Himalayan Medicinal Plants

Himalayan Medicinal Plants

Advances in Botany, Production & Research

Edited by

Nikhil Malhotra
ICAR-National Bureau of Plant Genetic Resources Regional Station, Shimla, India

Mohar Singh
ICAR-National Bureau of Plant Genetic Resources Regional Station, Shimla, India

Academic Press is an imprint of Elsevier
125 London Wall, London EC2Y 5AS, United Kingdom
525 B Street, Suite 1650, San Diego, CA 92101, United States
50 Hampshire Street, 5th Floor, Cambridge, MA 02139, United States
The Boulevard, Langford Lane, Kidlington, Oxford OX5 1GB, United Kingdom

Notices

Knowledge and best practice in this field are constantly changing. As new research and experience broaden our understanding, changes in research methods, professional practices, or medical treatment may become necessary.

Practitioners and researchers must always rely on their own experience and knowledge in evaluating and using any information, methods, compounds, or experiments described herein. In using such information or methods they should be mindful of their own safety and the safety of others, including parties for whom they have a professional responsibility.

To the fullest extent of the law, neither the Publisher nor the authors, contributors, or editors, assume any liability for any injury and/or damage to persons or property as a matter of products liability, negligence or otherwise, or from any use or operation of any methods, products, instructions, or ideas contained in the material herein.

Library of Congress Cataloging-in-Publication Data
A catalog record for this book is available from the Library of Congress

British Library Cataloguing-in-Publication Data
A catalogue record for this book is available from the British Library

ISBN: 978-0-12-823151-7

For information on all Academic Press publications visit our website at https://www.elsevier.com/books-and-journals

Publisher: Charlotte Cockle
Editorial Project Manager: Sara Valentino
Production Project Manager: Niranjan Bhaskaran
Cover Designer: Alan Studholme

Typeset by TNQ Technologies

Contents

Contributors

Ashrita
Biotechnology Division, CSIR-Institute of Himalayan Bioresource Technology, Palampur, Himachal Pradesh, India; Academy of Scientific and Innovative Research, CSIR-Institute of Himalayan Bioresource Technology, Palampur, Himachal Pradesh, India

Ashwani Bhardwaj
Defence Institute of High Altitude Research, Defence R & D Organization, Leh, Ladakh, India

Pushpender Bhardwaj
Defence Institute of High Altitude Research, Defence R & D Organization, Leh, Ladakh, India

Kirti Chawla
Plant Tissue Culture and Genetic Engineering, National Agri-Food Biotechnology Institute (NABI), Mohali, Punjab, India; Department of Biotechnology, Maharishi Markandeshwar (Deemed to be University), Mullana, Ambala, Haryana, India

Aditya Dogra
Department of Biotechnology, Shoolini Institute of Life Sciences and Business Management, Solan, Himachal Pradesh, India

Varun Garla
Department of Information Technology, Shoolini Institute of Life, Sciences and Business Management, Solan, Himachal Pradesh, India

Surendra Prakash Gupta
Department of Life Science, Shri Vaishnav Institute of Science, Shri Vaishnav Vidhyapeeth Viswavidhalaya, Indore, Madhya Pradesh, India

Vikrant Jaryan
Department of Botany, Sant Baba Bhag Singh University, Khiala, Jalandhar, Punjab, India

Anaida Kad
University Institute of Engineering and Technology (UIET), Panjab University, Chandigarh, India

Mamta Kashyap
Department of Biotechnology, Shoolini Institute of Life Sciences and Business Management, Solan, Himachal Pradesh, India

Anita Kumari
Department of Ornamental Plants and Agricultural Biotechnology, Agricultural Research Organization, Volcani Center, Rishon LeZion, Israel

Pankaj Kumar
Biotechnology Division, CSIR-Institute of Himalayan Bioresource Technology, Palampur, Himachal Pradesh, India

Pawan Kumar
Institute of Plant Science, Agricultural Research Organization (ARO), The Volcani Center, Rishon LeZion, Israel; Department of Biotechnology, Maharishi Markandeshwar (Deemed to be University), Mullana, Ambala, Haryana, India

Rahul Kumar
Faculty of Agricultural Sciences, DAV University, Jalandhar, Punjab, India

Varun Kumar
Department of Ornamental Plants and Agricultural Biotechnology, Agricultural Research Organization, Volcani Center, Rishon LeZion, Israel

Vishal Kumar
Govt. Senior Secondary School, Bhadwar, Kangra, Himachal Pradesh, India

Swaran Lata
Himalayan Forest Research Institute (HFRI), Conifer Campus, Shimla, Himachal Pradesh, India

Nikhil Malhotra
ICAR-National Bureau of Plant Genetic Resources Regional Station, Shimla, Himachal Pradesh, India

Ramgopal Mopuri
Institute of Animal Science, Agricultural Research Organization (ARO), The Volcani Center, Rishon LeZion, Israel

Avilekh Naryal
Defence Institute of High Altitude Research, Defence R & D Organization, Leh, Ladakh, India

Mahinder Partap
Biotechnology Division, CSIR-Institute of Himalayan Bioresource Technology, Palampur, Himachal Pradesh, India; Academy of Scientific and Innovative Research, CSIR-Institute of Himalayan Bioresource Technology, Palampur, Himachal Pradesh, India

Pratap Kumar Pati
Department of Biotechnology, Guru Nanak Dev University, Amritsar, Punjab, India

Archit Pundir
University Institute of Engineering and Technology (UIET), Panjab University, Chandigarh, India

Mohammed Saba Rahim
Agri-Biotechnology Division, National Agri-Food Biotechnology Institute, Mohali, Punjab, India

Shiv Rattan
Biotechnology Division, CSIR-Institute of Himalayan Bioresource Technology, Palampur, Himachal Pradesh, India

Joy Roy
Agri-Biotechnology Division, National Agri-Food Biotechnology Institute, Mohali, Punjab, India

Anil K. Sharma
Department of Biotechnology, Maharishi Markandeshwar (Deemed to be University), Mullana, Ambala, Haryana, India

Ashutosh Sharma
Faculty of Agricultural Sciences, DAV University, Jalandhar, Punjab, India

Himanshu Sharma
Agri-Biotechnology Division, National Agri-Food Biotechnology Institute, Mohali, Punjab, India

Indu Sharma
Department of Botany, Sant Baba Bhag Singh University, Khiala, Jalandhar, Punjab, India

Neha Sharma
Division of Crop Improvement, ICAR-Central Potato Research Institute, Shimla, Himachal Pradesh, India

Shivani Sharma
Himalayan Forest Research Institute (HFRI), Conifer Campus, Shimla, Himachal Pradesh, India

Vikas Sharma
Department of Molecular Biology and Genetic Engineering, Lovely Professional University, Jalandhar, Punjab, India

Vikas Sharma
Department of Botany, Sant Baba Bhag Singh University, Khiala, Jalandhar, Punjab, India

Kirti Shitiz
Defence Laboratory, Defence Research & Development Organization, Jodhpur, Rajasthan, India

Baldev Singh
Department of Biotechnology, Guru Nanak Dev University, Amritsar, Punjab, India

Jagdish Singh
Himalayan Forest Research Institute (HFRI), Conifer Campus, Shimla, Himachal Pradesh, India

Joginder Singh
Himalayan Forest Research Institute (HFRI), Conifer Campus, Shimla, Himachal Pradesh, India

Mohar Singh
ICAR-National Bureau of Plant Genetic Resources Regional Station, Shimla, Himachal Pradesh, India

Pradeep Singh
Department of Biotechnology, Guru Nanak Dev University, Amritsar, Punjab, India

Pramod Kumar Singh
Department of Biosciences, Christian Eminent College, Indore, Madhya Pradesh, India

Vijay Singh
Department of Botany, Mata Gujri College, Fatehgarh Sahib, Punjab, India

Archit Sood
Institute of Plant Sciences, Volcani Center, Agricultural Research Organization, Rishon LeZion, Israel

Hemant Sood
Department of Biotechnology and Bioinformatics, Jaypee University of Information Technology, Waknaghat, Solan, Himachal Pradesh, India

Ira Vashisht
Crop Genetics & Informatics Group, School of Computational and Integrative Sciences (SCIS), Jawaharlal Nehru University, New Delhi, India

Ashish R. Warghat
Biotechnology Division, CSIR-Institute of Himalayan Bioresource Technology, Palampur, Himachal Pradesh, India; Academy of Scientific and Innovative Research, CSIR-Institute of Himalayan Bioresource Technology, Palampur, Himachal Pradesh, India

CHAPTER

Introduction

1

Nikhil Malhotra, Mohar Singh

ICAR-National Bureau of Plant Genetic Resources Regional Station, Shimla, Himachal Pradesh, India

The Himalayan center of plant diversity is a narrow band of biodiversity lying on the southern margin of the Himalayas, the world's highest mountain range with elevations exceeding 8000 m (Barthlott et al., 2005). The Himalayan region, likewise other biomes of the world, is known since centuries for harboring a rich wealth of extremely valuable medicinal plants (Kala, 2005, 2010). The Indian Himalayas are home to more than 8000 species of vascular plants of which 1748 possess medicinal properties (Singh and Hajra, 1996; Samant et al., 1998; Joshi et al., 2016). At present, the trade of these plants from the Himalayas to the other parts of the world is speeding up due to increase in their demand, which subsequently affect the traditional collection practices of medicinal plants (Olsen, 1998; Bhat et al., 2013). Earlier, traditional healers mainly practiced the harvesting of medicinal plants but with the high demand at regional to international markets, many of the untrained collectors begin to participate in collection (Sharma and Kala, 2016), which has also resulted in adulteration of plant material (Sagar, 2014). Numerous wild and cultivated medicinal plants possessing bioactive compounds known as secondary metabolites have been utilized as curative agents since ancient times and medicinal plants have gained importance recently, not only as herbal medicines but also as natural ingredients for the cosmetic industries. As a result, a significant number of Himalayan plants now figure in the Red Data Book of rare, endangered, and threatened medicinal plants and require urgent conservation efforts (Rana and Samant, 2010; Goraya, 2011). Some of the important medicinal plants species of the Indian Himalayas include *Aconitum heterophyllum* (Atis), *Dactylorhiza hatagirea* (Salampanja), *Picrorhiza kurroa* (Kutki), *Podophyllum hexandrum* (Bankakdi), *Rauwolfia serpentina* (Sarpagandha), etc.

Secondary metabolites are a unique group of compounds produced by plants to protect against various biotic and abiotic factors (Kennedy and Wightman, 2011). These compounds, however, do not influence the primary metabolic activities such as growth and reproduction of plants (Arbona et al., 2013). The major classes of secondary metabolites include phenolics, alkaloids, tannins, saponins, lignins, glycosides, and terpenoids. Some of these compounds have become an integral part of plant−microbe interactions toward adapting to environmental irregularities. They regulate symbiosis, induce seed germination, and show allelopathic effect, i.e.,

Himalayan Medicinal Plants. https://doi.org/10.1016/B978-0-12-823151-7.00004-0

inhibit other competing plant species in their environment. Moreover, these compounds induce adverse physiological activities such as reduced digestive efficiency, reproductive failure, neurological problems, and gangrene and also possess high toxicity. The discovery of such unique compounds in the majority of Himalayan medicinal plants has inspired many scientific communities to explore their potential applications in various industries. The use of natural bioactive compounds and their products is thereby considered the most suitable source of alternative medicine. Thus, there is an unprecedented task to meet the increasing demand for plant secondary metabolites from flavor and fragrance, food, and pharmaceutical industries. However, their supply has become a major constraint since their large-scale cultivation is very limited. Moreover, it is difficult to obtain a constant quantity of compounds from cultivated plants as their yield fluctuates due to several factors including genotypic variations, geography, and edaphic conditions along with harvesting and processing methods.

Further, in situ and ex situ conservations are the most efficient methods to conserve genetic diversity of plants, including medicinal herbs. Traditionally, in situ conservation efforts have utilized the delineation of protected areas, whereas ex situ conservation efforts have included in vitro approaches and gene banks. However, conservation efforts have taken new dimensions with the advent of new technologies in recent years. As per the present scenario, these new approaches have integrated with traditional well-developed methods of conservation. The recent technological developments in high-throughput next-generation sequencing and other molecular biology techniques have provided greater opportunities to identify and characterize a large number of genes involved in important metabolite pathways. However, linking genotype to phenotype, predicting gene regulations, and ascertaining mutations require the utilization of vast genomic information and encompass the incorporation of intraspecific and environmental variability.

Unfortunately, there remains a paucity of information relating biological activities of essential metabolites with the ethnobotanical uses of the plants. In many cases, this may be due to the activity residing in nonvolatile components. Additionally, many researchers have neglected bioactivity screening related to ethnopharmacological uses. Thus, detailed work should be carried out to identify phytochemicals associated with biological activities, which support traditional uses of medicinal plants. Moreover, an integrated approach including conventional as well as emerging technologies should be utilized for the effective conservation of Himalayan herbs (Sharma and Kala, 2018). The latest ecological analysis methods coupled with whole-genome and transcriptome sequencing, metabolic engineering, and big data analytics should become an integral part of programs for the conservation and genetic improvement of the Himalayan plant wealth for future generations. Further, increased interdisciplinary collaboration and multiinstitutional focus for a resolute effort to this effect is urgently required. We encourage the preservation of traditional knowledge and uses of Himalayan medicinal plants and hope that additional steps should be undertaken to protect and maintain the Himalayan ecology.

References

Arbona, V., Manzi, M., de Ollas, C., Gómez-Cadenas, A., 2013. Metabolomics as a tool to investigate abiotic stress tolerance in plants. Int. J. Mol. Sci. 14, 4885–4911.

Barthlott, W., Mutke, J., Rafiqpoor, D., Kier, G., Kreft, H., 2005. Global centers of vascular plant diversity. Nova Acta Leopold. 92, 61–83.

Bhat, J., Kumar, M., Bussmann, R.W., 2013. Ecological status and traditional knowledge of medicinal plants in Kedarnath Wildlife Sanctuary of Garhwal Himalaya, India. J. Ethnobiol. Ethnomed. 9, 1–18.

Goraya, G., 2011. Conservation concerns for medicinal plants of Himachal Pradesh. ENVIS News Lett. Med. Plants 3, 15.

Joshi, R.K., Satyal, P., Setzer, W.N., 2016. Himalayan aromatic medicinal plants: a review of their ethnopharmacology, volatile phytochemistry, and biological activities. Medicines 3, 6.

Kala, C.P., 2005. Indigenous uses, population density, and conservation of threatened medicinal plants in protected areas of the Indian Himalayas. Conserv. Biol. 19, 368–378.

Kala, C.P., 2010. Medicinal Plants of Uttarakhand: Diversity Livelihood and Conservation. Biotech Books, Delhi, India, p. 188.

Kennedy, D.O., Wightman, E.L., 2011. Herbal extracts and phytochemicals: plant secondary metabolites and the enhancement of human brain function. Adv. Nutr. 2, 32–50.

Olsen, C.S., 1998. The trade in medicinal and aromatic plants from central Nepal to Northern India. Econ. Bot. 52, 279–292.

Rana, M.S., Samant, S.S., 2010. Threat categorisation and conservation prioritisation of floristic diversity in the Indian Himalayan region: a state of art approach from Manali wildlife sanctuary. J. Nat. Conserv. 18, 159–168.

Sagar, P.K., 2014. Adulteration and substitution in endangered, ASU herbal medicinal plants of India, their legal status, scientific screening of active phytochemical constituents. Int. J. Pharmaceut. Sci. Res. 5, 4023–4039.

Samant, S.S., Dhar, U., Palni, L.M.S., 1998. Medicinal Plants of Indian Himalaya: Diversity Distribution Potential Values. G.B. Pant Institute of Himalayan Environment and Development, Almora, India.

Sharma, N., Kala, C.P., 2016. Utilization pattern, population density and supply chain of *Rhododendron arboreum* and *Rhododendron campanulatum* in Dhauladhar mountain range of Himachal Pradesh, India. Appl. Ecol. Environ. Sci. 4, 102–107.

Sharma, N., Kala, C.P., 2018. Harvesting and management of medicinal and aromatic plants in the Himalaya. J. Appl. Med. Aromat. Plants 8, 1–9.

Singh, D.K., Hajra, P.K., 1996. Floristic diversity. In: Gujral, G.S., Sharma, V. (Eds.), Changing Perspective of Biodiversity Status in the Himalaya. British Council Division, British High Commission Publication, Wildlife Youth Services, New Delhi, India, pp. 23–38.

CHAPTER

Aconitum heterophyllum

2

Nikhil Malhotra[1], Shivani Sharma[2]

[1]*ICAR-National Bureau of Plant Genetic Resources Regional Station, Shimla, Himachal Pradesh, India;*
[2]*Himalayan Forest Research Institute (HFRI), Conifer Campus, Shimla, Himachal Pradesh, India*

2.1 Introduction

Out of many important medicinal plants cultivated in present times, *Aconitum* species finds a key position for their conservation and cultivation. The genus *Aconitum* belongs to the family Ranunculaceae. There are ~400 species of *Aconitum* occurring worldwide (Lane, 2004; Yin et al., 2019). In the northwest Himalayas, it is represented by 10 species and 2 varieties. Some of the important species of *Aconitum* are *Aconitum balfourii*, *Aconitum bisma*, *Aconitum carmichaeli*, *Aconitum chasmanthum*, *Aconitum deinorrhizum*, *Aconitum ferox*, *Aconitum japonicum*, *Aconitum napellus*, and *Aconitum violaceum* along with *A. heterophyllum*—the only nontoxic species of this genus (Chauhan, 2006; Buddhadev and Buddhadev, 2017). These herbaceous biennial plants are primarily natives of the mountainous parts of the Northern Hemisphere, growing in moisture retentive but well-drained soils on the mountain meadows (Tamura, 1995). These plants are tall, with erect stem being crowned by racemes of large and eye-catching blue, purple, white, yellow, or pink zygomorphic flowers with numerous stamens. The root is best harvested in the autumn as soon as the plant dies down and is dried for later use. In recent years, the demand for medicinal and aromatic plants has grown rapidly because of accelerated local, national, and international interest. *Aconitum* genus is the center of attraction in the field of herbal medicines because of its property of curing a wide range of diseases and, hence, the pressure on its natural habitat has increased. This is one of the most prized plant genuses which has been enlisted in the Red Data Book and is widely considered as a mystifying group due to fatal as well as therapeutic behavior (Tai et al., 2015). The pharmacological analysis of *Aconitum* species and their compounds have shown various therapeutic effects pertaining to cardiovascular and central nervous system (Dzhakhangirov et al., 1997; Friese et al., 1997; Ameri 1998; Polyakov et al., 2005) alongside anticancer (Solyanik et al., 2004), antimicrobial, and cytotoxic activities (Gavín et al., 2004; González et al., 2005). In recent years, a large number of studies have investigated the toxicological characteristics of *Aconitum*, its main alkaloids, and their derivatives (Xie et al., 2005; Fujita et al., 2007; Jaiswal et al., 2013, 2014). It has been observed that the whole plant of *Aconitum* is highly toxic with the concentration of toxic

Himalayan Medicinal Plants. https://doi.org/10.1016/B978-0-12-823151-7.00015-5

compounds higher in roots and flowers than in leaves and stems (Ding et al., 1993). The symptoms of toxicity affect mainly the central nervous system and the heart, with concomitant gastrointestinal signs. The cause of death is the development of ventricular tachyarrhythmia and heart arrest. No specific therapy exists for *Aconitum* poisoning, although cardiovascular supportive treatment is usually applied (Lin et al., 2004). The toxicity of *Aconitum* is mainly due to the diester diterpene alkaloids and monoester diterpene alkaloids such as deoxyaconitine, benzoylmesaconitine, jesaconitine, benzoylhypaconine, and benzoylaconine (Chinese Pharmacopoeia Commission, 2005; Srivastava et al., 2010; Nyirimigabo et al., 2015). Through various physical and chemical methods of treatment, highly toxic *Aconitum* alkaloids could be transformed into less toxic derivatives.

A. heterophyllum Wall, commonly known as "atis," is a rare diploid (2n = 16) Himalayan plant species found between 2400 and 3600 m amsl (Fig. 2.1). Ayurveda classical texts of 15th–16th century introduced "Abhava-Pratinidhi Dravya" concept, wherein it was categorized as an "abhava dravya" (unavailable drug). Its roots are ovoid-conical, tapering downward to a print, 2.0–7.5 cm long, 0.4–1.6 cm or more thick at its upper extremity, gradually decreasing in thickness toward tapering end, externally light ash-gray, white or gray-brown, while internally

FIGURE 2.1

Mature *Aconitum heterophyllum* plant.

starch white, external surface wrinkled marked with scars of fallen rootlet, and with a rosette of scaly rudimentary leaves on top. It is a cross-pollinated plant which flowers in the second year. The flowers are helmet shaped, bright blue or greenish blue in color and have a purple vein. For medicinal use, the roots from plants bearing fully developed tubers are collected (Kumar et al., 2016; Rajakrishnan et al., 2016). The tubers sometimes occur as a pair of mother and daughter tubers. Tuberization in *A. heterophyllum* is a distinctive process from young rootlet to fully mature storage roots which are committed to the storage of primary as well as secondary metabolites (Pal et al., 2015). *A. heterophyllum* has been listed as "critically endangered medicinal herb" by the International Union for Conservation of Nature and Natural Resources (IUCN, 1993; Nautiyal et al., 2002; CAMP, 2003; Srivastava et al., 2011), which has thereby prohibited the export of its plants, plant portions and their derivatives, and extracts obtained from the wild (Shah, 2005; Chinese Pharmacopoeia Commission, 2015). Owing to the huge cost for dried tuberous roots of *A. heterophyllum* (~₹10,000 per kg), and an ever-rising demand of raw material (>20 tons per year) (Aneesh et al., 2009; NMPB, 2015), overharvesting of its tubers has been facilitated over the years. This reckless collection has led to reduction in its population in natural habitat. Although efforts have been done to maintain its population in farm fields by conventional breeding and propagation methods (Fig. 2.2), nothing has been significantly achieved in R&D programs globally (Rawat et al., 2016). Nontoxic active components like atisine, hetisine, and heteratisine, collectively termed as "aconites," accumulating in tuberous roots of *A. heterophyllum* have wide pharmacological effects on immune, digestive, and nervous systems (Murti and Khorana, 1968; Pelletier et al., 1968; Mori et al., 1989; Rastogi and Mehrotra, 1991; Zhaohong et al., 2006; Nisar et al., 2009; Malhotra et al., 2014; Malhotra, 2017).

FIGURE 2.2

Field plantation of *Aconitum heterophyllum* at HFRI Farm, Shillaru, Himachal Pradesh, India.

Biotechnological interventions have substantially contributed in terms of higher aconites production and conservation in various *Aconitum* species, but the contemporary breakthroughs are still lacking in *A. heterophyllum*, besides a few research interventions made in the recent past. Although this plant has been circumspectly studied for its cultivation (Nautiyal et al., 2006; Srivastava et al., 2011), conservation and sustainable utilization (Pandey et al., 2005; Seethapathy et al., 2014; Kumar et al., 2016), cytology (Siddique et al., 1998; Rani et al., 2011; Jeelani et al., 2015), ecology (Nautiyal et al., 2002; Bhat et al., 2014; Jeelani et al., 2015), medicinal uses (Ukani et al., 1996; Nyirimigabo et al., 2015), phytochemical constituents (Gajalakshmi et al., 2011; Jaiswal et al., 2013; Jaiswal et al., 2014; Malhotra et al., 2014; Nagarajan et al., 2015a,b; Nyirimigabo et al., 2015; Kumar et al., 2016), reproductive biology (Siddique et al., 1998) along with reports on plant tissue culture (Giri et al., 1993, 1997; Jabeen et al., 2006; Solanki and Siwach, 2012), and OMICS-assisted approaches (Malhotra et al., 2014, 2016; Pal et al., 2015; Kumar et al., 2016), the comprehensive coverage of botany, production, and research advancements in *A. heterophyllum* have not been attempted till date. Thus, this chapter becomes very unique and important for the researchers and readers across the globe working on this high-value plant species.

2.2 Origin and distribution

Classification of the genus *Aconitum* has been extremely difficult because aconites are morphologically highly variable (Yang, 1990; Tamura, 1995; Luo, 2003). Numerous categorizations in this genus have been proposed (de Candolle, 1824; Nakai, 1953; Wang, 1965; Tamura, 1995), but due to the difference in explanation of features considered, these are still in great dispute. The major centers of *Aconitum* diversity are northwest and east Himalayas, southwest China, and Japan. Although the chloroplast DNA, nuclear ribosomal DNA (nrDNA), and nuclear internal transcribed spacer (ITS) sequence data have been used to study the phylogenetic relationships within *Aconitum* subgenus *Aconitum* (Kita et al., 1995; Kita and Ito, 2000; Luo et al., 2005) along with a study on chromosomal and molecular patterns (Mitika et al., 2007), significant information on adequate understanding of its phylogeny is still lacking. Moreover, separate studies for tracing the evolutionary history of each *Aconitum* species have not been done; therefore, no records are available for justifying the origin of *A. heterophyllum* also.

In India, *A. heterophyllum* is found and cultivated in the Himalayan states of Jammu and Kashmir, Ladakh, Himachal Pradesh, and Uttarakhand in the northwest along with Sikkim and Arunachal Pradesh in the east. It also occurs in Nepal, Bhutan, and parts of southwest China.

2.3 Medicinal properties

From ancient times, *A. heterophyllum* has been used in different formulations in the Indian Ayurvedic System for curing various diseases. Balachaturbhadra Churna, Caspa Drops, Chandraprabha Vati, Chaturbhadraka Vati, Chitrakadi vati, Kutajghan Vati, Livex, Panchatikta Guggulu Ghrita, Rasnerandadi Kwatha, Satyadi Yoga, Shaddharana churna, and Sudarshan Churna are some of the popular multidrug herbal formulations in which *A. heterophyllum* is used as one of the main ingredients (Lather et al., 2010; Nariya et al., 2011; Ajanal et al., 2012; Sojitra et al., 2013; Joshi et al. 2014, 2016; Kumar et al., 2014; Selvaraj et al., 2014; Chaudhary et al., 2015; Gupta et al., 2015; Dhamankar and Jadhav, 2016; Baishya et al., 2020). These drugs find common use in the treatment of diarrhea, fever, indigestion, inflammation, helminthiasis, hyperlipidemia, and other ailments. Some of the important medicinal properties of *A. heterophyllum* are listed in Table 2.1.

2.3.1 Antibacterial activity

Ahmad et al. (2008) isolated the new aconitine type norditerpenoid alkaloids, 6-dehydroacetylsepaconitine, and 13-hydroxylappaconitine from the tubers of *A. heterophyllum* along with the known alkaloids lycoctonine, delphatine, and lappaconitine, which were screened for antibacterial activity against different bacterial strains. They showed antibacterial activity against diarrhea causing gram-negative bacteria *Escherichia coli*, *Shigella flexneri*, *Pseudomonas aeruginosa*, and *Salmonella typhi*. This report strengthens the use of *A. heterophyllum* as an antimicrobial

Table 2.1 Medicinal properties of *Aconitum heterophyllum*.

Use	References
Abdominal distension	Imtiyaz et al. (2013)
Anti-Alzheimer's disease	Ahmad et al. (2017)
Antibacterial	Srivastava et al. (2011), Sinam et al. (2014)
Antidiabetic	Prasad et al. (2014), Nirja and Sharma (2016)
Antidiarrheal	Prasad et al. (2014), Paramanick et al. (2017)
Antihelminthic	Pattewar et al. (2012), Rungsung et al. (2013)
Antiinflammatory	Verma et al. (2010), Paramanick et al. (2017)
Antileucorrhea	Rana et al. (2013)
Antioxidant	Prasad et al. (2012), Rah et al. (2016)
Antiulcer	Rajakrishnan et al. (2020)
Aphrodisiac	Imtiyaz et al. (2013), Sojitra et al. (2013)
Arthritis	Lone and Bhardwaj (2013)
Hypolipidemic	Subash and Augustine (2012)
Immunomodulatory	Nagarajan et al. (2015b), Joshi et al. (2016)
Nephroprotective	Konda et al. (2016)

and/or antihelminthic agent. In another study by Sinam et al. (2014), the root alkaloid extract of *A. heterophyllum* showed antibacterial activity against *Bacillus subtilis*, *Bordetella bronchiseptica*, *Pseudomonas putida*, *Staphylococcus aureus*, and *Xanthomonas campestris*.

2.3.2 Antidiarrheal activity

The antidiarrheal activity of roots of *A. heterophyllum* may be attributed to an antisecretory and antienteropooling type effect as a result of reactivation of Na^+ and K^+ ATPase activity mediated through nitric oxide pathway. They cause either a decrease in mucosal secretion or increase in mucosal absorption, which allows the feces to become desiccated, thus retarding its movement through the colon (Prasad et al., 2014).

2.3.3 Antihelminthic activity

Aqueous and alcoholic extracts of tubers of *A. heterophyllum* gave encouraging results when evaluated against *Pheretima posthuma*, using piperazine citrate as standard. Time required for initial three paralytic attacks and deaths was used as parameters to evaluate the drug (Pattewar et al., 2012). It was revealed that a dose of 100% aqueous root extract was responsible for anthelmintic activity.

2.3.4 Antihyperlipidemic activity

The methanolic extract of tubers of *A. heterophyllum* had a hypolipidemic effect on diet-induced obese rats. It was observed that the pharmacological effect was due to the inhibition of hydroxymethylglutarate-Coenzyme A reductase and activation of lecithin-cholesterol acyltransferase. This resulted in lowering apolipoprotein B, total cholesterol, low-density lipoprotein cholesterol, and triglycerides in the blood serum along with the decrease in intestinal fat absorption and increase in apolipoprotein A with high-density lipoprotein cholesterol. These results supported the use of *A. heterophyllum* as an antihyperlipidemic agent (Subash and Augustine, 2012).

2.3.5 Antiinflammatory and antipyretic activity

For the assessment of antiinflammatory activity of *A. heterophyllum*, cotton-pellet–induced granuloma method was used. It was found that ethanolic extract of *A. heterophyllum* tuber had significant antiinflammatory activity, thereby providing scientific evidence for a traditional use as an antiinflammatory agent. Further, the antipyretic effects of roots of *A. heterophyllum* in the form of aqueous, chloroform, and hexane extracts were examined using the method of yeast-induced pyrexia, with aspirin as a standard antipyretic agent for comparison. These studies showed that the extracts were nontoxic with nonsignificant antipyretic activity (Verma et al., 2010).

2.3.6 Antioxidant activity

Prasad et al. (2012) demonstrated in vitro antioxidant activity of *A. heterophyllum* in different models which was attributed to low flavonoid and phenolic contents in its roots. Further, the root extracts of *A. heterophyllum* were tested for glycerol-induced acute renal failure in Wistar albino rats (Konda et al., 2016) and streptozotocin-induced diabetic rats (Rah et al., 2016), respectively. These studies revealed significant antioxidant property without any toxic effects.

2.3.7 Immunomodulatory activity

The immunomodulatory activity of ethanolic extract of *A. heterophyllum* tubers along with other medicines of the Ayurveda and Unani systems of medicine was investigated on delayed-type hypersensitivity, humoral responses to sheep red blood cells, skin allograft rejection, and phagocytic activity of the reticuloendothelial system in mice. It was found that the extract appeared to enhance the phagocytic function and inhibiting humoral component of the immune system. The results obtained from these preliminary studies showed that *A. heterophyllum* has immunomodulatory activity, which could possibly lead to synthesis of new immunomodulating agents of herbal origin (Atal et al., 1986; Gulati et al., 2002).

2.3.8 Nervous system stimulation

A. heterophyllum has the ability to make the sympathetic nervous system highly sensitive to physiological stimuli. It was found that while atisine had a hypotensive effect at every tested dose, the plant extract showed hypertensive properties. Hypertension produced by high doses of aqueous extract was attributed to the excitement of the sympathetic nervous system (Raymond-Hamet, 1954). Nisar et al. (2009) isolated two new diterpenoid alkaloids viz. heterophyllinine A and heterophyllinine B from the roots of *A. heterophyllum*, which were almost 13 times more selective in inhibiting the enzyme butyrylcholinesterase than acetylcholinesterase. These enzymes are involved in the transmission of nerve impulses.

2.4 Phytochemistry

A. heterophyllum is a rich source of alkaloids, flavonoids, free fatty acids, and polysaccharides (Rajakrishnan et al., 2016; Paramanick et al., 2017). The main alkaloid reported in *Aconitum* is aconitine that is highly toxic (O'Neil et al., 2001). However, among the reported species, *A. heterophyllum* is the only nontoxic species with therapeutic potential (Chauhan, 2006; Jaiswal et al., 2014; Malhotra et al., 2014). The pharmacological properties of *A. heterophyllum* are attributed to the nontoxic active constituents, i.e., aconites, including atisine which comprises the major alkaloid constituents of this plant species (Chatterjee and Prakash, 1994; Srivastava et al.,

2011). These constituents make it a safer herb to use when compared with other *Aconitum* species, since no purification process is mandatory for its purification or detoxification. Some of the important phytochemicals of *A. heterophyllum* are listed in Table 2.2.

The early investigations on the tubers of *A. heterophyllum*, beginning with 19th century works by Broughton, Wasowicz, and Wright, have been documented by Jowett (1896). Broughton was first to isolate atisine. Different salts (sulfate, hydrochloride, and platinichloride) were prepared from the alkaloid, and the molecular formula was deduced. Wasowicz showed that the aconitic acid is also present along with atisine besides suggesting slight modifications in the molecular formula of atisine. Wright proposed a new formula for atisine based on analysis of its aurichloride salt. Subsequently, investigations on the properties and composition of atisine and its salts were studied in great detail. No alkaloid other than atisine was found in such studies (Jowett, 1896). The structure of atisine and three other alkaloids hetisine, heteratisine, and benzoylheteratisine was confirmed by Jacobs and Craig (1942a,b).

Detailed studies on hetisine, atisine, and heteratisine helped in their structure elucidation (Solo and Pelletier, 1962; Aneja and Pelletier 1964; Pelletier and Parthasarathy 1965; Aneja et al., 1973). Further investigations on *A. heterophyllum* led to the isolation and structure elucidation of additional new diterpene alkaloids; atidine, F-dihydroatisine, hetidine, and hetisine as well as lactone alkaloids heterophyllisine, heterophylline, and heterophyllidine (Pelletier and Aneja, 1967; Pelletier et al., 1968). In 1982, a new entatisene diterpenoid lactone, atisenol, was isolated from

Table 2.2 Major phytochemicals of *Aconitum heterophyllum*.

Class	Chemical constituent	References
Alkaloid	Aconitine	O'Neil et al. (2001)
	Atidine	Pelletier et al. (1968)
	Atisine	Jacobs and Craig (1942a)
	Condelphine	Obaidullah et al. (2018)
	Heteratisine	Jacobs and Craig (1942b)
	Heterophylline-A	Obaidullah et al. (2018)
	Heterophylline-B	Obaidullah et al. (2018)
	Heterophyllidine	Pelletier et al. (1968)
	Hetisine	Pelletier et al. (1968)
	Isoatisine	Ahmad et al. (2017)
	Lappaconitine	Ahmad et al. (2008)
	6,15β-dihydroxylhetisine	Ahmad et al. (2017)
	13-hydroxylappoconitine	Nisar et al. (2009)
Glycoside	Steviol	Kumar and Chauhan (2016)
Terpenoid	Atisenol	Pelletier et al. (1982)

the tubers of *A. heterophyllum* (Pelletier et al., 1982). Moreover, the structure and, most importantly, the stereochemistry of atisine and related alkaloids were established by Dvornik and Edwards (1964).

From the reported literature, it is evident that alkaloids were the main focus of study in *A. heterophyllum.* Several pharmacological actions of *A. heterophyllum* have been attributed to their alkaloids (O'Neil et al., 2001). Later on, work on *Aconitum* alkaloids led to the isolation of two new aconitine-type norditerpenoid alkaloids 6-dehydroacetylsepaconitine and 13-hydroxylappaconitine along with known norditerpenoid alkaloids lycoctonine, delphatine, and lappaconitine (Ahmad et al., 2008). Although aconitine, which is the major alkaloid of other *Aconitum* species, is not a major constituent of *A. heterophyllum*, high-performance liquid chromatography (HPLC) studies carried out on the tubers from Kumaon and Garhwal regions of the Himalayas showed that aconitine is present in different populations and varies from 0.13% to 0.75% (dry weight basis) (Bahuguna et al., 2000; Pandey et al., 2008). HPLC studies on quantification of aconitine from tubers of *A. heterophyllum* from the Kashmir valley have reported 0.0014%–0.0018% aconitine (Jabeen et al., 2011). Similarly, a study by Bahuguna et al. (2013) reported higher content of atisine (0.35%) and aconitine (0.27%) in greenhouse-grown *A. heterophyllum* when compared with the naturally grown plants (0.19% and 0.16%, respectively). Further, a study by Malhotra et al. (2014) led to estimation of atisine in roots of plants of different age groups (1–3 years) which showed variation in atisine content. It increased from 0.14% in 1-year-old plants to 0.22% in 2-year-old plants and then decreased to 0.08% in the roots of 3-year-old plants. Atisine was not detected in shoots of *A. heterophyllum.* They also analyzed atisine/aconites content through HPLC and bromocresol green extraction method in 14 accessions of *A. heterophyllum* collected from different locations of Himachal Pradesh. The significant variation was observed among 14 accessions of *A. heterophyllum* as atisine content in roots ranged from 0.14% to 0.37% and total alkaloids (aconites) from 0.20% to 2.49%. Two accessions, namely AHCR and AHSR, showed the highest atisine content of 0.30% and 0.37%, as well as the highest total alkaloids content of 2.22% and 2.49%, respectively. Thereafter, steviol was quantified in roots of *A. heterophyllum.* It was found that high-content accession had 0.06% steviol which was sixfold greater as compared to roots of low-content accession (0.01%) (Kumar et al., 2016). Moreover, Kumar and Chauhan (2016) quantified steviol in leaves of *A. heterophyllum*, thereby providing a novel source for extraction of steviol which could benefit the harvesters to get additional economic returns on leaf biomass.

Soon after, extensive chromatographic separations along with mass and nuclear magnetic resonance spectroscopy analysis resulted in the isolation of three new diterpenoid alkaloids, 6β-methoxy, 9β-dihydroxylheteratisine, 1α,11,13β-trihydroxylhetisine, 6,15β-dihydroxylhetisine, and the known compounds namely atidine, heteratisine, hetisinone, iso-atisine, and 19-epi-isoatisine (Ahmad et al., 2017). Then, Obaidullah et al. (2018) isolated heterophylline-A and heterophylline-B, along with condelphine from the roots of the *A. heterophyllum.*

2.5 Adulteration and substitution

Natural sources of medicinal plants are unable to meet demand for popular herbal products. Populations of many species have limited distribution in their natural habitats, requiring conservation strategies for protection. Unavailability of such medicinal plants has led to arbitrary substitution and adulteration in raw drug market (Kumar 2014a,b; Mishra et al., 2015; Shanmughanandhan et al., 2016; Ichim, 2019). Adulteration is a practice of substituting the original crude drug partially or fully with other substances which are either free from or inferior in therapeutic and chemical properties or addition of low grade or entirely different drug similar to that of original drug substituted with an intention of enhancement of profits.

A. heterophyllum has been substituted by *Cyperus rotundus*, commonly called "Musta," in herbal-processing methods, thus affecting the quality of the herbal drug formulations (Venkatasubramanian et al., 2010; Adams et al., 2013; Kumar 2014b; Seethapathy et al. 2014, 2015). Being a low-cost substitute (~₹30–50 per kg), it shares similar biological functions like antidiabetic, antidiarrheal, antipyretic, and treatment of urinary tract infections (Mitra et al., 2003; Uddin et al., 2006; Venkatasubramanian et al., 2010; Nagarajan et al., 2015b). Seethapathy et al. (2014) differentiated *A. heterophyllum* and *C. rotundus* by using nrDNA ITS sequence–based sequence characterized amplified region (SCAR) markers for validating adulteration in herbal drugs. The former was not detected through SCAR markers, while the latter was identified in complex mixtures of DNA extracted from commercial formulations.

2.6 Omics-based advancements

Understanding the biology of aconites biosynthesis has provided insights about the sites of biosynthesis and accumulation of aconites in *A. heterophyllum*. Comparative genomics was utilized for cloning the 15 genes of aconites biosynthetic pathway along with expression analysis of those genes in relation to atisine/aconites content in *A. heterophyllum*. Multiple genes of mevalonic acid/methylerythritol 4-phosphate (MVA/MEP) pathways showed elevated expression in high atisine/aconites content accession as compared to low-atisine/aconites content accession. This was the first attempt toward molecular understanding of atisine/aconites biosynthesis in *A. heterophyllum*. Eight genes viz. 3-hydroxy-3-methylglutaryl-CoA synthase, 3-hydroxy-3-methylglutaryl-CoA reductase (HMGR), phosphomevalonate kinase, isopentenyl pyrophosphate isomerase, 1-deoxy-D-xylulose 5-phosphate synthase, 2-C-methylerythritol 4-phosphate cytidyltransferase, 1-hydroxy-2-methyl-2-(E)-butenyl 4-diphosphate synthase (HDS), and gerenyl diphosphate synthase with elevated expression in relation to aconites content could be used as suitable targets for developing gene markers for genetic improvement of *A. heterophyllum* (Malhotra et al., 2014).

With progress in modern technologies, transcriptomics has emerged as a powerful tool to capture traits of economic importance. The availability of whole transcriptome data could be used not only to discover candidate genes involved in tuberous root development and secondary metabolites production but also for understanding molecular basis of various biological processes in *A. heterophyllum* (Pal et al., 2015; Malhotra et al., 2016). Comparative next-generation sequencing transcriptome analysis between root and shoot tissues of *A. heterophyllum* predicted the candidate genes involved in the production of secondary metabolites. The in silico expression profiling for 15 genes identified 4 transcripts namely HDS, HMGR, mevalonate kinase, and mevalonate diphosphate decarboxylase with higher expression in root as compared to shoot transcriptomes. The pathway analysis performed for both the tissues suggested 341 and 329 mapped Kegg Orthologs (KOs) responsible for secondary metabolism in root and shoot transcriptomes, respectively, thereby attributing medicinal value to this plant species. In total, 77 interacting pathways associated with isoquinoline alkaloids biosynthesis were identified in root transcriptome of *A. heterophyllum* indicating how important primary and secondary metabolic pathways are connected with each other (Pal et al., 2015). Later, a study by Rai et al. (2017) has also revealed enrichment of essential biological processes and secondary metabolism in transcriptomes of *Aconitum carmichaelii* and *A. heterophyllum*.

A complete atisine biosynthetic pathway was also constructed connecting glycolysis, MVA/MEP, serine biosynthesis, and diterpene biosynthetic pathways in *A. heterophyllum* (Fig. 2.3). The study revealed phosphorylated pathway as a major contributor toward serine production in addition to repertoire of genes in glycolysis (glucose-6-phosphate isomerase, phosphofructokinase, aldolase, and enolase), serine biosynthesis (3-phosphoglycerate dehydrogenase and 3-phosphoserine aminotransferase), and diterpene biosynthesis (kaurene oxidase and kaurene hydroxylase) sharing a similar pattern of expression (2- to 4-folds) in roots compared to shoots vis-à-vis atisine content (0%–0.37%), thus suggesting their vital role in atisine biosynthesis (Kumar et al., 2016). Further, the biosynthetic machinery of tuberous roots was discerned to identify plausible key genes toward root biomass development by utilizing transcriptome datasets of *A. heterophyllum*. Four genes viz. ADP-glucose pyrophosphorylase, β-amylase, SRF receptor kinase (SRF), and expansin showed maximum contribution toward tuberous root development. There is possibility of altering the expression levels of these genes for improving tuberous root (biomass) yield for herbal drug industries. These results can be further explored to dissect the molecular regulation of tuberous root formation and growth in *A. heterophyllum* (Malhotra et al., 2016). Besides this, the role of ATP binding cassette transporters was also evaluated in tuberous roots (Malhotra, 2017).

2.7 Plant tissue culture–assisted progress

The use of plant tissue culture techniques has been employed for conservation of this medicinally important plant species. Plants of *A. heterophyllum* were obtained via somatic embryogenesis in callus derived from in vitro–raised leaf and petiole

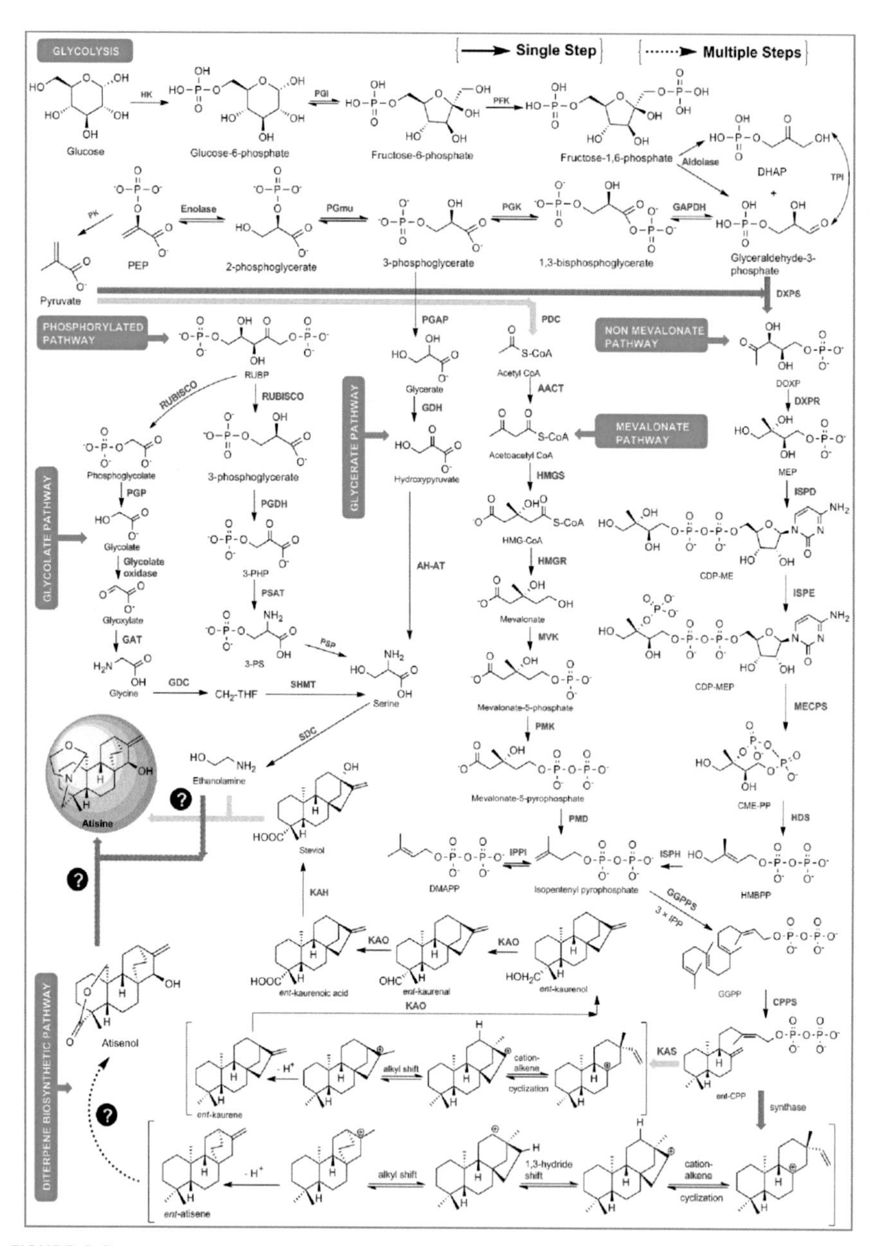

FIGURE 2.3

Metabolic pathway for atisine biosynthesis in *Aconitum heterophyllum* (Kumar et al., 2016).

explants (Giri et al., 1993). Then, a method for the production of hairy roots of *A. heterophyllum* was developed by Giri et al. (1997). Embryogenic callus cultures were successfully transformed using *Agrobacterium rhizogenes* strains viz. LBA 9402, LBA 9360, and A4 for the induction of hairy roots. Total alkaloid (aconites) content of transformed roots was found to be 2.96%, which was 3.75 times higher compared to 0.79% in the nontransformed (control) roots. Furthermore, thin layer chromatography analysis of aconites in the transformed roots revealed the presence of heteratisine, atisine, and hetidine in *A. heterophyllum*. In another study, a protocol was developed for in vitro shoot proliferation from callus cultures of *A. heterophyllum* (Jabeen et al., 2006).

Prolonged seed dormancy, high seedling mortality, and ecological constraints have made *A. heterophyllum* endangered. Some of the efforts made in this regard correspond to hot water treatment of seeds (Pandey et al., 2005), along with optimization of conditions for in vitro seed germination and shoot multiplication as reported by Solanki and Siwach (2012). In another study, Rana and Sreenivasulu (2013) found 80% ethanol induced seed germination in *A. heterophyllum*. They also identified 40 differentially expressed proteins from ethanol-treated and untreated seeds through comparative two-dimensional electrophoresis protein profiling. Further, a protocol for mass shoot multiplication was standardized by using different combinations of hormones (Mahajan et al., 2015). Later, Belwal et al. (2016) reported an optimized tissue culture protocol with reduced concentration of plant growth regulators and regeneration by shoot tips of *A. heterophyllum*. The clonal stocks obtained from this study were checked using intersimple sequence repeat markers for the production of true-to-type progenies.

2.8 Conclusion and future perspectives

The studies have suggested that the biosynthesis and accumulation of atisine/aconites occur entirely in the roots of *A. heterophyllum*. The presence of high atisine content in roots of 2-year-old plants would aid in the proper selection of raw material for the preparation of herbal drug formulations. The identification of elite chemotypes with high content of aconites will be helpful in commercial cultivation of genetically identical planting material of *A. heterophyllum*. This will further help in maintaining genetic uniformity, which would otherwise affect the amount of aconites and the quality of herbal drug efficacy. Also, for effectively resolving authentication challenges associated with the herbal market, DNA barcoding could be used in conjunction with need-based OMICS approaches. Further, the tuberous root formation in *A. heterophyllum* provides a unique system to explore mechanism of sink tissue formation and development vis-à-vis accumulation of medicinal metabolites. Lastly, the functional validation of key genes controlling aconites biosynthesis and tuberous root growth in *A. heterophyllum* would aid in designing a suitable genetic intervention strategy in this plant species. It is also presumed that genome sequencing along with CRISPR/Cas9 genome editing could be the next big steps toward better understanding of various biological as well as physiological processes in *A. heterophyllum*.

References

Adams, S.J., Kuruvilla, G.R., Krishnamurthy, K.V., Nagarajan, M., Venkatasubramanian, P., 2013. Pharmacognostic and phytochemical studies on Ayurvedic drugs *Ativisha* and *Musta*. Braz. J. Pharmacogn. 23, 398–409.

Ahmad, M., Ahmad, W., Ahmad, M., Zeeshan, M., Obaidullah, Shaheen, F., 2008. Norditerpenoid alkaloids from the roots of *Aconitum heterophyllum* Wall with antibacterial activity. J. Enzym. Inhib. Med. Chem. 23, 1018–1022.

Ahmad, H., Ahmad, S., Shah, S.A.A., Latif, A., Ali, M., Khan, F.A., Tahir, M.N., Shaheen, F., Wadood, A., Ahmad, M., 2017. Antioxidant and anticholinesterase potential of diterpenoid alkaloids from *Aconitum heterophyllum*. Bioorg. Med. Chem. 25, 3368–3376.

Ajanal, M., Gundkalle, M.B., Nayak, S.U., 2012. Estimation of total alkaloid in *Chitrakadivati* by UV-spectrophotometer. Ancient Sci. Life 31, 198–201.

Ameri, A., 1998. Effects of *Aconitum* alkaloids on the central nervous system. Prog. Neurobiol. 56, 211–235.

Aneesh, T.P., Hisham, M., Sonal Sekhar, M., Madhu, M., Deepa, T.V., 2009. International market scenario of traditional Indian herbal drugs – India declining. Int. J. Green Pharm. 3, 184–190.

Aneja, R., Pelletier, S.W., 1964. Diterpene alkaloids: structure of heteratisine. Tetrahedron Lett. 5, 669–677.

Aneja, R., Locke, D.M., Pelletier, S.W., 1973. The Diterpene alkaloids: structure and stereochemistry of heteratisine. Tetrahedron 29, 3297–3308.

Atal, C.K., Sharma, M.L., Kaul, A., Khajuria, A., 1986. Immunomodulating agents of plant origin. J. Ethnopharmacol. 18, 133–141.

Bahuguna, R., Purohit, M.C., Rawat, M.S., Purohit, A.N., 2000. Qualitative and quantitative variations in alkaloids of *Aconitum* species from Garhwal Himalaya. J. Plant Biol. 27, 179–183.

Bahuguna, R., Prakash, V., Bisht, H., 2013. Quantitative enhancement of active content and biomass of two *Aconitum* species through suitable cultivation technology. Int. J. Conserv. Sci. 4, 101–106.

Baishya, A., Das, B., Sarma, D., 2020. A detail pharmacognostic, physicochemical and phytochemical study of Satyadi yoga. J. Med. Plant Stud. 8, 07–10.

Belwal, N.S., Kamal, B., Sharma, V., Gupta, S., Dobriyal, A.K., Jadon, V.S., 2016. Production of genetically uniform plants from shoot tips of *Aconitum heterophyllum* Wall. – a critically endangered medicinal herb. J. Hortic. Sci. Biotechnol. https://doi.org/10.1080/14620316.2016.1184434.

Bhat, D., Joshi, G.C., Kumar, R., Tewari, L.M., 2014. Phytosociological features and threat categorization of *A. heterophyllum* Wall. ex Royle and *A. ferox* Wall. ex ser. in Kumaun Himalaya. J. Ecol. Nat. Environ. 6, 111–118.

Buddhadev, S.G., Buddhadev, S.S., 2017. A complete review on Ativisha – *Aconitum heterophyllum*. Pharma Sci. Monit. 8, 111–114.

CAMP, 2003. CAMP Workshop, 22–25th May, Shimla, Himachal Pradesh.

de Candolle, A.P., 1824. Aconitum. In: Prodromus Systematis Naturalis Regni Vegetabilis I. Treuttel and Würtz, Parisiis, pp. 56–64.

Chatterjee, A., Prakash, S.C., 1994. Treatise on Indian Medicinal Plants, first ed. Vedams Books International, New Delhi, pp. 111–121.

Chaudhary, S.A., Shingadiya, R.K., Patel, K.S., Kori, V.K., Rajagopala, S., Harisha, C.R., Shukla, V.J., 2015. Pharmacognostical & pharmaceutical evaluation of *Chaturbhadraka vati* — An Ayurvedic tablet. Pharma Sci. Monit. 6, 12—20.

Chauhan, N.S., 2006. Medicinal and Aromatic Plants of Himachal Pradesh, second ed. Indus Publishing Company, New Delhi.

Chinese Pharmacopoeia Commission, 2005. Pharmacopoeia of the People's Republic of China. China Medical Science and Technology Press, Beijing, China.

Chinese Pharmacopoeia Commission, 2015. Pharmacopoeia of the People's Republic of China. China Medical Science and Technology Press, Beijing, China.

Dhamankar, R.S., Jadhav, A.P., 2016. Evaluation of *Shaddharana churna* — An Ayurvedic formulation. Indian J. Nat. Prod. Resour. 7, 301—309.

Ding, L.S., Wu, F.E., Chen, Y.Z., 1993. Diterpenoid alkaloids from *Aconitum gymnandrum*. Acta Pharm. Sin. 28, 188—191.

Dvornik, D., Edwards, O.E., 1964. The structure and stereochemistry of atisine. Can. J. Chem. 42, 137—149.

Dzhakhangirov, F.N., Sultankhodzhaev, M.N., Tashkhodzhaev, B., Salimov, B.T., 1997. Diterpenoid alkaloids as a new class of antiarrhythmic agents: structure activity relationship. Chem. Nat. Compd. 33, 190—202.

Friese, J., Gleitz, J., Gutser, U.T., Heubach, J.F., Matthiesen, T., Wilffert, B., Selve, N., 1997. *Aconitum* sp. alkaloids: the modulation of voltage-dependent Na^+ channels, toxicity and antinociceptive properties. Eur. J. Pharmacol. 337, 165—174.

Fujita, Y., Terui, K., Fujita, M., Kakizaki, A., Sato, N., Oikawa, K., Aoki, H., Takahashi, K., Endo, S., 2007. Five cases of aconite poisoning: toxicokinetics of aconitines. J. Anal. Toxicol. 31, 132—137.

Gajalakshmi, S., Jeyanthi, P., Vijayalakshmi, S., Devi Rajeswari, V., 2011. Phytochemical constituent of *Aconitum* species — A review. Int. J. Appl. Biol. Pharmaceut. Technol. 2, 121—127.

Gavín, J.A., Reina, M., Medinaveitia, A., Guadaño, A., Santana, O., Martínez-Díaz, R., Ruiz-Mesía, L., Alva, A., Grandez, M., Díaz, R., Gavín, J.A., la Fuente, J.D., 2004. Structural diversity and defensive properties of norditerpenoid alkaloids. J. Chem. Ecol. 30, 1393—1408.

Giri, A., Paramir, S.A., Kumar, A.P.V., 1993. Somatic embryogenesis and plant regeneration from callus cultures of *Aconitum heterophyllum* Wall. Plant Cell Tissue Organ Cult. 32, 213—218.

Giri, A., Banerjee, S., Ahuja, P.S., Giri, C.C., 1997. Production of hairy roots in *Aconitum heterophyllum* Wall. using *Agrobacterium rhizogenes*. In Vitro Cell. Dev. Biol. Plants 33, 280—284.

González, P., Marín, C., Rodríguez-González, I., Hitos, A.B., Rosales, M.J., Reina, M., Díaz, J.G., González-Coloma, A., Sánchez-Moreno, M., 2005. In vitro activity of C_{20}-diterpenoid alkaloid derivatives in promastigotes and intracellular amastigotes of *Leishmania infantum*. Int. J. Antimicrob. Agents 25, 136—141.

Gulati, K., Ray, A., Debnath, P.K., Bhattacharya, S.K., 2002. Immunomodulatory Indian medicinal plants. J. Nat. Remedies 2, 121—131.

Gupta, R., Gupta, A., Singh, R.L., 2015. Hepatoprotective activities of triphala and its constituents. Int. J. Pharm. Rev. Res. 4, 34—55.

Ichim, M.C., 2019. The DNA-based authentication of commercial herbal products reveals their globally widespread adulteration. Front. Pharmacol. 10, 1227.

Imtiyaz, S., Tariq, M., Chaudhary, S.S., 2013. Aphrodisiacs used in Unani system of medicine. J. Biol. Sci. Opin. 1, 239–242.

IUCN, 1993. Draft IUCN Red List Categories. IUCN, Gland, Switzerland.

Jabeen, N., Shawl, A.S., Dar, G.H., Sultan, P., 2006. Callus induction and organogenesis from explants of *Aconitum heterophyllum* – Medicinal plant. Biotechnology 5, 287–291.

Jabeen, N., Rehman, S., Bhat, K.A., Khuroo, M.A., Shawl, A.S., 2011. Quantitative determination of aconitine in *Aconitum chasmanthum* and *Aconitum heterophyllum* from Kashmir Himalayas using HPLC. J. Pharm. Res. 4, 2471–2473.

Jacobs, W.A., Craig, L.C., 1942a. The aconite alkaloids VIII: on atisine. J. Biol. Chem. 143, 589–603.

Jacobs, W.A., Craig, L.C., 1942b. The aconite alkaloids IX: the isolation of two new alkaloids from *Aconitum heterophyllum*, heteratisine and hetisine. J. Biol. Chem. 143, 605–609.

Jaiswal, Y., Liang, Z., Yong, P., Chen, H., Zhao, Z., 2013. A comparative study on the traditional Indian Shodhana and Chinese processing methods for aconite roots by characterization and determination of the major components. Chem. Cent. J. 7, 169.

Jaiswal, Y., Liang, Z., Ho, A., Wong, L.L., Yong, P., Chen, H., Zhao, Z., 2014. Distribution of toxic alkaloids in tissues from three herbal medicine *Aconitum* species using laser microdissection, UHPLC–QTOF MS and LC–MS/MS techniques. Phytochemistry 107, 155–174.

Jeelani, S.M., Siddique, M.A.A., Rani, S., 2015. Variations of morphology, ecology and chromosomes of *Aconitum heterophyllum* Wall., an endangered Alpine medicinal plant in Himalayas. Caryologia 68, 294–305.

Joshi, A.J., Aparna, K., Rajagopala, S., Patel, K.S., Harisha, C.R., Shukla, V.J., 2014. A preliminary pharmacognostical and pharmaceutical evaluation of *Bala chaturbhadra Avaleha*. Ann. Ayurvedic Med. 3, 20–28.

Joshi, A.J., Aparna, K., Rajagopala, S., Shanthibhai, P.K., Channapa, R.H., Vinay, S.J., 2016. Comparative standardization of different market samples of Ayurvedic formulation – *Balachaturbhadra Churna*. Int. J. Green Pharm. 10, S65.

Joshi, A.J., Aparna, K., Rajagopala, S., Shanthibhai, P.K., Nariya, M., Ashok, B.K., 2016. Evaluation of immunomodulatory activity of *Balachaturbhadra Churna* – An Ayurvedic formulation. Indian J. Nat. Prod. Resour. 7, 293–300.

Jowett, H.A., 1896. Contributions to our knowledge of the aconite alkaloids, part XIII: on atisine, the alkaloid of *Aconitum heterophyllum*. J. Chem. Soc. Trans. 69, 1518–1526.

Kita, Y., Ito, M., 2000. Nuclear ribosomal ITS sequences and phylogeny in East Asian *Aconitum* subgenus *Aconitum* (Ranunculaceae), with special reference to extensive polymorphism in individual plants. Plant Systemat. Evol. 225, 1–13.

Kita, Y., Ueda, K., Kadota, Y., 1995. Molecular phylogeny and evolution of the Asian *Aconitum* subgenus *Aconitum* (Ranunculaceae). J. Plant Res. 108, 429–442.

Konda, V.G.R., Eerike, M., Raghuraman, L.P., Rajamanickam, M.K., 2016. Antioxidant and nephroprotective activities of *Aconitum heterophyllum* root in glycerol induced acute renal failure in rats. J. Clin. Diagn. Res. 10, FF01–FF05.

Kumar, S.P., 2014a. Adulteration and substitution in endangered, costly herbal medicinal plants of India, investigates their active phytochemical constituents. Int. J. Pharm. Therapeut. 5, 243–260.

Kumar, S.P., 2014b. Adulteration and substitution in endangered, ASU herbal medicinal plants of India, their legal status, scientific screening of active phytochemical constituents. Int. J. Pharmaceut. Sci. Res. 5, 4023–4039.

Kumar, V., Chauhan, R.S., 2016. Higher amount of steviol detected in the leaves of a non-toxic endangered medicinal herb, *Aconitum heterophyllum*. J. Plant Biochem. Biotechnol. https://doi.org/10.1007/s13562-016-0361-y.

Kumar, P.U., Balachandran, I., Rema Shree, A.B., 2014. Standardisation of quality parameter and quantification of 6-shogoal in *Chaturbhadra Kvatha Churna* — A polyherbal Ayurvedic formulation. Eur. J. Biomed. Pharmaceut. Sci. 1, 83—97.

Kumar, V., Malhotra, N., Pal, T., Chauhan, R.S., 2016. Molecular dissection of pathway components unravel atisine biosynthesis in a non-toxic *Aconitum* species, *A. heterophyllum* Wall. 3 Biotech 6, 106.

Kumar, V., Raina, R., Sharma, S., 2016. Pollination in *Aconitum heterophyllum* Wall. — A critically endangered temperate Himalayan medicinal plant species. Indian For. 142, 1191—1194.

Kumar, V., Raina, R., Sharma, Y., 2016. Studies on seed source variation and seedling vigour in *Aconitum heterophyllum* Wall. Med. Plants 8, 238—243.

Lane, B., 2004. The Encyclopaedia of forensic science. Med. Hist. 36, 53—69.

Lather, A., Gupta, V., Bansal, P., Singh, R., Chaudhary, A.K., 2010. Pharmacological potential of ayurvedic formulation: Kutajghan vati — A review. J. Adv. Sci. Res. 1, 41—45.

Lin, C.C., Chan, T.Y., Deng, J.F., 2004. Clinical features and management of herb-induced aconitine poisoning. Ann. Emerg. Med. 43, 574—579.

Lone, P.A., Bhardwaj, A.K., 2013. Potent medicinal herbs used traditionally for the treatment of arthritis in Bandipora, Kashmir. Int. J. Recent Sci. Res. 4, 1766—1770.

Luo, Y., 2003. Taxonomic Revision of *Aconitum* L. (Ranunculaceae) from Sichuan, with a Study on the Phylogeny of This Genus Based on Molecular Evidence (Ph.D. thesis). Institute of Botany, Chinese Academy of Sciences, China.

Luo, Y., Zhang, F-m, Yang, Q.-E., 2005. Phylogeny of *Aconitum* subgenus *Aconitum* (Ranunculaceae) inferred from ITS sequences. Plant Systemat. Evol. 252, 11—25.

Mahajan, R., Kapoor, N., Singh, I., 2015. Effect of growth regulators on in vitro cultures of *Aconitum heterophyllum*: an endangered medicinal plant. Int. J. Pure Appl. Biosci. 3, 50—55.

Malhotra, N., 2017. Unraveling Molecular Biology of Aconites Biosynthesis in a High Value Medicinal Herb *Aconitum heterophyllum* Wall (Ph.D. thesis). Jaypee University of Information Technology, India.

Malhotra, N., Kumar, V., Sood, H., Singh, T.R., Chauhan, R.S., 2014. Multiple genes of mevalonate and non-mevalonate pathways contribute to high aconites content in an endangered medicinal herb, *Aconitum heterophyllum* Wall. Phytochemistry 108, 26—34.

Malhotra, N., Sood, H., Chauhan, R.S., 2016. Transcriptome-wide mining suggests conglomerate of genes associated with tuberous root growth and development in *Aconitum heterophyllum* Wall. 3 Biotech 6, 152.

Mishra, P., Kumar, A., Nagireddy, A., Mani, D.N., Shukla, A.K., Tiwari, R., Sundaresan, V., 2015. DNA barcoding: an efficient tool to overcome authentication challenges in the herbal market. Plant Biotechnol. J. https://doi.org/10.1111/pbi.12419.

Mitika, J., Sutkowska, A., Ilnicki, T., Joachimiak, A.J., 2007. Reticulate evolution of high-Alpine *Aconitum* (Ranunculaceae) in the Eastern Sudetes and Western Carpathians (Central Europe). Acta Biol. Cracov. Ser. Bot. 49, 15—26.

Mitra, S.K., Sachan, A., Udupa, V., Seshadri, S.J., Jayakumar, K., 2003. Histological changes in intestine in semichronic diarrhoea induced by lactose enriched diet in rats: effect of Diarex-Vet. Indian J. Exp. Biol. 41, 211—215.

Mori, T., Ohsawa, T., Murayama, M., 1989. Studies on *Aconitum* species, VIII, components of "Kako-Bushi-Matsu". Heterocycles 29, 873–884.

Murti, B.S.R., Khorana, M.L., 1968. Identification of aconites and estimation of alkaloids. Indian J. Pharm. 30, 206–208.

Nagarajan, M., Kuruvilla, G.R., Subrahmanya Kumar, K., Venkatasubramanian, P., 2015a. Abhava pratinidhi dravya: a comparative phytochemistry of Ativisha, Musta and related species. J. Ayurveda Integr. Med. https://doi.org/10.4103/0975-9476.146550.

Nagarajan, M., Kuruvilla, G.R., Subrahmanya Kumar, K., Venkatasubramanian, P., 2015b. A review of pharmacology of Ativisha, Musta and their substitutes. J. Ayurveda Integr. Med. https://doi.org/10.4103/0975-9476.146551.

Nakai, T., 1953. A new classification of *Lycoctonum* and *Aconitum* in Korea, Japan, and their surrounding areas. Bull. Natl. Sci. Mus. 32, 1–53.

Nariya, M.B., Parmar, P., Shukla, V.J., Ravishankar, B., 2011. Toxicological study of *Balacaturbhadrika churna*. J. Ayurveda Integr. Med. 2, 79–84.

Nautiyal, B.P., Prakash, V., Bahuguna, R., Maithani, U.C., Bisht, H., Nautiyal, M.C., 2002. Population study for monitoring the status of rarity of three aconite species in Garhwal Himalaya. Trop. Ecol. 43, 297–303.

Nautiyal, B.P., Vinay, P., Chauhan, R.S., Purohit, H., Nautiyal, M.C., 2006. Cultivation of *Aconitum* species. J. Trop. Med. Plants 6, 193–200.

Nirja, R., Sharma, M.L., 2016. Antidiabetic and antioxidant activity of ethanolic extract of *Ajuga parviflora Benth.* (Lamiaceae) vern. Neelkanthi, Neelbati. Int. J. Pharmaceut. Sci. Rev. Res. 41, 232–238.

Nisar, M., Ahmad, M., Wadood, N., Lodhi, M.A., Shaheen, F., Choudhary, M.I., 2009. New diterpenoid alkaloids from *Aconitum heterophyllum* Wall: selective butyrylcholinestrase inhibitors. J. Enzym. Inhib. Med. Chem. 24, 47–51.

NMPB, 2015. National Medicinal Plants Board Annual Report. Available at: https://www.nmpb.nic.in/.

Nyirimigabo, E., Xu, Y., Li, Y., Wang, Y., Agyemang, K., Zhang, Y., 2015. A review on phytochemistry, pharmacology and toxicology studies of *Aconitum*. J. Pharm. Pharmacol. https://doi.org/10.1111/jphp.12310.

Obaidullah, Ahmad, M.N., Ahmad, W., Tariq, S.A., ur Rahman, N., Ahmad, S., Ahmad, M., 2018. Isolation and characterization of C19-diterpenoid alkaloids from the roots of *Aconitum heterophyllum* Wall. J. Med. Chem. Drug Des. 1, 101.

O'Neil, M.J., Smith, A., Hecklman, P.E., Obenchain, J.R., Gallipeau, J.A.R., D'Arecca, M.A., 2001. The Merck Index, thirteenth ed. Merck and Co. Inc., New Jersey, p. 118.

Pal, T., Malhotra, N., Chanumolu, S.K., Chauhan, R.S., 2015. NGS transcriptomes reveal association of multiple genes and pathways contributing to secondary metabolites accumulation in tuberous roots of *Aconitum heterophyllum* Wall. Planta 242, 239–258.

Pandey, S., Kushwaha, R., Prakash, O., Bhattacharya, A., Ahuja, P.S., 2005. Ex situ conservation of *Aconitum heterophyllum* Wall.—an endangered medicinal plant of the Himalaya through mass propagation and its effect on growth and alkaloid content. Plant Genet. Resour. 3, 127–135.

Pandey, H., Nandi, S.K., Kumar, A., Agnihotri, R.K., Palni, L.M., 2008. Acontine alkaloids from the tubers of *Aconitum heterophyllum* and *A. balfourii*: critically endangered medicinal herbs of Indian central Himalaya. Natl. Acad. Sci. Lett. 31, 89–93.

Paramanick, D., Panday, R., Shukla, S.S., Sharma, V., 2017. Primary pharmacological and other important findings on the medicinal plant "*Aconitum heterophyllum*" (Aruna). J. Pharmacopuncture 20, 089–092.

Pattewar, A.M., Pandharkar, T.M., Yerawar, P.P., Patawar, V.A., 2012. Evaluation of in-vitro antihelminthic activity of *Aconitum heterophyllum*. J. Chem. Biol. Phys. Sci. 2, 2401–2407.
Pelletier, S.W., Aneja, R., 1967. The Diterpene alkaloids: three new diterpene lactone alkaloids from *Aconitum heterophyllum Wall*. Tetrahedron Lett. 6, 557–562.
Pelletier, S.W., Parthasarathy, P.C., 1965. The diterpene alkaloids: further studies of atisine chemistry. J. Am. Chem. Soc. 87, 777–798.
Pelletier, S.W., Aneja, R., Gopinath, K.W., 1968. The alkaloids of *Aconitum heterophyllum* Wall: isolation and characterization. Phytochemistry 7, 625–635.
Pelletier, S.W., Ateya, A.M., Finer-Moore, J., Mody, N.V., Schramm, L.C., 1982. Atisenol: a new ent-atisene diterpenoid lactone from *Aconitum heterophyllum*. J. Nat. Prod. 45, 779–781.
Polyakov, N.E., Khan, V.K., Taraban, M.B., Leshina, T.V., Salakhutdinov, N.F., Tolstikov, G.A., 2005. Complexation of lappaconitine with glycyrrhizic acid: stability and reactivity studies. J. Phys. Chem. B 109, 24526–24530.
Prasad, S.K., Kumar, R., Patel, D.K., Sahu, A.N., Hemalatha, S., 2012. Physicochemical standardization and evaluation of in-vitro antioxidant activity of *Aconitum heterophyllum* Wall. Asian Pac. J. Trop. Biomed. 2, S526–S531.
Prasad, S.K., Jain, D., Patel, D.K., Sahu, A.N., Hemalatha, S., 2014. Antisecretory and antimotility activity of *Aconitum heterophyllum* and its significance in treatment of diarrhoea. Indian J. Pharmacol. 46, 82–87.
Rah, T.A., Hemalatha, S., Elanchezhiyan, C., Archunan, G., 2016. Ameliorative effect of *Aconitum heterophyllum* (Wall ex Royle) on, erythrocyte antioxidants and lipid peroxidation in stz- induced diabetic rats. Eur. J. Pharmaceut. Med. Res. 3, 280–286.
Rai, M., Rai, A., Kawano, N., Yoshimatsu, K., Takahashi, H., Suzuki, H., Kawahara, N., Saito, K., Yamazaki, M., 2017. De Novo RNA sequencing and expression analysis of *Aconitum carmichaelii* to analyze key genes involved in the biosynthesis of diterpene alkaloids. Molecules 22, 2155.
Rajakrishnan, R., Lekshmi, R., Samuel, D., 2016. Analytical standards for the root tubers of Ativisha – *Aconitum heterophyllum* Wall. ex Royle. Int. J. Sci. Res. Publ. 6, 531–534.
Rajakrishnan, R., Alfarhan, A.H., Al-Ansari, A.M., Lekshmi, R., Sreelakshmi, R., Benil, P.B., Kim, Y.-O., Tack, J.-C., Na, S.W., Kim, H.-J., 2020. Therapeutic efficacy of the root tubers of *Aconitum heterophyllum* and its substitute *Cyperus rotundus* in the amelioration of pylorus ligation induced ulcerogenic and oxidative damage in rats. Saudi J. Biol. Sci. 27, 1124–1129.
Rana, B., Sreenivasulu, Y., 2013. Protein changes during ethanol induced seed germination in *Aconitum heterophyllum*. Plant Sci. 198, 27–38.
Rana, C.S., Radha, B., Sharma, A., Dangwal, L.R., Tiwari, J.K., 2013. Herbal remedies for Leucorrhoea: a study from the Garhwal Himalaya, India. Glob. J. Res. Med. Plants Indig. Med. 2, 685–691.
Rani, S., Kumar, S., Jeelani, S.M., Kumari, S., Gupta, R.C., 2011. Meiotic studies in some members of Ranunculaceae from Western Himalaya (India). Caryologia 64, 405–418.
Rastogi, R., Mehrotra, B.N., 1991. Compendium of Indian Medicinal Plants, second ed. Central Drug Research Institute, Lucknow and National Institute of Science Communication, New Delhi.
Rawat, J.M., Rawat, B., Bhandari, A., Yadav, S., Mishra, S., Chandra, A., Mishra, S.N., 2016. *Aconitum* biotechnology: recent trends and emerging perspectives. Acta Physiol. Plant. 38, 280.

Raymond-Hamet, 1954. Hypertensive effect of *Aconitum heterophyllum* Wallich. Comptes Rendus Seances Soc. Biol. Ses Fil. 148, 1221–1224.

Rungsung, W., Dutta, S., Das, D., Hazra, J., 2013. A brief review on the botanical aspects and therapeutic potentials of important Indian medicinal plants. Int. J. Herb. Med. 1, 38–45.

Seethapathy, G.S., Balasubramani, S.P., Venkatasubramanian, P., 2014. nrDNA ITS sequence based SCAR marker to authenticate *Aconitum heterophyllum* and *Cyperus rotundus* in Ayurvedic raw drug source and prepared herbal products. Food Chem. 145, 1015–1020.

Seethapathy, G.S., Ganesh, D., Kumar, J.U.S., Senthilkumar, U., Newmaster, S.G., Ragupathy, S., Shaanker, R.U., Ravikanth, G., 2015. Assessing product adulteration in natural health products for laxative yielding plants, Cassia, Senna, and Chamaecrista, in Southern India using DNA barcoding. Int. J. Leg. Med. 129, 693–700.

Selvaraj, S., Ramanathan, R., Vasudevaraja, V., Rajan, K.S., Krishnaswamy, S., Pemiah, B., Sethuraman, S., Ramakrishnan, V., Krishnan, U.M., 2014. Transcriptional regulation of the pregnane-X receptor by the Ayurvedic formulation Chandraprabha Vati. RSC Adv. 4, 64967.

Shah, N.C., 2005. Conservation aspects of *Aconitum* species in the Himalayas with special reference to Uttaranchal (India). In: Newsletter of the Medicinal Plant Specialist Group of the IUCN Species Survival Commission, IUCN Species Survival Commission, Gland and Cambridge, pp. 9–14.

Shanmughanandhan, D., Ragupathy, S., Newmaster, S.G., Mohanasundaram, S., Sathishkumar, R., 2016. Estimating herbal product authentication and adulteration in India using a vouchered, DNA-based biological reference material library. Drug Saf. https://doi.org/10.1007/s40264-016-0459-0.

Siddique, M.A.A., Beigh, S.Y., Wafai, B.A., 1998. Patris (*Aconitum heterophyllum*) Ranunculaceae – An important endangered N.W. Himalayan drug plant – problems and prospects. Proc. Natl. Acad. Sci. India 12, 23–25.

Siddique, M.A.A., Wafai, B.A., Mehmooda, 1998. Breeding system analysis of an important medicinal plant (*Aconitum heterophyllum* Wall ex Royle) of Kashmir Himalaya, I, Cytogenetics, resource allocation and propagation. In: Govil, J.N. (Ed.), Glimpses in Plant Research Current Concepts of Multidiscipline Approach to the Medicinal Plants (Part I). Today and Tomorrow Printers and Publishers, New Delhi, pp. 179–189.

Sinam, Y.M., Kumar, S., Hajare, S., Gautam, S., Devi, G.A.S., Sharma, A., 2014. Antibacterial property of *Aconitum heterophyllum* root alkaloid. Int. J. Adv. Res. 2, 839–844.

Sojitra, J., Dave, P., Pandya, K., Parikh, V., Patel, P., Patel, G., 2013. Standardization study of poly herbal formulation – Caspa Drops. Int. J. Pharmaceut. Sci. Drug Res. 5, 113–119.

Solanki, P., Siwach, P., 2012. Optimization of conditions for in vitro seed germination and shoot multiplication of *Aconitum heterophyllum* Wall. Int. J. Med. Aromatic Plants 2, 481–487.

Solo, A.J., Pelletier, S.W., 1962. Aconite alkaloids: the structure of hetisine. J. Org. Chem. 27, 2702–2703.

Solyanik, G.I., Fedorchuk, A.G., Pyaskovskaya, O.N., Dasyukevitch, O.I., Khranovskaya, N.N., Aksenov, G.N., Sobetsky, V.V., 2004. Anticancer activity of aconitine-containing herbal extract BC1. Exp. Oncol. 26, 307–311.

Srivastava, N., Sharma, V., Kamal, B., Jadon, V.S., 2010. *Aconitum*: need for sustainable exploitation (with special reference to Uttarakhand). Int. J. Green Pharm. https://doi.org/10.4103/0973-8258.74129.

Srivastava, N., Sharma, V., Dobriyal, A.K., Kamal, B., Gupta, S., Jadon, V.S., 2011. Influence of pre-sowing treatments on *in vitro* seed germination of ativisha (*Aconitum heterophyllum* Wall) of Uttarakhand. Biotechnology 10, 215–219.

Srivastava, N., Sharma, V., Saraf, K., Dobriyal, A.K., Kamal, B., Jadon, V.S., 2011. In vitro antimicrobial activity of aerial parts extracts of *Aconitum heterophyllum* Wall. ex Royle. Indian J. Nat. Prod. Resour. 2, 504–507.

Subash, A.K., Augustine, A., 2012. Hypolipidemic effect of methanol fraction of *Aconitum heterophyllum* Wall ex Royle and the mechanism of action in diet-induced obese rats. J. Adv. Pharm. Technol. Res. 3, 224–228.

Tai, C.-J., El-Shazly, M., Wu, T.-Y., Lee, K.-T., Csupor, D., Hohmann, J., Chang, F.-R., Wu, Y.-C., 2015. Clinical aspects of *Aconitum* preparations. Planta Med. 81, 1017–1028.

Tamura, M., 1995. Ranunculaceae. In: Hiepko, P. (Ed.), Die Natürlichen Pflanzenfamilien, Zweite Auflage. Duncker und Humblot, Berlin, pp. 274–291.

Uddin, S.J., Mondal, K., Shilpi, J.A., Rahman, M.T., 2006. Antidiarrhoeal activity of *Cyperus rotundus*. Fitoterapia 77, 134–136.

Ukani, M.D., Mehta, N.K., Nanavati, D.D., 1996. *Aconitum heterophyllum* (ativisha) in ayurveda. Ancient Sci. Life 16, 166–171.

Venkatasubramanian, P., Kumar, S.K., Nair, V.S., 2010. *Cyperus rotundus*, a substitute for *Aconitum heterophyllum*: studies on the Ayurvedic concept of Abhava Pratinidhi Dravya (drug substitution). J. Ayurveda Integr. Med. 1, 33–39.

Verma, S., Ojha, S., Raish, M., 2010. Anti-inflammatory activity of *Aconitum heterophyllum* on cotton pellet-induced granuloma in rats. J. Med. Plants Res. 4, 1566–1569.

Wang, W.T., 1965. Notulae de *Ranunculaceis sinensibus* (II). Acta Phytotax. Sin. Addit. 1, 49–103.

Xie, Y., Jiang, Z.H., Zhou, H., Xu, H.X., Liu, L., 2005. Simultaneous determination of six *Aconitum* alkaloids in proprietary Chinese medicines by high-performance liquid chromatography. J. Chromatogr. A 1093, 195–203.

Yang, Q.E., 1990. Systematic Studies on the Genus *Aconitum* L. (Ranunculaceae) from Yunnan (Ph.D. thesis). Kunming Institute of Botany, Chinese Academy of Sciences, China.

Yin, T., Zhou, H., Cai, L., Ding, Z., 2019. Non-alkaloidal constituents from the genus *Aconitum*: a review. RSC Adv. 9, 10184.

Zhaohong, W., Wang, J., Xing, J., He, Y., 2006. Quantitative determination of alkaloids in four species of *Aconitum* by HPLC. J. Pharmaceut. Biomed. Anal. 40, 1031–1034.

CHAPTER

Arnebia euchroma

3

Kirti Chawla[1,4,#], Ramgopal Mopuri[2,#], Anil K. Sharma[4], Pawan Kumar[3,4]

[1]*Plant Tissue Culture and Genetic Engineering, National Agri-Food Biotechnology Institute (NABI), Mohali, Punjab, India;* [2]*Institute of Animal Science, Agricultural Research Organization (ARO), The Volcani Center, Rishon LeZion, Israel;* [3]*Institute of Plant Science, Agricultural Research Organization (ARO), The Volcani Center, Rishon LeZion, Israel;* [4]*Department of Biotechnology, Maharishi Markandeshwar (Deemed to be University), Mullana, Ambala, Haryana, India*

3.1 Introduction

Medicinal plants, referred to as "medicinal herbs", are receiving unrivaled attention across the world because they have enormous medicinal properties attributed to phytochemicals or natural products. One important medicinal plant found (3200–4500 height about mean sea level) in the northwestern regions of the Himalayas is *Arnebia euchroma* (Royle ex Benth), often known as ratanjot (Aswal and Mehrotra, 1994). Different types of secondary metabolites found in *A. euchroma* correspond to its medicinal value across the globe. The dried roots of this plant are used in industry to extract natural compounds for the preparation of drugs. Owing to various medicinal properties, the plant is overexploited, which has resulted in its population decline and in its critically endangered status (Kala, 2000). The plant species seems to be difficult to grow via conventional agricultural practices; thus, it has failed to fulfill growing industrial demand (Gupta et al., 2014). Therefore, plant cell culture technologies could be an alternative approach to fulfill the escalating demand for the extraction of natural metabolites of *A. euchroma* (Onrubia et al., 2012; Verpoorte et al., 2002; Sharma et al. 2008).

3.2 Physiological structure

It is an erect, hairy perennial herb 15–40 cm tall. Roots are thick and exclude a purple dye. Stems are generally one or several, and branched; they are erect, spreading pale yellow or white, and hirsute (Fig. 3.1). Leaves are linear with long bristly hair and are thorny, which protects the plant from grazing. Flowers are many-flowered

[#] Authors contributed equally.

Himalayan Medicinal Plants. https://doi.org/10.1016/B978-0-12-823151-7.00002-7

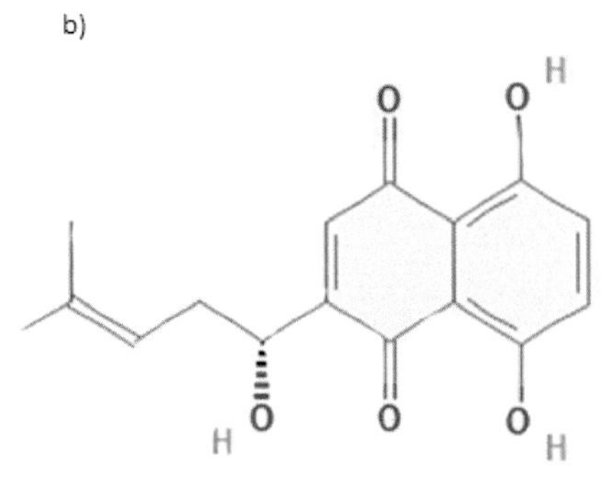

FIGURE 3.1

(a) *Arnebia* species in its natural habitat. (b) Structure of shikonin.

cymes with rounded and terminal pink-purple clusters that later turn blackish-purple and are almost stalkless. Flowers are heterostylous, hermaphrodite, and insect pollinated (Singh et al., 2017). The calyx is dense, pale yellow, and hirsute on both sides; the corolla is tubular-campanulate and is dark purple, often pale yellow or red tinged. Less fruit setting is observe in plants, because the plants are self-incompatible. Most of the seeds fall in the base vicinity of the plant and germinate close to each other, creating a dense population of the plant. The flowers bloom from Jun. to Aug. and seeds are ripe from Jul. to Sep. The plant is suitable for growth in sandy soils, rocky slopes, meadows, and gravelly marshes.

3.3 Medicinal properties of *Arnebia euchroma*

Dried *Arnebia* root powder is used to cure cough and lung problems as well as to treat menorrhagia. Mixed with apricot or mustard oil, the root fiber gives off a red color that can be used as a hair oil to control dandruff and hair loss. The roots can also be used to treat skin disorders such as burns (Pirbalouti et al., 2011). The plant is used as a colorant for designing cups and dyeing cloth, and in various dish preparations. The whole plant was used to improve soil fertility by mixing it with soil (Uniyal et al., 2002). Doctors in the western Himalayas use the root extract to make native medicines prescribed for blood purification and cough (Shen et al., 2002).

In China, drugs obtained from this plant are used as an anti-HIV medicine. They possesses an Me_2CO aqueous extract of *A. euchroma* ($\leq$20 μg/mL), which has inhibitory activity against HIV-infected H9 cells (Kashiwada et al., 1995). Traditionally, in China, the plant is used in medicines to cure smallpox, facilitate the passage of stool, and cool the blood (Chang et al., 1993), and it has contraceptive action (Ren et al., 2011; Shan Hai Chung I Yao ZaZhi 5, 3–5). Shikonin, the secondary metabolite produced by the plant, has anticancer (Sankawa et al., 1977), antibacterial (Gao, 1986), antifungal (Kyogoku et al., 1973), and antiviral properties. Shikonin also possess inhibitory activity against platelet aggregation (Chang et al., 1993). Moreover, the plant inhibits the growth of *Escherichia coli, Bacillus typhi, Staphylococcus aureus, Pseudomonas,* and *B. dysenteriae.* The root extracts of *A. euchroma* have a high wound-healing capacity (Aliasl et al., 2014). The plant is also useful for treating urinary disorders and kidney-associated complications (Hosseini et al., 2018). A clinical trial of *A. euchroma* extracts showed the potential ability to cure diabetes through its role in inhibiting adipocyte signaling and glucose sensitivity (Lee et al., 2010).

A. euchroma is traditional medicinal plant that grows in the Himalayan regions of India and some parts of China. Various parts, especially roots of this plant, were used in traditional medicine to cure many health problems (Ou et al., 2017). The crude extracts and bioactive compounds of *A. euchroma* have versatile biological activities attributed to different kinds of metabolites (Fig. 3.2). Phytochemicals isolated from *A. euchroma* are shikonin, alkannin, naphthoquinone, shikometabolin H (a new meroterpenoid), epoxyarnebinol, and iso-hexyl-naphthopurpurin had potential health benefits among the compounds that are available in the roots or other parts of *A. euchroma* (Fig. 3.3 and Table 3.1).

3.3.1 Anticancer effects

The phytocompound deoxyshikonin isolated from *A. euchroma* significantly downregulated the proteins of PI3K and the p-Pl3K/Akt/mTOR pathway in HT29 and DLD-1 cells. Acetylshikonin isolated from *A. euchroma* is a potential inhibitor of tumor growth in human lung adenocarcinoma cell A549 (Xiong et al., 2009). Preliminary clinical studies revealed that shikonin exerts additive and synergetic interactions in combination with potential pharmacological drugs used in cancer therapy (Boulos et al., 2019).

3.3.2 Antiinflammatory effects

The polysaccharides available in *A. euchroma* modulate body temperature, reduce the number of leukocytes, improve the complement system and lung permeability, and lower oxidative stress (Ou et al., 2017). In vivo studies of 10 mg/kg per day shikonin, a derivative of *Lithospermum* (the dry root of borage perennial, the herbaceous plant *A. euchroma*), inhibits inflammation and chondrocyte apoptosis thorough the PI3K/Akt pathway (Fu et al., 2016).

3.3.3 Antiobesity effects

The prevalence of obesity is a global health issue linked to many metabolic complications. One comorbidity is metabolic syndrome, which is correlated with body waist circumference and abdominal fat thickness. Methods are widely available to reduce fat thickness around the abdomen, such as liposuction, to remove fat in specific parts. External application of an ointment made with extracts of *A. euchroma* were reported to have potential efficacy in obese women, and to reduce body weight (2.96 kg), abdominal fat thickness (2.3 cm), and abdominal circumference (11.3 cm) (Siavash et al., 2016).

Phytochemical compounds found in *A.euchroma*

a) S-enantiomer

R= -H = Alkanin

b)R-enantiomer

R= -H = Shikonin

c) 1, 4 Naphthoquinone

d)Isohexyl naphthopurpurin

e)Deoxyshikonin

f) Acetylshikonin

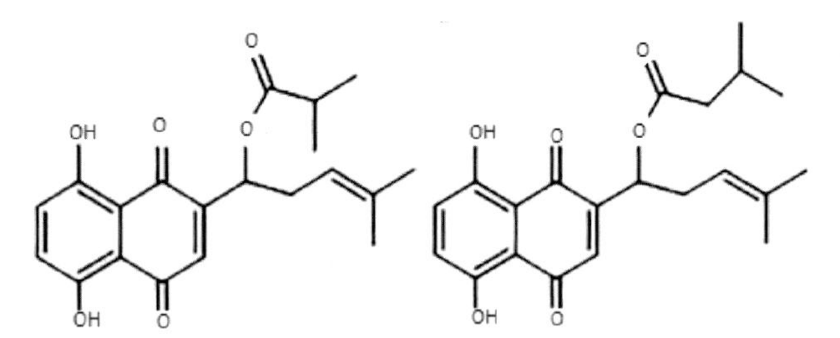

g)Isobutyrylshikonin

h)β,β′-dimethylacrylshikonin

FIGURE 3.2

Phytochemical compounds identified in *Arnebia euchroma*.

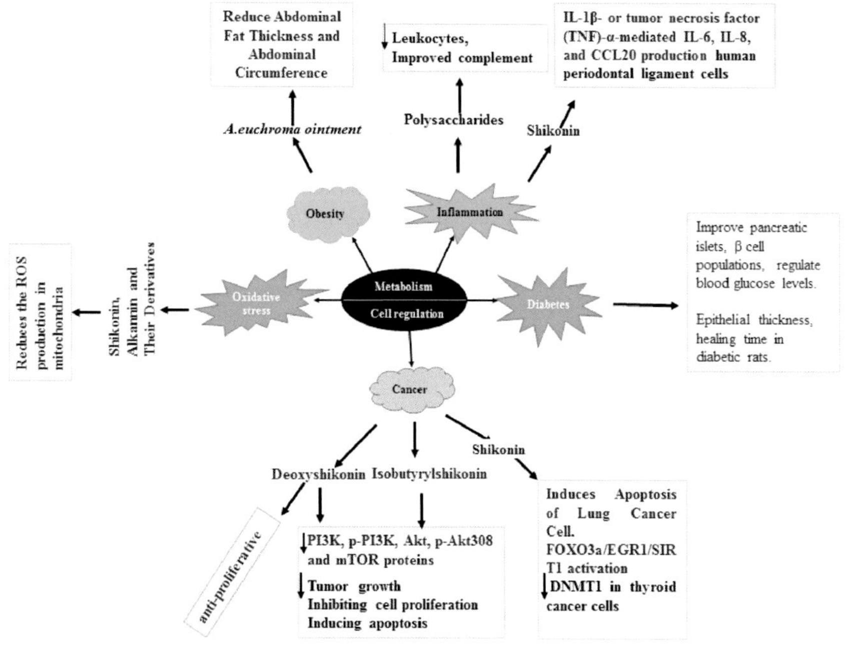

FIGURE 3.3

Health consequences according to dysregulated metabolic or cell function, and possible therapy by *Arnebia euchroma*. *IL*, interleukin; *ROS*, reactive oxygen species.

Table 3.1 Shikonin derivatives and their pharmacological properties.

Name	Properties	References
Shikonin	Antitumor, antipyretic, antifungal, antibacterial, wound healing, stimulation of peroxidase, secretion and induction of nerve growth factors	Ge et al. (2006), Dai et al. (2009), Andujar et al. (2012)
Teracrylshikonin	Antimicrobial	Shen et al. (2002), Sasaki et al. (2002)
β,β-Dimethyl acryl shikonin	Antimicrobial	Shen et al. (2002)
Deoxy alkannin, deoxy shikonin, or arnebin-7	Antidermatophytic, antibacterial, antitumor	Shen et al. (2002), Sasaki et al. (2002)
Isovaleryl alkannin dimer	Human lung cancer and human bronchial cancer	Liao et al. (2020)
β-Acetoxy-isovaleryl alkannin	Human lung cancer and human bronchial cancer	Liao et al. (2020)

3.3.4 Antidiabetic and diabetic wound-healing activity

A stereological study on rats orally administered *A. euchroma* extract at a dose of 100 or 300 mg kg/body weight resulted in improved pancreatic islet volume, β-cell population and regulated blood glucose levels (Noorafshan et al., 2017). *A. euchroma* also has potential applications for diabetic foot ulcers; significant effects were found for epithelial thickness and complete healing time (Isharif et al., 2019). The root phytochemical extracted by hexane and further formulated as an ointment had significant wound-healing activity (Akkol et al., 2009).

Healing of wounds is a complex process leading to the regeneration of damaged skin tissue. Through its fibroblast-regulating activity, a gel made from *A. euchroma* showed excisional wound-healing properties (Mohsenikia et al., 2017). The Sheng-ji Hua-Yu formulation, which has *A. euchroma* as one of the components, possesses diabetic potential wound healing though reepithelization (Kuai et al., 2018) (Fig. 3.4).

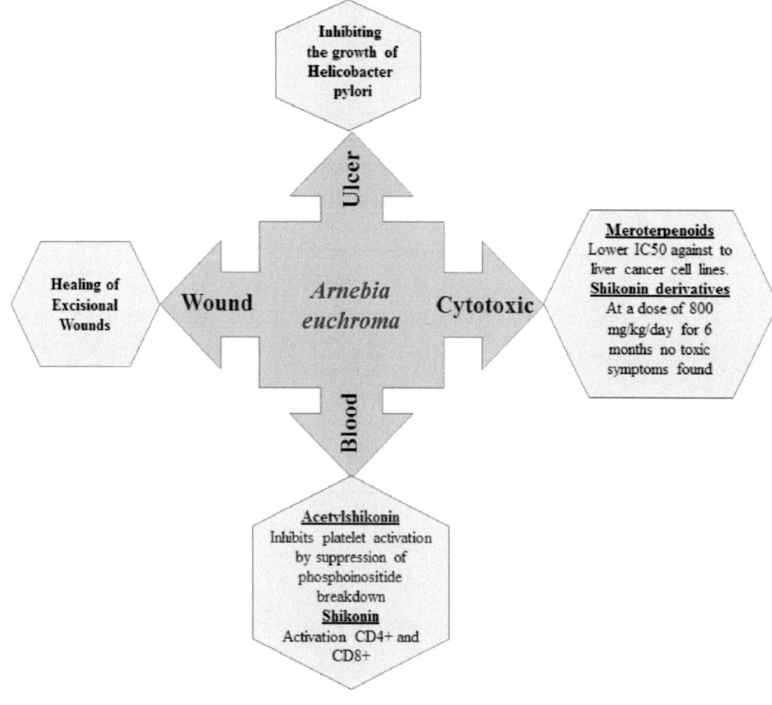

FIGURE 3.4

Cytotoxic, wound-healing, ulcer, and blood cells effect of *Arnebia euchroma*.

3.3.5 Cytotoxic activity of *Arnebia euchroma*

Cytotoxic studies are one of the most important parameters for assessing the dose concentration that is safe for respective species. The meroterpenoids isolated from *A. euchroma* gave potent IC_{50} activity against MMC-7721 (6.40 μM), HepG2 (3.86 μM), QGY-7703 (3.43 μM), and HepG2/ADM (11.31 μM) human liver cancer cell lines (Wang et al., 2018) (Fig. 3.3). Novel phytochemical compounds isolated from the roots were tested against cytotoxicity in different cancer cells (human leukemia cell CCRF-CEM, breast cancer cell MDA-MB-231, human glioblastoma cell U251, and colon cancer cell HCT 116); the propionyl alkannin had potent cytotoxic activity with low IC_{50} values (Damianakos et al., 2012). Use of the extract of *A. euchroma* against human gastric adenocarcinoma cells resulted in significant cytotoxic activity in a dose-dependent manner (Nadoushan et al., 2016).

3.4 In vitro production strategies

Because of its many medicinal properties, *A. euchroma* is being overexploited and therefore is considered a under critically endangered species. Moreover, cultivation of this plant through conventional agricultural practices is difficult because it is challenging to germinate from seed (Malik et al., 2011). Therefore, many strategies can be undertaken to increase the shikonin content, either through metabolic engineering of various genes involved in shikonin biosynthesis or by in vivo production of shikonin by plant cell culture.

Metabolic engineering is a conventional method of optimizing regulatory and genetic processes within the cell to increase the production of certain substances within the cell. In *A. euchroma*, geranyl pyrophosphate (GPP) and p-hydroxybenzoate (PHB) are basic precursors of the shikonin biosynthesis pathway. In GPP, it is formed by cytosolic the mevalonate (MVA) pathway; however, PHB is formed by the phenylpropanoid pathway. The MVA pathway is inhibited by mevinolin, an inhibitor of the HMGR enzyme. GPP and PHB are coupled by a reaction catalyzed by PHB to form m-geranyl-p-hydroxybenzoate (GHB). The deletion of mevinolin-expressing gene can increase the HMGR concentration, which can further increase the shikonin content and overexpression of GHB from more GHB, which can also increase the shikonin content. Overexpression of PHB can also increase the shikonin content However, to the best of the authors' knowledge, no such research has been done to date. Boehm et al. (2000) manipulated the shikonin biosynthesis pathway by introducing the bacterial gene *UbiA*. It catalyzes the formation of 3-geranyl-4-hydroxybenzoate (GBA), using GPP as substrate. This resulted in an increase in GBA concentration by a factor of 50 compared with control. However, a low increase (22%) in the shikonin content was observed in transformants compared with control. This indicates that overexpression of *UbiA* alone is insufficient to increase the shikonin content, and various other genes are responsible for increasing the shikonin content. The yield of shikonin can also be increased by irradiating cell suspension cultures at 2, 16, or 32 Gy. Gamma irradiation stimulates the activity of PHB, a key enzyme of shikonin biosynthesis. At 16 Gy gamma irradiation, the shikonin content increased by 400%, whereas at 2 and 32 Gy, the shikonin content increased by only 240% and 180%, respectively (Chung et al., 2006).

Another way to increase the shikonin content is by controlling the production of various plant metabolites by cell cultures. This not only relieves the pressure on the natural habitats of plant species, it provides suitable conditions for year-round metabolite production (Sood, 2020). The production of plant secondary metabolites can be enhanced by exposing cell cultures to biotic or abiotic elicitors. The effects of various fungal elicitors on shikonin production of *A. euchroma* were studied, prepared from *Aspergillus niger* and *Rhizopus oryzae* during suspension cultures in Erlenmeyer flasks. Addition of fungal elicitors to culture medium led to a rapid increase in shikonin content. A maximal concentration of 89.75 mg/L was achieved on the sixth day of addition compared with control (without elicitors). *R. oryzae* elicitors increased the shikonin content 2.24-fold higher than that of control. Moreover, when fungal elicitation and in situ hybridization were combined, the shikonin content increased by 6.15-fold compared with that of the control (Fu and Lu, 1999).

3.5 Physiobiological factors affecting shikonin production

Various physical factors such as light, temperature and pH, and nutritional factors such as carbon source and nitrogen concentration also affect the shikonin content. Gupta et al. (2016) showed that until the fourth day of growth, cell suspensions increased in shikonin content irrespective of light conditions. However, as the culture period progressed, there was a decrease in shikonin content under light conditions, but the shikonin content increased under dark conditions (Gupta et al., 2016). (Malik et al., 2011) showed that of the various temperature conditions, the maximum shikonin yield in cell cultures was at 25°C (Malik et al., 2011). Sucrose acts like a substrate; not only does it have a role in cell growth, it also influences the shikonin content. A maximum content of shikonin was observed at 6% sucrose concentration (Malik et al., 2011). The pH of the medium also has a pronounced effect on shikonin content. As the pH of the medium increased from 5.0 to 9.5, there was a simultaneous increase in shikonin content; a maximum content of shikonin was found at pH = 8.75 (Gupta et al., 2016; Malik et al., 2011). Nitrogen concentration also effects shikonin accumulation. It was shown that a high N concentration in a basal medium retards or inhibits shikonin production (Mizukami et al., 1977). Mizukami et al. (1977) showed that Fe as $FeSO_4$ and Ca as $CaCl_2$ ions at a high concentration inhibit shikonin content. Wang et al. (2014) formed a two-cell suspension line of *A. euchroma*: (1) red shikonin-proficient cell lines and (2) white shikonin-deficient cell lines. Administration of methyl-jasmonate (MeJA) to cell suspension cultures of *A. euchroma* resulted in the increased accumulation of shikonin derivatives, rosmarinic acid, and shikonofuran derivatives in red shikonin-proficient lines (RCL). Suspension culture experiments showed that MeJA leads to a rapid increase in the expression of five important genes (HMGR, GDPS, C_4H, $_4CL$, and PGT) of shikonin biosynthesis in white shikonin-proficient line. The total shikonin yield in RCL cell lines treated with MeJA increased by 1.52-fold compared with that of untreated cells. Kumar et al. (2014) showed the effect of salicylic acid treatment in the cell culture of *Arnebia* euchroma to elicit shikonin content.

3.6 Omics advancement to unravel the biosynthetic machinery of shikonin

In the modern era, the genomic resource has been lacking for *A. euchroma* to date. However, transcriptome resources of *A. euchroma* could reveal the molecular components (pathway genes, transcription factors, microRNAs, and transporters) associated with the biosynthesis of shikonin, a highly valuable metabolite. In other medicinal plant species such as *Podophyllum hexandrum, Swertia chirayita, Picrorhiza kurroa* etc., transcriptome resources has been used to identified the crucial molecular components including key genes, transcription factors and miRNAs associated with the biosynthetic pathways of secondary metabolite production (Kumar et al., 2017, 2018; Padhan et al., 2016; Pal et al., 2018; Shitiz et al., 2015). The MVA pathway, methylerythritol 4-phosphate pathway, and phenylpropanoid pathway contributed to the biosynthesis of shikonin and its derivatives (Fig. 3.5). A transcriptome study showed that p-hydroxybenzoate geranyl transferase (PGT) genes have a vital role in the biosynthetic pathways of shikonin and their derivatives. The transcriptome database of *A. euchroma* unraveled six PGTs that have crucial significance in shikonin biosynthesis. The higher expression of PGTs in the underground parts of *A. euchroma* is directly correlated with the higher content of shikonin (Liu et al., 2016). Similarly, 27 ERF transcription factor family genes identified from the transcriptome resource of *A. euchroma* have a significant role in shikonin biosynthesis (Xie et al., 2014). A study by Takanashi et al. (2019) demonstrated that some of the genes and enzymes (O-acetyltransferase, polyphenol oxidase, cannabidiolic acid synthase-like, and neomenthol dehydrogenase-like protein) possess an important role in shikonin biosynthesis through the use of proteomic and transcriptomic approaches. Wang et al. (2019) showed that CYP76B74 catalyzes the $3''$-hydroxylation of geranyl hydroquinone in the biosynthesis of shikonin, and thus provided an update on its biosynthetic machinery.

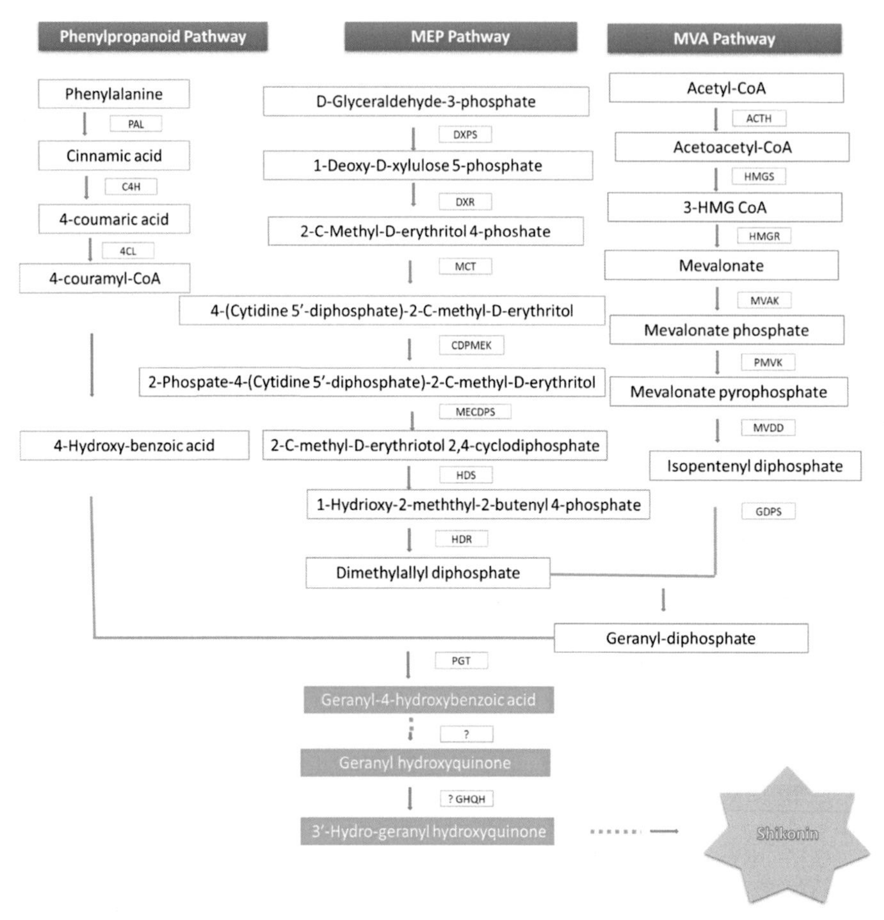

FIGURE 3.5

Biosynthetic pathway of shikonin. *4CL*, 4-coumaroyl-CoA ligase; *C4H*, cinnamic acid 4-hydroxylase; *CDPMEK*, 4-(cytidine 5′-diphospho)-2-C-methyl-D-erythritol 2-phosphokinase; *DXPS*, 1-deoxy-D-xylulose-5-phosphate synthase; *DXR*, 1-deoxy-D-xylulose-5-phosphate reductoisomerase; *GDPS*, geranyl-diphosphate synthase; *GHQH*, geranyl hydroquinone 3″-hydroxylase; *HDR*, hydroxy-2-methyl-2-(E)-butenyl-4-diphosphate reductase; *HDS*, 1-hydroxy-2-methyl-2-butenyl-diphosphate synthase; HMGR, 3-hydroxy-3-methylglutaryl-CoA reductase; *HMGS*, 3-hydroxy-3-methylglutaryl-CoA synthase; *IPPI*, isopentenyl pyrophosphate isomerase; *MCT*, 2-C-methyl-D-erythritol 4-phosphate cytidylyltransferase; *MECDPS*, 2-C-methyl-D-erythritol 2,4-cyclodiphosphate synthase; *MEP*, 2-C-methyl-D-erythritol-4-phosphate; *MVA*, mevalonate; *MVAK*, mevalonate 5-phosphokinase; *MVDD*, mevalonate diphosphate decarboxylase; *PAL*, phenylalanine ammonia lyase; *PGT*, 4-hydroxybenzoate-3-geranyltransferase; *PMVK*, 5-phosphomevalonate phosphokinase.

Adapted from Wang, S., Wang, R., Liu, T., Lv, C., Liang, J., Kang, C., Zhou, L., Guo, J., Cui, G., Zhang, Y., Werck-Reichhart, D., Guo, L., Huang, L., 2019. CYP76B74 catalyzes the 3″-hydroxylation of geranylhydroquinone in shikonin biosynthesis. Plant Physiol. doi:10.1104/pp.18.01056, Kumar, P., Kumar, V., Garlapati, V.K., 2016a. Biosynthesis and pharmacological evaluation of shikonin-A highly valuable metabolite of North-Western Himalayas: mini review. Med. Plants Int. J. Phytomed. Related Ind. 8, 267–274.

3.7 Conclusion and future perspectives

This chapter compiles information about the physiological attributes, in vitro production platform, biosynthetic pathways, medicinal properties, and physiobiological factors of *A. euchroma*, a highly valuable medicinal plant species from the northwestern Himalayas. Transcriptome resources are available to unravel molecular components (pathway genes, transcription factors (TFS), ABC transporters, etc.) associated with the biosynthesis of shikonin and its derivatives. Various elicitors have been tried in tissue culture practices to enhance the in vitro content of shikonin; however, the increased content is low. There is an urgent need for the full genome sequencing of *A. euchroma* so that the shikonin biosynthetic machinery might unravel and provide a robust platform to explore the crucial molecular elements. Plants of *A. euchroma* are highly used in industrial sectors to extract anticancer drugs, but the supply of raw material is low. Because of the high demand for *A. euchroma*, it is highly exploited and thus has been placed into the category of critical endangered medicinal plant species from the northwestern Himalayas. Thereby, there is an essential need to protect this highly valuable medicinal plant species in its natural habitat. Omics technologies include transcriptomics, metabolomics, and proteomics, which offer a platform to explore the genes—metabolite relationship to fulfil the gap in the biosynthesis of shikonin.

References

Akkol, E., Koca, U., Demet, Y., Toker, G., Yesilada, E., 2009. Exploring the wound healing activity of *Arnebia densiflora* (Nordm.) Ledeb. by in vivo models. J. Ethnopharmacol. 124 (1), 137—141.

Aliasl, J., Khoshzaban, F., Barikbin, B., Naseri, M., Kamalinejad, M., et al., 2014. Comparing the healing effects of *Arnebia euchroma* ointment with petrolatum on the ulcers caused by fractional CO2 laser: a single-blinded clinical trial. Iran. Red Crescent Med. J. 16 (10), e16239.

Andújar, I., Ríos, J.L., Giner, R.M., Miguel Cerdá, J., Recio, M.D.C., 2012. Beneficial effect of shikonin on experimental colitis induced by dextran sulfate sodium in BALB/c mice. Evid. base Compl. Alternative Med. 2012.

Aswal, B.S., Mehrotra, B.N., 1994. Flora of Lahaul-Spiti (a cold desert in North West Himalaya). M/s Bishen Singh Mahendra Pal Singh; First edition, pp. 10—15. In press.

Boehm, R., Sommer, S., Li, S.M., Heide, L., 2000. Genetic engineering on shikonin biosynthesis: expression of the bacterial ubiA gene in *Lithospermum erythrorhizon*. Plant Cell Physiol. 41 (8), 911—919.

Boulos, J.C., Rahama, M., Hegazy, M.F., Efferth, T., 2019. Shikonin derivatives for cancer prevention and therapy. Canc. Lett. 459 (10), 248—267.

Chang, Y.S., Kuo, S.C., Weng, S.H., Jan, S.C., Ko, F.N., Teng, C.M., 1993. Inhibition of platelet aggregation by shikonin derivatives isolated from Arnebia euchroma. Planta medica 59 (05), 401—404.

Chung, B.Y., Lee, Y.B., Baek, M.H., Kim, J.H., Wi, S.G., Kim, J.S., 2006. Effects of low-dose gamma-irradiation on production of shikonin derivatives in callus cultures of *Lithospermum erythrorhizon* S. Radiat. Phys. Chem. 75 (9), 1018–1023.

Dai, Q., Fang, J., Zhang, F.S., 2009. Dual role of shikonin in early and late stages of collagen type II arthritis. Mol. Biol. Rep. 36 (6), 1597–1604.

Damianakos, H., Kretschmer, N., Sykłowska-Baranek, K., Pietrosiuk, A., Chinou, I., 2012. Antimicrobial and cytotoxic isohexenylnaphthazarins from *Arnebia euchroma* (Royle) Jonst. (Boraginaceae) callus and cell suspension culture. Molecules 17, 14310–14322.

Fu, D., Shang, X., Ni, Z., Shi, H., 2016. Shikonin inhibits inflammation and chondrocyte apoptosis by regulation of the PI3K/Akt signaling pathway in a rat model of osteoarthritis. Exp. Ther. Med. 12 (4), 2735–2740.

Fu, X.Q., Lu, D.W., 1999. Stimulation of shikonin production by combined fungal elicitation and in situ extraction in suspension cultures of *Arnebia euchroma*. Enzym. Microb. Technol. 24 (5–6), 243–246.

Gao, J., 1986. A general review of resources, chemistry, pharmacology and clinical application of Gromwell. Zhongcaoyao 17, 28–31.

Ge, F., Wang, X., Zhao, B., Wang, Y., 2006. Effects of rare earth elements on the growth of *Arnebia euchroma* cells and the biosynthesis of shikonin. Plant Growth Regul. 48, 283–290.

Gupta, B., Chakraborty, S., Saha, S., Chandel, S.G., Baranwal, A.K., Banerjee, M., Chatterjee, M., Chaudhury, A., 2016. Antinociceptive properties of shikonin: in vitro and in vivo studies. Can. J. Physiol. Pharmacol. 94 (7), 788–796.

Gupta, K., Garg, S., Singh, J., Kumar, M., 2014. Enhanced production of napthoquinone metabolite (shikonin) from cell suspension culture of *Arnebia* sp. and its up-scaling through bioreactor. 3 Biotech 4 (3), 263–273.

Hosseini, A., Mirzaee, F., Davoodi, A., Jouybari, H.B., Azadbakht, M., 2018. The traditional medicine aspects, biological activity and phytochemistry of *Arnebia* spp. Med. Glas. 15 (1).

Isharif, A., Shafiei, E., Hosseinzadeh, M., 2019. Comparison of the effects of oral *Arnebia euchroma* and oral ANGIPARS on wounds in diabetic rats. Int. J. Pharmaceut. Res. 11 (3), 120–123.

Kala, C.P., 2000. Status and conservation of rare and endangered medicinal plants in the Indian trans- Himalaya. Biol. Conserv. 93, 371–379.

Kashiwada, Y., Nishizawa, M., Yamagishi, T., Tanaka, T., Nonaka, G.I., Snider, J.V., Lee, K.H., 1995. Sodium and potassium salts of caffeic acid tetramers from *Arnebia euchroma* as anti-HIV agents. J. Nat. Prod. 58, 392–400.

Kuai, L., Zhang, J., Deng, Y., Xu, S., Xu, X., Wu, M., Guo, D., Chen, Y., Wu, J., Zhao, X., Nian, H., Li, B., 2018. Sheng-ji Hua-yu formula promotes diabetic wound healing of re-epithelization via Activin/Follistatin regulation. BMC Compl. Alternative Med. 18, 32.

Kumar, P., Saini, M., Bhushan, S., Warghat, A.R., Pal, T., Malhotra, N., Sood, A., 2014. Effect of salicylic acid on the activity of PALand PHB geranyl transferase and shikonin derivatives production in cell suspension cultures of *Arnebia ecuhroma* RoyleJohnst- a medicinally important plant species. Appl. Biochem. Biotechnol. 173, 248–258.

Kumar, P., Kumar, V., Garlapati, V.K., 2016a. Biosynthesis and pharmacological evaluation of shikonin-A highly valuable metabolite of North-Western Himalayas: mini Review. Med. Plants Int. J. Phytomed. Related Ind. 8, 267–274.

Kumar, P., Sharma, R., Jaiswal, V., Chauhan, R.S., 2016b. Identification, validation, and expression of ABC transporters in *Podophyllum hexandrum* and their role in podophyllotoxin biosynthesis. Biol. Plant. 60 (3), 452–458.

Kumar, P., Jaiswal, V., Pal, T., Singh, J., Chauhan, R.S., 2017. Comparative whole-transcriptome analysis in Podophyllum species identifies key transcription factors contributing to biosynthesis of podophyllotoxin in *P. hexandrum*. Protoplasma 254 (1), 217–228.

Kumar, P., Padhan, J.K., Kumar, A., Chauhan, R.S., 2018. Transcriptomes of *Podophyllum hexandrum* unravel candidate miRNAs and their association with the biosynthesis of secondary metabolites. J. Plant Biochem. Biotechnol. 27 (1), 46–54.

Kyogoku, K., Terayama, H., Tachi, Y., Suzuki, T., Komatsu, M., 1973. Constituents of "shikon." II. Comparison of contents, constituents, and antibacterial effect of fat soluble fraction between "nanshikon" and "koshikon.". Shoyakugaku Zasshi 27, 31–36.

Lee, H., Kang, R., Yoon, Y., 2010. Shikonin inhibits fat accumulation in 3T3-L1 adipocytes. Phytother Res. 24, 344–351.

Liao, M., Zeng, C.F., Liang, F.P., 2020. Two new dimeric naphthoquinones from *Arnebia euchroma*. Phytochem. Lett. 37, 106–109.

Liu, T., Lv, C.G., Wang, S., Yang, W.Z., Guo, L.P., 2016. Transcriptome-based gene mining and bioinformatics analysis of p-hydroxybenzoate geranyl transferase genes in *Arnebia euchroma*. Zhongguo Zhongyao Zazhi 41 (8), 1422–1429. https://doi.org/10.4268/cjcmm201608090.

Malik, S., Bhushan, S., Sharma, M., Ahuja, P.S., 2011. Physico-chemical factors influencing the shikonin derivatives production in cell suspension cultures of Arnebia euchroma (Royle) Johnston, a medicinally important plant species. Cell Biol. Int. 35 (2), 153–158.

Mizukami, H., Konoshima, M., Tabata, M., 1977. Effect of nutritional factors on shikonin derivative formation in *Lithospermum* callus cultures. Phytochemistry 16 (8), 1183–1186.

Mohsenikia, M., Khakpour, S., Azizian, Z., Ashkani-Esfahani, S., Razavipour, S.T., Toghiani, P., 2017. Wound healing effect of *Arnebia euchroma* gel on excisional wounds in rats. Adv. Biomed. Res. 6, 2. https://doi.org/10.4103/2277-9175.199260.

Nadoushan, M.R.J., Karimi, M., Fattah, N., Zade, M.T., Nazarbeigi, S., 2016. Cytotoxicity effect of *Arnebia euchroma* against human gastric adenocarcinoma cell line (AGS). Trad. Intrgr. Med. 1 (4), 142–146.

Noorafshan, A., Ebrahimi, S., Esmaeilzadeh, E., Arabzadeh, H., Bahmani-Jahromi, M., Ashkani-Esfahani, S., 2017. Effect of *Arnebia euchroma* extract on streptozotocin induced diabetes in rats: a stereological study. Acta Endocrinol. 13 (3), 272–277. https://doi.org/10.4183/aeb.2017.272.

Onrubia, M., Cusidó, R.M., Ramirez, K., Hernández-Vázquez, L., Moyano, E., Bonfill, M., Palazon, J., 2012. Bio processing of plant in vitro systems for the mass production of pharmaceutically important metabolites: paclitaxel and its derivatives. Curr. Med. Chem. In press.

Ou, Y., Jiang, Y., Li, H., Zhang, Y., Lu, Y., Chen, D., 2017. Polysaccharides from *Arnebia euchroma* ameliorated endotoxic fever and acute lung injury in rats through inhibiting complement system. Inflammation 40 (1), 275–284. https://doi.org/10.1007/s10753-016-0478-0.

Pal, T., Padhan, J.K., Kumar, P., Sood, H., Chauhan, R.S., 2018. Comparative transcriptomics uncovers differences in photoautotrophic versus photoheterotrophic modes of nutrition in relation to secondary metabolites biosynthesis in *Swertia chirayita*. Mol. Biol. Rep. 45 (2), 77–98.

Padhan, J.K., Kumar, P., Sood, H., Chauhan, R.S., 2016. Prospecting NGS-transcriptomes to assess regulation of miRNA-mediated secondary metabolites biosynthesis in *Swertia chirayita*, a medicinal herb of the North-Western Himalayas. Med. Plants Int. J. Phytomed. Related Ind. 8 (3), 219–228.

Pirbalouti, A.G., Azizi, S., Koohpayeh, A., Golparvar, A., 2011. Evaluation of the burn healing properties of Arnebia Euchroma Rolye (Johnst) in diabetic rats. Int. Conf. BioSci. BioChem. Bio. In press.

Ren, S., Wang, X., Ma, B., Yuan, Q., Zhang, H., Yu, X., Hao, A., 2011. Arnebia preventing the expression of Muc1 protein decrease results in anti-implantation in early pregnant mice. Contraception 83 (4), 378–384.

Sankawa, U., Ebizuka, Y., Miyazaki, T., Isomura, Y., Otsuka, H., Sibata, S., Inomata, M., Fukuoka, F., 1977. Antitumor activity of shikonin and its derivatives. Chem. Pharm. Bull. 25, 2392–2397. In press.

Sasaki, K., Abe, H., Yoshizaki, F., 2002. In vitro antifungal activity of naphthoquinone derivatives. Biol. Pharm. Bull. 25, 669–670.

Sharma, N., Sharma, U.K., Malik, S., Bhushan, S., Kumar, V., Verma, S.C., Sharma, N., Sharma, M., Sinha, A.K., 2008. Isolation and purification of acetyl shikonin and b-acetoxy isovaleryl shikonin from cell suspension cultures of *Arnebia euchroma* (Royle) Johnston using rapid preparative HPLC. J. Separ. Sci. 31, 629–635.

Shen, C.-C., Syu, W.-J., Li, S.-Y., Lin, C.-H., Lee, G.-H., Sun, C.-M., 2002. Antimicrobial activities of naphthazarins from *Arnebia euchroma*. J. Nat. Prod. 65, 1857–1862.

Shitiz, K., Sharma, N., Pal, T., Sood, H., Chauhan, R.S., 2015. NGS transcriptomes and enzyme inhibitors unravel complexity of picrosides biosynthesis in *Picrorhiza kurroa* Royle ex. Benth. PLoS One 10 (12), e0144546.

Siavash, M., Naseri, M., Rahimi, M., 2016. *Arnebia euchroma* ointment can reduce abdominal fat thickness and abdominal circumference of overweight women: a randomized controlled study. J. Res. Med. Sci. 21, 63. https://doi.org/10.4103/1735-1995.187347.

Singh, H., Chauhan, R., Raina, R., 2017. Population structure, ecological features and associated species of *Arnebia euchroma*. J. Pharmacogn Phytochem 6, 2005–2007.

Sood, H., 2020. Production of medicinal compounds from endangered and commercially important medicinal plants through cell and tissue culture technology for herbal industry. In: Bioactive Compounds. IntechOpen.

Takanashi, K., Nakagawa, Y., Aburaya, S., Kaminade, K., Aoki, W., Saida-Munakata, Y., Sugiyama, A., Ueda, M., Yazaki, K., 2019. Comparative proteomic analysis of *Lithospermum erythrorhizon* reveals regulation of a variety of metabolic enzymes leading to comprehensive understanding of the shikonin biosynthetic pathway. Plant Cell Physiol. 60 (1), 19–28.

Uniyal, S.K., Awasthi, A., Rawat, G.S., 2002. Current status and distribution of commercially exploited medicinal and aromatic plants in upper Gori valley, Kumaon, Himalaya, Uttranchal. Curr. Sci. 82 (10), 1246–1252. In press.

Verpoorte, R., Contin, A., Memelink, J., 2002. Biotechnology for the production of plant secondary metabolites. Phytochem. Rev. 1 (1), 13–25.

Wang, Y., Zhu, Y., Xiao, L., Ge, L., Wu, X., Wu, W., Wan, H., Zhang, K., Li, J., Zhou, B., Tian, J., Zeng, X., 2018. Meroterpenoids isolated from *Arnebia euchroma* (Royle) Johnst. and their cytotoxic activity in human hepatocellular carcinoma cells. Fitoterapia 131, 236–244. https://doi.org/10.1016/j.fitote.2018.11.005.

Wang, S., Guo, L.P., Xie, T., Yang, J., Tang, J.F., Li, X., Wang, X., Huang, L.Q., 2014. Different secondary metabolic responses to MeJA treatment in shikonin-proficient and shikonin-deficient cell lines from *Arnebia euchroma* (Royle) Johnst. Plant Cell Tissue Organ Cult. 119 (3), 587–598.

Wang, S., Wang, R., Liu, T., Lv, C., Liang, J., Kang, C., Zhou, L., Guo, J., Cui, G., Zhang, Y., Werck-Reichhart, D., Guo, L., Huang, L., 2019. CYP76B74 catalyzes the 3″-hydroxylation of geranylhydroquinone in shikonin biosynthesis. Plant Physiol. https://doi.org/10.1104/pp.18.01056.

Xie, T., Wang, S., Huang, L., Wang, X., Kang, L.P., Guo, L.P., 2014. Transcriptome-based bioinformatics analysis of *Arnebia euchroma* ERF transcription factor family. Zhongguo Zhongyao Zazhi 39 (24), 4732–4739.

Xiong, W., Luo, G., Zhou, L., Zeng, Y., Yang, W., 2009. In vitro and in vivo antitumor effects of acetyl shikonin isolated from *Arnebia euchroma* (Royle) Johnst (Ruanzicao) cell suspension cultures. Chin. Med. 4, 14.

CHAPTER

Dactylorhiza hatagirea

4

Archit Sood

Institute of Plant Sciences, Volcani Center, Agricultural Research Organization, Rishon LeZion, Israel

4.1 Introduction

Dactylorhiza hatagirea (D. Don) Soo, an orchid prevalent to North-Western Himalayan region (3000–4200 m above sea level), is a highly valued medicinal herb of family Orchidaceae. It is commonly called as "Marsh Orchis." There is existence of different names of *D. hatagirea*, region wise. Angmo-Lakpa, Salem Panja, Hatajari, and Munjataka are some of the specific names attributed to *D. hatagirea* in India. It is a terrestrial herb which can reach up to the height of 50–60 cm. Rhizome of *D. hatagirea* is tuberous with 2 – 5 finger-like lobes. Its stem is slender with the presence of leaves all over the surface (Selvam, 2012). Leaves generally vary from 4 to 6 in number and are erect, subsessile, and lanceolate to rectangular in shape (Figs. 4.1 and 4.2). Inflorescences of *D. hatagirea* are spicate racemes with flowers in proportionate density. Its flowers are generally purple in appearance and fruits are ellipsosid. Many minute seeds are present which generally give dusty

FIGURE 4.1

Dactylorhiza hatagirea population (plants with violet color flowers) in natural habitat at Lahaul and Spiti District, Himachal Pradesh, India.

Himalayan Medicinal Plants. https://doi.org/10.1016/B978-0-12-823151-7.00003-9

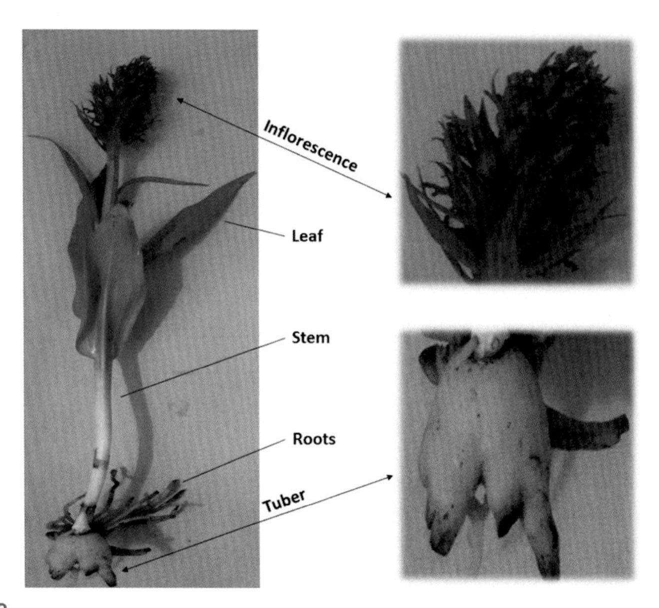

FIGURE 4.2

Partwise differentiation of *Dactylorhiza hatagirea*.

appearance (www.bsienvis.nic.in.) The flowering in *D. hatagirea* generally starts at the onset of July and remains up to the end of August (Chaurasia et al., 2007).

4.2 Origin and distribution

The distribution of *D. hatagirea* encompasses Asian countries viz., Afghanistan, Pakistan, India, China, and Bhutan. As far as India is concerned, this orchid is found in the Himalayan states of Jammu and Kashmir (former state), Himachal Pradesh, Uttarakhand, and Sikkim (Dhar and Kachroo, 1983; Samant et al., 2001). Inhabitation of *D. hatagirea* includes alpine meadows, humus rich soil surrounded by grasses, near to snowy streams along with other herbs (Bhatt et al., 2005; Pant and Rinchen, 2012). Unscientific exploitation of *D. hatagirea* is in practice on very large scale due to its high demand aimed at therapeutic uses. The specific pollinator prerequisite conditions and particular mycorrhizal associations for growth impart *D. hatagirea* a slow-growing herb. The cultivation of *D. hatagirea* through seeds is very problematic as they are very minute structures and also having very low viability. Its cultivation can be achieved through divisions of the tuber part which already contain the stem part with bud. In India, only few centers in Himalayan states have been succeeded in making a repository of germplasm of *D. hatagirea* for use by the growers as well as scientific community. It has been reported that taxonomy of orchids like *D. hatagirea* is generally considered as more complicated due to morphological variations among species and more levels of hybridization amid species (Pillon et al., 2007).

4.3 Medicinal uses

For millenniums, many orchid species have been used as a source of medicine to treat various disorders and ailments. Orchids have also been used in Chinese medicine system for many years. The Book of Herbs (CE 500) encompassed the application of many medicinal orchids such as *Dendrobium* species (State Pharmacopeia Committee of China, 2010). Many species from genus *Dactylorhiza* have been used in combating major diseases. The medicinal uses of *D. hatagirea* range across various medicine systems like Ayurveda, Unani, Tibetan, and Folk medicine. Mainly tuber part of this plant is being used as it has neurostimulant, astringent, antibacterial, aphrodisiac, immunomodulator, demulcent, and nutritional properties (Vij et al., 1992; Thakur and Dixit, 2007). The mucilaginous substance present in the tubers of *D. hatagirea* is also used to treat aliments like diarrhea, chronic fever, and dysentery (Selvam, 2012). *D. hatagirea* is generally considered as medicinal orchid as there are various reports on health benefits associated with it. Many studies have been performed in the past which demonstrate its potential in treating various disorders and health ailments. The roots and tubers of *D. hatagirea* have been found to be aphrodisiac in nature which denotes its potential in refining the sexual behavior and performance (Thakur and Dixit, 2007). The antioxidant activity and ameliorative effect of this plant extracts have been nominated as herbal cure for sexual dysfunction (Thakur et al., 2008). Recently, Sirohi and Sagar (2019) have reported the neuropharamcological properties linked with *D. hatagirea* by demonstrating the effect of hydroalcoholic extracts of roots and tubers. Also other parts of plant like leaves have shown to be associated with antidiabetic activity. The methanol extracts of leaves of *D. hatagirea* have shown the antidiabetic activity in various cell lines tested (Alsawalha et al., 2019).

4.4 Dactylorhin

Due to a strong urge of this orchid herb in medicinal industry, its annual demand is very high, i.e., more than 5000 tonnes. Its economical importance can be assessed by the fact that the value for its dry tubers ranges across Rs 1800–Rs 2000 per Kilogram (Kala, 2004). Overall, the mature tuber consists of mucilage (up to 45%), glucosides, starch, volatile oil, albumen, phosphate, and chloride (Chaurasia et al., 2007). The main medicinal properties of *D. hatagirea* are generally attributed toward the presence of a glucoside, dactylorhin (Fig. 4.3) in the underground part, i.e., tuber. Dactylorhin is the main active chemical constituent in the mature tubers of plants in the family Orchidaceae. Overall, five distinct dactylorhin compounds (dactylorhin A, dactylorhin B, dactylorhin C, dactylorhin D, and dactylorhin E) have been documented from *Dactylorhiza* species and some other medicinal orchids so far.

4.4.1 Biosynthesis of dactylorhin

Although dactylorhin is considered as the main medicinal constituent of *D. hatagirea*, the pathway completing its biosynthesis in plant tissues remains still

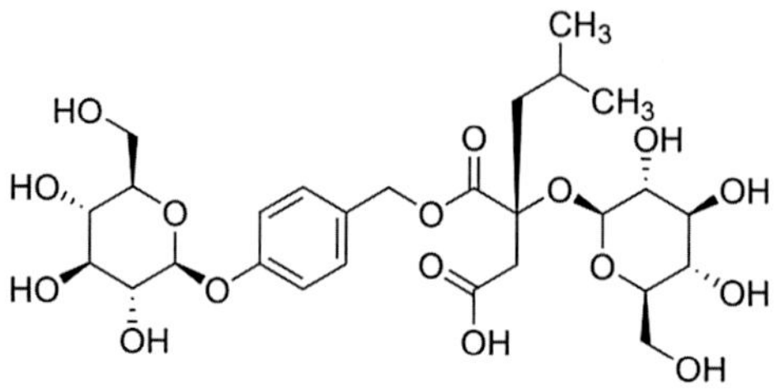

FIGURE 4.3

Chemical structure of dactylorhin.

to be elucidated. There are few reports on the isolation and quantification of different dactylorhin compounds such as dactylorhin A, dactylorhin B, dactylorhin C, dactylorhin D, and dactylorhin E along with other compounds such as dactyloses (Kizu et al., 1999). Three dactylorhin compounds, i.e., dactylorhin A, dactylorhin B, and dactylorhin E have been also isolated from other orchids viz., *Gymnadenia conopsea* R. Br. and *Coeloglossum viride* along with other compounds such as loroglossin and militarine (Li et al., 2009). Using RNA-Seq approach, a comparative transcriptome-based characterization has been performed to elucidate the biosynthetic pathway of dactylorhin employing different conditions through plant cell culture techniques (Sood).

4.5 Biotechnological interventions in *Dactylorhiza hatagirea*

Whenever there is debate on the role of biotechnology for scientific improvement of orchids, a large repository for the same always exists. A very informative report on the use of modern biotechnological interventions has been documented where Hossain et al. (2013) have signified the areas like *in vitro* propagation for mass production and conservation, functional genomics, genetic transformation, as well as the role of pharmaceutical biotechnology. For genus *Dactylorhiza* also, there are adequate number of reports on the application of biotechnology-based tools for various studies in other species such as *Dactylorhiza fuchsii*, *Dactylorhiza incarnata*, *Dactylorhiza majalis*, and *Dactylorhiza maculate* (Hedren et al., 2008; Balao et al., 2017; Devos et al., 2005). However, very limited biotechnological interventions have been carried out on *D. hatagirea* overall as compared to various other herbs endemic to same growing region, on national and international levels, to date.

4.5.1 Tissue culture strategies for conservation of *Dactylorhiza hatagirea*

Tissue culture is an imperative step for conservation of terrestrial orchids at *ex situ* level (Jakobsone et al., 2007). It is important to know about the morphological and physiological characteristics for germination of particular orchid species to establish effective tissue culture approach.

Like other orchid species, genus *Dactylorhiza* has not been studied with respect to employing various tissue culture strategies for mass propagation. However, some prominent tissue culture approaches have been applied up to some extent for conservation of critically endangered medicinal orchid, *D. hatagirea* which in turn linked to low rate of success. This could be attributed to many limitations linked with this orchid herb such as association of fungus and presence of mucilaginous substances in underground parts such as tubers as well as roots. *In vitro* propagation of *D. hatagirea* was also achieved successfully using various plant growth regulators (Giri and Tamta, 2012). Different explants such as green pod, shoot bud, leaf segment, and tuber segment were used for *in vitro* plantlet regeneration (Fig. 4.4). However, the success rate was very low. It was observed that Murashige Skoog (MS) medium added with peptone having concentration of 1.0 g/L, morphoino-ethane sulfonic acid with 1.0 g/L concentration, and activated charcoal in ratio of 0.1% showed effectiveness in terms of producing protocorm-like bodies and the plantlet formation.

If given proper environment and suitable conditions, seed embryos can also develop into entities like protocorm for mass propagation. In a different study, a positive attempt was made to culture immature seed embryos of *D. hatagirea* to develop into protocorms and also in shoot regeneration (Warghat et al., 2014). Out of various media combinations, MS medium with 3 mg/L of indole butyric acid and 1 mg/L of kinetin produced significant results in terms of production of shoots and roots from the culture of protocorms with leaf primordia. Further propagation of *in vitro* plants was also significant in the potting mixture of cocopeat + perlite + vermiculite (1:1:1).

Also, for enhancing the rate of callus induction, *in vitro* propagation of tubers/roots was also achieved where young shoots from mature plants were taken as explants (Fig. 4.5) (Sood). The shoots explants were cultured in various media combinations with various plant growth regulators (Table 4.1), and it was observed that they served as an alternate for better callus induction as compared to direct callus formation.

4.5.2 Role of molecular markers

There is a sharp decline in the natural populations of *D. hatagirea* due to which this species has been recorded as critically endangered species in CAMP (Conservation Assessment and Management Plan), endangered by CITES (Convention on International Trade in Endangered Species) under appendix II and critically rare by IUCN (International Union for Conservation of Nature and Natural Resources) (Uniyal et al., 2002; Bhatt et al., 2005). There is scarcity of information available with respect to study the genetic diversity and population structure of *D. hatagirea* at molecular- or DNA-based markers. It is anticipated that efficient and robust molecular markers should be used for the conservation and genetic improvement programs at very significant level.

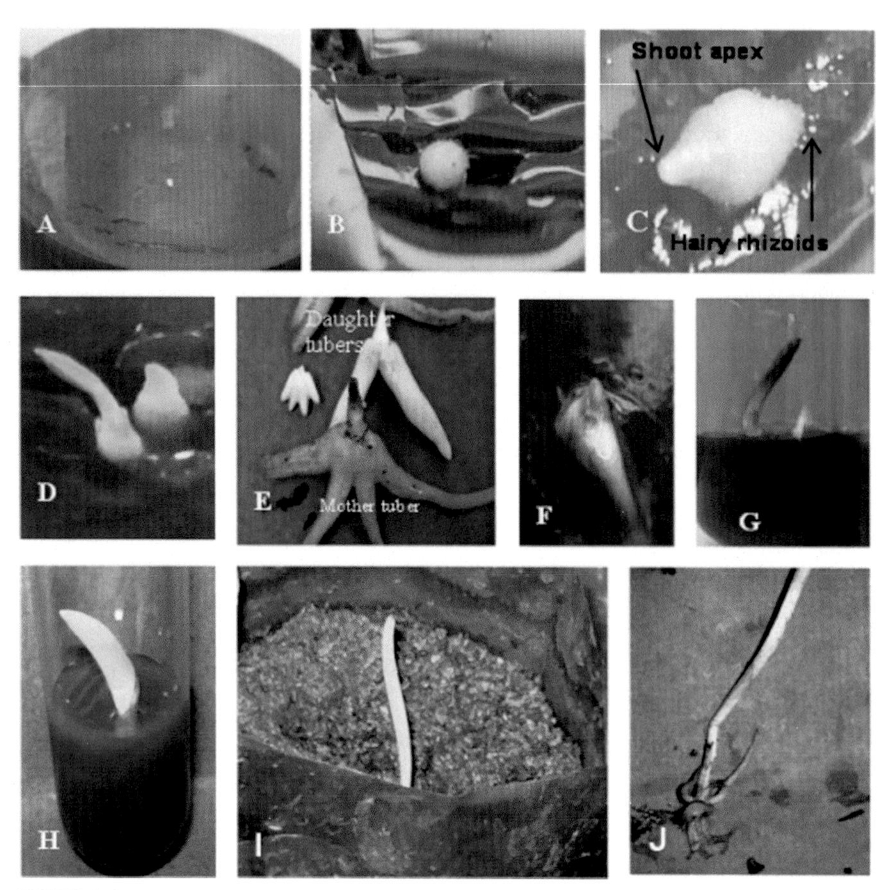

FIGURE 4.4

(a) Seeds swelling after 9 weeks of inoculation. (b) Development of protocorm after 20 weeks of seed germination. (c) Enlarged view of protocorm showing distinct shoot apex and rhizoids after 22 weeks of seed germination. (d) In vitro–raised plantlets after 91 weeks of seed inoculation on MS medium. (e) Mother tuber and daughter tubers. (f) Daughter tubers having shoot buds used as an explants. (g) and (h) Shoots developed from the sprouted buds and attained an average length of 2.72 cm on MS medium supplemented with TDZ (10.0 μM). (i) Sprouted plantlet inside the poly house after 12 weeks of planting. (j) Root emergence in apical segments of tuber after 16 weeks of planting. *μM*, Micro molar; *MS*, Murashige and Skoog; *TDZ*, Thiadiazuron.

Courtesy: Giri, D., Tamta, S., 2012. Propagation and conservation of Dactylorhiza hatagirea *(D. Don) Soo, an endangered alpine orchid. Afr J Biotechnol 11 (62), 12586–12594.*

Inter Simple Sequence Repeats markers were employed to estimate the level of genetic diversity among *D. hatagirea* populations in cold desert of Ladakh region of Indian Himalayas (Fig. 4.6) (Warghat et al., 2013). It was observed that a moderate level of genetic variations among populations exists. Also, random amplified

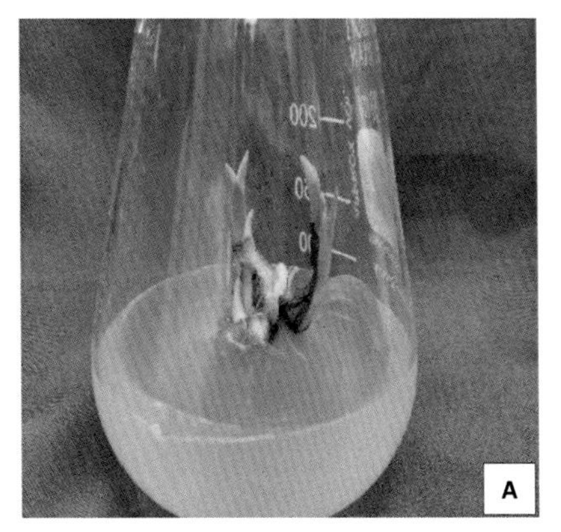

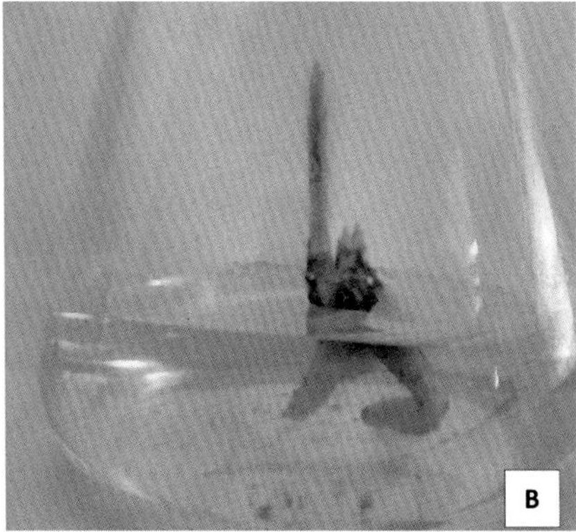

FIGURE 4.5

(a) *In vitro* plantlet regeneration and (b) tuber/root induction in *Dactylorhiza hatagirea*.

Table 4.1 Media combinations used for *in vitro* plant propagation and tuber/root induction.

Media composition	Response
BM2	−
BM2 + IBA (1 mg/L)	+
MS + IBA (1 mg/L)	−

BM2, *orchid medium with vitamins, sucrose, casein hydrolysate, 6-BAP;* IBA, *indole-3-butyric acid;* mg/L, *milligram per liter;* MS, *Murashige and Skoog basal medium.*

polymorphic DNA markers were used to assess the genetic diversity among nine populations of *D. hatagirea* in Ladakh region which revealed the occurrence of moderate genetic variations among population (Fig. 4.7) (Warghat et al., 2012).

Overall low level of genetic diversity among populations supports the theory that endangered or threatened species with narrow level of distribution are usually depauperate. The low genetic variations are generally linked with the less ability of the species to cope up with environmental fluctuations as well as at higher risk of being extinct. This can also lead to increase the phenomenon of inbreeding (Frankham et al., 2010).

For efficient markers to be used in conservation genetic studies among closely related species, some features like absence of genotypic linkage disequilibrium within populations and also cross-specific amplification are anticipated. Later in 2014, a study reported development of simple sequence repeat (SSR) markers to assess the genetic diversity within and between Chinese populations of *D. hatagirea* (Lin et al., 2014). This resulted in the identification of 14 SSR markers of different repeats which can be employed to for population-level studies to assess genetic diversity among *D. hatagirea* and its closely related other orchid species as

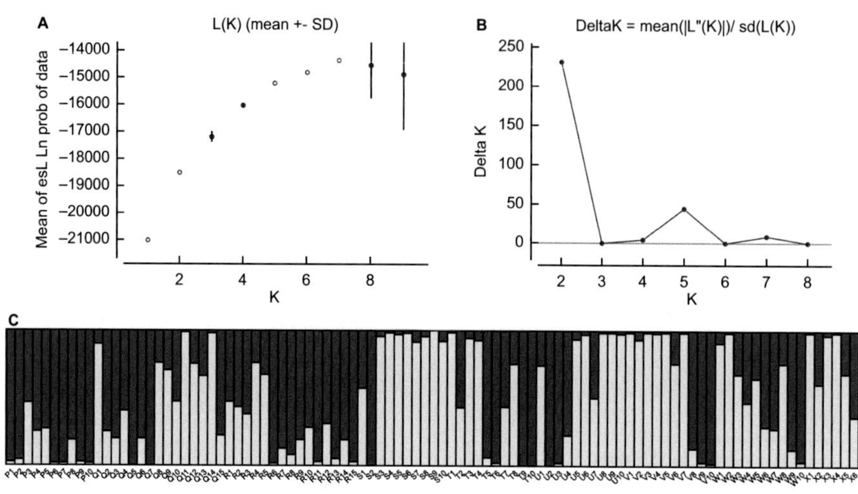

FIGURE 4.6

Structure package–based analysis of *Dactylorhiza hatagirea* population (Based on ISSR analysis). (a) The relationship between K and Ln P D; (b) the relationship between K and dK; (c) the grouping when K = 2. Location code: P—Bogdang; Q—Skampuk; R—Skurru; S—Hunder; T—Turtuk; U—Tirith; V—Sumur; W—Changlung; X—Staksha.

Courtesy: Warghat, A.R., Bajpai, P.K., Srivastavaa, R.B., et al., 2013. Population genetic structure and conservation of small fragmented locations of Dactylorhiza hatagirea *in Ladakh region of India. Sci Hortic 164, 448–454.*

well. Keeping in mind the same objective for Indian populations from Western Himalayan region, (Sharma et al. (2015)) reported characterization of novel polymorphic SSR markers, out of which seven were further identified to be linked with studying the extent of genetic diversity and population genetics in *D. hatagirea*. Later, a different marker approach, i.e., isoenzymes, was also used along with morphological and biochemical parameters to study variability among populations of *D. hatagirea* which indicates a significant level of intrapopulation diversity (Chauhan et al., 2014).

There is an immense need to employ more robust and specific molecular markers along with screening of large number of populations for all the habitats of *D. hatagirea* in Himalayan region to enhance the conservation management strategies on a large scale.

4.5.3 Role of "omics" technologies

"Omics" refers to the systematic analysis at specific level in any living system. In the past, various "omics" technologies have paved the way forward for understanding of organism at cellular, gene, protein, and metabolite level (Witzel et al., 2015). During the last decade, plant science has witnessed the use of "omics" approaches at very

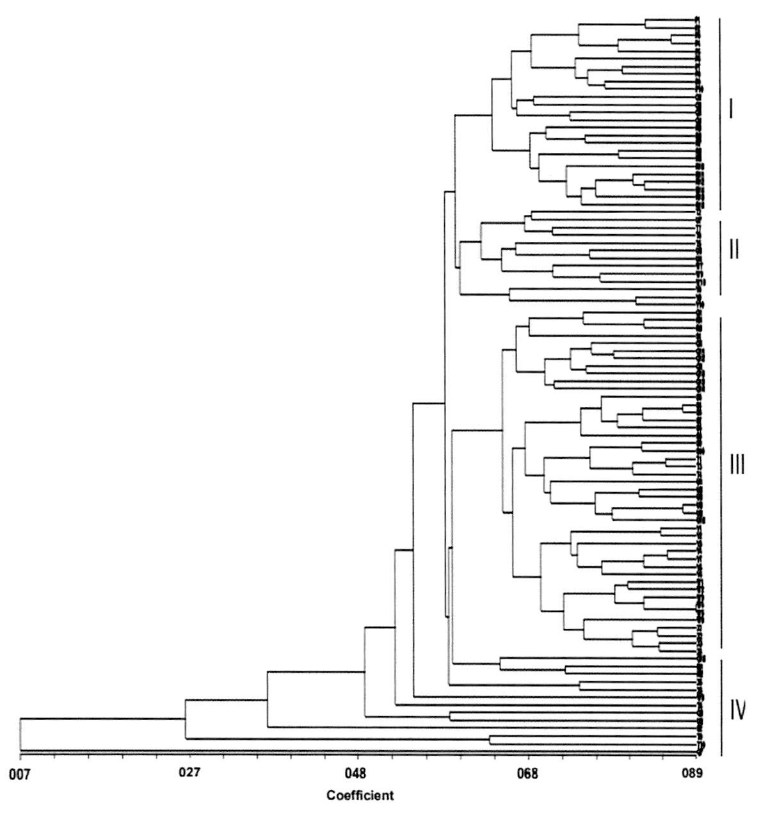

FIGURE 4.7

Dendrogram of 96 individuals of 9 population of *Dactylorhiza hatagirea* based on UPGMA analysis of RAPD polymorphism.

Courtesy: Warghat, A.R., Bajpai, P.K., Murkute, A.A., Sood, H., Chaurasia, O.P., Srivastava, R.B., 2012. Genetic diversity and population structure of Dactylorhiza hatagirea *(Orchidaceae) in cold desert Ladakh region of India. J Med Plants Res 6, 2388–2395.*

large and in an efficient manner and perhaps the most implemented for it. Modern-day "omics" approaches are anticipated for full-level understanding of unexplored mechanisms linked with various pathways and processes associated with plants. The most potential application of "omics" for plant research currently is genomics which has enabled techniques like genetic transformation, genome-assisted breeding, and development of transgenic/genome-edited plants (Hruz et al., 2008; Tsai et al., 2017). More or less it is imperative that techniques like next-generation sequencing (NGS) have commissioned the role of "omics" for betterment of an organism at various developmental stages and processes.

4.5.4 Next-generation sequencing to characterize transcriptome

For genus *Dactylorhiza*, the use of ultramodern and advanced NGS platforms is well documented for applications like genome and transcriptome sequencing to characterize chloroplast genome, phylogenomic relationship, and ecological divergence among various species (Balao et al., 2017; Brandrud et al., 2020; May et al., 2019; Paun et al., 2011).

As *D. hatagirea* is concerned, there are very few reports on employing NGS platforms to study any genomic and transcriptomic characterization. Recently, Nova-Seq sequencing platform from Illumina Inc. was employed to perform RNA-Seq analysis of various *in vitro* cultured and wild tissues of *D. hatagirea* to elucidate biosynthetic pathway of important secondary metabolite, i.e., dactylorhin (Fig. 4.8) (Sood). Also another study reported the use of another NGS platform, i.e., Genome Analyzer IIx (Illumina Inc.) to elucidate biosynthetic mechanism of various secondary metabolites such as resveratrol and stilbenes in the tuber of *D. hatagirea* (Dhiman et al., 2019). Also molecular cues linked with freezing stress and other environmental factors were identified using this transcriptomic characterization.

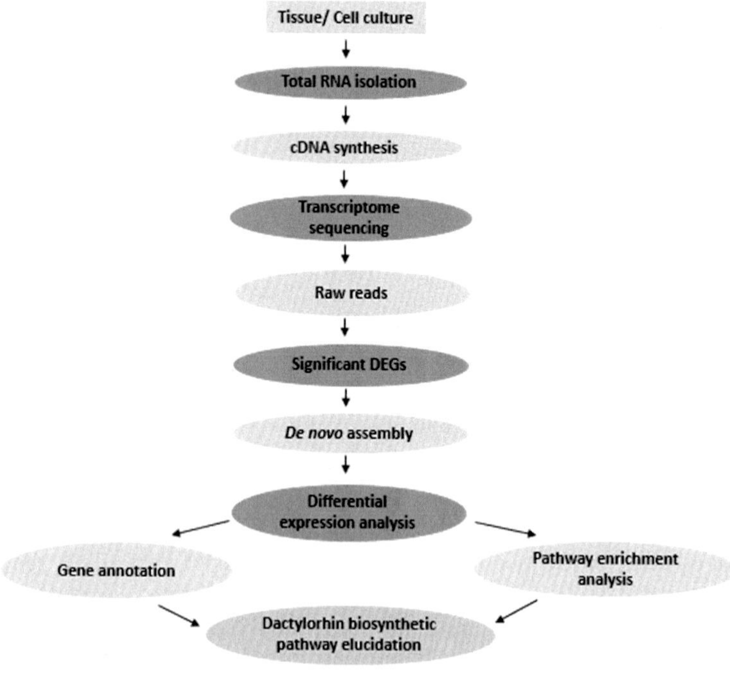

FIGURE 4.8

Methodology followed to characterize transcriptome for the elucidation of dactylorhin biosynthesis pathway in *Dactylorhiza hatagirea*. *cDNA*, Complementary deoxy ribonucleic acid; *DEGs*, Differentially expressed genes; *RNA*, Ribonucleic acid.

4.6 Conclusion and future anticipation

It is summarized that there is an immense need to conserve this miraculous and perhaps the only medicinal orchid, *D. hatagirea*, confined to Indian Western Himalayas along with other medicinal herbs. More biotechnology-based interventions are needed to apply in an effective manner so that maximum medicinal benefits could be harnessed from *D. hatagirea* and also other medicinal orchids. More spatiotemporal-enabled transcriptomic characterization, development of novel molecular markers, standardization of tissue culture protocols, and more importantly, the elucidation of genome sequence of *D. hatagirea* will facilitate the overflow of fruitful information which can be applied for novel insights on factors like its adaptation in the extreme conditions, conservation, and also to gain maximum medicinal benefits. Along with these, Government agencies and departments should prioritize more management strategies to conserve the natural wealth in the form of various still unexplored medicinal herbs in the Indian Himalayan region so that maximum benefits can be garnered for the betterment of mankind. There is also a demanding need for the whole scientific fraternity to broad their objectives toward effective and fruitful research for medicinal plants confined to the Himalayas. Also consciousness for conservation of this and other Himalayan herbs through involvement of local growers and farmers is much anticipated.

Acknowledgments

Author acknowledges Science and Engineering Research Board (SERB, Department of Science and Technology (DST)), Government of India, New Delhi, for providing research grant in the form of National-Post Doctoral Fellowship vide letter no. PDF/2016/002430. Author is also grateful to Dr. Sanjay Kumar (Director, CSIR-IHBT Palampur, India) and Dr. Ashish Rambhau Warghat (Scientist, CSIR-IHBT Palampur, India) for providing the necessary facilities and infrastructure to carry out the research work.

References

Alsawalha, M., Al-Subaei, A.M., Al-Jindan, R.Y., Bolla, R.S., 2019. Anti-diabetic activities of *Dactylorhiza hatagirea* leaf extract in 3T3-L1 cell line model. Pharm. Mag. 15 (64), 212–217.

Balao, F., Trucchi, E., Wolfe, T.M., et al., 2017. Adaptive sequence evolution is driven by biotic stress in a pair of orchid species (*Dactylorhiza*) with distinct ecological optima. Mol. Ecol. 26 (14), 3649–3662.

Bhatt, A., Joshi, S.K., Gairola, S., 2005. *Dactylorhiza hatagirea* (D.Don) Soo-a west Himalayan Orchid in peril. Curr. Sci. 89, 610–612.

Brandrud, M.K., Baar, J., Lorenzo, M.T., et al., 2020. Phylogenomic relationships of diploids and the origins of allotetraploids in *Dactylorhiza* (Orchidaceae). Syst. Biol. 69 (1), 91–109.

Chauhan, R.S., Nautiyal, M.C., Vashistha, R.K., et al., 2014. Morphobiochemical variability and selection strategies for the germplasm of *Dactylorhiza hatagirea* (D. Don) Soo: An endangered medicinal orchid. J. Botany 2014, 5. Article ID 869167.

Chaurasia, O.P., Ahmed, Z., Ballabh, B., 2007. In: Ethnobotany and Plants of Trans-Himalaya. Satish Serial Publishing House, Delhi, p. 544.

Devos, N., Oh, S.H., Raspé, O., Jacquemart, A.-L., Manos, P.S., 2005. Nuclear ribosomal DNA sequence variation and evolution of spotted marsh-orchids (*Dactylorhiza maculata* group). Mol. Phylogenet. Evol. 36, 568–580.

Dhar, U., Kachroo, P., 1983. Alpine Flora of Kashmir Himalaya. Scientific Publishers, Jodhpur, India.

Dhiman, N., Sharma, N.K., Thapa, P., et al., 2019. *De novo* transcriptome provides insights into the growth behaviour and resveratrol and trans-stilbenes biosynthesis in *Dactylorhiza hatagirea* - An endangered alpine terrestrial orchid of western Himalaya. Sci. Rep. 9 (1), 13133.

Frankham, R., Ballou, J.D., Briscoe, D.A., 2010. Introduction to Conservation Genetics. Cambridge University Press, Cambridge, UK.

Giri, D., Tamta, S., 2012. Propagation and conservation of *Dactylorhiza hatagirea* (D. Don) Soo, an endangered alpine orchid. Afr. J. Biotechnol. 11 (62), 12586–12594.

Hedrén, M., Nordström, S., Ståhlberg, D., 2008. Polyploid evolution and plastid DNA variation in the *Dactylorhiza incarnata/maculata* complex (Orchidaceae) in Scandinavia. Mol. Ecol. 17, 5075–5091.

Hossain, M.M., Kant, R., Van, P.T., Winarto, B., Zeng, S., Teixeira da Silva, J.A., 2013. The application of biotechnology to orchids. Crit. Rev. Plant Sci. 32 (2), 69–139.

Hruz, T., Laule, O., Szabo, G., et al., 2008. Genevestigator V3: A reference expression database for the meta-analysis of transcriptomes. Adv. BioInform. 2008, 5. Article ID 420747.

https://www.bsienvis.nic.in. (Accessed 11 July 2020).

Jakobsone, G., Dapkūnienė, S., Cepurīte, B., Belogrudova, I., 2007. The conservation possibilities of endangered orchid species of Latvia and Lithuania. Monographs of botanical gardens (European botanic gardens together towards the implementation of plant conservation strategies). Balt. Bot. Gard. 1, 65–68.

Kala, C.P., 2004. Assessment of species rarity. Curr. Sci. 86, 1058–1059.

Kizu, H., Kaneko, E., Tomimori, T., 1999. Studies on nepalese crude drugs XXVI, chemical constituents of panch aunle, the roots of *Dactylorhiza hatagirea* D. DON. Chem. Pharm. Bull. 47 (11), 1618–1625.

Li, M., Guo, S.X., Wang, C.L., Xiao, P.G., 2009. Quantitative determination of five glycosyloxybenzyl 2-isobutylmalates in thetubers of *Gymnadenia conopsea* and *Coeloglossum viride* var. bracteatum by HPLC. J. Chromatogr. 47 (8), 709–713.

Lin, P., Zeng, L., Yang, Z., et al., 2014. Development and characterization of polymorphic microsatellitemarker for *Dactylorhiza hatagirea* (D. Don) Soo. Conserv. Gen. Res. 6, 29–31.

May, M., Novotna, A., Minasiewicz, J., et al., 2019. The complete chloroplast genome sequence of *Dactylorhiza majalis* (Rchb.) P.F. Hunt et Summerh. (Orchidaceae). Mitochondrial DNA Part B 4 (2), 2821–2823.

Pant, S., Rinchen, T., 2012. *Dactylorhiza hatagirea*: A high value medicinal orchid. J. Med. Plant. Res. 6, 3522–3524.

Paun, O., Bateman, R.M., Fay, M.F., et al., 2011. Altered gene expression and ecological divergence in sibling allopolyploids of *Dactylorhiza* (Orchidaceae). BMC Evol. Biol. 11, 113.

Pillon, Y., Fay, M.F., Hedrén, M., et al., 2007. Evolution and temporal diversification of western European polyploid species complexes in *Dactylorhiza* (Orchidaceae). Taxon 56, 1185–1208.

Samant, S.S., Dhar, U., Rawal, R.S., 2001. Himalayan Medicinal Plants Potential and Prospects. Gyanodaya Prakashan, Nainital.

Selvam, A.B.D., 2012. Pharmacognosy of Negative Listed Plants, pp. 59–68.

Sharma, S., Sharma, V., Chhabra, M., et al., 2015. Characterization of novel polymorphic microsatellite markers in *Dactylorhiza hatagirea*: A critically endangered orchid species from western Himalayas. Conserv. Gen. Res. 7, 285–287.

Sood, A, et al., In preparation. Unpublished.

Sirohi, B., Sagar, R., 2019. Effect of hydroalcoholic extract of *Dactylorhiza hatagirea* roots & *Lavandula stoechas* flower on thiopental sodium induced hypnosis in mice. JDDT 9, 414–417.

State Pharmacopeia Committee of China, 2010.

Thakur, M., Dixit, V.K., 2007. Aphrodisiac activity of *Dactylorhiza hatagirea* (D.Don) Soo in male albino rats. Evid. Based Complement Alternat. Med. 4, 29–31.

Thakur, M., Bhargava, S., Dixit, V.K., 2008. Evaluation of antioxidant activity and ameliorative effect of *Dactylorhiza hatagirea* on sexual dysfunction in hyperglycemic male rats. Planta Med. 74 - PA313.

Tsai, A.Y.L., Chan, K., Ho, C.Y., et al., 2017. Transgenic expression of fungal accessory hemicellulases in *Arabidopsis thaliana* triggers transcriptional patterns related to biotic stress and defense response. PLoS One 12 (3), e0173094.

Uniyal, S.K., Awasthi, A., Rawat, G.S., 2002. Current status and distribution of commercially exploited medicinal and aromatic plants in upper Gori Valley, Kumaon Himalaya, Uttaranchal. Curr. Sci. 82, 1246–1252.

Vij, S.P., Srivastav, R.C., Mainra, A.K., 1992. On the Occurrence of *Dactylorhiza hatagirea* (D. Don) Soo in Sikkim. Orch. News, pp. 8–9.

Warghat, A.R., Bajpai, P.K., Murkute, A.A., Sood, H., Chaurasia, O.P., Srivastava, R.B., 2012. Genetic diversity and population structure of *Dactylorhiza hatagirea* (Orchidaceae) in cold desert Ladakh region of India. J. Med. Plant. Res. 6, 2388–2395.

Warghat, A.R., Bajpai, P.K., Srivastavaa, R.B., et al., 2013. Population genetic structure and conservation of small fragmented locations of *Dactylorhiza hatagirea* in Ladakh region of India. Sci. Hortic. 164, 448–454.

Warghat, A.R., Bajpai, P.K., Srivastatva, R.B., Chaurasia, O.P., 2014. In vitro protocorm development and mass multiplication of an endangered orchid, *Dactylorhiza hatagirea*. Turk. J. Bot. 38 (4), 737–746.

Witzel, K., Neugart, S., Ruppel, S., et al., 2015. Recent progress in the use of 'omics technologies in brassicaceous vegetables. Front. Plant Sci. 6, 244.

CHAPTER

Fritillaria roylei

5

Pankaj Kumar[1], Ashrita[1,2], Mahinder Partap[1,2], Ashish R. Warghat[1,2]

[1]*Biotechnology Division, CSIR-Institute of Himalayan Bioresource Technology, Palampur, Himachal Pradesh, India;* [2]*Academy of Scientific and Innovative Research, CSIR-Institute of Himalayan Bioresource Technology, Palampur, Himachal Pradesh, India*

5.1 Introduction

Fritillaria roylei Hook. is an important Himalayan medicinal herb that belongs to family Liliaceae, commonly known by different names such as kakoli, ksirakakol, Payasya, Ksirasukla, and Himalayan Fritillary (Mehrotra and Ojha, 2006; Joshi et al., 2007; Chauhan et al., 2011; Bisht et al., 2016; Kumar et al., 2020a). This valuable herb is in continuous demand due to the fact that it is one of the top 18 active species, with a global market value of USD 400 million and a local Indian market value of approximately Rs 15,000 per kg/dry bulb (Wang et al., 2017; Luo et al., 2018; Kumar et al., 2020a). It inhabits in alpine slopes, shrubberies, and grows well in light sandy or medium loam well-drained acidic soils. *F. roylei* is perennial, glabrous, an erect bulbous herb (15–60 cm height), mottled stem with leaf number ranged from 7 to 11 (5–7 cm long), opposite/alternate/in whorls of three or four; leaf morphology varies from linear lanceolate and obtuse to acute acuminate. Flowers are solitary or two to three in raceme: nodding, bell shaped; color varies from yellowish-green checkered with dull purple and narrow ovate petals (4–5 cm long). Fruit are broadly oblong, obtusely angled, 6-winged capsules. Small globose bulbs are usually covered with membranous scales (bulb scales) (Goraya et al., 2013). Plant morphological variations are possibly due to habitat variations, agro-environment, altitude, ecological niche, and genetic variability. Conventional propagation procedures in *F. roylei* are generally hindered due to prolonged life cycle, i.e., ~80–90 days above the ground and ~270–280 days beneath the ground due to geophytic nature (Carosso et al., 2011; Petric et al., 2012). Bulbous *F. roylei* plants are naturally propagated by seed and vegetatively via daughter bulbs. Dormancy occurs in the winters because of the low temperature that eventually allows vegetative period and flowering during the spring. Flowering occurs in June–July, whereas fruiting occurs in July–august. Seeds in each valve are arranged in two rows. Contingent to agro-climatic environments, single mother bulb can only produce two–three bulblets. However, seed propagation is also very low due to weak seedlings and takes about 4 to 6 years of the growing period from the initial

Himalayan Medicinal Plants. https://doi.org/10.1016/B978-0-12-823151-7.00010-6

phase to maturity. Premature/early snowfall sometimes also attributed to reproductive-phase obstruction that hinders seed regeneration and maturation (Petric et al., 2012; Bisht et al., 2016; Kumar et al., 2020a). Due to the immense pharmaceutical/herbal potential of bulbous *F. roylei* that is mainly attributed due to steroidal alkaloids resulted in overexploitation from its wild habitats that leads to rare, endangered, and threatened species category (CITES; Ved and Goraya, 2008; Chauhan et al., 2011). Till date, only a few reports are available regarding conservation as well as cultivation strategies for *F. roylei* viz., bulblets regeneration (Joshi et al., 2007; Kumar et al., 2020a) and seed germination (Chauhan et al., 2011). However, detailed scientific information regarding trade volume, climate change impact, genetic diversity, and species germination behavior is lacked. Keeping in view, *F. roylei* deserves attention both for its economic and ecological values, and its rehabilitation in the natural habitat is the key concern because of critically endangered status. Therefore, captive cultivation strategies, status survey, conservation measures, i.e., in situ (protection of existing sites) and *ex situ* (germplasm bank establishment), fast-track medicinal plants cultivation, and traditional knowledge validation are the need of the hour to fulfill the demand–supply gap in bioactive molecule production and also for extensive applications in traditional medicine system. Biotechnological interventions such as plant cell and tissue culture approaches, metabolic engineering, molecular taxonomy, molecular phylogeny, and omics sciences (genomics, transcriptomics, proteomics, metabolomics, and phenomics) provide novel insights into metabolite upscaling vis-s-vis secondary metabolism biosynthetic pathways elucidation (Kumar et al., 2020b). Molecular bioprospection also plays a vital role for exploring the unexplored traits of *F. roylei* that are in supply crisis or nearly in extinction stage for their commercial application in synthetic biology, biocatalysis, and for the development of new biocatalyst to validate traditional knowledge-based herbal formulations.

5.2 Origin and distribution

Northern Hemisphere temperate region is the natural habitat of *Fritillaria* species (Hao et al., 2015). However, center of diversity for the genus *Fritillaria* is reported in the East Mediterranean region with primary evolutionary center Iran (Kamari and Phitos, 2006). Bisht et al. (2016) reviewed an article on *F. roylei* and reported the highest number of *Fritillaria* taxa in Turkey (33), China (30), Greece (24), California (18), and India (6). Out of six, *F. roylei* and *Fritillaria cirrhosa* are predominantly present at an altitude 2800–4000 m above sea level of western Himalayan region of India. *F. roylei* is distributed over the Himalayan states of Jammu and Kashmir (Minimarg and Gurez valley), Himachal Pradesh (Chitkul Kinnaur), Parju and Talra (Chhajpur), Rohtang slopes, Pangi valley, Chanshal, Mural Danda, Bharmour (Fig. 5.1A–F), and Uttarakhand (Rudranath, Tungnath, Valley of Flowers, Dayara, Dronagiri, Govind NP, Khatling, Kedarnath, and Punchchuli area) (Goraya et al., 2013; Kumar et al., 2020a).

FIGURE 5.1

Fritillaria roylei plant collection sites (A,B,C,D); biology, and steroidal alkaloids, i.e., sipeimine, peimine, and peimisine (E,F).

5.3 Phytochemistry

Bulbous *F. roylei* mainly constitutes steroidal alkaloids viz., sipeimine ($C_{27}H_{43}NO_3$), average mass 429.635 Da, Synonym: [Cevan-6-one, 3,20-dihydroxy-,(3β,5α,17β); (3b)-3,20-dihydroxycevan-6-one; raddeamine; peiminine; imperialin; imperialine; kashmirine, verticinone, fritillarine, Zhebeinone], peimine ($C_{27}H_{45}NO_3$, average mass 431.651 Da, Synonym: [Cevane-3,6,20-triol, (3β,5α,6α,22β), verticine, dihydroisoimperialine, wanpeinine A, zhebeinine], and peimisine ($C_{27}H_{41}NO_3$) average mass 427.619 Da, Synonym: [Veratraman-6(5H)-one, 17,23-epoxy-3-hydroxy-,(3β)-, Ebeiensine, Peimissine] (Chi et al., 1940; Wu, 1944; Chou, 1947; Jiang et al., 2001; Chatterjee et al., 1976; Goraya et al., 2013; Bisht et al., 2016; Kumar et al., 2020a). The bulb of *F. roylei* also constitutes propeimin ($C_{29}H_{48}O_3N$) and sterols ($C_{27}H_{46}O$) as neutral constituents (Bisht et al., 2016) (Fig. 5.1).

5.4 Therapeutics potential/biological significance

Bulbs of *F. roylei* are used for the treatment of chronic respiratory disorders such as asthma, tuberculosis, etc., in traditional Indian system of medicine (ISM) from Vedic times (Joshi et al., 2007; Chauhan et al., 2011; Kumar et al., 2020a). In ISM, it is also used as aphrodisiac, healing wounds, corms, rheumatism, burns, and stomach problems (Bisht et al., 2016). Traditionally, bulbs of this herb are boiled with orange peel for asthma and tuberculosis treatment, whereas bulb powder mixed in milk is used as stimulant/tonic for body weakness in some parts of India such as Uttarakhand and

Jammu and Kashmir (Shaheen et al., 2012; Bisht et al., 2013). Similarly, in traditional Chinese medicine, bulbous *Fritillaria* is also used as a source of the expectorant drug, antitussive (cough suppressant), and for treatment of bronchial disorder and pneumonia (Hao et al., 2015; Luo et al., 2018; Kumar et al., 2020b). Different medicinal attributes such as antiasthmatic, antitussive, antirheumatic, antitumor, antiulcers, antihypertensive, galactagogue, hemostatic, febrifuge, oxytocic, ophthalmic, antimicrobial, and anti-viral properties are reported in bulbous *Fritillaria* (Bisht et al., 2016).

5.5 In vitro conservation/morphogenesis

In vitro plant morphogenesis refers to the capacity of cultured plant explants and cells resulting in the development of discrete organs, whole plants, and mass of undifferentiated cells (callus). It has provided opportunities for numerous applications of in vitro plant biology studies of biochemistry, basic botany, propagation, conservation, breeding, and development of transgenic crops. Researchers have been trying for years to conserve and devise a sustainable conservation protocol for bulbous *Fritillaria*. Rate of vegetative propagation in *Fritillaria* is very low, since one mother bulb can produce two to three daughter bulbs depending on ecological niche conditions and cultivation procedures (Ulug et al., 2010). Propagation through seed is even more difficult than vegetative propagation. The seedling can take 4–6 years to become mature plants (Petric et al., 2012; Kumar et al., 2020a). To overcome these situations, various efforts for in vitro regeneration or morphogenesis have been attempted since the early 1980s on different medicinal *Fritillaria* species. Tissue culture techniques provide an alternative and sustainable protocol over to vegetative propagations. Through plant tissue culture technology, initial explants, i.e., bulb scales, bulb segments (transverse or vertical cuts), or whole bulbs, were used as explants for micropropagation of *Fritillaria*. Calli, somatic embryos, and bulblets have been induced and regenerated using different concentrations and combinations of plant regulators under optimized culture conditions such as photoperiod, light flux, humidity, and temperature for the effective morphogenetic response of *Fritillaria* species (Petric et al., 2012). Petric et al. (2012) reviewed an article on in vitro morphogenesis of *Fritillaria* species and provided detailed scientific insight into propagation and somatic embryogenesis using bulb and bulb scale culture, stem and inflorescence culture, zygotic embryo culture, leaf base culture, root culture, androgenesis, plant acclimatization, and in vitro metabolite production particularly alkaloids of *Fritillaria* species. *Fritillaria* species can be regenerated through both direct and through intermediate callus phase, indirect organogenesis. However, the success of in vitro regeneration depends on various factors such as selection of explants, age of explant, type and composition of media, specific growth regulators, etc. Almost all explants of *Fritillaria* plant have been used for direct organogenesis in various species such as bulb scale and leaf of *Fritillaria camtschatcensis* (Otani and Shimanda, 1997), seedlings of *F. cirrhosa*

(Chang et al., 2020), bulb scales of *F. cirrhosa* (Wang et al., 2010), bulb scale, shoot part, and flower part of *Fritillaria imperialis* (Petric et al., 2012), petals of *F. imperialis* (Mohammadi- Dehcheshmeh et al., 2008), etc. For indirect organogenesis, mostly bulbs, bulb scales, and immature embryos have been used (Ozcan et al., 2007; Joshi et al., 2007; Kumar et al., 2020a). The exposure to low temperatures is also reported to play central role in *in vitro* morphogenesis in many *Fritillaria* species during different growth phases of tissue culture. For some species, the multiplication rate can be increased when initial explants are stored under low-temperature conditions for several weeks before in vitro morphogenesis induction. Gao et al. (1999) investigated the effect of various media on the growth rate and regeneration yield in *Fritillaria unibracteata*. They concluded Murashige and Skoog (MS) medium as the most suitable medium for organ culture. Apart from MS medium, Linsmaier and Skoog medium was also used for micropropagation of *F. camtschatcensis* (Otani and Shimanda, 1997) and N6 medium was used for bulblet regeneration of *Fritillaria alburyana* (Ozcan et al., 2007). Varying concentration of sugar was also evaluated and sugar toxicity was reported on increasing sugar substitutes (Otani and Shimanda, 1997). NAA in varying concentration of 0.1−4.0 mg/L is the most commonly used auxin for regeneration as well as callus induction. Other than this, IAA and IBA also performed well for plant regeneration (Wang et al., 2010). In cytokinin, KN (0.1−2.0 mg/L) and BAP (0.5−4.0 mg/L) produce a good regenerative response and induced roots formation as well. Other than these, zeatin, picloram, and TDZ were also used. The cultures performed optimum in 20−25 ± 2°C in light as well as dark culture conditions. Chilling treatment of 5−10°C was used to break dormancy for improved regeneration as well as acclimatization (Peak and Murthy, 2002; Petric et al., 2012). In general, the explants took 2−5 months to in vitro regeneration and callus proliferation depending upon species and explant selection and culture conditions. The regenerated bulblets show 60% −90% regeneration frequency (Otani and Shimada, 1997; Wang et al., 2010) with 90% survival rate in some species (Chen et al., 2000). A reliable and efficient in vitro plant regeneration protocol was optimized in *F. roylei*. Joshi et al. (2007) and Kumar et al. (2020a) also reported 83.30% and 79.67% explant survival using 4% sodium hypochlorite for 20 min and 0.1% mercury chlorite treatment for 15 min in *F. roylei* bulb scale sterilization, respectively. Joshi et al. (2007) reported high frequency (95.8%) in vitro bulblet regeneration in *F. roylei* after 8 weeks of culture using MS medium with 5.0 μM kinetin and 2.0 μM NAA and found average number of bulblet per explant (10.1 ± 0.63 bulblets).

Recently, Kumar et al. (2020a) reported efficient in vitro regeneration in *F. roylei* using bulb as explants and also callus induction for metabolite accumulation and enhancement. Maximum percent response (77.78%) was reported using 1.0 mg/L KIN + 0.5 mg/L NAA and 68.89% using 1.5 mg/L BAP + 0.5 mg/L NAA was observed. Direct regeneration of shoot buds from the surface of cultured bulb scales was observed after 6 weeks of incubation at 25 ± 2°C and 16/8 h photoperiod. Furthermore, Kumar et al. (2020a) also carried out the phytochemical analysis in different parts of *F. roylei* such as bulb, stem, leaf, and floral bud and also in in vitro–raised cultures, i.e., callus and directed in vitro–regenerated plantlets. These findings concluded that apart from bulb, leaf, stem, and floral buds could also be used as a potential alternative source because of the presence of steroidal alkaloids for plant regeneration studies (i.e., metabolite enhancement using cell culture) and for pharmaceutical/herbal usage. The establishment of in vitro model in *F. roylei* for mass scale cultivation, conservation, metabolite production, and scale-up ultimately resulted in the promotion of herbal cultivation and significantly utilized bioresources for catalyzing bioeconomy in a sustainable and eco-friendly manner.

5.6 Omics advancements

The omics resources mainly include genome, transcriptome, proteome, and metabolome studies which provide deeper insights into broader aspects of molecular biology in a precise biological sample in a nonbiased means. Omics studies are neoteric wizardry in the field of modern science with advent in 1990s. Recently, next-generation sequencing (NGS) technologies have directed the attention toward the genomes of many plant species to be sequenced, assembled, and analyzed. The size of genome may differ >40 times within a single genus of plants due to the presence of a large number of transposable elements and ploidy level, which resulted in the variation in genome size between closely related species. In genus *Fritillaria*, vast diversity in genome size had been reported between 30.15 and 85.38 Gb (Leitch et al., 2007; Ambrozova et al., 2011). The size of the nuclear genome for *Fritillaria assyriaca* is reported to be 125 Gbp (Bennett and Smith, 1991). Apart from nuclear and mitochondrial genome, plants also contain chloroplast (CP) genome, and unlike those, the CP genome is highly conserved in most angiosperms. The chloroplast genome of *Fritillaria* has also been vigorously analyzed in recent years. Single Molecule, Real-Time sequencing technology has been reported as a rapid method to obtain a complete CP genome (Li et al., 2016). This technology has been utilized to study the CP genome of different *Fritillaria* species viz., *Fritillaria hupehensis*, *Fritillaria. taipaiensis*, and *F. cirrhosa* (Li et al., 2014). The CP genome has annotated through DOGMA (Dual Organaller GenoMe Annotator; http://dogma.ccbb.utexas.edu/), and tRNA genes are detected by tRNAscan-SE Search Server (http://lowelab.ucsc.edu/tRNAscan-SE). Then circular CP genome map was constructed by OGDRAW (https://chlorobox.mpimp-golm.mpg.de/OGDraw.html). The study reported 20 and 70 putative single nucleotide polymorphisms (SNPs) in *F. taipaiensis* and *F. cirrhosa*, respectively. SNPs in CP genome were reported to have functional insight into evolutionary ecology studies. Li et al. (2016) followed the similar approach with *F. unibracteata* var. *wabuensis*. They reported its CP genome encodes 133 genes: 88 protein coding, 37 tRNA (covering all 20 amino acids), and 8 rRNA genes. Park et al. (2017) deduced the CP genome of *Fritillaria ussuriensis* and *F. cirrhosa* and performed a comparative analysis with other species of *Fritillaria*. They reported that the lengths of genome were 151,524 and 151,083 bp, respectively, and encodes 111 genes in *F. ussuriensis*, and 112 in *F. cirrhosa* comprised 77 protein-coding regions in *F. ussuriensis* and 78 in *F. cirrhosa*. Compared with different *Fritillaria* species' CP genomes revealed seven highly divergent regions in the coding regions (*matK*, *rpoC1*, *rpoC2*, *ycf1*, *ycf2*, *ndhD*, and *ndhF)* and in intergenic spacers regions. Recently, Li et al. (2018) analyzed the phylogenetic relation by comparing CP genome among 13 species of genus *Fritillaria* namely *Fritillaria pallidiflora*, *Fritillaria tortifolia*, *Fritillaria walujewii*, *Fritillaria verticillata*, *Fritillaria karelinii*, *Fritillaria meleagroides*, *Fritillaria yuminensis*, *F. ussuriensis*, *F. cirrhosa*, *Fritillaria hupensis*, *F. taipaiensis*, *F. unibracteata*, and *Fritillaria thunbergii* and identified highly variable CP DNA sequences. Ten highly divergent regions were identified in CP genome, which could be valuable in phylogenetic and population genetic studies. These omics-based studies offer detailed scientific information on characteristics of *Fritillaria* species and further enhance the understanding of evolution in *Fritillaria* species.

Molecular bioprospection for the elucidation of plant secondary metabolic pathways and to co-relate the metabolite content, transcriptomics-based studies have been largely exploited in recent years (Kumar et al., 2020b). By functional analysis and validation of identified key regulatory genes, transcription factors, transporter genes, and many more involved in particular metabolic processes and pathways can be further used for targeted metabolite enhancement using plant genetic engineering and metabolic engineering approaches. In the past few years, de novo transcriptome analysis of *Fritillaria imperialis* and *F. cirrhosa* has been performed for the identification of involved putative genes in steroidal alkaloid biosynthesis (Eshaghi et al., 2019; Zhao et al., 2018). The findings suggested that nonmevalonate pathway, i.e., 2-C-methyl-D-erythritol 4-phosphate/1-deoxy-D-xylulose 5-phosphate pathway, acts as a source of isoprene precursor. Eshaghi et al. (2019) identified 10 unique genes involved in steroidal alkaloid biosynthesis in *F. imperialis* and suggested that squalene synthase gene may perform key enzymatic role in this biosynthetic process. Other than this, the important role of cytochrome P450 family genes has also been suggested in steroidal alkaloid biosynthesis. A metabolomics study has been also performed in *F. cirrhosa* at different growth stages (within 7 years) using ultrahigh-performance liquid chromatography–quadrupole time-of-flight mass spectrometry (Geng et al., 2018). They characterized marker compounds in plant growth stages and reported higher alkaloid content in *Fritillaria* bulbs at the early stages of development. Recently, Wang et al., 2020, performed metabolomics study using a hybrid method of matrix-assisted laser desorption/ionization mass spectrometry and multivariate statistical analysis using bulb of five different *Fritillaria* species, i.e., *F. cirrhosa* Bulbous, *F. hupehensis*, *F. ussuriensis*, *Fritillaria pallidiflora*, and *F. thunbergii.* Metabolomics-based approaches help in the differentiation of herbal/medicinal plants such as *Fritillaria* species to avoid any adulteration that is very much essential to its clinical usage.

5.7 Conclusions and future perspectives

By taking consideration of high demand of *F. roylei*, detailed scientific research is needed about its propagation, captive cultivation, policy interventions for conservation strategies, habitat survey, different biotic and abiotic stress pressures, climate change, habitat degradation, anthropogenic activities, diversity studies, functional genomics, and system biology. The illegal and nonscientific collection should be strictly prohibited from its natural habitat and also proper policy procedures, guidelines, training on scientific methods of plant growth and behavior, development, conservation, management, and on harvesting should be mandatory to the individuals' involved in the harvesting of this medicinal herb.

Acknowledgments

The author, PK, thankfully acknowledges National Post-Doctoral Fellowship Award (GAP-0231, PDF/2017/000309), Science and Engineering Research Board, Department of Science & Technology, Government of India, New Delhi, India. The authors are thankful to the Director, CSIR-IHBT, for providing necessary facilities under Phytopharmaceutical mission (HCP-0010) and Biotechnological interventions for sustainable bioeconomy generation through characterization, conservation, prospection, and utilization of Himalayan Bioresource (MLP-0201).

References

Ambrozová, K., Mandáková, T., Bures, P., Neumann, P., Leitch, I.J., Koblízková, A., Macas, J., Lysak, M.A., 2011. Diverse retrotransposon families and an AT-rich satellite DNA revealed in giant genomes of *Fritillaria lilies*. Ann. Bot. 107 (2), 255–268.

Bennett, M.D., Smith, J.B., 1991. Nuclear DNA amounts in angiosperms. Philos. Trans. R. Soc. London, Ser. A B 334, 309–345.

Bisht, V.K., Kandari, L.S., Negi, J.S., Bhandari, A.K., Sundriyal, R.C., 2013. Traditional use of medicinal plants in district Chamoli, Uttarakhand, India. J. Med. Plants Res. 7, 918–929.

Bisht, V.K., Negi, B.S., Bhandari, A.K., Kandari, L.S., 2016. *Fritillaria roylei* Hook. in Western Himalaya: species biology, traditional use, chemical constituents, concern and opportunity. J. Med. Plants Res. 10 (6–7), 375–381.

Carasso, V., Hay, F.R., Probert, R.J., Mucciarelli, M., 2011. Temperature control of seed germination in *Fritillaria tubiformis* subsp. moggridgei (Liliaceae) a rare endemic of the South-West Alps. Seed Sci. Res. 21 (1), 33–38.

Chang, H., Xie, H., Lee, M., et al., 2020. In vitro propagation of bulblets and LC–MS/MS analysis of isosteroidal alkaloids in tissue culture derived materials of Chinese medicinal herb *Fritillaria cirrhosa* D. Don. Bot. Stud. 61 (9) https://doi.org/10.1186/s40529-020-00286-2.

Chatterjee, A., Dhara, K.P., Pascard, C., Prange, T., 1976. Kashmirine, a new steroidal alkaloid from *Fritillaria roylei*, Hook (Liliaceae). Tetrahedron Lett. 33, 2903–2904.

Chauhan, R., Nautiyal, M.C., Vashistha, R.K., Prasad, P., Nautiyal, A.R., Kumar, A., Teixeira da Silva, J.A., 2011. Morpho-biochemical variability and selection strategies for the germplasm of *Fritillaria roylei* Hook. (Liliaceae) – An endangered medicinal herb of Western Himalaya, India. J. Plant Breed Crop Sci. 3 (16), 430–434.

Chen, U.C., Tai, C.D., Chen, C.C., Tsay, H.S., 2000. Studies on the tissue culture of *Fritillaria hupehensis* Hsiao et K. C. Hsia III. Influence of medium component and light treatment on in vitro rooting and acclimatization of bulblet. Agric. Sci. China 49 (4), 39–47.

Chi, Y.F., Kao, Y.S., Chang, K.J., 1940. The alkaloids of *Fritillaria roylei* II. Isolation of peiminine. J. Am. Chem. Soc. 62, 2896–2897.

Chou, T.Q., 1947. Some minor alkaloids of Pei-Mu, *Fritillaria roylei*. J. Am. Chem. Soc. 36, 215–217.

Eshaghi, M., Shiran, B., Fallahi, H., Ravash, R., Deri, B.B., 2019. Identification of genes involved in steroid alkaloid biosynthesis in *Fritillaria imperialis* via de novo transcriptomics. Genomics 111 (6), 1360–1372.

Gao, S.L., Zhu, D.N., Cai, Z.H., Jiang, Y., Xu, D.R., 1999. Organ culture of a precious Chinese medicinal plant *Fritillaria unibracteata*. Plant Cell Tissue Organ Cult. 59, 197–201.

Geng, Z., Liu, Y., Gou, Y., Zhou, Q., He, C., Guo, L., Zhou, J., Xiong, L., 2018. Metabolomics study of cultivated bulbus *Fritillaria cirrhosa* at different growth stages using UHPLC-QTOF-MS coupled with multivariate data analysis. Phytochem. Anal. 29 (3), 290–299.

Goraya, G.S., Jishtu, V., Rawat, G.S., Ved, D.K., 2013. Wild Medicinal Plants of Himachal Pradesh: An Assessment of Their Conservation Status and Management Prioritisation. HPFD, Shimla, pp. 79–80.

Hao, Da-C., Gu, X.J., Xiao, P.G., 2015. Medicinal Plants: Chemistry, Biology and Omics. Elsevier, pp. 137–170.

Jiang, R.W., Ma, S.C., But, P.P.H., Dong, H., Mak, T.C.W., 2001. Sipeimine, a steroidal alkaloid from *Fritillaria roylei* Hooker. Acta Crystallogr. C 57, 170–171.

Joshi, S.K., Dhar, U., Andola, H.C., 2007. In vitro bulblet regeneration and evaluation of *Fritillaria roylei* hook. – A high value medicinal herb of the Himalaya. Acta Hortic. 756, 75–84.

Kamari, G., Phitos, D., 2006. Karyosystematic study of *Fritillaria messanensis* (Liliaceae). Willdenowia 36, 217–233.

Kumar, P., Partap, M., Ashrita, Rana, D., Kumar, P., Warghat, A.R., 2020a. Metabolite and expression profiling of steroidal alkaloids in wild tissues compared to bulb derived in vitro cultures of *Fritillaria roylei* – High value critically endangered Himalayan medicinal herb. Ind. Crops Prod. 145, 111945.

Kumar, P., Shaunak, I., Verma, M.L., Verma, M.L., Chandel, A.K., 2020b. Biotechnological application of health promising bioactive molecules. In: Biotechnological Production of Bioactive Compounds. Elsevier, pp. 165–189.

Leitch, I.J., Beaulieu, J.M., Cheung, K., Hanson, L., Lysak, M.A., Fay, M.F., 2007. Punctuated genome size evolution in Liliaceae. J. Evol. Biol. 20 (6), 2296–2308.

Li, Q., Li, Y., Song, J., Xu, H., Xu, J., Zhu, Y., Li, X., Gao, H., Dong, L., Qian, J., Sun, C., Chen, S., 2014. High-accuracy de novo assembly and SNP detection of chloroplast genomes using a SMRT circular consensus sequencing strategy. New Phytol. 204 (4), 1041–1049.

Li, Y., Li, Q., Li, X., Song, J., Sun, C., 2016. Complete chloroplast genome sequence of *Fritillaria unibracteata* var. wabuensis based on SMRT sequencing technology. Mitochondrial DNA A, DNA Mapp. Seq. Anal. 27 (5), 3757–3758.

Li, Y., Zhang, Z., Yang, J., Lv, G., 2018. Complete chloroplast genome of seven *Fritillaria* species, variable DNA markers identification and phylogenetic relationships within the genus. PloS One 13 (3), e0194613.

Luo, D., Liu, Y., Wang, Y., Zhang, X., Huang, L., Duan, B., 2018. Rapid identification of *Fritillaria cirrhosa* bulbs and its adulterants by UPLC-ELSD fingerprint combined with chemometrics methods. Biochem. Systemat. Ecol. 76, 46–51.

Mehrotra, N.N., Ojha, S.K., 2006. Ayurvedic Rasayana Therapy and Rejuvenation (Kaya-Kalp) Current R&D Highlights. January–March 6–10.

Mohammadi-Dehcheshmeh, M., Khalighi, A., Naderi, R., Saradi, M., Ebrahimie, E., 2008. Petal: a reliable explant for direct bulblet regeneration of endangered wild population of *Fritillaria imperialis* L. Acta Physiol. Plant. 30, 395–399.

Otani, M., Shimada, T., 1997. Micropropagation of *Fritillaria cumtschatcensis* (L.) Ker-Gawl., "Kuroyuri", vol. 5. Bulletin of RIAR, Ishikawa Agricultural College, pp. 39–44.

Ozcan, S., Parmaksiz, I., Mirici, S., Çoçu, S., Uranbey, S., Ipek, A., Sancak, C., Sarihan, E., Gurbuz, B., Sevimay, C.S., Arslan, N., 2007. Efficient in vitro bulblet regeneration from immature embryos of endemic and endangered geophyte species in *Sternbergia, Muscari* and *Fritillaria* genera. In: Xu, Z., Li, J., Xue, Y., Yang, W. (Eds.), Biotechnology and Sustainable Agriculture 2006 and Beyond. Springer, Berlin, Germany, pp. 381–383.

Paek, K.Y., Murthy, H.N., 2002. High frequency of bulblet regeneration from bulb scale sections of *Fritillaria thunbergii*. Plant Cell Tissue Organ Cult. 68, 247–252.

Park, I., Kim, W.J., Yeo, S.M., Choi, G., Kang, Y.M., Piao, R., Moon, B.C., 2017. The complete chloroplast genome sequences of *Fritillaria ussuriensis* maxim. and *Fritillaria cirrhosa* D. Don, and comparative analysis with other *Fritillaria* species. Molecules 22 (6), 982.

Petric, M., Subotic, A., Trifunovic, M., Jevremovic, S., 2012. Morphogenesis in vitro of *Fritillaria* spp. Floriculture Ornamental Biotech. 6 (1), 78–89.

Shaheen, H., Shinwari, Z.K., C!ureshi, R.A., Ullah, Z., 2012. Indigenous plant resources and their utilization practices in village populations of Kashmir Himalayas. Pakistan J. Bot. 44, 739–745.

Ulug, B.V., Korkut, A.B., Sisman, E.E., Vuz, M., 2010. Research on propagation methods of Persian lily bulbs (*Fritillaria persica* Linn) with various vegetative techniques. Pakistan J. Bot. 42, 2785–2792.

Ved, D.K., Goraya, G.S., 2008. Demand and Supply of Medicinal Plants in India. NMPB, New Delhi & FRLHT, Bangalore, India.

Wang, D., Chen, X., Atanasov, A.G., Yi, X., Wang, S., 2017. Plant resource availability of medicinal *Fritillaria* species in traditional producing regions in Qinghai-Tibet Plateau. Front. Pharmacol. 8, 502.

Wang, Z., Xie, H., Ren, J., Chen, Y., Li, X., Chen, X., Chan, T., Wah, D., 2020. Metabolomic approach for rapid differentiation of *Fritillaria* bulbs by matrix-assisted laser desorption/ionization mass spectrometry and multivariate statistical analysis. J. Pharm. Biomed. https://doi.org/10.1016/j.jpba.2020.113177.

Wang, Y.H., Dai, Y., He, Z.S., Sun, Y.X., Yan, S.J., Xu, S.J., Wang, X.R., 2010. The effects of in vitro culture conditions on regeneration of *Fritillaria cirrhosa*. Zhong Yao Cai 33 (6), 854–856.

Wu, Y.H., 1944. The constituents of *Fritillaria roylei*. J. Am. Chem. Soc. 66, 1778–1780.

Zhao, Q., Li, R., Zhang, Y., Huang, K., Wang, W., Li, J., 2018. Transcriptome analysis reveals in vitro-cultured regeneration bulbs as a promising source for targeted *Fritillaria cirrhosa* steroidal alkaloid biosynthesis. 3 Biotech 8 (4), 191.

CHAPTER

Picrorhiza kurroa 6

Neha Sharma

Division of Crop Improvement, ICAR-Central Potato Research Institute, Shimla, Himachal Pradesh, India

6.1 Introduction

The genus *Picrorhiza* belongs to the family Plantaginaceae. *Picrorhiza kurroa* Royle ex Benth, *Picrorhiza scrophulariiflora* Pennell, and *Picrorhiza tungnathii* are the three species of this genus. *P. kurroa* is an important medicinal herb which has been utilized in ancient medicinal system like Ayurveda as well as in pharmaceutical companies for the treatment of various ailments. It is a diploid species (2n = 34) with an estimated genome size of 3452.34 Mbp. It is known by various names such as "Kutki" in Hindi and Nepali, "Katuki" in Bengali, "Karu" in Punjabi, "Putising" in Dzongkha, "Katurohini" in Sanskrit, "Hellebore" in English, "Hunhunglien" in Chinese, "Kaur" in Kashmir Himalayas, "Kadu" in Himachal Himalayas, and "Kadvi" in Uttarakhand Himalayas (Mondal et al., 2013). This medicinal herb is perennial and generally grows in rocky slopes and crevices of Himalayan region. The plant is self-regenerating in nature through seeds and rhizomes. Stolons mature into a rhizome with formation of new shoots and roots, which after detachment from mother plant give rise to independent plants (Fig. 6.1A). Shoots are of 5–9 cm in length from joints of rhizomes. Leaves are basal, spathulate to narrow elliptic, serrated, and 7–11 cm in length. Roots are inflexible, almost 6–10 inches in length, creeping, and bitter in taste. Flowers are sessile, zygomorphic, bilipped, bisexual, pale or purplish blue in color, occurring on a long spike and appear in June through August (Fig. 6.1B). Manual harvesting of the plants takes place in October through December (Raina et al., 2010). Various pharmaceutically important compounds like picroside I (P-I), picroside II (P-II), kutkoside, vanillic acid (VA), apocynin, androsin, cucurbitacins, picrotin, picrotoxinin, etc., are present in this medicinal herb. P-I and P-II are the main bioactive constituents of this species, which are used in various herbal formulations for the treatment of liver diseases, fever, allergy, hepatitis B, leukoderma, gastrointestinal, urinary disorders, etc (Bhandari et al., 2009; Mondal et al., 2013). They also possess antioxidant, immunomodulatory, antimalarial, antiinflammatory, anticancerous, neuroprotective, antiasthmatic, and antidiabetic properties (Sah and Varshney, 2013). In terms of the economic value of traded materials, this plant comes in list of top 15 plant species traded in India. The demand of *Picrorhiza* is increasing each year and has raised up to 415 MT/

Himalayan Medicinal Plants. https://doi.org/10.1016/B978-0-12-823151-7.00011-8

FIGURE 6.1

Mature *Picrorhiza kurroa* plants showing different organs (A) and flowering (B).

year (Ved and Goraya, 2008). Price of this plant material varied from Rs. 220 to 340 per kg which will be much higher today (Shitiz et al., 2013). In order to get 1 Kg dry weight of rhizomes and stolons of *P. kurroa*, it is estimated that a total of 300–400 individual plants are uprooted and maximum demand is fulfilled from wild or natural sources. Poor seed germination and since most extraction takes place prior to seed set, the most preferred way of propagation for this plant is through rhizomes. During harvesting, due to lack of knowledge, whole plants are uprooted, wherein rhizomes are kept while remaining plant material including young buds are thrown, which hinders its propagation in wild. Therefore, owing to extensive and unorganized collection, *P. kurroa* has been listed as an endangered plant species by International Union for Conservation of Nature (Nayar and Sastri, 1990). It is also placed in the negative list of exports by Ministry of Commerce, GOI (Raina et al., 2010).

6.2 Origin and distribution

The genus *Picrorhiza* and the species *P. kurroa* were depicted for the first time on a drawing published by Royle (Royle, 1835–1940a). The first written description of genus and species was given by Bentham in "Scrophularineae Indicae" (Bentham, 1835). Therefore, the name of the species is *P. kurroa* Royle ex Bentham. According to Royle, the generic name has been derived from the bitter root, where "Picros" means bitter, while "rhiza" means root which is used in native medicine (Royle, 1835–1940b). Smith and Cave collected and identified *P. kurroa* for the first time at the base of the Zemu Glacier in Sikkim at 4300 m altitude (Smith and Cave, 1911).

P. kurroa grows naturally in dry western Himalayan region and distributed abundantly from Kashmir to Kumaon (Pennell, 1943), Pakistan to Uttarakhand regions (Polunin and Stainton, 1990), Garhwal to Bhutan, north Burma, west China, and southeast Tibet at 3000–5000 m altitudes (Table 6.1) (Chettri et al., 2005).

Table 6.1 Distribution of *Picrorhiza kurroa* in the Himalayan region.

State	Location	Reference
Jammu and Kashmir	Kolohoi, Zojpal, Sonsa Nag	Coventry (1927)
	Burzil Pass, Kamri Pass, Mir Panzil Pass, Deosai road, Nafran, Pahlgam, Sonamarg, Tragbal, Zoji La Pass	Pennell (1943)
	Lipper Valley, northwest of Kashmir Valley	Singh and Kachroo (1976)
	Simthan, Jammu	Sharma and Kachroo (1981)
	Pir Panjal Range, Kishanganga valley, Upper Lidder valley	Kapahi et al. (1993)
Himachal Pradesh	Chhitkul, Chandrokani, Changla Pass, Pieri Pass, Kinnauar, Kugti Pass	Pennell (1943)
	Manimahesh, Rohtang Pass, Lahaul and Spiti Forest Division	Uniyal et al. (1982)
	Lahaul valley, Bharmour valley Kinnaur, Kulu, Rohru, Dhauladhar valley, Pangi valley,	Chauhan (1988), Kapahi et al. (1993)
	Chepuwa, Talla Johar, Eastern Kumaon	Mehta et al. (1994)
	Chhakinal watershed, District Kullu	Dobriyal et al. (1997)
	Great Himalayan National Park, Dhauladhar Wildlife Sanctuary, Kugti Wildlife Sanctuary	Singh and Rawat (2000), Uniyal et al. (2006)
Uttarakhand	Harsil, Raithal, Sukhi, Sayara, Tehri-Garhwal in Bhagirathi Valley, Bhilangna Valley, Sahastra Tal, Damodar Valley, Deodi Ramani, Gangotri, Tungnath, Valley of Flowers, Glacial Valley	Gupta (1989)
	Kedarnath, Har-ki-dun, Ponwati, Tali, Harsil, Gangotri in Garhwal hills, Pithoragarh District of Kumaon Hills	Kapahi et al. (1993), Kapil (1995)
	Valley of Flowers National Park, Kedarnath Wildlife Sanctuary, Nanda Devi Biosphere Reserve, alpine ranges of Ralam, Dhauli, Kali Valleys	Uniyal et al. (2002), Semwal et al. (2007)

6.3 Phytochemistry of *Picrorhiza kurroa*

The phytochemistry of *P. kurroa* has been extensively studied, and up to now 132 active ingredients have been identified from different parts of the plant such as roots, stem, leaf, and seeds. Diverse classes of secondary metabolites like iridoid glycosides, cucurbitacins, and phenolics are isolated from this important medicinal herb. Various iridoids glycosides extracted from *P. kurroa* include picroside I, II, III, V, minecoside, catalpol, kutkoside, 6-feruloylcatalpol, pikuroside, and

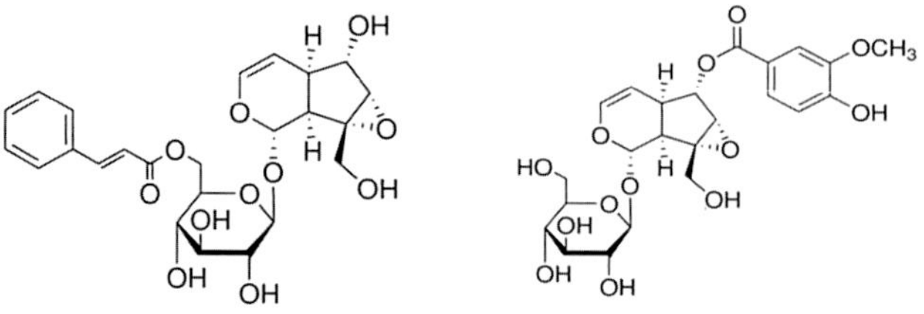

Picroside I

Picroside II

FIGURE 6.2

Chemical structures of Picroside-I and Picroside-II.

veronicoside (Sah and Varshney, 2013). P-I and P-II are the main bioactive constituents of *P.* kurroa, having similar basic structure except P-I having cinnamate moiety and P-II having vanillate moiety (Fig. 6.2). It has been observed that P-I is produced in shoots, P-II in roots, and both P-I and P-II are biosynthesized or accumulated in stolons/rhizome tissues (Pandit et al., 2013a). These are used in various herbal formulations such as Picroliv, Katuki, Arogya, Kutaki, Livocare, Livomap, Livomyn, Livplus, Pravekliv, and Vimliv for the treatment of various disorders. Picroliv is the standard herbal preparations of *P. kurroa* which is a mixture of iridoid glycoside containing 60% P-I and kutkoside in the ratio of 1:1.5 (Pandit et al., 2013b). Picrolax is a form of *P. kurroa* extract, which is recommended for laxative purposes and confers 45 mg P-I (0.02%) and 17.6 mg P-II (0.01%) for every 1.585g of supplement (Upadhyay et al., 2013). The extracts of *P. kurroa* have also been found to contain cucurbitacins such as cucurbitacin B, D, R, arvenin III, etc. These are triterpenoid compounds, possessing a wide range of biological activities and are bitter and toxic in nature. Further, recognized class viz. phenolics containing various compounds like apocyanin, androsin, ellagic acid, picein, and VA are also present in *P. kurroa* extracts (Sah and Varshney, 2013).

6.4 Medical significance

P. kurroa is traditionally well recognized in Indian Ayurvedic system for treatment of liver disorders and has also been implicated in curing other ailments like asthma, gastrointestinal and urinary problems, leukoderma and snake bites, etc (Sah and Varshney, 2013). It is used as a primary ingredient in an Ayurvedic preparation viz. Arogyavardhini to treat liver ailments. In Nepal and Bhutan, it is recommended as a curative agent against hepatitis, jaundice, colds, coughs, and fever. It is also used to treat high blood pressure, eye disease, intestinal pain, bile disease, gastritis, sore throats, and also acts as an antidote against scorpion bite. Similarly, in Chinese medicinal system, this important plant is known to cure jaundice, digestive disorders, diarrhea, and dysentery (Mulliken, 2000).

The traditional uses of *P. kurroa* have been investigated by the several pharmacological studies. Its extract has various pharmaceutical properties including hepatoprotective (Sinha et al., 2011), antiinflammatory (Zhang et al., 2012), antioxidant (Rajkumar et al., 2011), antimalarial (Irshad et al., 2011), anticancerous (Joy et al., 2000), antidiabetic (Kumar et al., 2017a), immunomodulatory (Sane et al., 2011), free radical scavenging (Chander et al., 1992), antineoplastic (Rajkumar et al., 2011), anticholestatic (Verma et al., 2009), antiallergic and antianaphylactics (Baruah et al., 1998), and antihepatitis B surface antigen activities (Mehrotra et al., 1990). The major properties have been described below:

Hepatoprotective: *P. kurroa* extracts have been reported to possess hepatoprotective action against various toxins such as amanita poisoning, carbon tetrachloride (Lee et al., 2007), galactosamine (Dwivedi et al., 1992), ethanol (Sinha et al., 2011), aflatoxin-B1 (Dwivedi et al., 1993), acetaminophen (Kumar and Shukla, 2017), and thioacetamide (Dwivedi et al., 1991), in both in vitro and in vivo experiments. A hydroalcoholic extract of *P. kurroa* has been found effective against nonalcoholic fatty liver disease by reversal of the fatty infiltration of the liver and by decreasing the quantity of hepatic lipids (Shetty et al., 2010). It is also known to supply advanced neutraceutical activity for superior hepatoprotection by improving intestinal absorption (Jia et al., 2015). The active constituent picroliv isolated from *P. kurroa* has been shown to possess hepatoprotective activity by suppressing the deleterious effects of alcohol like cholestasis, altered serum, and liver markers along with reduced levels of alcohol-metabolizing enzymes in rat hepatocytes (Saraswat et al., 1999).

Antioxidant: *P. kurroa* extract has significant antioxidant potential due to its free radicals scavenging activity which prevents various diseases. Ray et al. (2002) have suggested that oral administration of ethanol extract of rhizome of *P. kurroa* (20 mg/kg body weight) enhanced the healing of stomach wall of indomethacin-induced gastric ulcerated rats by an in vivo free radical scavenging action. Rajkumar et al. (2011) have studied the antioxidant effectiveness of plant extracts by using radical scavenging assays, ferric-reducing antioxidant property, and thiobarbituric acid assay for analyzing inhibition of lipid peroxidation. Further, Krupashree et al. (2014) have analyzed that DPPH radical scavenging and metal chelating activities in *P. kurroa* extract with IC50 of 75.16 ± 3.2 and 55.5 ± 4.8 μg/mL and also showed inhibition of H_2O_2-induced plasmid DNA damage and AAPH-induced oxidation of bovine serum albumin and lipid peroxidation of rat hepatic tissues. Furthermore, antioxidant potential has also been explored in leaves, roots, and rhizomes of this important plant (Kant et al., 2013; Sharma et al., 2018).

Anticarcinogenic: Different studies have been reported on the anticancer activity of extracts of *P. kurroa* in *in vitro* as well as in vivo systems. Antitumor and anticarcinogenic activities of *P. kurroa* extract have been verified in mice (Joy et al., 2000). Oral administration of *P. kurroa* extract inhibited the tumor

incidence and tumor-related deaths in sarcoma-induced mice. The extract reduced transplanted solid tumors induced by Dalton's lymphoma ascites tumor cell lines and enhanced the life span of ascites tumor–bearing mice. It also restricted yeast topoisomerase I and II enzyme activity when tested on *Saccharomyces cerevisiae* mutant cell cultures. Rajeshkumar and Kuttan (2001) have reported the anticancerous activity of picroliv in N-nitrosodiethylamine–induced hepatic cancer in mice. Oral dose of 100–200 mg/kg body weight of picroliv in BALB/c mice decreased the sarcoma and papilloma resulted from 20-methylcholanthrene and 7, 12 dimethylbenz[a]anthracene. Anticarcinogenic response of picroliv has also been reported in rats administered with 1, 2-dimethylhydrazine hydrochloride (Rajeshkumar and Kuttan, 2003). *P. kurroa* extract containing iridoid glycosides P-I, kutkoside, and kutkin have been studied for their antiinvasion activity in human breast cancer cells (MCF-7). The extract showed cell toxicity at 50 and 100 μg/mL and decreased the activity of matrix metalloproteins involved in breast cancer progression (Rathee et al., 2013). Soni and Grover have reviewed the therapeutic potential of picrosides against different types of cancers and highlighted the effect of picrosides on different anticancer activities like free radical scavenging activity, metal ion chelator, detoxifying activity, cell cycle regulation, and apoptotic induction (Soni and Grover, 2019).
Antidiabetic: Aqueous extract of *P. kurroa* over the course of 14 days in diabetic rats induced with streptozotocin–nicotinamide was found to improve blood glucose and insulin levels in dose-dependent manner (Husain et al., 2009). Kumar et al. (2017a,b) have observed that hydroalcoholic extract of *P. kurroa* rhizome induced the insulin secretion and regenerated β-cell in streptozotocin-induced diabetic rats. They have also found that the extract showed antihyperglycemic effects along with improvement in hepatic and renal functions against oxidative damage.
Antimalarial and antimicrobial: There are various reports on antimalarial and antimicrobial potential of *P. kurroa.* Singh and Banyal (2011) and Banyal et al. (2014) have tested the effectiveness of antimalarial properties of *P. kurroa* against *Plasmodium berghei.* Extracts of dried stolons of this plant displayed wide range of antimicrobial activity against various pathogenic microbes including *Gloeocercospora sorghi*, *Erwinia chrysanthemi*, *Rhizoctonia solani*, *Fusarium oxysporum*, and *Sporisorium scitamineum* (Laxmi and Preeti, 2015). Further, antifungal properties were examined in alcoholic extract obtained from roots of *P. kurroa* and found effective against *Candida tropicalis*, *Candida albicans*, *Penicillium marneffi*, and *Trichophyton rubrum* (Shubha et al., 2016).

6.5 Tissue culture status of *Picrorhiza kurroa*

In vitro plant tissue culture techniques are a useful tool for conservation and rapid propagation of high-value medicinal plants along with biosynthesis of important secondary metabolites for their commercialization and sustainable utilization in herbal industries. Initial studies on in vitro multiplication of *P. kurroa* were attempted using different explants on different media combination supplemented with varying concentrations of growth hormones such as culturing of shoot tips on Murashige and Skoog (MS) medium supplemented with kinetin (KN) (3–5 mg/L), stem cuttings with 6-benzyladenine (BA) (0.11–2.25 mg/L) alone or in combination with indole-3-acetic acid (0.02–0.2 mg/L) or gibberellic acid (GA3) (0.03–0.35 mg/L), and runners along with axillary shoots with BA at a lower concentration (0.23 mg/L) (Lal et al., 1988; Upadhyay et al., 1989; Chandra et al., 2006). Low-cost micropropagation protocol for *P. kurroa* has been developed by Sood and Chauhan (2009a). Indirect shoot organogenesis with callus formation was reported on MS media supplemented with 2, 4-dichlorophenoxyacetic acid (2 mg/L) and indole-3-butyric acid (IBA) (0.5 mg/L) using leaf, nodal segments, and root explants followed by shoot regeneration on MS medium containing BA (2 mg/L) + KN (3 mg/L) (Sood and Chauhan, 2009b), while direct shoot organogenesis has been done on Gamborg's B5 medium supplemented with 3 mg/L KN and 1 mg/L IBA using leaf explants (Bhat et al., 2012). Further, somatic embryos were developed on MS medium containing 0.11 mg/L thidiazuron and 0.5 mg/L IBA using nodal explants of *P. kurroa* (Sharma et al., 2010). Lately, Sharma et al. (2015) have demonstrated seaweed extract as an economic alternative to MS media for large-scale propagation of *P. kurroa* under in vitro conditions. They have observed enhanced plant biomass, plant length, number of shoots, root length, and number of roots of this plant on MS medium supplemented with seaweed extract alone or in combination with growth hormones.

Secondary metabolites are synthesized in tissue-specific manner and cell cultures offer a suitable biological system for biosynthesis of secondary metabolites of commercial importance. Sood and Chauhan (2010) have reported the biosynthesis of P-I in shoot cultures, while absenteeism of P-II in *P. kurroa* plants grown under in vitro conditions. Picrosides production is regulated by light and temperature, and high picrosides content has been observed in shoots grown at low temperature (15°C) and under illumination as compared to 25°C and dark conditions (Kawoosa et al., 2010). Sharma et al. (2016a) have investigated the P-I biosynthesis in different morphogenetic stages including plant segment, callus initiation, callus mass, shoot primordia, multiple shoots, and fully developed stages of *P. kurroa*. Reduction in P-I amount was observed from 2.51 μg/mg to nondetectable level during dedifferentiation in callus initiation and callus mass stages; however, these levels increased consistently with the progression of redifferentiation during shoot primordia, multiple shoots, and fully developed stages, thereby suggesting shoots as the site of P-I biosynthesis rather than undifferentiated callus cultures. On the contrary, Ganeshkumar et al. (2017) have detected P-I and P-II contents in 16-week-old callus cultures of *P. kurroa*. This might be due to the fact that the biosynthetic routes for secondary metabolites production can be cells/tissues specific and can vary based on genotype, developmental stage, altitude, and other environmental factors. Further, Sharma et al. (2015) have enhanced the P-I production by two- to four-folds at 15°C and

25°C upon seaweed extract treatment. Other elicitors including methyl jasmonate, sodium nitroprusside, and abscisic acid have also been tested and seaweed extract has shown highest P-I production in *in vitro* grown plants of *P. kurroa* (Sharma et al., 2016b). The effect of various elicitors like hydrogen peroxide, abscisic acid, methyl jasmonate, and salicylic acid was also studied, and enhanced picrosides accumulation was observed upon hydrogen peroxide and abscisic acid treatment, whereas methyl jasmonate and salicylic acid application decreased the content in *P. kurroa* (Bhat et al., 2014). Furthermore, Kumar et al. (2016) have studied that the exogenous feeding of precursors viz. cinnamic acid and catalpol influenced the synthesis of P-I and showed that cinnamic acid and cinnamic acid + catalpol stimulated P-I production with 1.6-fold and 4.2-fold, respectively, in shoot cultures of *P. kurroa*. Kannojia et al. (2018) reported enhancement in P-II content upon copper sulfate and silver nitrate treatment on leaves of *P. kurroa* and found maximum increment in the total picroside content with 4 μM of silver nitrate after 72 h of treatment. Apart from elicitor treatment and precursor feeding, several studies related to genetic transformation for secondary metabolite production have been attempted in *P. kurroa* (Verma et al., 2007; Bhat et al., 2012). Hairy root lines of *P. kurroa* through *Agrobacterium rhizogenes* for picrotin and picrotoxinin production have been established (Mishra et al., 2011; Rawat et al., 2013). Yield enhancement strategies for production of picroliv from hairy root cultures of *P. kurroa* have also been reported (Verma et al., 2015). Therefore, these reports on mass propagation and secondary metabolite production under in vitro conditions can act as a platform for reducing pressure on wild population of this valuable medicinal plant along with ensuring the supply of raw material to industries.

6.6 Omics advancements in picrosides biosynthesis

The biosynthesis and accumulation of secondary metabolites are controlled by structural and regulatory genes in different plant species. The development of "omics" technologies has improved our knowledge in understanding the complex biosynthetic pathways involved in secondary metabolite production. Various studies have been taken up to understand the molecular basis of picrosides biosynthesis in *P. kurroa*. Comparative genomics approach was used to clone several genes including 3-hydroxy-3-methylglutaryl coenzyme A reductase (HMGR), 1-deoxy-D-xylulose-5-phosphate synthase (DXPS), 1-deoxy-D-xylulose-5-phosphate reductoisomerase, 4-diphosphocytidyl-2-C-methyl-D-erythritol kinase, 4-hydroxy-3-methylbut-2-enyl diphosphate reductase (HDS), acetyl-CoA acetyltransferase, isopentenyl pyrophosphate isomerase, geranyl diphosphate synthase, phenylalanine ammonia lyase (PAL), 2-C-methylerythritol 4-phosphate cytidyl transferase, 2-C-methylerythritol-2,4-cyclophosphate synthase, 1-hydroxy-2-methyl-2-(E)-butenyl 4-diphosphate synthase (HDS), hydroxymethyl glutaryl CoA synthase, and phosphomevalonate kinase (PMK) of methylerythritol pathway (MEP), mevalonate (MVA), and phenylpropanoid pathways involved in picrosides production in

P. kurroa (Kawoosa et al., 2010; Pandit et al., 2013b; Singh et al., 2013; Bhat et al., 2014). High-throughput de novo transcriptome sequencing and analyses were done in plants grown at two temperature conditions 15°C and 25°C using Illumina sequencing technology to investigate temperature-mediated molecular changes in *P. kurroa* (Gahlan et al., 2012). Various studies have been taken up in which expression analysis of various genes of these pathways was done vis-à-vis picrosides content in different tissues, developmental stages, elicitor treatments to provide the insight into the metabolic basis controlling the biosynthesis of P-I and P-II in *P. kurroa* (Singh et al., 2013; Pandit et al., 2013b; Sharma et al., 2015, 2016b). These studies have identified key genes involved in biosynthesis of picrosides in *P. kurroa*. Further, bioretrosynthetic approach was used to propose a plausible complete biosynthetic pathway for the first time in *P. kurroa* by assembling end products to their precursors and involved four different modules, i.e., MEP, MVA, shikimate/phenylpropanoid, and iridoid pathways for P-I and P-II production (Kumar et al., 2013). Furthermore, Shitiz et al. (2015) have made an important endeavor toward picrosides biosynthesis in *P. kurroa* by completely elucidating the P-I biosynthetic pathway using NGS transcriptomes and enzyme inhibitor studies (Fig. 6.3).

NGS transcriptome resources obtained from varying conditions of picrosides content were also mined for different molecular factors like miRNAs, kinases, and transcription factors regulating the secondary metabolite production. Validation of miRNA and expression analysis by qRT-PCR, and 5′ RACE revealed that miRNA-4995 has a regulatory role in P-I biosynthesis (Vashisht et al., 2015). Key transcription factors were identified through in silico transcript abundance, and qPCR analysis was done for estimation of their gene expression levels in different tissues and genotypes varying for picrosides content in *P. kurroa* (Vashisht et al., 2016). Vashisht et al. (2018) have found upregulated expression of 16 kinases genes involved in plant—pathogen interactions, abiotic stress, wounding, hormonal response, and carbohydrate metabolism in high P-I—accumulating conditions, indicating their possible role in eliciting P-I biosynthesis in *P. kurroa*. In another study, Shitiz et al. (2017) have analyzed the picrosides content and genetic profiles of 26 accessions of *P. kurroa* obtained from different locations in north-western Himalayas and observed total picrosides content (P-I and P-II) ranged from 3.7% to 10.9% in dry rhizomes. They have also studied the genetic diversity and correlated picrosides content with their genetic factors and genetic profiles using simple sequence repeats identified from *P. kurroa* transcriptomes. These markers would be helpful in the development of DNA diagnostics for the authentication of quality plant material.

The precursors for secondary metabolites are provided by primary metabolic pathways such as photosynthesis, glycolysis, pentose phosphate pathway, TCA/Citric acid cycle, etc. Therefore, proper understanding of picrosides biosynthesis requires the coordination of primary and secondary metabolic pathways by tracking events occurring at the molecular level. A study was carried out by Kumar et al. (2015) to determine the gene expression patterns along with P-I content at different stages of *P. kurroa* growth viz. 0, 10, 20, 30, and 40 days of culture in vitro in order to

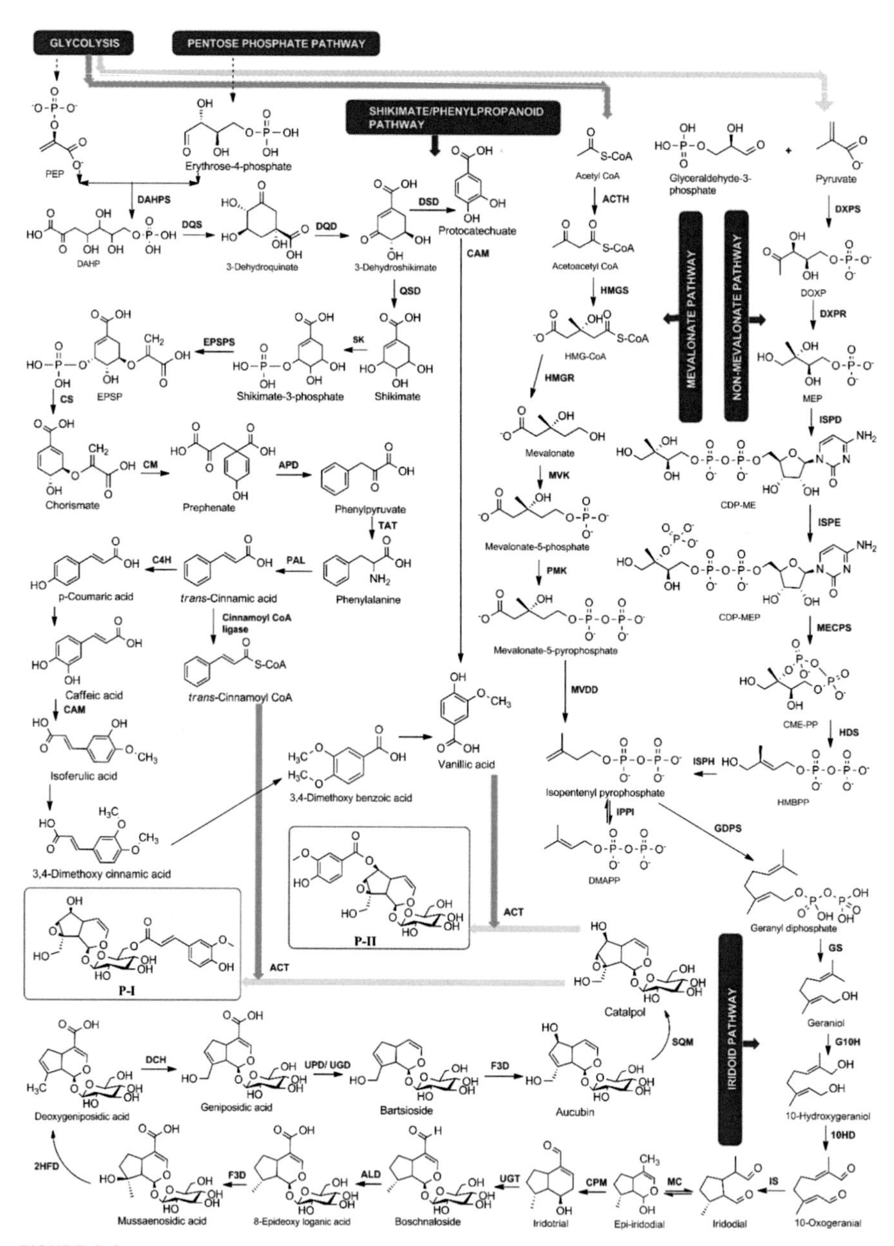

FIGURE 6.3

Complete biosynthetic pathway for Picroside-I and Picroside-II of *Picrorhiza kurroa*.

Adapted from Shitiz, K., Sharma, N., Pal, T., Sood, H., Chauhan, R.S., 2015. NGS transcriptomes and enzyme inhibitors unravel complexity of picrosides biosynthesis in Picrorhiza kurroa Royle ex. Benth. PLoS One 10 (12), e0144546.

identify regulatory steps that provide insight into the metabolic basis controlling the biosynthesis of P-I. Sharma et al. (2016a) have revealed that P-I biosynthesis was developmentally regulated during different stages of differentiation in *P. kurroa.* Expression analysis of multiple genes of primary and secondary metabolic pathways at different morphogenetic stages of *P. kurroa* confirmed their involvement in P-I biosynthesis vis-à-vis shoot development. Temperature influenced P-I production by regulating all integrating pathways of secondary metabolism in this plant species. Genes such as HMGR, PMK, DXPS, GS, G10H, DAHPS, and PAL showed 47- to 87-folds high expression in shoots of FD stage of *P. kurroa* at 15°C compared to 25°C. Recently, Kumar et al. (2017b) have attempted to integrate metabolomics with NGS transcriptomes to decipher metabolic route of P-II biosynthesis using different accessions containing differential P-II content (1.3%–2.6%) and found that P-II is biosynthesized via degrading ferulic acid to produce VA. They have also shown that P-I and P-II are biosynthesized by independent mechanisms in different compartments of *P. kurroa*, i.e., shoots and stolons, respectively.

Proteomic tools also aided in better understanding of various physiological processes occurring in plants. SDS-PAGE and 2-D gel electrophoresis along with mass spectrometry revealed altered proteins belonging to several functional categories including stress response, signaling pathways, photosynthesis, cell cycle, transport, transcription, and translation factors. Few imperative enzymes were identified by differential protein expression using SDS-PAGE followed by MALDI-TOF/TOF MS in *P. kurroa* under picrosides accumulating and nonaccumulating conditions (Sud et al., 2013, 2014).

6.7 Conclusions and future perspective

P. kurroa is an important medicinal herb with a rich source of hepatoprotective picrosides and other secondary metabolites. Accessions with high picrosides content should be used for mass propagation and production of quality raw material to meet the demand of pharmaceutical industry. Cell culture techniques can provide a platform for large-scale in vitro production of P-I in *P. kurroa*, thereby, relieving the pressure from its natural habitat and providing a sustainable strategy for the conservation of this endangered plant species. Omics-based approaches are utilized to elucidate the complexity of P-I and P-II biosynthesis across different tissues and accessions, which enlightened the path for its in vitro production. Expression analysis of the biosynthetic pathway genes in relation to picrosides content and identification of key genes revealed that picrosides biosynthesis is being regulated at various control points in different modules of the biosynthetic route. These genes can be taken up as candidates for optimizing flux levels for enhanced production of picrosides. The knowledge of complete pathway and corresponding genes would be helpful in understanding molecular basis of picrosides biosynthesis as well as planning genetic improvement strategies for enhancing picrosides production in *P. kurroa*.

References

Banyal, H.S., Rani, Devi, N., 2014. *Picrorhiza kurrooa* Royal ex Benth exhibits antimalarial activity against *Plasmodium berghei* Vincke and Lips, 1948. Asian J. Biol. Sci. 7, 72–75.

Baruah, C.C., Gupta, P.P., Nath, A., Patnaik, G.K., Dhawan, B.N., 1998. Anti-allergic and anti-anaphylactic activity of picroliv-a standardized iridoid glycoside fraction of *Picrorhiza kurroa*. Pharmacol. Res. 38 (6), 487–492.

Bentham, G., 1835. Scrophularineae Indicae. James Ridgway and Sons, London, p. 47.

Bhandari, P., Kumar, N., Singh, B., Gupta, A.P., Kaul, V.K., Ahuja, P.S., 2009. Stability-indicating LC–PDA method for determination of picrosides in hepatoprotective Indian herbal preparations of *Picrorhiza kurroa*. Chromatographia 69, 221–227.

Bhat, W.W., lattoo, S.K., Rana, S., Razdan, S., Dhar, N., Dhar, R.S., Vishwakarma, R.A., 2012. Efficient plant regeneration via direct organogenesis and Agrobacterium tumefaciens-mediated genetic transformation of Picrorhiza kurroa: an endangered medicinal herb of the alpine Himalayas. In vitro Cellular and Developmental Biology-Plant 48 (3), 295–303.

Bhat, W.W., Razdan, S., Rana, S., Dhar, N., Wani, T.A., Qazi, P., Vishwakarma, R., Lattoo, S.K., 2014. A phenylalanine ammonia-lyase ortholog (PkPAL1) from *Picrorhiza kurrooa* Royle ex. Benth: molecular cloning, promoter analysis and response to biotic and abiotic elicitors. Gene 547 (2), 245–256.

Chander, R., Kapoor, N.K., Dhawan, B.N., 1992. Picroliv: picroside I and kutkoside from *Picrorhiza kurrooa* are scavengers of superoxide anions. Biochem. Pharmacol. 44, 180–183.

Chandra, B., Palni, L.M.S., Nandi, S.K., 2006. Propagation and conservation of *Picrorhiza kurroa* Royle ex Benth.: an endangered Himalayan medicinal herb of high commercial value. Biodivers. Conserv. 15, 2325–2338.

Chauhan, N.S., 1988. Endangered Ayurvedic pharmacopoeial plant resources of Himachal Pradesh. In: Kaushik, P. (Ed.), Indigenous Medicinal Plants – Including Microbes and Fungi. Today and Tomorrow's Printers and Publishers, New Delhi, pp. 199–205.

Chettri, N., Sharma, E., Lama, S.D., 2005. Non-timber forest produces utilization, distribution and status in a trekking corridor of Sikkim. India, Lyonia 8, 89–101.

Coventry, B.O., 1927. Reprint 1984. Wild Flowers of Kashmir, vol. 2. B. Singh, M.P. Singh, Dehradun, pp. 89–90.

Dobriyal, R.M., Singh, G.S., Rao, K.S., Saxena, K.G., 1997. Medicinal plant resources in Chhakinal watershed in the Northwestern Himalaya. J. Herbs, Spices, Med. Plants 5, 15–27.

Dwivedi, Y., Rastogi, R., Sharma, S.K., Garg, N.K., Dhawan, B.N., 1991. Picroliv affords protection against thioacetamide induced hepatic damage in rats. Planta Med. 57, 25–28.

Dwivedi, Y., Rastogi, R., Garg, N.K., Dhawan, B.N., 1992. Picroliv and its components kutkoside and picroside I protect liver against galactosamineinduced damage in rats. Pharmacol. Toxicol. 71, 383–387.

Dwivedi, Y., Rastogi, R., Mehrotra, R., Garg, N.K., Dhawan, B.N., 1993. Picroliv protects against aflatoxin B1 acutehepatotoxicity in rats. Pharmacol. Res. 27, 189–199.

Gahlan, P., Singh, H.R., Shankar, R., Sharma, N., Kumari, A., Chawla, V., Ahuja, P.S., Kumar, S., 2012. De novo sequencing and characterization of *Picrorhiza kurrooa* transcriptome at two temperatures showed major transcriptome adjustments. BMC Genom. 13, 126.

Ganeshkumar, Y., Ramarao, A., Veeresham, C., 2017. Picroside I and picroside II from tissue cultures of *Picrorhiza kurroa*. Pharmacogn. Res. 9 (1), S53–S56.

Gupta, R.K., 1989. The Living Himalayas, vol. 2. Today and Tomorrow's Printers and Publishers, New Delhi, p. 296.

Husain, G.M., Singh, P.N., Kumar, V., 2009. Antidiabetic activity of standardized extract of *Picrorhiza kurroa* in rat model of NIDDM. Drug Discovery Therapeut. 3 (3), 88–92.

Irshad, S., Mannan, A., Mirza, B., 2011. Antimalarial activity of three pakistani medicinal plants. Pak. J. Pharm. Sci. 24 (4), 589–591.

Jia, D., Barwal, I., Thakur, S., Subhash, C.Y., 2015. Methodology to nanoencapsulate hepatoprotective components from *Picrorhiza kurroa* as food supplement. Food Biosci. 9, 28–35.

Joy, K.L., Rajeshkumar, N.V., Kuttan, G., Kuttan, R., 2000. Effect of *Picrorrhiza kurroa* extract on transplanted tumours and chemical carcinogenesis in mice. J. Ethnopharmacol. 71, 261–266.

Kannojia, G., Helena, D.S., Kumari, A., Gaur, A.K., 2018. Supplementing effect of abiotic elicitor on picroside II content on in vitro Culture of *Picrorhiza kurroa* Royle ex Benth. Genomics Gene Ther. Int. J. https://doi.org/10.23880/ggtij-16000110.

Kant, K., Walia, M., Agnihotri, V.K., Pathania, V., Singh, B., 2013. Evaluation of antioxidant activity of *Picrorhiza kurroa* (leaves) extracts. Indian J. Pharmaceut. Sci. 75, 324–329.

Kapahi, B.K., Srivastava, T.N., Sarin, Y.K., 1993. Description of *Picrorhiza kurroa*, a source of the Ayurvedic drug Kutaki. Int. J. Pharmacogn. 31, 217–222.

Kapil, R.S., 1995. Studies on Ayurvedic plants. J. Non Timber For. Prod. 2, 32–36.

Kawoosa, T., Singh, H., Kumar, A., Sharma, S.K., Devi, K., Dutt, S., Vats, S.K., Sharma, M., Ahuja, P.S., Kumar, S., 2010. Light and temperature regulated terpene biosynthesis: hepatoprotective monoterpenepicroside accumulation in *Picrorhiza kurroa*. Funct. Integr. Genom. 10, 393–404.

Krupashree, K., Kumar, K.H., Rachitha, P., Jayashree, G.V., Khanum, F., 2014. Chemical composition, antioxidant and macromolecule damage protective effects of *Picrorrhiza kurroa* Royle ex Benth. South Afr. J. Bot. 94, 249–254.

Kumar, P., Shukla, S.K., 2017. Hepatoprotective efficacy of *Picrorhiza kurroa* in experimentally induced hepatotoxicity in cockerels. Int. J. Curr. Microbiol. Appl. Sci. 6 (4), 2614–2622.

Kumar, V., Sood, H., Chauhan, R.S., 2013. A proposed biosynthetic pathway of picrosides linked through the detection of biochemical intermediates in the endangered medicinal herb *Picrorhiza kurroa*. Phytochem. Anal. 24, 598–602.

Kumar, V., Sharma, N., Shitiz, K., Singh, T.R., Tandon, C., Sood, H., Chauhan, R.S., 2015. An insight into conflux of metabolic traffic leading to picroside I biosynthesis by tracking molecular time course changes in a medicinal herb, *Picrorhiza kurroa*. Plant Cell Tissue Organ Cult. 123 (2), 435–441.

Kumar, V., Sharma, N., Sood, H., Chauhan, R.S., 2016. Exogenous feeding of immediate precursors reveals synergistic effect on picroside-I biosynthesis in shoot cultures of *Picrorhiza kurroa* Royle ex Benth. Sci. Rep. 6, 29750.

Kumar, S., Patial, V., Soni, S., Sharma, S., Pratap, K., Kumar, D., Padwad, Y., 2017a. *Picrorhiza kurroa* enhances β-Cell mass proliferation and insulin secretion in streptozotocin evoked β-cell damage in rats. Front. Pharmacol. 8, 1663–9812.

Kumar, V., Bansal, A., Chauhan, R.S., 2017b. Modular design of picroside-II biosynthesis deciphered through NGS transcriptomes and metabolic intermediates analysis in naturally variant chemotypes of a medicinal herb, *Picrorhiza kurroa*. Front. Plant Sci. 8, 564.

Lal, N., Ahuja, P.S., Kukreja, A.K., Pandey, B., 1988. Clonal propagation of *Picrorhiza kurroa* Royle ex Benth. by shoot tip culture. Plant Cell Rep. 7, 201–205.

Laxmi, V., Preeti, C., 2015. Antimicrobial activity of dried stolon extracts of *Picrorhiza kurroa* Royle ex. Benth. – An endemic and endangered Himalayan herb. ENVIS Bull. Himalayan Ecol. 23, 127–132.

Lee, H.S., Keum, K.Y., Ku, S.K., 2007. Effects of *Picrorrhiza rhizoma* water extracts on the subacute liver damages induced by carbon tetrachloride. J. Med. Food 10, 110–117.

Mehrotra, R., Rawat, S., Kulshrestha, D.K., Patnaik, G.K., Dhawan, B.N., 1990. In vitro studies on the effect of certain natural products against hepatitis B virus. Indian J. Med. Res. 92, 133–138.

Mehta, I.S., Joshi, G.C., Basera, P.S., 1994. The folklore medicinal plants of Talla Johar of Eastern Kumaun (Central Himalaya). In: Gupta, B.K. (Ed.), Higher Plants of Indian Subcontinent, vol. 3, pp. 125–133.

Mishra, J., Bhandari, H., Singh, M., Rawat, S., Agnihotri, R.K., Mishra, S., Purohit, S., 2011. Hairy root culture of *Picrorhiza kurroa* Royle ex Benth.: a promising approach for the production of picrotin and picrotoxinin. Acta Phsiol. Plant. 33, 1841–1846.

Mondal, T.K., Bantawa, P., Sarkar, B., Ghosh, P., Chand, P.K., 2013. Cellular differentiation, regeneration, and secondary metabolite production in medicinal *Picrorhiza* spp. Plant Cell Tissue Organ Cult. 112, 143–158.

Mulliken, T.A., 2000. Implementing CITES for Himalayan medicinal plants *Nardostachys grandiflora* and *Picrorhiza kurroa*. In: TRAFFIC B, vol. 18, pp. 63–72.

Nayar, M.P., Sastri, A.R.K., 1990. Red Data Plants of India. CSIR Publication, New Delhi, p. 271.

Pandit, S., Shitiz, K., Sood, H., Chauhan, R.S., 2013a. Differential biosynthesis and accumulation of picrosides in an endangered medicinal herb *Picrorhiza kurroa*. J. Plant Biochem. Biotechnol. 22, 335–342.

Pandit, S., Shitiz, K., Sood, H., Naik, P.K., Chauhan, R.S., 2013b. Expression pattern of fifteen genes of non-mevalonate (MEP) and mevalonate (MVA) pathways in different tissues of endangered medicinal herb *Picrorhiza kurroa* with respect to picrosides content. Mol. Biol. Rep. 40 (2), 1053–1063.

Pennell, F.W., 1943. The Scrophulariaceae of the Western Himalayas. Bishen Singh Mahendra Pal Singh, Dehra Dun, pp. 63–66 (Reprint 1997).

Polunin, O., Stainton, A., 1990. Flowers of the Himalaya, sixth ed. Oxford University Press, Delhi, p. 295.

Raina, R., Mehra, T.S., Chand, R., Sharma, Y.P., 2010. Reproductive biology of *Picrorhiza kurroa* – A critically endangered high value temperate medicinal plant. J. Med. Aromatic Plants 1 (2), 40–43.

Rajeshkumar, N.V., Kuttan, R., 2001. Protective effect of picroliv, the active constituent of *Picrorhiza kurroa*, against chemical carcinogenesis in mice. Teratog. Carcinog. Mutagen. 21 (4), 303–313.

Rajeshkumar, N.V., Kuttan, R., 2003. Modulation of carcinogenic response and antioxidant enzymes of rats administered with 1,2-dimethylhydrazine by Picroliv. Cancer Lett. 191 (2), 137–143.

Rajkumar, V., Guha, G., Kumar, R.A., 2011. Antioxidant and antineoplastic activities of *Picrorhiza kurroa* extracts. Food Chem. Toxicol. 49, 363–369.

Rathee, D., Thanki, M., Bhuva, S., Anandjiwala, S., Agrawal, R., 2013. Iridoid glycosides-kutkin, picroside I, and kutkoside from *Picrorrhiza kurroa* Benth inhibits the invasion

and migration of MCF-7 breast cancer cells through the down regulation of matrix metalloproteinases. Arab. J. Chem. 6 (1), 49–58.
Rawat, J.M., Rawat, B., Mehrotra, S., 2013. Plant regeneration, genetic fidelity, and active ingredient content of encapsulated hairy roots of *Picrorhiza kurroa* Royle ex Benth. Biotechnol. Lett. 35, 961–968.
Ray, A., Chaudhuri, S.R., Majumdar, B., Bandyopadhyay, S.K., 2002. Antioxidant activity of ethanol extract of rhizome of *Picrorhiza kurroa* on indomethacin induced gastric ulcer during healing. Indian J. Clin. Biochem. 17, 44–51.
Royle, J.F., 1835–1940a. Illustrations of the Botany and Other Branches of the Natural History of the Himalayan Mountains, and of the Flora of Cashmere. Volume Plates, Reprint 1970. Today and Tomorrow's Printers and Publishers, New Delhi, p. 71.
Royle, J.F., 1835–1940b. Illustrations of the Botany and Other Branches of the Natural History of the Himalayan Mountains, and of the Flora of Cashmere. Volume Text. Reprint 1970. Today & Tomorrow's Printers & Publishers, New Delhi, p. 291.
Sah, J.N., Varshney, V.K., 2013. Chemical constituents of *Picrorhiza* genus: a review. Am. J. Essent. Oil Nat. Prod. 1 (2), 22–37.
Sane, S.A., Shakya, N., Gupta, S., 2011. Immunomodulatory effect of picroliv on the efficacy of paromomycin and miltefosine in combination in experimental visceral leishmaniasis. Exp. Parasitol. 127, 376–381.
Saraswat, B., Visen, P.K.S., Patnaik, G.K., Dhawan, B.N., 1999. Ex vivo and in vivo investigations of picroliv from *Picrorhiza kurroa* in an alcohol intoxication model in rats. J. Ethnopharmacol. 66, 263–269.
Semwal, D.P., Saradhi, P.P., Nautiyal, B.P., Bhatt, A.B., 2007. Current status, distribution and conservation of rare and endangered medicinal plants of Kedarnath Wildlife Sanctuary, Central Himalayas, India. Curr. Sci. 92 (12), 1733–1738.
Sharma, B.M., Kachroo, P., 1981. Flora of Jammu and Plants of Neighbourhood, vol. 1. Bishen Singh, Mahendra Pal Singh, Dehra Dun, p. 243.
Sharma, N., Chauhan, R.S., Sood, H., 2015. Seaweed extract as a novel elicitor and medium for mass propagation and picroside-I production in an endangered medicinal herb *Picrorhiza kurroa*. Plant Cell Tissue Organ Cult. 122 (1), 57–65.
Sharma, S., Katoch, V., Rathour, R., Sharma, T.R., 2010. In vitro propagation of endangered temperate Himalayan medicinal herb Picrorhiza kurroa Royle ex Benth using leaf explants and nodal segments. J. Plant Biochem. Biotechnol. 19, 111–114.
Sharma, N., Chauhan, R.S., Sood, H., 2016a. Discerning picroside-I biosynthesis via molecular dissection of in vitro shoot regeneration in *Picrorhiza kurroa*. Plant Cell Rep. 35 (8), 1601–1615.
Sharma, N., Kumar, V., Chauhan, R.S., Sood, H., 2016b. Modulation of picroside-I biosynthesis in grown elicited shoots of *Picrorhiza kurroa* in vitro. J. Plant Growth Regul. 35 (4), 965–973.
Sharma, R., Deb, K., Ambwani, T.K., Ambwani, S., 2018. Preliminary phytochemical screening and antioxidative potential of *Picrorhiza kurroa* Royle ex. Benth. Bull. Environ. Pharmacol. Life Sci. 7 (11), 134–139.
Shetty, S.N., Mengi, S., Vaidya, R., Vaidya, A.D.B., 2010. A study of standardized extracts of *Picrorhiza kurroa* Royle ex Benth in experimental nonalcoholic fatty liver disease. J. Ayurveda Integr. Med. 1 (3), 203–210.
Shitiz, K., Pandit, S., Chauhan, R.S., Sood, H., 2013. Picrosides content in the rhizomes of *Picrorhiza kurroa* Royle ex Benth traded for herbal drugs in the markets of North India. Int. J. Med. Aromatic Plants 3 (2), 226–233.

Shitiz, K., Sharma, N., Pal, T., Sood, H., Chauhan, R.S., 2015. NGS transcriptomes and enzyme inhibitors unravel complexity of picrosides biosynthesis in *Picrorhiza kurroa* Royle ex. Benth. PLoS One 10 (12), e0144546.

Shitiz, K., Pandit, S., Chanumolu, S., Sood, H., Singh, H., Singh, J., Chauhan, R.S., 2017. Mining simple sequence repeats in *Picrorhiza kurroa* transcriptomes for assessing genetic diversity among accessions varying for picrosides contents. Plant Genet. Resour. 15 (1), 79–88.

Shubha, K.S., Sumana, K., Lakshmidevi, L., 2016. Antifungal activity of *Solanum xantocarpum* Sch and Wend and *Picrorhiza kurroa* Royle ex Benth against some clinical dermatophytes. Int. J. Curr. Microbiol. Appl. Sci. 5, 236–244.

Singh, V., Banyal, H.S., 2011. Antimalarial effects of *Picrorhiza kurrooa* Royle ex Benth extracts on *Plasmodium berghei*. Asian J. Exp. Biol. Sci. 2 (3), 529–532.

Singh, G., Kachroo, P., 1976. Forest Flora of Srinagar and Plants of Neighbourhood. Reprint 1994. Bishen Singh, Mahendra Pal Singh, Dehradun, p. 21.

Singh, S.K., Rawat, G.S., 2000. Flora of Great Himalayan National Park, Himachal Pradesh. Bishen Singh, Mahendra Pal Singh, Dehradun.

Singh, H., Gahlan, P., Kumar, S., 2013. Cloning and expression analysis of ten genes associated with picrosides biosynthesis in *Picrorhiza kurroa*. Gene 515, 320–328.

Sinha, S., Bhat, J., Joshi, M., Ghaskadbi, S., 2011. Hepatoprotective activity of *Picrorhiza kurroa* Royle Ex. Benth extract against alcohol cytotoxicity in mouse liver slice culture. Int. J. Green Pharm. 5 (3), 244–253.

Smith, W.W., Cave, G.H., 1911. The Vegetation of the Zemu and Llonakh Valleys of Sikkim, vol. 4. Recordings Bot. Survey India, pp. 141–260.

Soni, D., Grover, A., 2019. "Picrosides" from Picrorhiza kurroa as potential anti-carcinogenic agents. Biomedicine & Pharmacotherapy 109, 1680–1687.

Sood, H., Chauhan, R.S., 2009a. Development of a low cost micropropagation technology for an endangered medicinal herb (*Picrorhiza kurroa*) of North-Western Himalayas. J. Plant Sci. 4, 21–31.

Sood, H., Chauhan, R.S., 2009b. High frequency callus induction and plantlet regeneration from different explants of *Picrorhiza kurroa*-a medicinal herb of Himalayas. Afr. J. Biotechnol. 8, 1965–1972.

Sood, H., Chauhan, R.S., 2010. Biosynthesis and accumulation of a medicinal compound, picroside-I, in cultures of *Picrorhiza kurroa* Royle ex Benth. Plant Cell Tissue Organ Cult. 100, 113–117.

Sud, A., Chauhan, R.S., Tandon, C., 2013. Identification of imperative enzymes by differential protein expression in *Picrorhiza kurroa* under metabolite accumulating and non-accumulating conditions. Protein Pept. Lett. 20, 826–835.

Sud, A., Chauhan, R.S., Tandon, C., 2014. Mass spectrometric analysis of differentially expressed proteins in an endangered medicinal herb, *Picrorhiza kurroa*. Biomed Res. Int. https://doi.org/10.1155/2014/326405.

Uniyal, M.R., Bhat, A.V., Chaturvedi, P.N., 1982. Preliminary observations on medicinal plants of Lahaul Spiti forest division in Himachal Pradesh. Bull. Med. Ethno. Bot. Res. 3, 1–26.

Uniyal, S.K., Awasthi, A., Rawat, G.S., 2002. Current status and distribution of commercially exploited medicinal and aromatic plants in upper Gori valley, Kumaon Himalaya, Uttaranchal. Curr. Sci. 82 (10), 1246–1252.

Uniyal, S.K., Singh, K.N., Jamwal, P., Lal, B., 2006. Traditional use of medicinal plants among the tribal communities of Chhota Bhangal, Western Himalaya. J. Ethnobiol. Ethnomed. 2, 14.

Upadhyay, R., Arumugam, N., Bhojwani, S.S., 1989. In vitro propagation of *Picrorhiza kurroa* Royle ex Benth. – An endangered species of medicinal importance. Phytomorphology 39, 235–242.

Upadhyay, D., Dash, R.P., Anandjiwala, S., Nivsarkar, M., 2013. Comparative pharmacokinetic profiles of picrosides I and II from kutkin, *Picrorhiza kurroa* extract and its formulation in rats. Fitoterapia 85, 76–83.

Vashisht, I., Mishra, P., Pal, T., Chanumolu, S., Singh, T.R., Chauhan, R.S., 2015. Mining NGS transcriptomes for miRNAs and dissecting their role in regulating growth, development, and secondary metabolites production in different organs of a medicinal herb, *Picrorhiza kurroa*. Planta 241 (5), 1255–1268.

Vashisht, I., Pal, T., Sood, H., Chauhan, R.S., 2016. Differential metabolites content transcriptomes pinpoint key transcription factors regulating picrosides biosynthesis in a medicinal herb, *Picrorhiza kurroa* Royle ex. Benth. Mol. Biol. Rep. 43 (12), 1395–1409.

Vashisht, I., Pal, T., Bansal, A., Chauhan, R., 2018. Uncovering interconnections between kinases vis-à-vis physiological and biochemical processes contributing to picroside-I biosynthesis in a medicinal herb, *Picrorhiza kurroa* Royle ex. Benth. Acta Physiol. Plant. 40, 115.

Ved, D.K., Goraya, G.S., 2008. Demand and Supply of Medicinal Plants in India. Foundation for Revitalisation of Local Health Traditions (Bangalore, India), National Medicinal Plants Board. Bishen Singh, Mahendra Pal Singh.

Verma, P.C., Rahman, L.Q., Negi, A.S., Chand, J.D., Khanuja, S.P.S., Banerjee, S., 2007. *Agrobacterium rhizogenes*-mediated transformation of *Picrorhiza kurroa* Royle ex Benth.: establishment and selection of superior hairy root clone. Plant Biotechnol. Rep. 1, 169–174.

Verma, P.C., Basu, V., Gupta, V., Saxena, G., ur Rahman, L., 2009. Pharmacology and chemistry of a potent hepatoprotective compound picroliv isolated from the roots and rhizomes of *Picrorhiza kurroa* Royle ex Benth. (Kutki). Curr. Pharmaceut. Biotechnol. 10 (6), 641–649.

Verma, P.C., Singh, H., Negi, A.S., Saxena, G., Rahman, L., Banerjee, S., 2015. Yield enhancement strategies for the production of picroliv from hairy root culture of *Picrorhiza kurroa* Royle ex Benth. Plant Signal. Behav. 10 (5), e1023976.

Zhang, D.K., Yu, J.J., Li, Y.M., Wei, L.N., Yu, Y., Feng, Y.H., Wang, X., 2012. A *Picrorhiza kurroa* derivative, picroliv, attenuates the development of dextran-sulfate-sodium-induced colitis in mice. Mediat. Inflamm. 751629.

CHAPTER

Podophyllum hexandrum

7

Jagdish Singh, Joginder Singh, Swaran Lata

Himalayan Forest Research Institute (HFRI), Conifer Campus, Shimla, Himachal Pradesh, India

7.1 Introduction

North Western Himalayas harbors a large number of economically important plants, which includeseveral species of medicinal value. The ever-increasing demand, particularly in view of worldwide shift for the drugs of herbal origin over synthetic counterparts, has led to overexploitation of medicinal plants. Apart from healthcare, medicinal plants are mainly the alternate income generating source of underprivileged communities; therefore, strengthening this sector may benefit and improve the living standard of poor people. *Podophyllum hexandrum* Royle belongs to family Berberidaceae, and it is a perennial herb bearing the common name Himalayan Mayapple. The genus *Podophyllum* is also called as Mayapple because its fruits ripen in spring. The name *Podophyllum* is taken from "podos" a foot and "phyllon" a leaf and refers to the resemblance of the leaves to a duck's foot. The genus *Podophyllum* is represented by three species *P. hexandrum*, *P. peltatum*, and *P. sikkimensis* in Himalayan zone of India. *P. hexandrum* Royle also known as the Indian Podophyllum is a perennial herb, growing on the lower slopes of the Himalayas in scrub and forest from Afghanistan eastward to central China commonly distributed in Himalayan region of Asian continent and popularly known as Himalayan Mayapple, while *P. peltatum*, commonly distributed in Atlantic North America and popularly called as American Mayapple.

The rhizomes of *P. hexandrum* are known to contain several lignans. The lignans occurring in *Podophyllum* possess antitumor properties. *Podophyllum* being the most active cytotoxic herb contains 4.3% podophyllotoxin on a dry weight basis. The whole plant has also got great importance in traditional systems of medicines including Ayurveda, Unani, and Tibetan systems for curing several diseases. The rhizomes and roots of *Podophyllum* species have gained much importance throughout the world as being the main source or the starting material for the alkaloid podophyllotoxin and its semisynthetic compounds, the etoposide, teniposide, and etoposide phosphate since their use in treatment of specific types of cancers. These compounds are useful in the treatment of refractory testicular carcinomas, nonlymphocytic leukemias and non-Hodgkin's lymphomas and etoposide in the treatment of lung cancer in addition to its therapeutic value against the AIDS-associated Kaposi sarcoma. It has served as a commercial source of podophyllotoxin and related

Himalayan Medicinal Plants. https://doi.org/10.1016/B978-0-12-823151-7.00001-5

aryltetralin lignans and several other bioactive constituents. It also has numerous applications in modern medicine by virtue of its free radical scavenging capacity. In United States Bristol Co. and in Switzerland Sandoz prepared hundreds of semi-synthetic compounds. Out of these only above three are widely used as antitumor agents with minimal toxic or side effects. Destruction of plant populations due to overexploitation or natural calamities affects drug supply and the content of bioactive secondary metabolite in the plant. Therefore, immediate thrust has to be given for generating the reliable conventional protocols of mass cultivation of *P. hexandrum*. Moreover, wild populations may be represented by various genotypes growing under different environmental conditions which may affect drug profile leading to problems in the purity of the final product. Thus, cultivation of suitable clones would ensure a reliable supply of the material with consistent quality. Thus, there is a need to understand and conserve the genetic diversity of this important medicinal plant. At present two species of *Podophyllum*, viz. the American *Podophyllum* or Mayapple (*P. peltatum*) and the Himalayan *Podophyllum* (*P. hexandrum*), are the main sources of podophyllotoxin. Podophyllotoxin is an expensive starting compound for the chemical synthesis of its derivatives. *P. hexandrum* is now being considered as a rare and threatened species mainly due to the large-scale removal of its underground parts that still continues at rates well over natural regeneration. Attempts have been made to conserve this plant through in vitro propagation and artificial breaking of seed dormancy.

P. hexandrum is native to the temperate region of Himalayan countries like Afghanistan, Pakistan, India, Nepal, Bhutan, and in Southwest China (Airi et al., 1997; Gupta and Sethi, 1983; Giri and Narasu, 2000a,b; Chaudhary et al., 1998). It is categorized as globally rare plant in IUCN red list and has endangered status in India (Chaudhary et al., 2014). It is distributed in very restricted pockets in the Himalayan zone at altitudes ranging from 2000 to 4000 m a.s.l (Bhadula et al., 2000), mostly found in Alpine region. Extraction of the whole *P. hexandrum* plant in order to harvest the roots and rhizome containing the useful drug has resulted into a decline of this species and even disappearance in some regions. The population of these plants throughout its range was observed to be very sparse, in decline, and receding toward higher elevations (Rao, 1998). It is variously considered as endangered (Bhadula et al., 2000) and critically endangered (CAMP, 1998). It has declined considerably as a result of exploitation to meet the increasing demands of the pharmaceutical industry (Bhadula et al., 2000). Export of this species from India has been prohibited, although illegal removal continues (Nadeem et al., 2000).

In recent years, the frequency of *P. hexandrum* in nature has declined because these plant species are collected in large quantities from wild to meet the ever-increasing demand of pharmaceutical industry (Sultan et al., 2006).

P. hexandrum (Himalayan Mayapple) was known as Aindri ("a divine drug") in ancient times. It has been reported to be used through the ages and in modern times as a cure for allergic and inflammatory conditions of the skin; biliary fever; burning sensation; cold; constipation; cancer of the brain, bladder, and lung; erysipelas; Hodgkin's disease; insect bites; mental disorders; monocytoid leukemia non-Hodgkin's lymphoma; rheumatism; septic wounds; plague; and venereal warts. However, currently the commercial source of podophyllotoxin is the rhizomes and roots of *P. hexandrum.* Availability of podophyllotoxin isolated from plant has its limitations, due to scarce occurrence of the plant because of intense collection from nature and lack of organized cultivation (Gupta and Seth, 1983). Himalayan Mayapple is overexploited because the whole plant is uprooted for extraction of podophyllotoxin which is a serious issue from conservation point of view (Canel et al., 2000). In view of overexploitation due to its medicinal importance, a conservation strategy should be devised to increase its population number by propagating the plant through seed in a short period thereby compensating or reducing the harvesting pressure on rhizome. Exploitation of *Podophyllum* from the wild is prohibited for export from India under CITES (Convention on International Trade in Endangered Species of wild flora and fauna). Only cultivated/artificially propagated plant species is allowed for export under cover of CITES export permit and Legal Procurement Certificates or certificate of cultivation from the designed authorities. *P. hexandrum* needs study of its variability and population under different locations

Mature plant of *P. hexandrum.*

Profuse rooting in *P. hexandrum.*

with scientific basis and its *ex situ* and *in situ* conservation. National Medicinal Plant Broad, India, has initiated efforts toward conservation of rare, endangered, and threatened medicinal plants throughout the country (Kaul et al., 1998).

7.2 Taxonomy

Flower of P. hexandrum.

P. hexandrum is believed to have originated in Himalayan regions. According to Bentham and Hooker's System of Classification, the genus *Podophyllum* is classified as follows:

Kingdom	Plantae
Subkingdom	Phanerogamia
Division	Angiospermae
Class	Dicotyledons
Subclass	Polypetalae
Order	Ranunculales
Family	Berberidaceae
Genus	*Podophyllum*
Species	*hexandrum* Royle.

Vernacular Names:

Hindi	Bakra, Chimara, Papra, Papri
Himachali	Ban-kakri, Kakariya
Uttaranchal (Bhotia)	Ghi-cupra
Punjabi	Bankakri
Urdu	Ban kakri
Kashmiri	Baruwangan, Nirbishi, Papra, Bhavanbakra
English	Mayapple
Ladakhi	Ol-Mo-Se, Demo-khushu
Kinnauri	Ululu

7.3 Adulterant

The rhizomes in trade are also adulterated with the rhizome of wild growing *Ainsliaea latifolia* belonging to Asteraceae family (Puri and Jain, 1988).

7.4 Geographical distribution and status

It occurs in forests and on open slopes from 2400 to 4500 m (Polunin and Stainton, 1984). This species has been considered a rare and threatened species, and removal rates exceed natural regeneration rates (Nadeem et al., 2000). Occurrence is reported in Uttarakhand (Rao, 1998), and more particularly in Kumaon (Airi et al., 1997), and Garhwal (Bhadula et al., 2000). It is also reported in Himachal Pradesh, where it is found sporadically in Rohru, Kullu, Kangra, Chamba, Nichar, and Lahaul and Spiti forest divisions (Chauhan, 1999). It is distributed in restricted pockets of the Himalayas ranging from 2000 to 4000 m (Bhadula et al., 2000). The population of these plants throughout its range was observed to be very sparse, in decline and receding toward higher elevations (Rao, 1998). It is variously considered as endangered (Bhadula et al., 2000) or critically endangered (CAMP, 1998). It has declined considerably as a result of exploitation to meet the increasing demand of the pharmaceutical industry (Bhadula et al., 2000). *Podophyllum* takes years for establishment and lives long in a suitable habitat (Facciola, 1990). According to reports, about 37.3 tones of rhizomes of *P. hexandrum* were uprooted during 1995–2000 in Himachal Pradesh (Ali and Sharma, 2013). *P. hexandrum* has been overharvested to meet the demands of pharmaceutical industries. The entire plant is harvested due to presence of podophyllotoxin in its rhizome. That is why the species has acquired the endangered status (Gupta and Sethi, 1983; Bhadula et al., 2000; Airi et al., 1997). Leaves of *P. hexandrum* contain higher amounts of podophyllotoxin in comparison to *P. peltatum*.

Podophyllum hexandrum in Suru Valley, Ladakh.

Nag et al. (2013) reported that *P. hexandrum* has a wide region of distribution, within it appears mostly in valleys with secondary vegetation. In any population, the plant shows a kind of clumped distribution pattern. Earlier, *P. hexandrum* was used in folk medicine by local healers in small quantities, but commercialization of the plant for its medicinal attributes in recent years has increased demand and consequent exploitation. The size of the wild populations has been declining owing to overexploitation, habitat fragmentation, long dormancy, and low rate of natural regeneration. The population size of *P. hexandrum* in the Himalayas is very low (40–700 plants per location) and is declining rapidly each year. Some populations in certain pockets have virtually disappeared owing to anthropogenic activities and overexploitation (Bhadula et al., 2000). Therefore, *P. hexandrum* has been classified as an endangered species in India since 1987 (Nayar and Sastry, 1987).

Singh et al. (2018) reported that most of the identified habitat for *P. hexandrum* in Himachal Pradesh and Ladakh (J&K) was moist and rocky moist having gentle to steep slope which shows habitat specificity of species. According to literature also the species prefers moist humid conditions and soils with rich in organic matter. Several species like *Achillea millefolium, Bergenia ciliata, Dactylorhiza hatagirea, Heracleum candicans, Salix* spp., *Rosa webbiana, Primula* spp., *Aconitum heterophyllum, Angelica glauca*, etc., were commonly associated with *P. hexandrum* in different identified sites. *Aesculus indica, Salix* spp., *Quercus semecarpifolia, Pinus wallichiana, Betula utilis, Abies pindrow, Juniperus* spp., *Rhododendron campanulatum*, etc., were the prominent main tree species found in different identified sites. They have further identified 41 different sites from Himachal Pradesh and Ladakh, UT, the details of sites are given in Table 7.1:

Table 7.1 *Podophyllum hexandrum* in different geographical locations of Himachal Pradesh (HP) and Ladakh.

Sr. No.	State/UT	District	Forest division	Source/Sites	Latitude (N)	Longitude (E)	Altitude (m)
1.	HP	Kullu	Naggar	Bansherudhar	32°10′795″	77°12′994″	2866
2.	HP	Lahaul and Spiti	Keylong	Chimrit	32°43′249″	76°40′870″	2718
3.	HP	Chamba	Killar	Karyuni	33°03′271″	76°26′300″	3064
4.	HP	Chamba	Killar	Hudanbhaturi	33°05′980″	76°29′223″	3580
5.	HP	Chamba	Killar	Suralbhaturi	33°08′594″	76°27′015″	3320
6.	HP	Kenner	Yangapa	Shelti (Yangapa)	31°39′583″	77°07′32″	2985
7.	HP	Lahaul and Spiti	Udaipur	Trilokinath	32°06′741″	76°41′471″	2951
8.	HP	Kinnaur	Nichar	Chotkhanda	31°308′77″	77°58′704″	3400
9.	HP	Kinnaur	Sangla	Chitkul	31°20′422″	78°26′702″	3450
10.	HP	Kullu	Manali	Kothi	32°19′206″	77°12′234″	3036
11.	HP	Kangra	Bir	Palachak	32°08′076″	76°45′416″	2760
12.	HP	Shimla	Rampur	Jurassi	N 31°18′583″	77°42′092″	2540
13.	HP	Kullu	Sainj	Khoruthach	31°45′976″	77°26′732″	2580
14.	HP	Chamba	Bhandal	Thathidhar	32°49′905″	75°58′600″	2899
15.	HP	Kullu	Anni	Jalorijot	N 31°32′236″	77°22′391″	3140
16.	HP	Chamba	Bharmour	Deol	32°18′210″	76°33′015″	3225
17.	HP	Kullu	Patlikul	Shangchar	32°10′502″	77°08′576″	3056
18.	HP	Kullu	Naggar	Parot	32°11′654″	77°17′012″	3032
19.	HP	Kullu	Naggar	Kulri (Bansheru)	32°10′64″	77°13′415″	3079
20.	HP	Kullu	Manali	Gulaba	32°19′114″	77°12′310″	3664
21.	HP	Kullu	Kullu	Majhoni (Tirthan)	31°45′640″	77°27′870″	3460
22.	HP	Kinnaur	Sarahan	Katgaon	31°33′061″	77°58′489″	2750
23.	HP	Chamba	Khajjiar	Kalatop	32°33′052″	76°01′168″	2356
24.	HP	Shimla	Rohru	Janglik	1°91′4116″	78°00′992″	2848

Continued

Table 7.1 *Podophyllum hexandrum* in different geographical locations of Himachal Pradesh (HP) and Ladakh.—*cont'd*

Sr. No.	State/UT	District	Forest division	Source/Sites	Latitude (N)	Longitude (E)	Altitude (m)
25.	HP	Shimla	Rohru	Maila (Diudi)	31°22′014″	78°01′129″	2990
26.	HP	Chamba	Bhandal	Dharwal	32°48′829″	75°56′766″	2810
27.	HP	Lahaul and Spiti	Keylong	Kangsar	32°32′721″	76°58′463″	3240
28.	HP	Shimla	Theog	Deha	32°08′715″	76°45′526″	2675
29.	HP	Lahaul and Spiti	Lahaul	Kardang	32°33′918″	77°01′088″	3148
30.	HP	Kullu	Sainj	GHNP	31°45′970″	E 77°26′728″	2582
31.	HP	Shimla	Rampur	Chottkanda (Panvi)	31°18′583″	77°42′092″	2835
32.	HP	Kangra	Bir	Joardu forest	32°08′715″	77°45′526″	2675
33.	Ladakh	Kargil	Sankoo	Prachik-2	34°04′129″	75°58′667″	3497
34.	Ladakh	Kargil	Sankoo	Beemaputri	34°12′353″	75°55′699″	3231
35.	Ladakh	Kargil	Sankoo	Panikher	34°07′119″	75°58′002″	3306
36.	Ladakh	Kargil	Sankoo	Shangara	34°12′984″	75°058′219″	3148
37.	Ladakh	Leh	Sankoo	Thangbu-1	34°12′337″	75°55′743″	3228
38.	Ladakh	Leh	Sankoo	Thangbu-II	34°12′457″	75°55′748″	3235
39.	Ladakh	Kargil	Sankoo	UMBA1	34°18′118″	75°55′709″	3129
40.	Ladakh	Kargil	Sankoo	UMBA 2	34°18′096″	75°55′677″	3142
41.	Ladakh	Kargil	Sankoo	Suru (Sangrol)	34°14′479″	75°58′466″	3085

Source: Singh, J., Singh, J., Tewari, V.P., 2018. Screening and evaluation of superior chemotypes of Podophyllum hexandrum royle from different geographical locations of North-west Himalayas. J. Plant Chem. Ecophysiol. 3 (1), 1–7.

7.5 Morphology

It is an erect, smooth, glabrous, somewhat fleshy, 15–40 cm tall herb with perennial roots and rhizomes. Mostly found in aggregation, which gives the plant a shrubby look. It bears one to four reproductive and five to eight vegetative shoots that appear in April–May. Whereas the vegetative shoot bears a single leaf, the reproductive shoots generally bear two and sometimes three to four palmate, orbicular, and deeply lobed (three to five) leaves. A single white or pink supraaxillary flower emerges at the middle of the flower-bearing shoot. The cup-shaped bisexual, actinomorphic, and gamosepalous flower (2–3 cm long) bears three deciduous petaloid sepals but six obovate/oblong polypetalous petals, six stamens with anthers opened as lateral slits, and a superior multilocular ovary with marginal placentation. The ovoid, pulpy, many seeded berry is borne on a pedicel is an erect, smooth, glabrous, somewhat fleshy. Fruit a berry, 2.5–5 cm, oblong ovoid or oblong-ellipsoid, scarlet, or red when ripe and many seeded. Seeds small and light black embedded in the red pulp. Rhizome fleshy, short, horizontally creeping with long dense fibrous roots, the underground part is known as rhizomatus root. Each node has the tendency to develop into a new plant offshoot (Collet, 1921; Hooker, 1875; Airi et al., 1997). Its flowering and fruiting period is May–September.

P. hexandrum in Lahaul Valley (Himachal Pradesh).

Bud initiation in *P. hexandrum.*

7.6 Chemical constituents

American *Podophyllum* contains 4%–5% podophyllum resin, whereas Indian spp. contains 7%–16%. The variation in percentage of resin is attributed to seasonal differences, different sites of growth, and age of the plant (Purohit et al., 1999). Sharma et al. (2000a,b) demonstrated much higher podophyllotoxin content in *P. hexandrum* collected from Jalori Pass (8000 ft) than in the Palampur (4000 ft) counterpart. Similarly active constituents of *P. hexandrum* (resin and podophyllotoxin) and *A. heterophyllum* alkaloids and aconitine from both wild (alpine region) and cultivated (subalpine region) plants were analyzed. Active constituents of both the plant species were significantly lower in cultivated plants than in wild plants. However, differences in actual amounts of each type of active constituent were small. These differences, which were consistent for all active constituents, could have resulted from moving the plants from an alpine to a subalpine environment or from growing wild rather than having been cultivated. Methanolic extracts of *P. hexandrum* were assessed for phytochemical components and results revealed that both the plant extracts contained glycosides, flavonoids, saponins, and terpenes. Alkaloids were absent in *P. hexandrum*. The absences of carbohydrates were detected in *P. hexandrum*.

Chattopadhyay et al. (2001) reported that podophyllotoxin was produced by cell culture of *P. hexandrum* under in vitro culture conditions. A maximum of 4.26 mg/L of podophyllotoxin was produced when *P. hexandrum* was cultivated in 3 liter stirred tank bioreactor. Woerdenbag et al. (1990) reported that cell suspension cultures, derived from roots of *P. hexandrum*, accumulate podophyllotoxin. Although some efforts have been done for production of podophyllotoxin by plant cell cultures using bioreactors, the quantity is not sufficient to meet the demand (Saurabh et al., 2002).

Prakash and Bisht (2010) reported variation in respect of isoenzyme and polypeptides in seeds of *P. hexandrum* having two and three leaves. The material was collected from two populations one growing at Harkidun (3000 m), district Uttarkashi, and the other in Valley of Flowers (3300m), district Chamoli, and cultivated at lower altitude Pothivasa (2200 m), district Rudraprayag of Uttarakhand. Several low molecular polypeptides were prominent in all the populations. While the presence of high molecular weight polypeptides was observed only in naturally grown population having three leaves, some specific bands of isoenzyme were observed in population cultivated at lower altitude.

Wani et al. (2012) assessed methanolic extracts of the two medicinal plants viz. *Rheum emodi* and *P. hexandrum* for phytochemical components. The results revealed that both the plant extracts contained glycosides, flavonoids, saponins, and terpenes. Alkaloids were present at low quantity in *R. emodi* but were absent in *P. hexandrum*. The absences of carbohydrates were detected in both the extracts. Proteins were present in low quantity in *R. emodi* but abundantly present in *P. hexandrum*.

The Indian *P. hexandrum* is superior to its American species, namely, *P. peltatum*, in terms of higher podophyllotoxin content (4% in the dried roots in comparison to only 0.25% for *P. peltatum*) (Hollithuis, 1988). Chaudhry et al. (2014) investigated *Podophyllum* species and revealed presence of a number of compounds like podophyllin, podophyllotoxin, quercetin, 4-demethylpodophyllotoxin, podophyllotoxin glucoside, 4-dimethyl podophyllotoxin glucoside, kaempferol, icropodophylotoxin, deoxypodophyllotoxin, picropodophylotoxin, sopicropodophyllone, 4-Methyl deoxypodophyllotoxin,-peltatin and S- peltatin. Nag et al. (2013) reported that rhizomes and roots of *P. hexandrum* contain antitumor lignans such as podophyllotoxin, 4-dimethyl podophyllotoxin, and podophyllotoxin 4-O-glucoside (Tyler et al., 1988; Broomhead and Dewick, 1990). Of these lignans, podophyllotoxin is the most important for its use in the semisynthesis of anticancer drugs, etoposide and teniposide (Issel et al., 1984). Podophyllotoxin content of Himalayan Mayapple is quite high (4.3%) compared to that of *P. peltatum* (0.25%), the most common species in the America (Jackson and Dewick, 1984). Hameed et al. (2014) had done comparative chemotaxonomic to investigate the phylogenetic relationship of different accessions within the *Podophyllum* species. Chemical profiles demonstrated that all *P. hexandrum* accessions collected from different geographical regions are chemically diverse. Chemotaxonomic data showed that chemical characters of the investigated species were able to generate essentially the same relationship as revealed by RAPD analysis. The study has revealed that maximum amount of the podophyllotoxin (5.97%) and podophyllotoxin b-D glycoside (5.72%) was present in the *Podophyllum* population collected from Keller (Shopian) and Khilanmarg (Gulmarg) area of Jammu and Kashmir, respectively. Mengfei et al. (2012) explained biochemical composition and antioxidant capacity of extracts from *P. hexandrum* rhizome. The rhizome extracts had greater antioxidant capacity than the petiole extracts in DPPH and FRAP assays. About 16 kinds of main reactive oxygen components were identified in the extracts.

Podophyllum herb contains 4.3% podophyllotoxin on a dry weight basis and it contains three times more podophyllotoxin than in American species *P. peltatum* (Fay and Ziegler, 1985; Nag et al., 2013). Podophyllotoxin content in *P. hexandrum* has been reported between 1.03 and 6.13% in 30 different geographical sources from Himachal Pradesh and Ladakh (Singh et al., 2018).

7.7 Molecular advancements

There is urgent need to conserve genetic diversity of this valuable medicinal plant, which may become extinct if reckless exploitation continues. Estimation of the level and distribution of genetic variation in endangered species is a primary objective implementation of conservation programs (Fritsch and Rieseberg, 1996). Therefore, it is necessary to evaluate the genetic variation from different regions for identification of elite germplasm with high genetic variability that can be used in conservation. Chaudhary et al. (2014) observed that *Podophyllum* is becoming rare and is at the risk of danger for being extinction. This exerted huge pressure on the population may result in the extinction of species. So efforts should be made to conserve germplasm of such valued species. For this certain protective measures should be taken. Exploitation in wild should be prohibited at Government level in order to conserve this plant in its natural habitat. Xiao et al. (2006) said that estimate of genetic diversity could provide a basis for conservation and utilization of the endangered *P. hexandrum*. The results of the study from Western Sichuan Province, China, showed that there is low genetic diversity at the species level and that genetic differentiation among populations was more obvious than within populations. Based on field survey of seven natural populations, they found that the habitats of all populations have been destroyed by heavy deforestation and extensive habitat loss over the past few decades. These factors, together with overcollecting, have led to a decreasing in population size and subsequent inbreeding depression. They also suggested initiating management program of *ex situ* conservation of the species. Their study also revealed that some populations harbor specific locally adapted genotypes that are suited to particular environments. Artificially propagated plants recruited from local seed sources are more likely to exhibit increased fitness over nonlocal genotypes in particular environments. Consequently, they suggest that care should be taken to separate seedlings from different populations and reintroduce seedlings only into their original parental localities. With combined and sustained efforts, they hope that the genetic diversity of the important medicinal and endangered species of *P. hexandrum* in western Sichuan Province will be guaranteed.

P. hexandrum is described as self-incompatible but some researchers believe that colonies in the wild may come from single seedling. Thus, one genotype grows in clonal patches (Laverty and Plowright, 1988). In contrast, Policansky (1983) reported that Mayapple colonies comprised more than one genotype and interpopulation crosses. This evidence suggests that Mayapple is at least partially self-incompatible. Siddique et al. (1990) described the *P. hexandrum* (2n = 2x = 12) for first time and found the haploid set comprise one metacentric, three submetacentric, and two acrocentric chromosomes. On the basis of studies on chromosome component of *P. hexandrum* from actively dividing root tip cells of germinating seed she also revealed that the chromosomes could be grouped into six duplets on the basis of overall length, centromeric position, and details of NOR (nucleolar organizing region). Metaphase chromosome in *P. hexandrum* (2n = 12) is large in size, indicating toward a large genome size, which is found to be 32.25 pg (2C content), using *Vicia faba* L. 'Inovec (2c = 26.90) as internal standard (Dolezel et al., 2007). Nag and Rajkumar (2011) found the karyotype formula $6m + 2m + 2\ st + 2\ t$ for *P. hexandrum* with secondary constriction in the chromosomes 1 and 7. Callus tissues of *P. hexandrum* have unique potential of generating variation in in vitro (Kumar and Mathur, 2004). Arumugam and Bhojwani (1994) reported stability in chromosome number in 18-month-old callus culture of *P. hexandrum* with all cells showing the diploid chromosome number, while both numerical and structural variation were observed by them in 3-year-old calli. The haploid set comprised of one metacentric, three submetacentric, and two acrocentric chromosomes, with the metacentric chromosomes being the longest and the acrocentric ones being the smallest in the complement; four chromosomes has secondary constrictions. Numerical chromosome variation during callus culture is evidently attained by some kind of endore-duplication and anaphase nondisjunction (Mukhopadhyay and Sharma, 1990). Genome multiplication is affected by certain spindle anomalies giving rise to cells with abnormal chromosome numbers, the frequency of which increases with time as a result of multiplication of division errors. Mixoploid nature of the source explants and culture conditions have also been suggested as some of the other causes leading to chromosomal variation in in vitro (Phillips et al., 1994). RAPD markers have been used to differentiate *P. hexandrum* populations from Chamba and Kullu district of Himachal Pradesh (Sharma et al., 2000a,b) and in discriminating seven *P. peltatum* accessions collected from three different sites in Lafayette County, Mississippi (Lata et al., 2002). Habitat fragmentation poses major threats to endangered plant species by reducing population size and increasing geographic isolation (Young et al., 1996). The increased risk of extinction of many endangered species including medicinal is often associated with small and isolated populations, population genetics is highly relevant to development of conservation strategies of endangered species (Ellstrand and Elam, 1993; Ottewell et al., 2015). Genetic variability has also been reported in *P. hexandrum*. Singh et al. (2018) screened population of the *P. hexandrum* from different geographical locations of Himachal Pradesh and Ladakh and observed considerable variations in morphological characters *viz*., plant height, leaf shape, fruit weight, and color. At least four distinct morphological variants with 1, 2, 3, and 4 leaves have been reported (Purohitet et al.,1999). Polypeptide patterns and esterase enzyme analysis have indicated the existence of high inter- and intrapopulation variation in

P. hexandrum from the Gharwal Himalayas (Bhadula et al., 2000). Based on high active ingredient content the HFRI has identified the superior chemo types of *P. hexandrum* by screening different populations from Himachal Pradesh and Ladakh. Purohit et al. (1999) also recorded significant variation in podophyllotoxin content among different populations from Himachal Pradesh and Ladakh. Populations of *P. hexandrum*, collected from alpine region, have been observed highest podophyllotoxin content. Substantial decrease in podophyllotoxin and resin contents was observed when plants collected from higher altitudes were planted at low altitudes (Sharma et al., 2000a,b), whereas minor difference in podophyllotoxin content of *P. hexandrum* was observed between wild plants and those cultivated in the region of natural habitat (Prasad, 2000).

Further, the availability of genomics, transcriptomics, and metabolomics resources can provide a powerful tool for identification, validation, characterization, and functional analyses of molecular components in order to infer their role in regulation of metabolic processes in plant species. Kumar et al. (2015, 2016, 2017, 2018) identified and validated crucial molecular components (pathway genes, transcription factors, ABC transporters, and miRNAs) contributing to podophyllotoxin biosynthesis machinery in *P. hexandrum*. The research work laid the foundation for future research aimed to escalate the production of bioactive compounds in *Podophyllum* species and eventually will be beneficial for the industries and society.

7.8 Good agricultural and collection practices

Within the overall context of quality assurance, the WHO guidelines on good agricultural and collection practices (GACPs) for medicinal plants are primarily intended to provide general technical guidance on obtaining medicinal plant materials of good quality for the sustainable production of herbal products classified as medicines. They apply to the cultivation and collection of medicinal plants, including certain postharvest operations. Raw medicinal plant materials should meet all applicable national and/or regional quality standards.

The main objectives of these guidelines are as follows:

- To contribute to the quality assurance of medicinal plant materials used as the source for herbal medicines, which aims to improve the quality, safety and efficacy of finished herbal products.
- To guide the formulation of national and/or regional GACP guidelines and GACP monographs for medicinal plants and related standard operating procedures.
- To encourage and support the sustainable cultivation and collection of medicinal plants of good quality in ways that respect and support the conservation of medicinal plants and the environment in general.

Extraction of the whole *P. hexandrum* plant in order to harvest the roots and rhizome, containing the useful drug, Indian *Podophyllum*, has led to a decline of this species and even disappearance in some regions. It is variously considered as endangered and critically endangered. It has declined considerably as a result of exploitation to meet the increasing demands of the pharmaceutical industry. Export of this species from India has been prohibited, although illegal removal continues. Medicinal plants are valued for their various active ingredient contents which are used by the pharmaceutical companies for various formulations. Therefore, it is of paramount importance that first of all we should identify the superior stock (high active ingredient contents) from the natural condition and then multiply the same for commercial cultivation. Development of cost-effective propagation methods will further help to mass multiplication of elite planting material thus making sure for the availability of quality planting stock to go for commercial cultivation of the species. Because the species is already endangered, and exploitation of its underground parts continues to exceed the rate of natural regeneration, it needs immediate attention for conservation. In an endeavor to development of the species the Institute has made an effort to develop appropriate propagation methods and identified the superior chemo types of *P. hexandrum* by screening different populations from Himachal Pradesh and Ladakh Valley (J&K). Various other institute and agencies are also actively involved in development of agro-techniques of the species.

7.8.1 Soil and fertilization

The enrichment of soil with nutrients is one of the most important tools to obtain satisfactory productions (Carrubba and Alessandra, 2015). The practice of use of fertilizers is also growing in medicinal and aromatic plant species. The use of fertilizers was defined often indispensable in order to obtain large yields. At the same time, however, the use of correct types and quantities of fertilizers was termed necessary. Generally an increased level of nutrients induces an enhancement of plant biomass, but when the goal of cultivation is different from herbage yield, i.e., when a special plant parts, i.e., seeds, roots, flowers, are of interest, or when the quality features are especially important, the outcome of fertilization may be dramatically different. A fine-tuned fertilization practice is therefore necessary, and forms, rates, and times of distribution of fertilizers must be accurately planned and managed. Nutrient availability in soils is related to several soil characters, both physical and chemical. Soil reaction (pH) and parent material are the major factors moderating the content and availability of mineral elements in the soil (Carrubba and Alessandra, 2015).

In natural condition, *P. hexandrum* has been found growing on open slopes and under forest covers, along sides of *Nallah*, along the borders of cultivated fields and species prefers moist habitat. The soil in natural habitat of the species is generally black and rich in organic matters (Sharma and Sharma, 2018; Singh et al., 2018). Under cultivated conditions, application of farmyard manures and humus increases plant growth and yield. Mixing of well-rotted and sieved farmyard manure should be done at the end of October or the beginning of November. Generally 50–60 quintal farmyard manure is required for 1 hectare of land (Singh et al., 2018; Sharma and Singh, 2014).

7.8.2 Propagation methods

Nursery Raising and Management: Generally light and well-drained soils are good for nursery raising. Before sowing of seeds, the beds should be properly ploughed and be mixed with farmyard manures at 04 kg/m^2. Seeds are generally sown during the month of September–November and start germinating by March–April. Regular watering (at least twice/week) and weeding is required for proper and healthy growth of nursery seedlings. Owing to its slow growth the seedlings become ready for transplantation in the next year (Sharma and Singh, 2014).

Propagation Methods: *P. hexandrum* Royle can be propagated through seeds and rhizomes (Nautiyal and Nautiyal, 2003; Qazi et al., 2011; Sharma and Singh, 2014; Sharma and Sharma, 2018). Seeds show erratic and poor germination under natural conditions. Sometimes the seeds germinate after remaining dormant for 1 or 2 years. The main reason for poor seed germination seems to be postharvest care of seeds. Nautiyal et al. (1987) observed no germination in seeds extracted from fresh berries; however, germination was recorded when fresh berries were used as such for germination. Seeds washed with water also showed better germination than unwashed seeds (Bhadula et al., 2000). Purohit and Nautiyal (1988) found inhibitory effect of cotyledons on plumule development in *P. hexandrum*. Bhadula et al. (2000) observed inhibitory effect of fruit pulp on seed germination and removal of seed coat was found helpful in enhancing seed germination. The propagation of species through seeds and rhizomes is described below.

Mature seed capsule of *P. hexandrum*.

Seeds inside the capsule.

Propagation Through Seeds: The seeds should be collected during the month of June to July after it gets ripened. Each berry of fruit of *P. hexandrum* contains small dark brown seeds ranging between 30 and 69 depending upon the size of fruit and locality. Seeds should be separated from pulp, washed under running tap water for 20 min, and dried under shade and stored at 4°C until used. The seeds should be sown in June–July immediately after collection. Seeds with hot water treatment for 24 h give a better germination percentage. They remain dormant for 9–10 months and germination starts in following spring after the melting of snow. Seed germination has been achieved with GA_3 200 ppm within 20 days under lab conditions. In Hiko trays under polyhouse condition seed germination took place within 55 days, while under nursery condition it took almost 210 days. In a field trial it has been observed that potting mixture of humus:soil:sand in the ratio of 2:2:1 found to be good for germination percentage and growth. Transplantation of seedling in the field at a spacing of 30 × 30 cm should be carried out when seedlings attain height of about 6–10 cm. About 4–5 Kg seeds per ha are required for direct sowing (Sharma and Singh, 2014).

Seeds of *P. hexandrum.*

Vegetative Propagation: The multiplication of *P. hexandrum* is also done through rhizome cuttings. The youngest top portion of the rhizome cuttings of 1.0–2.5 cm in length, bearing leaf primordium, leads to better sprouting in *P. hexandrum* when planted in June–July in well-prepared soil at a spacing of 30–30 cm (Nautiyal and Nautiyal, 2003). Treatment of apical segments with indole-3-butyric acid (IBA) or α-naphthalene acetic acid (NAA) increases rooting percentage and also results in multiple root formation (Nadeem et al., 2000). GA_3 also shows marked effect in inducing uniform sprouting and flowering in rhizomes of *P. hexandrum* grown at lower altitudes (Pandey et al., 2001).

Rhizome cuttings of *P. hexandrum.*

Rhizome cuttings of 1.0–2.5 cm in length should be taken from the youngest tip portion. The rhizome cuttings planted from May to the beginning of July gives the best results. To improve vegetative multiplication, rhizome segments should be treated with NAA or IBA before planting; as rooting percentage was observed almost double with IBA 100 ppm. The growth of plants is slow and it takes up 4–5 years to produce rhizome suitable for exploitation (Sharma and Singh, 2014).

7.8.3 Transplantation and planting density

Under natural conditions *P. hexandrum* prefers organic matter–rich light soil and adequate moisture. The plant also prefers open slopes as well as partial shade condition for luxuriant growth (Singh et al., 2018). The rhizomatous system of the *P. hexandrum* is deep and therefore soil must be at least 60 cm deep, free from stones/pebbles, etc., for its optimum growth and good yield. Hence, these points should be kept in mind before selection of cultivation site (Sharma and Singh, 2014). Due to slow growth of seedling, generally plants raised through vegetative propagation are preferred as later takes less time for maturity. For the field transplanting there should be 30 cm spacing between plant to plant and same distance for row to row. The planting materials, i.e., fresh rhizome of size 1–2 cm with growing apical buds, made from uprooted rhizome in March–April to June July should be planted 4–6 cm deep in the soil(Nautiyal and Nautiyal, 2003; Sharma and Singh, 2014).

Transplanting of *Podophyllum. hexandrum* in field beds.

7.8.4 Irrigation

Irrigation depends upon soil type and weather conditions. At the time of seedling transplantation and plantation through rhizome cuttings, the crop requires watering at alternate day during April—May. After proper establishment of plants, the crop requires watering twice in a week. The crop undergoes dormant during frozen and prolonged winters (November—March) and therefore there is no requirement of watering (Sharma and Singh, 2014).

7.8.5 Crop protection and maintenance

P. hexandrum is 3—6 years crop depending upon the mode of propagation followed. Weeds growing along with crop affect the growth and yield of the plant. Hence, regular weeding is required for optimum growth and yield. Waterlogging conditions should be avoided. In case of any insect pests attack, neem-based insecticide should be used. However, no major insect pests have been observed under nursery and cultivation condition (Sharma and Singh, 2014).

7.8.6 Maturity of crop and harvesting

The crop grown through rhizomes becomes ready for harvest after 3 or 4 years of plantation. However, crop raised from seed matures in 5—6 years after planting. The rhizomes may be harvested at the time of senescence (October—November) or at the time sprouting (April—May). The rhizomes harvested in spring (April—May) are reported to contain higher resin content than those obtained in autumn (October—November). Freshly harvested rhizomes are reported to contain higher quantities of active ingredients (*a.i.*) and quantity of *a.i.* reduces after prolonged storage.

7.8.7 Yield

The general cost norms for medicinal plants cultivation for the year 2017—18 as per NMPB are as given below.

Sr. No.	Expenses per acre	Rupees
1.	Cost norms (preparation of farm, hoeing and weeding, and irrigation)	58,564.00
2.	Other expenses	10,000.00
3.	Harvesting, drying, and transport expenses	5000.00
	Total	73,564.00

Sharma and Singh (2014) computed the economics of *P. hexandrum*. The details are as given below.

1.	Estimated production of dry roots and rhizomes after 3 years(quintals/acre)	2.0
2.	Latest rates of market (Rs./kg)	450.00
3.	Income from dried roots (Rs.)	90,000.00
4.	The number of new plants produced In the nursery from seeds—20000 × 3 years	60,000.00
5.	Income from selling plants (Rs. 5.0/plant) in Rs.	30,000.00
6.	Total income (Rs.)	1,20,000.00
7.	Total expenses (Rs.)	73,564.00
8.	Net profit after 3 years (Rs.)	46,436.00

7.8.8 Postharvest management

Fresh rhizomes should be immediately washed thoroughly with running water soon after harvesting of the crop, so that all the foreign material is removed. For the ease of drying the properly washed rhizomes are chopped with the help of sharp stainless steel blade into small pieces of size 2–4 cm. Postharvest management practices are very important to maintain the quality of crop. Rhizomes are the main parts which contain the drugs and the trade depends upon the size and quality of rhizomes. Hence, it is of paramount significance that scientific postharvest management practices should be followed to maintain the potency and vigor of drugs.

Drying and Storage: The chopped rhizomes immediately need to be properly dried. These fresh small pieces may be dried under partial shade which takes about 08–12 days for proper drying. Damp place for drying should be strictly avoided and there should be proper ventilation, failing which results in fungal infestation of the rhizomes, which will severely affect the quality of marketable product. The fresh rhizomes can be dried in a dryer under hot circulating air at 45–50°C for about 20–24 h. The dried rhizome should have minimum 10%–12% moist moisture content. After drying the rhizomes are properly packed in gunny bags and kept in cool places for storage. Storage in polythene bags should be avoided (Sharma and Singh, 2014).

7.9 Medicinal uses

It has been reported to be used through the ages and in modern times as a cure for allergic and inflammatory conditions of the skin; biliary fever; burning sensation; cold; constipation; cancer of the brain, bladder, and lung; erysipelas; Hodgkin's disease; insect bites; mental disorders; monocytoid leukemia non-Hodgkin's lymphoma; rheumatism; septic wounds; plague; and venereal warts (Duke and Ayensu, 1985; Leander and Rosen, 1988; Allevi et al., 1993; Pugh et al., 2001; Sharma et al., 2006). In recent times, the rhizomes and roots of *Podophyllum* species have gained much importance throughout the world as being the main source or the starting material for the alkaloid podophyllotoxin and its semisynthetic compounds, the etoposide, teniposide, and etoposide phosphate. It is used in the treatment of cancer, and especially in the treatment of ovarian cancer (Howes, 2001; Board, 2003; Farkya et al., 2004). The root and rhizome contains several lignans like podophyllotoxin, podophyllin, and berberine which possess antitumor activities like inhibitor of microtubule assembly, used in the treatment of lung cancer, testicular cancer, neuroblastoma, hepatoma, and other tumors (Chattopadhyay et al., 2002). It also shows antiviral activities by interfering with some critical viral processes (Giri and Narasu, 2000a,b). The whole plant, but especially the root, is cholagogue, cytostatic, and purgative. Podophyllin is considered a cholagogue, purgative, alternative, emetic, and a bitter tonic.

Ethnobotanical Uses: *P. hexandrum* plays very important role in traditional systems of medicines viz., Ayurveda, Unani, and Amchi system. All the parts, i.e., rhizomes, roots, leaves, and fruits, have medicinal properties. Ripe fruit of *P. hexandrum* are edible and used as a cough remedy (Chatterjee, 1952; Singh et al., 2017). The dry roots are powdered and taken orally with lukewarm water once a day for stomachache (Vidyarthi et al., 2014). The species is used for indigestion by Gaddis in Kangra and Chamba District of Himachal Pradesh, India. The decoction of the root is given in diarrhea and hepatic disorder. The *Podophyllum* is used as a purgative and also for treatment of vaginal warts (Anonymous, 1999). Tea prepared from roots is effective in controlling constipation. Root paste is applied on ulcer, cuts and wounds and roots are also used in jaundice, syphilis, fever and liver ailments (Bhattacharjee, 2001).

7.10 Conclusion

P. hexandrum is a very important temperate medicinal plant and in great demand. Due to overexploitation of the species, it has been categorized as globally rare plants in IUCN red list and has endangered status in India (Chaudhary et al., 2014). Both *in-situ* and *ex-situ* measures should be taken to protect and conserve this very important species. But before cultivation there is urgent need to identify the superior genetic stock of the species by screening different geographical locations of its natural habitat. To promote commercial cultivation of the species, it has to ensure the supply of quality planting stock to farmers. To increase the bargaining power and high economic return cultivation of the species through cluster formation is highly recommended.

References

Airi, S., Rawal, R.S., Dhar, U., Purohit, A.N., 1997. Population studies on *Podophyllum hexandrum* Royle – A dwindling, medicinal plants of the Himalaya. Plant Genet. Resour. Newsl. 110, 29–34.

Ali, M., Sharma, V., 2013. An urgent need for conservation of *Podophyllum hexandrum* (Himalayan Mayapple) – An economical important and threatened plant of cold Desert of Ladakh. Indian J. Biol. Chem. Res. 30 (2), 741–747.

Allevi, P., Anstrain, M., Claffrede, P., Begati, E., Macdonald, P., 1993. Sterio selective glycosylation of *Podophyllum* lignans – A new simple synthesis of etoposide. J. Org. Chem. 58, 4175–4178.

Anon, 1999. Indian Herbal Pharmacopoeia, vol. II. Ebenezer Printing House, Mumbai, pp. 110–113.

Arumugam, N., Bhojwani, S.S., 1994. Chromosome variation in callus and regenerants of *Podophyllum hexandrum* (Podophyllaceae). Caryologia 47, 249–256.

Bhadula, S.K., Singh, A., Lata, C.P., Purohit, A.N., 2000. Genetic resources of *Podophyllum hexandrum* Royale, an endangered medicinal species from Garhwal Himalaya, India. Int. Plant Genet. Res. Newsl 106, 26–29.

Bhattacharjee, S.K., 2001. Hand Book of Medicinal Plants. Jaipur Printer Publishers, Jaipur, p. 25.

Board, N., 2003. Herbs Cultivation and Their Utilization. Asia Pacific Business Press Kamal Nagar, Delhi, India, pp. 329–333.

Broomhead, A.J., Dewick, P.M., 1990. Tumor inhibitory aryltralin lignans in *Podophyllum hexandrum versipelle, Diphylleia cymosa* and *Diphylleia grayi*. Phytochemistry 29, 3831–3837.

CAMP, 1998. Selected medicinal plants of northern, northeastern & central India. In: Conservation Assessment and Management Plan (CAMP). Hosted by Forest Department of Uttar Pradesh, Lucknow, 21–25 January, 1998.

Canel, C., Moraes, R.M., Dayan, F.E., Ferreira, D., 2000. Molecules of interest – Podophyllotoxin. Photochemistry 54, 115–120.

Carrubba, Alessandra, 2015. Sustainable fertilization in medicinal and aromatic plants. In: AkosMathe (Ed.), Medicinal and Aromatic Plants of the World – Scientific, Production, Commercial and Utilization Aspects. Springer Dordrecht Heidelberg, New York London.

Chatterjee, R., 1952. Indian *Podophyllum*. Econ. Bot. 6, 342–354.

Chattopadhyay, S., Srivastava, A.K., Bhojwani, S.S., Bisaria, V.S., 2001. Development of suspension culture of *Podophyllum hexandrum* for the production of podophyllotoxin. Biotechnol. Lett. 23, 2063–2069.

Chattopadhyay, S., Srivastava, A.K., Bhojwani, S.S., Bisaria, V.S., 2002. Production of podophyllotoxin by plant cell cultures of *Podophyllum hexandrum* in Bioreactor. J. Biosci. Bioeng. 93, 215–220.

Chaudhary, S., Khalil, B., Yamin, Arshad, M., 2014. *Podophyllum hexandrum*: an endangered medicinal plant from Pakistan. Pure Appl. Biol. 3, 19–24.

Chauhan, N.S., 1999. Medicinal and Aromatic Plants of Himachal Pradesh. Indus Publishing Company, New, Delhi, p. 632.

Choudhary, D.K., Kaul, B.L., Khan, S., 1998. Cultivation and conservation of *Podophyllum hexandrum* – An overview. J. Med. Aromat. Plant Sci. 20, 1071–1073.

Collett, H., 1921. Flora Simlensis: A Hand Book of Flowering Plants of Shimla and the Neighborhood. Oxford University Press, p. 637.

Dolezel, J., Greihuber, J., Suda, J., 2007. Estimation of nuclear DNA content of plants using flow cytometry. Nat. Protoc. 2 (9), 2233–2244.

Duke, J.A., Ayensu, E.S., 1985. Medicinal Plants of China. Reference Publications, Inc. Algonac., Michigan, USA, p. 705. ISBN 0-917256-20-24.

Ellstrand, N.C., Elam, D.R., 1993. Population genetic consequences of small population size: implications for plant conservation. Annu. Rev. Ecol. Evol. Systemat. 24, 217–242.

Facciola, S., Cornucopia, 1990. A Source Book of Edible Plants. Kampong Publication, CA. ISBN 0-9628087-0-9.

Farkya, S., Bisaria, V.S., Srivastava, A.K., 2004. Biotechnological aspects of the production of the anticancer drug podophyllotoxin. Appl. Microbiol. Biotechnol. 65, 504–519.

Fay, D.A., Ziegler, H.W., 1985. Botanical source differentiation of *Podophyllum* resin by high performance liquid chromatography. J. Liq. Chromatogr. 8, 1501–1506.

Fritsch, P., Rieseberg, L.H., 1996. The use of random amplified polymorphic DNA (RAPD) in conservation genetics. In: Smith, T.B., Wayne, F.R.K. (Eds.), Molecular Genetic Approaches in Conservation. Oxford University Press, New York, USA, pp. 54–73.

Giri, A., Narasu, M.L., 2000a. Production of Podophyllotoxin from *Podophyllum hexandrum:* a potential natural product for clinically useful anticancer drugs. Cytotechnology 34, 17–26.

Giri, A., Narasu, M.L., 2000b. Transgenic hairy roots: recent trends and applications. Biotechnol. Adv. 18, 1–22.

Gupta, R., Sethi, K.L., 1983. Conservation of medicinal plant resources. In: Jain, S.K., Mehra, K.L. (Eds.), Conservation of Tropical Plant Resources, pp. 101–107.

Hameed, I., Ullah, A., Murad, W., Khan, S., 2014. *Podophyllum hexandrum* Royle. Sch. J. Agric. Sci. 4 (6), 331–338.

Hollithuis, J.J.M., 1988. Etoposide and tenopside, Bioanalysis, metabolism and clinical pharmacokietics. Pharm. World Sci. 10, 101–116.

Hooker, J.D., 1875. Flora of British India. L. Reeve and Co. Ltd., England, pp. 112–113.

Howes, F.N., 2001. Vegetable Gums and Resins. Reprint. Jodhpur, Scientific, p. 190.

Issel, B.F., Muggia, F.M., Carter, S.K., 1984. Etoposide (VP-16) – Current Status and New Developments. Academic Press, Orlando, USA.

Jackson, D.E., Dewick, P.M., 1984. Biosynthesis of *Podophyllum* lignanas –I. Interconversion of aryltralin lignan in *Podophyllum hexandrum*. Phytochemistry 23, 1037–1042.

Kaul, M.K., Sharma, P.K., Singh, V., 1998. Crude Drugs of Zanskar (Ladakh): Conservation Assessment and Management Plan – Workshop-Kullu.

Kumar, P.S., Mathur, V.L., 2004. Chromosomal instability in callus culture of *Pisumsativum*. Plant Cell Tissue Organ Cult. 78, 267–271.

Kumar, P., Pal, T., Sharma, N., Kumar, V., Sood, H., Chauhan, R.S., 2015. Expression analysis of biosynthetic pathway genes vis-à-vis podophyllotoxin content in *Podophyllum hexandrum* Royle. Protoplasma 252 (5), 1253–1262.

Kumar, P., Sharma, R., Jaiswal, V., Chauhan, R.S., 2016. Identification, validation and expression of ABC transporters in *Podophyllum hexandrum* Royle and their role in podophyllotoxin biosynthesis. Biol. Plant. 60, 452–458.

Kumar, P., Jaiswal, V., Pal, T., Singh, J., Chauhan, R.S., 2017. Comparative whole-transcriptome analysis in *Podophyllum* species identifies key transcription factors contributing to biosynthesis of podophyllotoxin in *P. hexandrum*. Protoplasma 254 (1), 217–228.

Kumar, P., Padhan, J.K., Kumar, A., Chauhan, R.S., 2018. Transcriptomes of *Podophyllum hexandrum* unravel candidate miRNAs and their association with the biosynthesis of secondary metabolites. J. Plant Biochem. Biotechnol. 27, 46–54.

Lata, H., Moraes, R.M., Douglas, A., Scheffler, B.R., 2002. Assessment of genetic diversity in *Podophyllum peltatum* by molecular markers. In: Janick, J., Whipkey, A. (Eds.), Trends in New Crops and New Uses. ASHS Press, Alexandria, VA, pp. 537–544.

Laverty, T.M., Plowright, R.C., 1988. Fruits and seed set in mayapple (*Podophyllum peltatum*): influence of intraspecific factors and local enhancement near *Pedicularis Canadensis*. Can. J. Bot. 66, 173–178.

Leander, K., Rosen, B., 1988. Medicinal Uses for Podophyllotoxin, vol. 4, p. 216.

Mengfei, L., Lanlan, Z., Delong, Y., Tiantian, L., Wei, L., 2012. Biochemical composition and antioxidant capacity of extracts from *Podophyllum hexandrum* rhizome. BMC Compl. Alternative Med. 12, 263.

Mukhopadhyay, S., Sharma, A.K., 1990. Chromosome number and DNA content in callus culture of *Costus speciosus* (Koen.). Sm. Genetica 80, 109–114.

Nadeem, M., Palini, L.M.S., Purohit, A.N., Pandey, H., Nandi, S.K., 2000. Propagation and conservation of *Podophyllum hexandrum* Royale: an important medicinal herb. Biol. Conserv. 92 (1), 121–129.

Nag, A., Rajkumar, S., 2011. Chromosome identification and karyotype analysis of *Podophyllum hexandrum* Roxb. exKunth using FISH. Physiol. Mol. Biol. Plants 17 (3), 313–316.

Nag, A., Bhardwaj, P., Ahuja, P.S., Sharma, R.K., 2013. Identification and characterization of novel UniGene-derived microsatellite markers in *Podophyllum hexandrum* (Berberidaceae). J. Genet. 92, e4–e7.

Nautiyal, M.C., Nautiyal, B.P., 2003. Agrotechniques for High Altitude Medicinal and Aromatic Plants. Bishen Singh and Mahendra Pal Singh, Dehradun, pp. 134–142.

Nautiyal, M.C., Rawat, A.S., Bhadula, S.K., Purohit, A.N., 1987. Seed germination in *Podophyllum hexandrum*. Seed Res. 15 (2), 206–209.

Nayar, M.P., Sastry, A.R.K., 1987. Red Data Book of Indian Plants, vols. I–III. BSI, Calcutta India.

Ottewell, K.M., Bickerton, D.C., Byrne, M., Lowe, A.J., 2015. Bridging the gap: a genetic assessment framework for population-level threatened plant conservation prioritization and decision-making. Divers. Distrib. 22, 174–188.

Panday, H., Nanda, S.K., Chandra, B., Nadeem, M., Palni, L.M.S., 2001. GP3 induced flowering in *Podophyllum hexandrum* Royle: a rare alpine medicinal herb. Acta Physiol. Plant. 23, 487.

Phillip, R.L., Kaeppler, S.M., Olhaft, P., 1994. Genetic instability of plant tissue cultures: breakdown of normal controls. Proc. Natl. Acad. Sci. U. S. A. 91, 5222–5226.

Policansky, D., 1983. Patches, clones and self-fertility of mayapples (*Podophyllum peltatum* L.). Rhodora 85, 253–256.

Polumin, O., Stainton, A., 1984. Flowers of the Himalaya. Oxford University Press, Oxford.

Prakash, V., Bisht, H., 2010. Genetic polymorphism in inter population variation of *Podophyllum hexandrum* Royle – An endangered medicinal plant of Himalaya, India. Rep. Opin. 2 (7), 55–58.

Prasad, P., 2000. Impact of cultivation on active constituents of the medicinal plants *Podophyllum hexandrum* and *Aconitum heterophyllumin* in Sikkim. Plant Genet. Resour. Newsl. 124, 33–35.

Pugh, N., Khan, I.A., Moraes, R.M., Pasco, D.S., 2001. Podophyllotoxin lignans enhance IL-1β but suppress TNF-α mRNA expression in LPS-treated monocytes. Immunopharmacol. Immunotoxicol. 23, 83–95.

Puri, H.S., Jain, S.P., 1988. *Ainsliaea latifolia*: an adulterant of Indian *Podophyllum*. Planta Med. 54 (3), 269.

Purohit, A.N., Nautiyal, M.C., 1988. Inhibitory effect of cotyledons on plumule development in two alpine rosettes. Can. J. Bot. 66, 205–206.

Purohit, M.C., Bahuguna, R., Maithani, U.C., Purohit, A.N., Rawat, M.S.M., 1999. Variation in podophylloresin and podophyllotoxin contents in different populations of *Podophyllum hexandrum*. Curr. Sci. 77 (8), 1078–1080.

Qazi, P., Rashid, A., Shawal, S.A., 2011. *Podophyllum hexandrum*: a versatile medicinal plant. Int. J. Pharm. Pharmaceut. Sci. 3, 261–268.

Rao, R.R., 1998. Assessment of biodiversity and infraspecific variation among two reputed medicinal plant genera *Aconitum* L. (Ranuculaceae) and *Podophyllum* L. (Podophyllaceae) in the Himalaya. ENVIS Bullet. 6 (2), 16–19.

Saurabh, C., Ashok, K.S., Bhojwani, S., Virendra, S.B., 2002. Production of podophyllotoxin by plant cell cultures of *Podophyllum hexandrum* in bioreactor. J. Biosci. Bioeng. 93, 215–220.

Sharma, A., Sharma, P., 2018. The Himalayan May apple (*Podophyllum hexandrum*): a review. Asian J. Adv. Basic Sci. 6 (2), 42–51.

Sharma, S., Singh, J., 2014. Atish, Choura, Vankakri Ki Kheti. HFRI Booklet under NMPB project.

Sharma, K.D., Singh, B.M., Sharma, T.R., Katoch, M., Guleria, S., 2000a. Molecular analysis of variability in *Podophyllum hexandrum* Royle – An endangered medicinal herb of Northwestern Himalaya. Plant Genet. Resour. Newsl. 124, 57–61.

Sharma, T.R., Singh, B.M., Sharma, N.R., Chauhan, R.S., 2000b. Identification of high podophyllotoxin producing biotypes of *Podophyllum hexandrum* Royle from North-western Himalaya. J. Plant Biochem. Biotechnol. 9, 49–51.

Sharma, R.K., Sharma, S., Sharma, S.S., 2006. Seed germination behaviour of some medicinal plants of Lahaul and Spiti cold desert (Himachal Pradesh): implications for conservation and cultivation. Curr. Sci. 90, 1113–1118.

Siddiue, M.A.A., Wafai, B.A., Dhar, U., 1990. Chromosome complement and nucleolar organization in *Podophyllum hexandrum*. Genetica 82, 59–62.

Singh, J., Singh, J., Kumar, N., Jishtu, V., Sharma, S., Dhupper, R., 2017. Ethno-medicinal plants used by indigenous people of Kanda Range, Chopal forest division, Himachal Pradesh. World J. Pharm. Pharmaceut. Sci. 7 (1), 697–710.

Singh, J., Singh, J., Tewari, V.P., 2018. Screening and evaluation of superior chemotypes of *Podophyllum hexandrum* Royle from different geographical locations of North-west Himalayas. J. Plant Chem. Ecophysiol. 3 (1), 1–7.

Sultan, P., Shawl, A.S., Ramteke, P.W., Jan, A., Nahida, C., Neelofer, J., Shabir, S., 2006. In vitro propagation for mass multiplication of *Podophyllum hexandrum*: a value medicinal plant. Asian J. Plant Sci. 5 (2), 179–184.

Tyler, V.E., Brady, L.R., Roberts, J.E., 1988. Pharmacology, ninth ed. Lea and Febiger, Philadelphia.

Vidyarthi, S., Samant, S.S., Sharma, P., 2014. Traditional and indigenous uses of medicinal plants by local residents in Himachal Pradesh, North western Himalaya, India. J. Biodiversity Sci. Ecosyst. Serv. Manag. 9 (3), 185–200.

Wani, S.A., Ashfaq, M., Shah, K.W., Singh, D., 2012. Phytochemical screening of methanolic extracts of *Podophyllum hexandrum* Royle and *Rheum emodi* Wall. J. Curr. Chem. Pharm. Sci. 2 (2), 125–128.

Woerdenbag, H.J., Uden, W.V., Frijlink, H.W., Lerk, C.F., Pras, N., Malingre, T.M., 1990. Increased podophyllotoxin production *in Podophyllum hexandrum* cell suspension cultures after feeding coniferyl alcohol as a β-cyclodextrin complex. Plant Cell Rep. 9 (2), 97–100.

Xiao, M., Li, Q., Wang, L., Guo, L., Li, J., Tang, L., Chen, F., 2006. ISSR analysis of the genetic diversity of the endangered species *Sinopodophyllym hexandrum* (Royle) Ying from western Sichuan Province, China. J. Integr. Plant Biol. 48 (10), 1140–1146.

Young, A., Boyle, T., Brown, T., 1996. The population genetic consequences of habitat fragmentation for plants. Trends Ecol. Evol. 11, 413–418.

Further reading

Hossain, M.A., Arefin, M.K., Khan, B.M., Rahman, M.A., 2005. Effects of seed treatments on germination and seedling growth attributes of Horitaki (*Terminalia chebula*) in the nursery. Research (2), 135–141.

Phillips, R., Foy, N., 1990. Herbs Pan Books Ltd. London. ISBN 0-330-30725-8.

Sharma, A., Singh, R., 2016. A breathtaking herbal plan of Himalaya – *Podophyllum hexandrum*. Int. J. Appl. Pure Sci. Agric. 2 (8), 192–194.

Tabassum, S., Bibi, Y., Zahara, K., 2014. A review on conservation status and pharmacological potential of *Podophyllum hexandrum*. Int. J. Biosci. 5 (10), 77–86.

CHAPTER

Rauwolfia serpentina 8

Kirti Shitiz[1], Surendra Prakash Gupta[2]

[1]*Defence Laboratory, Defence Research & Development Organization, Jodhpur, Rajasthan, India;*
[2]*Department of Life Science, Shri Vaishnav Institute of Science, Shri Vaishnav Vidhyapeeth Viswavidhalaya, Indore, Madhya Pradesh, India*

8.1 Introduction

Rauwolfia serpentina (L.) Benth. ex Kurz. is a medicinal shrub that is a member of the dogbane or *Apocynaceae* family. The plant has mentioned in Indian manuscripts as long ago as 1000 BC and is also known as *Sarpagandha* and *Chandrika*. Higher plants as source of medicinal compounds have continued to play a dominant role in the maintenance of human health since ancient times. Over 50% of all modern drugs are of natural product origin and they play an important role in drug development programs in the pharmaceutical industry. In fact, plants produce a diverse range of bioactive molecules making them a rich source of different types of medicine.

Plant-derived substances have recently become of great interest owing to their versatile applications. Medicinal plants are the richest bioresource of drugs of traditional system of medicine, modern medicine, pharmaceutical intermediates, and chemical entities for synthetic drugs (Ncube et al., 2008).

The genus *Rauwolfia* was named in honor of the 16th-century German physician Dr. Leonhard Rauwolf, who studied plants while traveling in India. *R serpentina* was selected for study due to its long, tapering, snakelike roots. The Indian political leader Mahatma Gandhi was known to employ *Rauwolfia*, reportedly using the root to make a tea that he consumed in the evening to help relax after a busy, overstimulated day (Jerie, 2007). *R serpentina* was used in folk medicine in India for centuries to treat a wide variety of sicknesses, including snake and insect bites, febrile conditions, malaria, abdominal pain, and dysentery. It was also used as a uterine stimulant, febrifuge, and cure for insanity (Tyler et al., 1988).

The Indian physician Rustom Jal Vakil is considered responsible for introducing *Rauwolfia* to Western medicine. He collected data on patients treated with *Rauwolfia* for 10 years, from 1939 to 1949. In 1949, he published a watershed paper on the antihypertensive properties of *R. serpentina* in the British Medical Journal. He presented his detailed results from treating 50 patients who had high blood pressure with the root of *Rauwolfia*. The results were remarkable and significant. By 1949, more than 90% of Indian physicians were using *Rauwolfia* in the treatment of high blood

Himalayan Medicinal Plants. https://doi.org/10.1016/B978-0-12-823151-7.00009-X

pressure. After Vakil's original paper, more than 100 scientific articles were published throughout the world (Vakil, 1949, 1955).

Scientific interest in medicinal plants has burgeoned in recent times due to increased efficiency of new plant-derived drugs and rising concerns about the side effects of modern medicine. Most of modern researchers on herbal medicine have hinged around traditional folklore medicine. The modern medicine has brought with it an array of drugs which are quite safer for human consumption.

The quality control of medicinal plant materials which are used worldwide as folk medicine or raw materials for the pharmaceutical industry has always been one of the main concerns of the World Health Organization (WHO). Therefore, WHO organized the meetings of experts from various countries to establish internationally accepted guidelines for assessing the quality of medicinal plants so that they can be used by the regulatory agency in each country to set up the national quality specifications of medicinal plant materials that all parties involved in the production of herbal medicine need to understand in order to manufacture good quality, effective, and safe products (Nascimento et al., 2000).

8.2 Origin and distribution

More than 100 species are included in the *Rauwolfia* genus, and they are found in tropical and subtropical regions of the world, including Europe, Africa, Asia, Australia, and the Central and South Americas. *R. serpentina* is native to the moist, deciduous forests of southeast Asia, including India, Burma, Bangladesh, Sri Lanka, and Malaysia. The plant usually grows to a height between 60 and 90 cm and has pale green leaves that are 7–10 cm long and 3.5–5.0 cm wide. Generally, it grows in the region with annual rainfall of 200–250 cm at 1000 m altitude (Brijesh, 2011).

In Indian subcontinent prospective, it is cultivated on a small scale in India and Bangladesh. In India, it is found in the central region, i.e., between Sirmor and the Gorakhpur district of Uttar Pradesh, in shady, moist, or sometimes swampy localities. In the east Bihar, North Bengal, and Assam as well as in Khasi, Jaintia, and Gharo Hills, and plant is encountered more commonly on the forest margin of mixed deciduous forests. In the Western Ghats, it occurs more frequently in Goa, Coorg, the North Kanara and Shimoga districts of Karnataka and Palghat, Calicut, and Trichur in Kerala. In Orissa, Andhra Pradesh, and Himachal Pradesh, the areas comprising the catchments of the river Godavari are the richest (Dey and De, 2012).

It is a compound commonly used in Asian medicine, including traditional Ayurveda medicine recognized for its many therapeutic effects, especially in the treatment of diarrhea, malaria, hypertension, and male infertility. Its roots are used to treat snake-bite, rheumatism, hypertension, insanity, epilepsy and eczema. Leaf extract is known to remove cornea's opacity. Its active components are alkaloids and about 50 have been identified, although the primary psychoactive components appear to be reserpine, rescinnamine, and deserpidine (Dey and De, 2010).

The roots of this plant have been used for centuries in ayurvedic medicines under the name sarpagandha and nakuli for the treatment of mental disorders. Thus, the overall effect is inhibitory and sedative. Its root is ground into a powder and packaged in this form or sold in tablets or capsules. It has been stated that the drug is useful in mental disease, epilepsy, sleeplessness, and several other ailments (Ojha and Mishra, 1985). Due to its medicinal values, the root of this plant has been popular both in India and Malaya Peninsula, from ancient times as an antidote to the stings of insects and poisonous reptile. It has also been used as febrifuge and stimulant to uterine contraction for insomnia and most of all for insanity.

8.2.1 Selected vernacular names

It is most commonly known as "***Sarpagandha***" or snake smell or repellent. The plant is also known by various common names in different places as given below:

Hindi	Chandrabhaga, Chota-chand, Sarpagandha
English	Devil-pepper or Indian snakeroot
Assamese	Arachoritita
Bengali	Chandra
Canarase	Sarpagandhi, Shivanabhiballi, Sutranavi, Patalagandhi
Gujarati	Amelpodee
Kannada	Sutranabhu
Malayalam	Amalpori, Cuvanna amalpori
Marathi	Adkai, Harki
Oriya	Patalagarur, Sanochado
Sanskrit	Sarpagandha, Chandrika, Patalguruda
Tamil	Chevanamalpodi
Telugu	Patalaguni, Patalagaruda, Sarpagandha
Urdu	Asrel
Arunachal Pradesh	Bhungmaraja
Chinese	Lu fu mu
Latin	*Rauwolfia serpentina*

8.2.2 Brief description of *Rauwolfia* species

Several species of the *Rauwolfia* genus have been studied from the Indo–Malaysian region and from Africa. Some of them are subdivided into subspecies and varieties. There are approximately 85 species of the genus *Rauwolfia* mainly found in tropical regions. *R. serpentina*, *Rauwolfia caffra*, *Rauwolfia tetraphylla, Rauwolfia micrantha*, and *Rauwolfia vomitoria* are some of the important species of *Rauwolfia*. The Reserpine content in the root of *R. serpentina* has attracted worldwide attention for drug development. Summarized description of morphological characteristics of *Rauwolfia species* is given in Table 8.1.

Table 8.1 Morphological features of some *Rauwolfia* species.

Traits	*Rauwolfia serpentina*	*Rauwolfia caffra*	*Rauwolfia tetraphylla*	*Rauwolfia micrantha*	*Rauwolfia densiflora*	*Rauwolfia perakensis*	*Rauwolfia vomitoria*
Appearance	Shrub	Woody shrub	Shrub	Woody shrub	Erect herb	Shrub	Large shrub
Heights	60–90 cm	35 cm	1 m	600 m	3–6 m	4 cm in length	8m
Leaves	7–10 cm	6–7 cm	7–25 cm	8–13 cm	7–10 cm	7–25 cm.	3.5–27 cm
Petiole	1–1.5 cm	0.5–6.0 cm	2–5 mm	1 cm	0.5–1.5 cm	10–15 mm	0.5–3.5 cm
Inflorescence	Corymbose cymes	Umbel	Reticulate venation	Corymbose panicle	Umbellate	Cyme	Congested cyme
Flower	Irregular	Bisexual	White colored	Cymes	White	Corymbose cymes	Bisexual
Fruits	Drupe-shiny black	Drupe	Syncarpous	Seeded drupe	Ellipsoidal drupe	Indehiscent pods	Ellipsoid drupe
Distribution	Indian subcontinent	Tanzania to South Africa	Native to Tropical America	Native to Southwestern India	Western and the Eastern Ghats of India	Malaya region	Native to Africa

8.2.3 *Rauwolfia serpentina* (L.) Benth. ex. Kurz.

R. serpentina (L.) Benth. ex. Kurz. is an under shrub, belongs to Apocynaceae family of dicotyledonous. It is indigenous to India and other tropical countries of Asia and is naturalized in distribution. Typical morphological characteristics of the plant include small in appearance, completely erect in size with presence of glabrous shrub, where the approximate height includes 30–60 cm. Also the presence of whorled type leaves possess length 7.5–17.5 cm, with lanceolate or oblanceolate appearance in shape, acute or acuminate aperture, along with characteristic tapering at the petiole. Moreover, the plant bears flowers with white to pinkish appearance, peduncles of 5.0–7.5 cm in length, pedicels and calyx red, with calyx lobes of 2.5 mm long and lanceolate. The roots of the plants are 5–15 cm long and 3–20 mm in diameter, subcylindrical to tapered structure. Moreover, the plants include tortuous or curved shaped structure, rarely branched appearance, occasionally bearing twisted rootlets. Moreover, the external appearance of the plant indicates light brown to grayish yellow and grayish brown color.

8.3 Medical significance of *Rauwolfia*

Plants are utilized therapeutically in different countries and are a source of numerous potent and powerful drugs. Medicinal plants are used by 80% of the world's population as the main accessible medicines especially in developing nations. *Rauwolfia* can be regarded as a typical drug of ayurvedic medicaments.

R. serpentina has a broad range of therapeutic spectrum, mainly effective in the treatment of hypertension and psychotic disorders like schizophrenia, anxiety, epilepsy, insomnia, insanity, and furthermore, utilized as a sedative, a hypnotic drug. *Rauwolfia* has been studied widely in researches as a treatment for autistic children between the ages of 3.5 and 9 years (Lehman et al., 1957).

The plant is accounted for a large number of therapeutically useful indole alkaloids and these alkaloids are extensively situated in the roots. Alkaloids of this plant have a great therapeutic significance to treat cardiovascular diseases, hypertension (Silja et al., 2008), arrhythmia, breast cancer, and human promyelocytic leukemia (Itoh et al., 2005).

The *Rauwolfia* root has been consumed since the pre-Vedic period as a medication in India, to treat snake bites and fever and bug stings (Thakar, 2010). Its roots are utilized as an esteemed medicine for blood pressure, insomnia anxiety, excitement, schizophrenia, insanity, epilepsy, hypochondria, and other disorders of the central nervous system (Singh et al., 2010; Agrawal and Mishra, 2013).

The root was believed to stimulate uterine contraction and suggested for the use in childbirth. However, the juice of the leaves has been used as a remedy for the opacity of the cornea. *Rauwolfia*'s juice and extract acquired from the root can be used for treating gastrointestinal and circulatory illnesses. The juice of tender leaves and root extracts are used to treat liver pain, stomach pain, dysentery, and to eliminate intestinal worms.

The extract is likewise utilized to treat cancer which is one of the leading causes of death. Plant extract has been reported to use in treatment of prostate cancer and AIDS (Dey and De, 2011). Extracts from the root and bark of the plant are enriched with compounds of β-carboline alkaloid family of which the main constituent is *alstonine*. This compound has been reported to reduce tumor cell growth in mice inoculated with YC8 lymphoma cells or Ehrlich ascetic cells. The plant extract has antiprostate cancer activity in both in vitro and in vivo model systems which, based upon analyses of gene expression patterns of treated prostate cancer cells, may be modulated by its effects on DNA damage and cell cycle control signaling pathways.

8.4 Phytochemical constituents of *Rauwolfia*

The phytochemical analysis of *R. serpentina* has numerous medicinal values. Alkaloids are huge cluster of phytochemicals which contain a heterocyclic nitrogen ring. Till date, over 6000 basic nitrogen containing organic compounds have been isolated, which are now classified under different categories of alkaloids. Out of this huge number, around 15% of compounds have been isolated from vascular terrestrial plants of 150 different families. The pure alkaloids are used as analgesic, antispasmodic, and bactericidal effects (Okwu and Okwu, 2004). The medicinal value of plants lies in the bioactive phytochemical constituents that produce definite physiological effects on human body. These natural compounds formed the base of modern drugs as we use today (Koche et al., 2010). Herbal medicines are becoming popular in modern world as people resort to natural therapies. Natural products isolated from higher plants and microorganisms have been providing novel clinically active drugs (Niraimathi et al., 2012). The plant contains more than 70 distinct alkaloids which belong to the monoterpenoid indole alkaloid (MIA) family. The major alkaloids are reserpine, ajmaline, ajmalicine, ajmalimine, deserpidine, indobine, indobinine, reserpiline, rescinnamine, rescinnamidine, serpentine, serpentinine, and yohimbine (Howes and Louis, 1990; Srivastava et al., 2006). Chemical structures of major alkaloids present in *R. serpentina* are presented in Fig. 8.1. Different types of alkaloids in *R. serpentina* along with their physical and medicinal properties are given in Table 8.2.

8.4.1 Reserpine

Reserpine is a pure crystalline single alkaloid. It is a white-to-yellow powder that becomes darker when exposed to light. It is odorless, insoluble in water, slightly soluble in alcohol, and freely soluble in acetic acid. It has a chemical formula of $C_{33}H_{40}N_2O_9$, a molecular mass of 609 g, and a bitter taste. Reserpine (3,4,5-trimethyl benzoic acid ester of reserpic acid, an indole derivative of 18-hydroxy yohimbine type) is used in hypersensitive reactions and also act as natural tranquilizer (Banerjee and Modi, 2010). Reserpine can be used in the antihypertensive actions by act on peripheral nervous system by binding to catecholamine storage

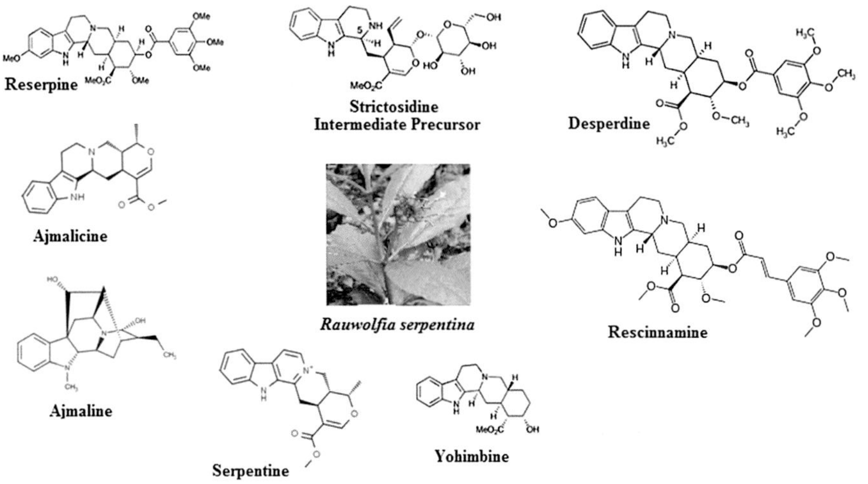

FIGURE 8.1

Chemical structures of some alkaloids present in *Rauwolfia serpentina*.

vesicles present in the nerve cell (Ellenhorn and Barceloux, 1988; Gilman et al., 1990). It is useful for the treatment of hypertension, cardiovascular diseases, and neurological diseases (Weiss and Fintelmann, 2000; Pullaiah, 2002).

The mechanism of action of reserpine is well researched and well documented. Reserpine binds to protein receptors called vesicular monoamine transporters (VMATs) in the organelle membranes of specialized secretory vesicles of presynaptic neurons. Reserpine prevents intracellular neurotransmitters from binding to VMAT proteins and stops secretory vesicles from uptaking neurotransmitters. Ultimately, use of reserpine provides that no or few neurotransmitters are released from the presynaptic neuron (Nammi et al., 2005). As a result, no or only slight promulgation of the nerve impulse occurs in the postsynaptic neuron.

8.4.2 Biosynthesis of reserpine

Alkaloids have a strong genetic—physiological function and background in the organisms which produce them. The biogenesis of alkaloids is therefore a part of the total genetic—functional strategy of such metabolisms. In the genus *Rauwolfia*, MIAs are formed via complex biosynthetic sequences. The characteristics of *R. serpentina* are only the most prominent representatives of the aforementioned class of specialized metabolites, collectively forming several hundred different MIA structures within the plant genera. MIAs are medicinally important class of compounds abundant in the roots of *Rauwolfia* species (Apocynaceae). MIAs have the pentacyclic ring system consisting of monoterpene and indole moieties which are biologically derived from secologanin and tryptophan, respectively. MIAs are widespread within the Apocynaceae, underscoring the importance of numerous members of the botanical family as sources of high-value compounds boasting pharmacological potential.

Table 8.2 Phytochemicals constituents of *Rauwolfia* and their medicinal properties.

Alkaloids	Molecular formula	Nature	Functions	Melting point	Category of drug	References
Reserpine	$C_{33}H_{40}N_2O_9$	Indole alkaloid, soluble in chloroform	Antipsychotic, antihypertensive	264.5°C	Indole alkaloids antihypertensive agent, adrenergic uptake inhibitor, antipsychotic agent, hypotensive agent, membrane transport modulator, neurotransmitter agent (OCT2 substrates)	Weiss and Finelman (2000), Pullaiah (2002), Nammi et al. (2005), Banerjee and Modi (2010)
Ajmaline	$C_{20}H_{26}N_2O_2$	Alkaloid, miscible in water	Antiarrhythmic	158°C	Indole alkaloids antiarrhythmics (class I and III), cardiovascular system, membrane transport modulators, secologanin tryptamine alkaloids, sodium channel blocker, voltage-gated sodium channel blocker	Brugada et al. (2003)
Rescinnamine	$C_{35}H_{42}N_2O_9$	Weakly basic indole alkaloids	Antihypertensive	238°C	Alkaloid angiotensin converting enzyme inhibitor cardiovascular system	Kolh et al. (1954)
Serpentine	$C_{20}H_{21}N_2O_3$	Basic anhydronium alkaloids	Tranquilizer	153°C	Type II topoisomerase inhibitor	Dassonneville et al. (1999)
Deserpidine	$C_{32}H_{38}N_2O_8$	Ester alkaloid	Antipsychotic and antihypertensive	230.5°C	Inhibitor of the ATP/Mg^{2+} pump	Varchi et al. (2005)
Yohimbine	$C_{21}H_{26}N_2O_3$	Indoloquinolizidine alkaloid	Selective alpha-adrenergic antagonist, aphrodisiac	241°C	Treatment of erectile dysfunction	Morales (2000)
Ajmalicine	$C_{21}H_{24}N_2O_3$	Indoline alkaloids	Vasodilator	250°C	Vasodilator agent, antihypertensive agent	Wink and Roberts (1998)

The shikimate pathway consists of a sequence of seven metabolic steps, in which phosphoenolpyruvate and erythrose 4-phosphate are converted to chorismate, the precursor of the aromatic amino acids and many aromatic secondary metabolites. Tryptophan decarboxylase catalyzes the conversion of L-tryptophan to L-tryptamine. Common to the biosynthesis of all MIAs is the formation of their terpenoid precursor, secologanin intermediate of MEP pathway (Stockigt and Zenk, 1977). Its subsequent ligation to tryptamine yields the universal intermediate, Strictosidine. There are different alkaloids produced due to enzyme strictosidine synthase (STR) in *R. serpentina*, mainly the indole alkaloids are shown in Fig. 8.2. The general role played by intermediate strictosidine in the biosynthesis of all MIAs is firmly established (Kutchan et al., 1988; Bracher and Kutchan, 1992).

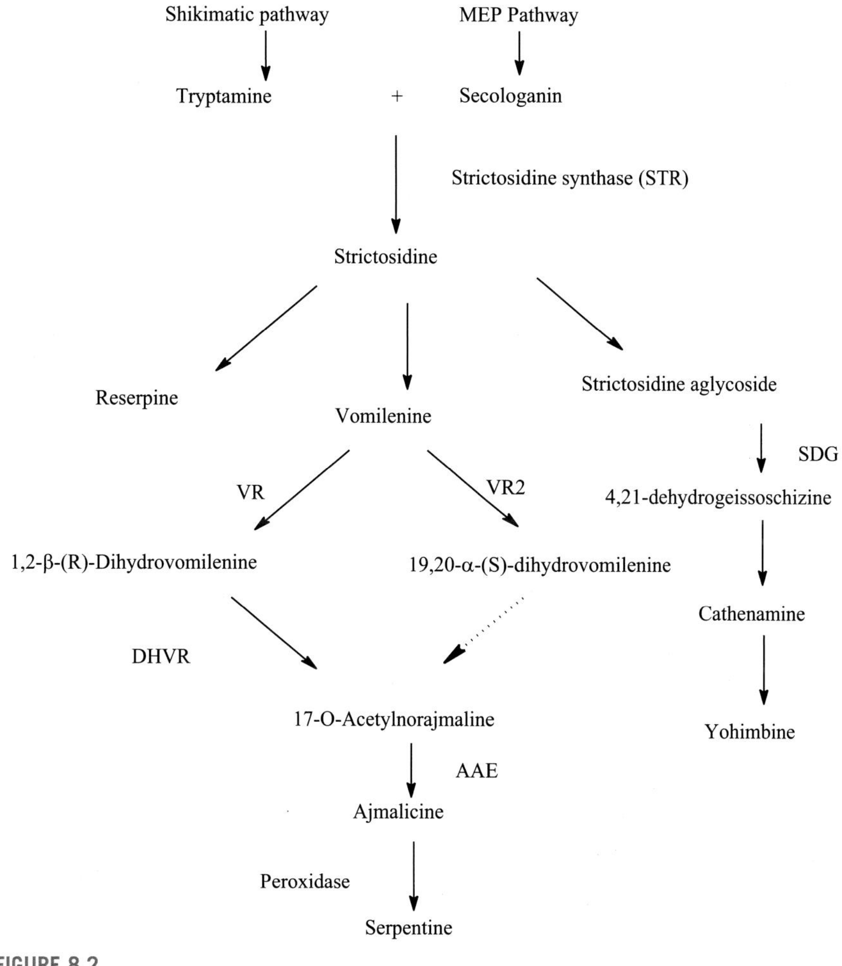

FIGURE 8.2

Enzyme-catalyzed biosynthesis of reserpine alkaloids via strictosidine intermediate.

8.4.3 Biosynthesis of strictosidine

The biological material can be the key to successfully investigate a biosynthetic pathway at the molecular level. This is especially true for pathways operating in higher plants due to their slow growth characteristics. This "upstream pathway" was elucidated in detail on enzymatic and genetic level (Geu-Flores et al., 2012; Asada et al., 2013; Salim et al., 2013, 2014; Miettinen et al., 2014). However, the manifold reactions, spearheaded by deglucosylation of strictosidine, providing the abundance of MIA carbon skeletons, are still elusive. Further, of the various "downstream pathways" leading to the pharmacologically important structures, only a few are known. This condensation is catalyzed by the enzyme STR (Treimer and Zenk, 1979) through a stereoselective Pictet–Spengler reaction mechanism in order to yield a β-carboline product (Maresh et al., 2008). Strictosidine β-glucosidase (Luijendijk et al., 1998) cleaves the glucose moiety of strictosidine to produce an unstable aglycone molecule which spontaneously leads to the formation of a series of reactive intermediates that serve as starting materials for the biosynthesis of different MIA backbones/groups.

8.4.4 Ajmaline

Ajmaline is a class I antiarrhythmic agent, it is highly useful in diagnosing Brugada syndrome (hereditary cardiac disorder) (Brugada et al., 2003), and differentiating between subtypes of patients with this disease (Paul et al., 2003). These agents are primarily classified into four major groups on the basis of their mechanism of action, i.e., sodium channel blockade, beta-adrenergic blockade, repolarization prolongation, and calcium channel blockade. Ajmaline is a sodium channel blocker that shows instant action when given intravenously, which makes it ideal for diagnostic purposes (Dobbels et al., 2016). It has been reported to stimulate respiration and intestinal movements. The action of ajmaline on systemic and pulmonary blood pressure is similar as of serpentine.

8.4.5 Biosynthesis of ajmaline

Simplified pathway leading from strictosidine via vomilenine to ajmaline. Two routes from vomilenine to 17-O-acetylnorajmaline can be postulated, depending on which of the two double bonds is reduced, first: reduction of the indolenine ring in 1,2-position (VR) followed by reduction of the 19,20-double bond (DHVR), as proposed by Gao et al. (2002) and von Schumann et al. (2002). The next step along the ajmaline biosynthetic pathway involves hydrolysis of the 17-O-acetylated norajmaline by the enzyme acetylajmalan esterase (AAE). The last step involves an S-adenosyl-L-methionine–dependent methyltransferase (Stöckigt et al., 1983) which catalyzes the indoline nitrogen methylation of norajmaline to produce ajmaline.

8.4.6 Biosynthesis of serpentine

Serpentine are indole alkaloids of considerable medicinal importance. Serpentine is an inhibitor of topoisomerase (Type II) as well as antipsychotic properties (Dassonneville et al., 1999; Santos et al., 2017). In vacuole, an enzyme (PER) peroxidase is responsible for oxidation of ajmalicine to serpentine (O'Connor and Maresh, 2006).

8.4.7 Biosynthesis of yohimbine

Yohimbine is an indoloquinolizidine alkaloid, used as a selective alpha-adrenergic antagonist or alpha-blocker in the blood vessels for the treatment of erectile dysfunction in man (Goldberg and Robertson, 1983; Morales, 2000). Its action on peripheral blood vessels is weaker as compared to reserpine. Yohimbine has a mild antidiuretic action, probably via stimulation of hypothalmic center and release of posterior pituitary hormone. Antagonism at these receptors relaxes smooth muscle and lowers blood pressure.

Biosynthetic pathway of Yohimbine is involved homoallylic isomerization of the keto dehydrogeissoschizine followed by 1,4 conjugate addition. However, detailed mechanism of biosynthetic route of Yohimbine is yet to be identified due to unidentification of enzymes in deglycosylated strictosidine.

8.5 Omics strategies and advancements

Plants are a rich source of assorted specialized metabolites that have been utilized for thousands of years as scents, flavoring agents, pigments, insect repellents, and therapeutic compounds (Facchini et al., 2012). These secondary metabolites are considered to be the fundamental approach used by plants to acclimate and persist in various ecological niches and environmental conditions and to tackle biotic as well as abiotic intrusions in their natural habitat (Furstenberg-Hagg et al., 2013; Weng, 2014).

The specialized plant secondary metabolites are characterized by complex chemical structures which are derived from simpler precursors, suggesting the involvement of complicated biosynthetic machinery and regulatory processes that have been strongly favored and developed via natural selection (Moore et al., 2014).

Therefore, mapping plant genes to a biosynthetic pathway necessitates the tedious and time-consuming experimental approach of figuring out one gene at a time. Such identification becomes even more challenging due to genetic redundancy and tight genetic regulation; therefore, to confirm or verify the function of a plant gene, multiple lines of evidence are required. Since secondary metabolic pathways and their regulation include extremely complex frameworks with interconnected components, studies comprising association-based analysis within multiple elements can serve as an important alternate for plant-derived metabolic pathway discovery (Steuer, 2007; Sweetlove et al., 2008; Tomar and De, 2013).

In recent years, "Omics"-based strategies have gained a great deal of popularity among plant biologists and research community as an approach for functional characterization of target plant genes and to investigate systems response under specific conditions (Muranaka and Saito, 2013; Saito, 2013; Wurtzel and Kutchan, 2016). Qualitative and quantitative analysis of various elements of a biological system, such as specialized metabolites, transcript expression, and protein levels while capturing the spatiotemporal responses, provides imperative insights into various ongoing processes and interactions/associations within them. While each omics datasets gives a broad overview of the static or dynamic condition of a biological system, the integration of various datasets provides an effective and efficient means to strengthen genuine observations and reduces the probability of false positives/negatives (Moreno-Risueno et al., 2010; Deshmukh et al., 2014; Rai et al., 2016). The representation of different omics approaches and strategies used in *R. serpentina* has been depicted in Fig. 8.3.

8.5.1 Genomic technologies and genetic markers

Whole-genome sequences serves as an imperative resource for understanding the total biosynthetic potential of a medicinal plant. It also facilitates the development of herbal medicines and selection of cultivars with desired agricultural traits and high levels of secondary metabolites having pharmaceutical and medicinal importance (Hao and Xiao, 2015; Unamba et al., 2015). Total reliance on medicinal plants for extraction of essential bioactive compounds has proved to be an impractical and unsustainable approach (Chang and Keasling, 2006; Facchini et al., 2012); therefore,

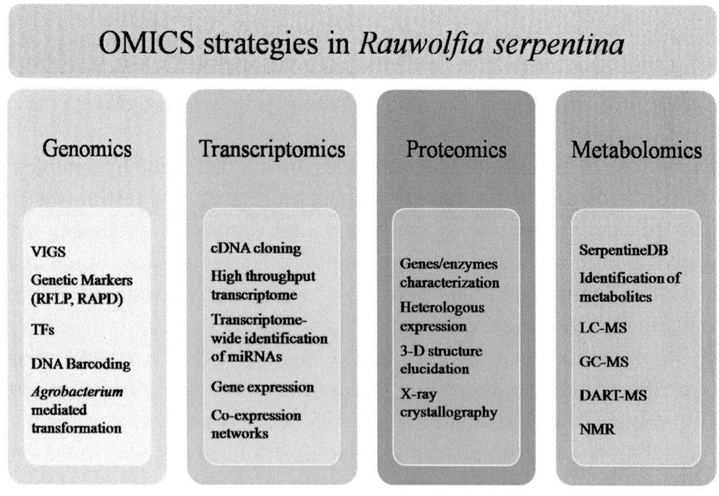

FIGURE 8.3

The representation of different omics approaches and strategies used in *Rauwolfia serpentina*.

knowledge and understanding of biosynthetic pathway components offers an opportunity to develop alternate sources for obtaining these important compounds.

Genomic sequences give vital information on plant origin and evolution, inheritable characters, physiological and developmental schematics, epigenetic regulation, and metabolic potential, which serves as the basis for interpreting genetic as well as chemodiversity at the molecular level (Dhanapal and Govindaraj, 2015; Hao and Xiao, 2015; Unamba et al., 2015).

Genomic sequences helps in the development of easily available and robust functional genomic resources, like full-length complementary DNA (cDNA) clones, tagged mutant lines, and facile and rapid transformation method. It also provides essential information on multiple homologous genes of specialized biosynthetic pathways, thereby improving capability and possibilities to carry out gene knockout experiments to decipher their functionality.

Elucidation of the MIA biosynthesis has recently advanced in Apocynaceae family through simultaneous development of transcriptomic resource analyses and reverse genetics strategies achieved by means of virus-induced gene silencing (VIGS). Most of these tools have been basically adapted for *Catharanthus roseus*; however, the VIGS technique has been barely used on other *Apocynaceae* species. *Rauwolfia* species not only constitutes a significant source of explicit and valuable secondary metabolites such as reserpine, ajmaline, ajmalicine, etc., but are also well-established models for understanding alkaloid metabolism, and as such would certainly benefit from an effective VIGS procedure. It has been demonstrated that biolistic-mediated VIGS technique can be efficiently used for gene silencing in both *R. serpentina* and *R. tetraphylla* taking advantage of a recently modified inoculation method in tobacco rattle virus (TRV) vectors via particle bombardment (Corbin et al., 2017). TRV vectors, namely pTRV1 and pTRV2-MCS encoding the two genomic components of TRV obtained from Arabidopsis Biological Resource Centre (http://www.arabidopsis.org), were used to generate silencing constructs and for propagating the virus within *Rauwolfia* plantlets. After standardizing bombardment conditions while minimizing transformed plantlet injury, gene downregulation was observed with an approximately 70% decrease in expression by silencing phytoene desaturase gene in both *Rauwolfia* species. This established gene silencing methodology will thus contribute as a valuable tool in identification and characterization of alkaloid biosynthesis genes in these prevalent *Rauwolfia* species as well as other closely related medicinal plant species.

DNA-based molecular markers have been widely utilized in recent years for assessment of genetic diversity among the germplasm in various medicinal plant species. Nair et al. (2014) investigated genetic diversity in populations of *R. serpentina* based on random amplified polymorphic DNA (RAPD) markers. Within population a high genetic diversity and among population high genetic differentiation was revealed which was suggested to be caused both by habitat fragmentation of the lower size populations as well as the low gene flow level among them. The findings of this study showed that RAPD-based assessment of genetic diversity in *R. serpentina* seemed to be adequately informative and powerful. The information on genetic diversity and structure among populations of *R. serpentina* would be helpful in developing appropriate conservation strategies and breeding programs.

Comparative genomics has been described as one of the most effective and successful methods for characterization of gene functions, identification of biomarkers, and investigation of evolutionary relatedness in humans (Moreno et al., 2008) as well as plants (Moreno et al., 2008; Michael and Jackson, 2013). Pathania et al. (2016) carried out comparative co-expression analysis of *R. serpentina* and *C. roseus* which revealed evolutionary complexity in secondary metabolism. Comparative analysis approach is instigated for comparing two or more organisms in order to identify similarities among them as well as to investigate mechanisms responsible for diversification in various key biological phenomena such as photosynthesis, reproduction and defense response, metabolic pathways, and many more. The basis of comparative analysis is that biologically significant processes remain conserved across different organisms in comparison to nonrelevant associations which decline with evolutionary time scale (Hansen et al., 2014). Comparative co-expression analysis technique can be used to determine genes playing role in differential tissue-specific and species-specific biosynthesis of metabolites. High-throughput expression data availability and the use of computational analysis approach for integration of expression data prompted to determine candidate genes/biomarkers involved in variation of MIAs between *R. serpentina* and *C. roseus*. Network-based approach was used to carry out differential expression analysis for identification of candidate genes accountable for species-specific production of metabolites in these medicinal plants. The key genes of MIA biosynthesis contributing toward diversification of metabolites have been identified using this approach.

Gene expression is considered as an intricate phenomenon being regulated by a set of proteins known as transcription factors (TFs), which are responsible for activating or repressing various genes (Mitsuda and Ohme-Takagi, 2009). This gene regulation by TFs takes place via a set of highly synchronized internal and/or external signals. Some TFs also interact with one another to regulate genes (Yang et al., 2012). Identification of TFs regulating secondary metabolism in *R. serpentina* has been carried out by Pathania and Acharya (2016). Unraveling the interaction between genes and various TFs is essential for gaining complete understanding and knowledge of secondary metabolism in plants. It is necessary to identify transcriptional regulators along with their targets (genes) involved in secondary metabolites biosynthesis, to gain in-depth insights of metabolic pathways associated with them. The authors employed an integrative approach using omics data in order to identify TFs with unknown functionality and illustrated their roles in regulating valuable metabolites along with metabolic traits. Identification of TFs functionality was carried out by implementing gene co-expression network analysis, which could not have been possible to annotate using any other simple methods. The TF families, WRKY and AP2-EREBP, were identified to be playing regulatory roles in regulating alkaloids biosynthesis in *R. serpentina*. TFs regulating interconnecting steps between primary and secondary metabolism in terpene indole alkaloids biosynthesis were also identified. Certain TFs were also suggested to control specialized metabolites synthesis and predicted that associated genes may be physically clustered in genome. These TFs may control the major alkaloids biosynthesis under influence of various environmental stimuli. This study has established an integrated genome- and systems-wide basic structure for identification of TFs and elucidation of their intricate regulatory behavior, linked with secondary metabolites synthesis in *R. serpentina*.

DNA barcoding is a widely used and promising approach in genomics for identification and authentication of medicinal plants, especially in case of endangered species. Various ongoing DNA barcoding projects targeting to identify all biological species and the great reduction in costs have encouraged molecular diagnostic techniques to become more and more common as an effective tool in applied sciences particularly in wildlife forensics. Forensic identification of *R. serpentina* using this genomic technique has been carried out by Eurlings et al. (2013). The capability of molecular identification was evaluated by looking for species-specific polymorphisms in DNA. DNA polymorphisms were identified by investigating the applicability of fast mutating regions of chloroplast DNA (trnL—trnF intergenic spacer, rps16 intron region, and rpl16 intron region) to distinguish between *R. serpentina* and other species in *Rauwolfia* genus closely related to *R. serpentina*. The rps16 intronic region was observed to be providing highest polymorphism at the species level among all three chloroplast DNA regions after sequence comparison of limited number of *Rauwolfia* species. Therefore, only this genetic locus was targeted for developing a marker for identification of species. Two species-specific indels were found for *R. serpentina* in the rps16 intronic region. The DNA barcoding method developed was tested for reproducibility, specificity, sensitivity, and stability. This genomic-based DNA diagnostic tool for molecular identification is suggested to be highly proficient; however, it can also be complemented with mass spectrometry as an additional identification tool since alkaloid composition in *Rauwolfia* species also appears to be highly species specific (Woodson et al., 1957).

Another molecular marker technique restriction fragment length polymorphism (RFLP) has also been used in *Rauwolfia* species for investigating variability of ribosomal RNA (rRNA) genes (Andreev et al., 2005). Ribosomal DNA repeats which are high copy number and tandemly organized in the genome, consisting of regions with diverse functions and various mutation rate, frequently serve as a suitable model for studying evolutionary relationships and for identification of genetic changes. The most conserved sequences of 18S—25S rRNA genes are the regions which contain ITS-1, 5.8S rDNA, ITS-2, and 25S rDNA and 3′-part of transcribed area with 18S rRNA gene while nontranscribed integenic spacer is the most unstable region. Analysis of cultured tissues along with intact plants of certain *Rauwolfia* species for 18S—25S and 5S rRNA genes was carried out to compare sequence variability occurring as a consequence of species evolution in nature and also tissue culture—induced variations. The RFLP in both the regions was found in intact plants of *Rauwolfia* species as well as long-term tissue cultures of *R. serpentina*. In addition, quantitative alterations in 18S–25S rRNA genes were also observed in long-term tissue cultures. These outcomes demonstrate that the observed rDNA variability in both types of plants, intact and tissue cultured, is attributed to the variations in the same regions of rRNA genes.

8.5.2 In vitro biotechnological interventions

Recently, Mukherjee et al. (2019) reviewed and illustrated various in vitro biotechnological interventions on the *Rauwolfia* genus which gives significant ideas on the ongoing research activities and imminent prospective related to in vitro propagation, hairy root culture, synthetic seed production, somatic embryogenesis, polyploidy induction, secondary metabolite quantitation, germplasm conservation, and the genetic improvement of the genus.

Genetic variability is one of the essential factors to be considered in tissue culture. To ensure the genetic homogeneity of the in vitro cultured plants, assessment of clonal fidelity has to be done. This assessment in order to ascertain the uniformity of tissue cultured plants is done at the genetic level by means of suitable genetic markers. PCR amplification–based DNA markers, like RAPD and inter simple sequence repeats (ISSRs), are used most commonly for the assessment of clonal fidelity (Williams et al., 1990; Zietkiewics et al., 1994). Clonal fidelity assessments have been performed on in vitro raised plantlets of *R. serpentina* via direct regeneration (Goel et al., 2010; Saravanan et al., 2011; Faisal et al., 2012a; Senapati et al., 2014; Rana et al., 2015), indirect regeneration (Saravanan et al., 2011; Rohela et al., 2013), as well as encapsulated explants (Faisal et al., 2012b, 2013; Gantait et al., 2017). Pandey et al. (2014) investigated the long-term stability of hairy root culture of *R. serpentina* on the basis of biomass and TIA production along with studying the expression of *rol* B and *rol* C genes. Many studies have been reported which show stimulatory effect of various Ri T-DNA *rol* genes in enhanced secondary metabolites production in hairy root clones. Among these genes, *rol* B and *rol* C have gained maximum attention (Bulgakov, 2008). Consistency in expression of pRi T-DNA gene is essential for production of secondary metabolites in long-term hairy root cultures. Another PCR-based marker, start codon targeted (ScoT) along with RAPR and ISSR markers, has recently been used for the assessment of genetic fidelity of acclimated plantlets in *R. tetraphylla* (Rohela et al., 2019).

For in vitro production and enhancement of secondary metabolites like reserpine, ajmaline, ajmalicine, yohimbine, serpentine, etc., in *R. serpentina*, hairy root culture technology has been employed. Enhancement of secondary metabolites content has been achieved through genetic transformation in *Rauwolfia* species. Genetic transformation aims to increase the content of secondary metabolites in plants by producing those specific plant organs and tissues which are abundant in secondary metabolites, such as roots in case of *R. serpentina*. Mass production of *Rauwolfia* roots is generally achieved through *Agrobacterium rhizogenes*, which contains root-inducing plasmid (Ri plasmid) and carries out the genetic transformation of the explant cells for producing adventitious roots. The plant parts after genetic transformation are then further cultured and maintained under artificial controlled tissue culture conditions. Different *Agrobacterium* strains have been employed for transformation experiments, namely, strain ATCC 15834 (Benjamin et al., 1993; Sudha et al., 2003), strain A4 (Goel et al., 2010), and strain LBA 9402 (Ray et al., 2014). An overall enhancement in alkaloid content has been observed in most of the transformation experiments conducted. For example, higher ajmalicine content of 0.006 mg/g dry weight and ajmaline content of 0.01 mg/g dry weight has been reported by Sudha et al. (2003). Enhanced reserpine content of 0.0858% dry weight and 2.54 ± 0.66 mg/g dry weight has also been reported by Goel et al. (2010) and Ray et al. (2014), respectively.

Elicitor-based enhancement of secondary metabolites has also been achieved in *R. serpentina*. Enhancement of ajmaline and ajmalicine in hairy root cultures using biotic (mannan from *Saccharomyces cerevisiae*) and abiotic (NaCl) elicitors through *Agrobacterium*-mediated transformation has been reported by Srivastava et al. (2016). Ajmalicine production was observed to be enhanced by up to 14.8-fold on elicitation with NaCl and ajmaline concentration increased upto 2.9-fold with mannan after 1 week of treatment. Effect of another abiotic plant growth regulator, melamine salt of bis (methylol) phosphinic acid (melaphene) on alkaloid synthesis, especially ajmaline has been investigated in *R. serpentina* calli (Kozlova et al., 2005). In presence of melaphene an increase of 80% in ajmaline content was observed and total alkaloids content increased by 220%. Melaphene was observed to exert significant effects on the growth and metabolism of *R. serpentina* callus during initial stages of culturing.

Elicitation is one of the most important approaches for enhancing bioactive compound levels through plant tissue culture. Further, identification of genes upregulated in response to elicitor treatment followed by cloning those genes through genetic engineering techniques, the plants with desired and increased amount of secondary metabolites could be generated.

8.5.3 Transcriptomic resources

Transcriptomics refers to global analysis of transcriptome profile and represents the initial level of the vigorous state of a biological system which provides an indication of probable ongoing processes at a specific time and biological state. Expression analysis under differential conditions/stresses allows the know-how of plant response machinery and regulation of various biological processes. Differential expression studies have resulted in functional characterization of genes and enzymes playing role in specialized metabolites biosynthesis. The major revolution in transcriptome analysis started with RNA sequencing (RNA-seq) based on high-throughput next-generation sequencing (NGS). RNAseq-based profiling of transcriptome does not necessitate prior genome information in comparison to microarray analysis, thereby serving as a vital method for accelerating transcriptome profiling of nonmodel plants leading to functional characterization of gene transcripts of medicinal plants as well (Wang et al., 2009; Jain, 2012; Xiao et al., 2013;

Han et al., 2016; Rai et al., 2017). RNA-seq–based profiling of transcriptome concomitant with co-expression analysis has widely been used in a number of studies for identification of enzymes involved in the biosynthesis of medicinally important phytochemicals.

Considering the importance of *R. serpentina* attributed to its pharmaceutically and economically important secondary metabolites, its transcriptomic resources were generated using high-throughput NGS technology and submitted to National Center for Biotechnology Information database (Gongora-Castillo et al., 2012). These transcriptomic resources and expression abundance matrices thus generated provided a rich resource of information for understanding specialized plant metabolism, and promoted realization of novel systems for production of plant-derived pharmaceutical compounds.

A major breakthrough in area of transcriptomics in case of *R. serpentina* has been made by Prakash et al. (2015) with transcriptome-wide identification of microRNAs (miRNAs) and prediction of their prospective targets. MicroRNAs or miRNAs are short noncoding RNAs of around 19–24 nucleotides in length and are considered as potential regulators of gene expression at both transcriptional as well as posttranscriptional level. Plant miRNAs show nearly perfect to completely perfect complementarity to their target mRNA transcript(s) thereby regulating the concerned gene either through translational degradation or repression (Aukerman and Sakai, 2003; Schwab et al., 2005). Prakash et al. (2015) identified and characterized 15 conserved miRNAs from *R. serpentina* belonging to 13 different families through in silico analysis of nucleotide datasets available. Gene Ontology (GO) annotation revealed that predicted miRNA targets included TFs and genes involved in various biological processes like primary metabolism, secondary metabolism, growth, development, stress response, and diseases resistance. Some miRNAs were suggested to target genes of secondary metabolic pathways of pharmaceutical importance, such as anthocyanin and alkaloids biosynthesis. These findings provided first-hand information of miRNAs and their targets in *R. serpentina* and also contributed in gaining insights into miRNA-mediated gene regulation mechanisms in plants. The identification of miRNAs and their respective targets in transcriptomic sequences of *R. serpentina* suggests that various regulatory genes and biosynthetic pathways are under miRNA control which could be for genetic manipulation of secondary metabolism for increased production of pharmaceutically and economically important alkaloids.

Another study in transcriptomics led to the discovery of two important candidate genes responsible for indole and side chain N-methylation of ajmaline using previously characterized c-tocopherol-like N-methyltransferase (c-TLMT) as a query sequence against the *R. serpentina* root transcript database (http://www.phytome_tasyn.ca) (Cazares-Flores et al., 2016).

8.5.4 Characterization and functional expression of specialized metabolism genes

Proteins are considered as the work-horses of a cell which are involved in functioning and regulation of every biological process. Synthesis of proteins and its modification is a highly regulated and multiple-step process and serves as the basis for tight homeostasis by which any biological system is characterized. Information about the dynamic state of protein localization and abundance, protein—protein interaction, protein—nucleotide interaction, as well as posttranslational modification gives significant clues in understanding cell functionality and regulation.

In recent years, advancements in protein identification using mass spectrometry (MS) with enhanced resolution and accuracy as a result of cumulative improvements in instrumentation technology, sample preparation along with computational analysis, have made it possible to identify and quantify almost all the expressed proteins (Di Girolamo et al., 2013; Hu et al., 2015).

Furthermore, continuous addition in RNA-seq—based transcriptome profile numbers and de novo transcriptomic assemblies for different plant species has made available a rich database resource for proteome studies. Thus, it has led to protein identification in nonmodel plant species with unknown genomic sequences (Lopez-Casado et al., 2012).

Proteomics have been used successfully by various researchers for identification of candidate proteins or enzymes involved in biosynthesis of specialized metabolites (especially alkaloids) in medicinal plants. The formation of secologanin, the terpenoid precursor, is common to the synthesis of all MIAs. On subsequent ligation with tryptamine, secologanin produces the universal intermediate known as strictosidine. In-depth knowledge of the biosynthetic pathway and characterization of its corresponding genes is a prerequisite for any metabolic engineering intervention for enhanced metabolites production. In *Rauwolfia* genus, MIAs are produced through complex biosynthetic sequences. In spite of the enormous wealth of information regarding biochemistry and molecular genetics underneath these processes, few reaction steps remain elusive.

Many research groups have contributed in elucidation of the biosynthetic pathway enzymes/genes in *Rauwolfia* species. Various enzymes involved in MIA biosynthesis pathway in *R. serpentina* have been identified and characterized in the past. Polz et al. (1986) isolated, purified, and characterized 2β(R)-17-0-acetylajmalan: acetyltransferase from cell suspension cultures of *R. serpentina*, an acetylesterase involved in biosynthesis of ajmaline alkaloid. The enzyme catalyzes a late biochemical step in ajmaline biosynthesis. It carries out hydrolysis of ajmalan group 17-O-acetylated alkaloids thereby producing the concerned deacetylated compounds. This esterase exhibited an exceptionally high substrate selectivity and entirely takes acetylated ajmaline derivatives having naturally occurring 2β(R)-configuration.

Complementary DNA (cDNA) cloning and expression of various genes from *R. serpentina* has been carried out in recent years. Kutchan (1989) successfully cloned and carried out heterologous expression of enzymatically active stricosidine synthase from *R. serpentina* in *Escherichia coli*. This enzyme carries out the catalyzation reaction involving stereospecific condensation of trytamine and secologanin thereby producing strictosidine, the main intermediate in biosynthesis of indole alkaloids. STR is the first enzyme belonging to plant secondary metabolism to be successfully expressed in a microorganism.

Ikeda et al. (1990) identified a dehydrogenase enzyme, acyclic monoterpene primary alcohol: $NADP^+$ oxidoreductase from *R. serpentina* cultured cells. After purification its physicochemical and enzymological properties were studies. It is an important enzyme in monoterpene alcohol biosynthesis in plants which has been reported to catalyze dual reaction steps in the oxidation of 10-hydroxygeraniol to produce 10-oxogeranial via 10-oxogeraniol or 10-hydroxygeranial.

Full-length cDNA cloning and heterologous expression of strictosidine glucosidase (SG) from *Rauwolfia* in *E. coli* was carried out by Gerasimenko et al. (2002). Strictosidine glucosidase catalyzes the second step in biosynthetic pathway of different MIA classes. The kinetic parameters of purified heterologously expressed enzyme were evaluated.

Bayer et al. (2004) reported the heterologous expression and molecular analysis of vinorine synthase (VS). It catalyzes acetyla-CoA or CoA-mediated reversible formation of vinorine (or 11-methoxy-vinorine) and 16-*epi*-vellosimine (or gardneral) alkaloids. Heterologous expression in *E. coli* yielded functional enzyme and further its molecular characterization was done using site-directed mutagenesis, to gain insights into this unknown acetyltransfer reaction mechanism.

Gao et al. (2002) detected 1,2-dihydrovomilenine reductase in *Rauwolfia* cell suspension cultures. This enzyme converts specifically 2β(R)-1,2-dihydrovomilenine into 17-O-acetylnorajmaline (close precursor for biosynthesis of ajmaline) through an NADPH-dependent reaction. The enzyme was purified using combination of various chromatographic techniques viz. HPLC, affinity chromatography, and ion exchange chromatography and its enzymatic properties were studied. This identification of this enzyme closed the gap at the end of ajmaline pathway and holds an explicit position in ajmaline biosynthesis.

Functional expression of AAE from *Rauwolfia* in a plant-virus expression system has been reported (Ruppert et al., 2005a). This esterase enzyme plays an important role in late stage of ajmaline synthesis. A full-length cDNA clone of AAE was generated on the basis of partial peptide sequences of AAE extracted and purified from cell suspension cultures of *Rauwolfia*. AAE was functionally expressed in the leaves of *Nicotiana benthamiana* Domin employing a viral proviral system based on *Agrobacterium*-mediated delivery of assembly of the DNA modules (Marillonnet et al., 2004). The functional expression of AAE failed in *E. coli* but was made possible in *N. benthamiana* by employing virus-based novel expression system.

All the major biosynthetic reaction steps involved in ajmaline biosynthesis in *Rauwolfia* have been established till now at the enzyme level. Out of the well-investigated 10 steps at the enzymatic level, functional cloning of six of the enzymes had been carried out, namely, STR (strictosidine synthase), SG (strictosidine glucosidase), PNAE (polyneuridine aldehyde esterase), VS (vinorine synthase), CPR (cytochrome P450 reductase), and AAE (acetylajmalan acetylesterase). An account of cDNA cloning of major enzymes of the ajmaline biosynthesis pathway in *R. serpentina* has been presented in a review article (Ruppert et al., 2005b). Due to the availability of cDNAs of these enzymes, their detailed molecular analysis became attainable. Many of the enzymes like STR, SG, and VS have been demonstrated to be produced in *E. coli* at significant levels; however, vomilenine reductase and 1,2-dihydrovomilenine reductase remained to be heterologously expressed.

Nearly half of the enzymes of ajmaline biosynthesis pathway have been heterologously overexpressed in *E. coli*, yeast, or the recently developed novel virus–based plant expression system such as *N. benthamiana* (Marillonnet et al., 2004, 2005; Ruppert et al., 2005b). Functional heterologous expression led to the production of proteins in highly purified form in milligram quantities sufficient for crystallization of the *Rauwolfia* enzymes and X-ray analysis for three-dimensional structure elucidation. STR, the enzyme which generates the ultimate biogenic precursor of all MIAs, has been crystallized and its three-dimensional structure has been elucidated in complex with its substrates (Ma et al., 2004; Koepke et al., 2005). This was the first enzyme from the novel STR1 family to be analyzed structurally in detail. Its three-dimensional structure shows a six-bladed beta-propeller fold which represents a novel structure in plant proteins (Ma et al., 2006). Another enzyme, vinorine synthase, located in the middle of multistep ajmaline pathway, has been characterized structurally in detail (Ma et al., 2005). Earlier, purification and partial amino acid sequencing of vinorine synthase was reported by Gerasimenko et al. (2004). Now, the acetyl CoA–dependent vinorine synthase may act as structural basis for elucidation of three-dimensional structure of other members of the enzyme family. Additionally, strictosidine glucosidase (SG), the next enzyme which follows directly after STR in indole alkaloid biosynthesis, has also been recently overexpressed in *E. coli*, crystallized and structure has been elucidated through X-ray analysis (Gerasimenko et al., 2002; Barleben et al., 2005). This is one of the two important glucose-hydrolyzing enzymes taking part in *Rauwolfia* alkaloid metabolism. The second one is raucaffricine glucosidase (RG), the enzyme catalyzing the deglucosylation of substrate raucaffricine. This hydrolysis results in the formation of aglycone vomilenine which is a direct intermediate in the pathway leading to target compound ajmaline. Purification and partial amino acid sequencing of raucaffricine-O-β-D-glucosidase from *Rauwolfia* was reported by Warzecha et al. (1999). Raucaffricine glucosidase from *R. serpentina* has been overexpressed in *E. coli* expression system and its purification and crystallization was carried out (Ruppert et al., 2006). Further, preliminary X-ray analysis of the recombinant enzyme was also performed. In future, the detailed three-dimensional structure evaluation will provide better understanding of the reaction mechanism of raucaffricine glucosidase.

Concerning the number of identified enzymatic steps and downstream pathways genes, *C. roseus* and *R. serpentina* are, largely, the best studied systems in medicinal plants. In *Rauwolfia*, all nine important biosynthetic steps leading to the formation of ajmaline (from strictosideine, the parent precursor) have been identified by analysis of purified enzymatic fractions. Further, six *Rauwolfia* genes specifically involved in MIA biosynthesis have been cloned (as previously mentioned, Ruppert et al., 2005b), including two involved in side reactions in ajmaline biosynthesis, significantly increasing our knowledge on specialized metabolic route of this plant. Moreover, information is also available on three-dimensional structure of five of the identified enzymes (Panjikar et al., 2012), thereby allowing mechanistic studies and mutagenesis experiments aimed at enhancing their catalytic properties, as demonstrated for STR (Bernhardt et al., 2007).

In ajmaline biosynthesis, reduction of vomilenine to acetylnorajmaline is an important step. Previous studies suggested the involvement of two different enzymes consecutively reducing the aforementioned intermediate (Stockigt and Zenk, 1995). Later on, two NADPH-dependent reductases were isolated and characterized from *R. serpentina* cell cultures (Gao et al., 2002; von Schumann et al., 2002). On the basis of obtained data, it was postulated that an enzyme named vomilenine reductase (VR) was accountable for initial reduction of vomilenine resulting in the formation of 1,2-dihydrovomilenine which served as a substrate for 1,2-dihydrovomilenine reductase (DHVR) to yield acetylnorajmaline. Both enzymes were found to be highly substrate specific in planta. Geissler et al. (2016) recently carried out the molecular cloning and functional characterization of three different cinnamyl alcohol dehydrogenase (CAD)-like reductases from *R. serpentina* cell cultures and *R. tetraphylla* roots. The enzymes were heterologously expressed in *E. coli* as recombinant proteins and purified. Functional analysis of the purified enzymes was carried out using a set of MIAs as potential substrates. One of the CAD enzymes was identified to be involved in lignin formation. Other two reductases included isoenzymes derived from orthologous genes of *Rauwolfia* species. This study suggested that a cinnamyl alcohol dehydrogenase (CAD)-like reductase is responsible for structural diversity of MIAs in *Rauwolfia*.

Recently, benzenoid/phenylpropanoid meta/para-O-methyltransferase enzymes have been identified and isolated from *R. serpentina* roots (Wiens and Luca, 2016). Functional expression and biochemical characterization of methyltransferases was carried out. Two O-methyltransferases were reported, the first one is a nonspecific caffeic acid 4-O-methyltransferase which methylates caffeic and sinapic acids at the para position, whereas the second one is caffeic acid 3-O-methyltransferase, alike other previously characterized enzymes in this family. These enzymes are suggested to be involved in the assembly of trimethoxycinnamic acid and trimethoxybenzoic acid which are further used in the biosynthesis of acyl substituted MIAs such as reserpine and rescinamine in *R. serpentina*.

Cazares-Flores et al. (2016) reported the identification, molecular cloning, and functional expression of two *R. serpentina* cDNAs, norajmaline N-methyltransferase (NNMT) and ajmaline N_{β}-methyltransferase (ANMT) in *E. coli*. These enzymes belong to the family of newly discovered γ-tocopherol-like N-methyltransferases (γ-TLMT) and are responsible for N-methylation of indole and side chain of ajmaline. These enzymes were expressed as recombinant proteins which showed remarkably high specificity for molecules having ajmalan-type backbone and also strict regiospecific N-methylation. Additionally, gene transcript and enzyme activity was observed to be enhanced in *R. serpentina* roots which correlates with ajmaline accumulation in roots. This study identified the enzyme involved in the formation of N_β-methylajmaline thereby elucidating the final step of the ajmaline biosynthesis pathway.

Recent progress in ajmaline pathway elucidation was made by Dang et al. (2017) with the discovery of vinorine hydroxylase (VH), a cytochrome P450 enzyme, thereby providing a missing enzymatic step in ajmaline biosynthesis. Like all other MIAs, ajmaline is also produced through the intermediate strictosidine, which is subjected to a number of catalytic steps in order to produce structurally diverse metabolites of the MIA family (O'Connor and Maresh, 2006). In ajmalan alkaloid pathways in plants, the intermediate vomilenine can either be transformed into ajmaline or instead can be isomerized to form perakine which is an alkaloid having structurally distinct scaffold. Dang et al. (2017) reported the identification and characterization of a cytochrome P450 enzyme, vinorine hydroxylase, which hydroxylates vinorine to produce vomilenine, and occurs as a mixture of quickly interconverting epimers. This study was carried out by utilizing available *Rauwolfia* transcriptome (Medicinal Plant Genomics Resource (MPGR), http://medicinalplantgenomics.msu.edu/index.shtml). Surprisingly, this cytochrome P450 enzyme also catalyzes the nonoxidative isomerization of vomilenine, the ajmaline precursor to perakine. This unexpected dual enzymatic activity of vinorine hydroxylase thus provides control of the bifurcation of these pathway branches in alkaloid biosynthesis in *R. serpentina*. This finding highlights the surprising catalytic versatility that has evolved over time in plant pathways.

Later, Dang et al. (2018) presented the discovery of a new cytochrome P450 gene from *R. serpentina* encoding sarpagan bridge enzyme (SBE). SBE has been reported to catalyze a key cyclization reaction in the biosynthesis of MIAs and marks the entry of strictosidine, the central MIA intermediate, into sarpagan, ajmalan, and ajmaline alkaloid classes (Namjoshi and Cook, 2016). Though the enzymatic activity of SBE was identified over 20 years ago in plant extracts (Schmidt and Stöckigt, 1995) but the corresponding gene had not been discovered. Recently, a cytochrome P450 gene encoding SBE activity has been identified (Dang et al., 2018). Heterologous expression of SBE was carried out using *Agrobacterium*-mediated transient expression in *N. benthamiana*. This enzyme carries out the conversion of geissoschizine (derived from strictosidine) to sarpagan alkaloid polyneuridine aldehyde. Interestingly, it was also demonstrated that SBE also performs the aromatization of ajmalicine to serpentine.

The major enzymes involved in MIA biosynthesis in *R. serpentina* along with omics studies carried out has been mentioned in Table 8.3.

8.5.5 Metabolomic advancements and identification of metabolites

Metabolomics is a modern advanced omics approach which holds enormous potential in comprehensive profiling of plant secondary metabolites (Ellis et al., 2007). Within past decade metabolomics has become a fundamental part of functional genomics. Motivated by technological advances, primarily in enhanced mass resolution and sensitivity, chromatography and other analytical techniques, high-

Table 8.3 Enzymes involved in MIA biosynthesis in *Rauwolfia serpentina*.

S. No.	Gene/enzyme	Omics studies/data	References
1	Strictosidine synthase (STR)	cDNA cloning, characterization, and heterologous expression in *Escherichia coli*, insect, and yeast, 3D structure elucidation	Kutchan (1989), Ma et al. (2004), Koepke et al. (2005)
2	Strictosidine glucosidase (SG)	cDNA cloning, characterization, and heterologous expression in *E. coli*, 3D structure elucidation	Gerasimenko et al. (2002), Barleben et al. (2005)
3	Sarpagan bridge enzyme/ geissoschizine dehydrogenase (SBE)/(GDH)	Cloning, characterization, and expression in *Nicotiana benthamiana*	Schmidt and Stöckigt (1995), Dang et al. (2018)
4	Polyneuridine aldehyde esterase (PNAE)	cDNA cloning, characterization, and heterologous expression in *E. coli*, 3D structure elucidation	Dogru et al. (2000), Yang et al. (2009)
5	Vinorine synthase (VS)	cDNA cloning, characterization, and heterologous expression in *E. coli*, 3D structure elucidation	Bayer et al. (2004), Ruppert et al. (2005b), Ma et al. (2005)
6	Vinorine hydroxylase (VH)/CYP reductase	cDNA cloning, characterization, and heterologous expression in *E. coli*	Ruppert et al. (2005b)
7	Vomilenine reductase (VR, DHVR)	cDNA cloning and characterization, heterologous expression in *E. coli*	Geissler et al. (2016)
8	Dihydrovomilenine reductase (DHVR)	Detection, isolation, and purification	Gao et al. (2002)
9	Acetylajmalan esterase (AAE)	cDNA cloning, characterization, and heterologous expression in *E. coli*	Ruppert et al. (2005a)
10	Norajmaline Nα-methyltransferase (NAMT)	cDNA cloning, characterization, and heterologous expression in *E. coli*	Cazares-Flores et al. (2016)
11	Raucaffricine synthase (RS)	Not characterized yet	–
12	Raucaffricine glucosidase (RG)	cDNA cloning, characterization, and heterologous expression in *E. coli*, preliminary X-ray analysis	Ruppert et al. (2006)
13	Perakine reductase (PR)	cDNA cloning, characterization, and heterologous expression in *E. coli,* 3D structure elucidation	Sun et al. (2008), Sun et al. (2012)

throughput analysis, and availability of increasing bioinformatics databases, tools, and other resources, metabolomics is being widely utilized as the major tool for systems biology, functional genomics, and molecular breeding (Hall et al., 2002; Bino et al., 2004; Baran et al., 2009; Quanbeck et al., 2012; Rhee, 2013).

Metabolomics-based approaches have been the main contributor in exploring metabolite diversity and pharmaceutically active compounds present in various medicinal plant species. Metabolomics has become a very useful tool in drug discovery by allowing identification and profiling principal bioactive compounds in medicinal plants (Gahlaut et al., 2013). Multidisciplinary aspects of metabolomics have been explored. Metabolomics can be investigated in different areas such as drug discovery and development, evaluation of plant medicines using high-throughput screening, and many others (Ulrich-Merzenich et al., 2007).

Metabolome fingerprinting may prove to be very useful in the area of herbal medicines for drug discovery and development, analysis of gene function, systems biology, and various diagnostic tools through different advanced hyphenated technologies. A study comprising the characterization of set of defined metabolites is referred as "targeted" metabolomics and generally associate NMR-MS techniques, which is useful for these types of analysis (Dudley et al., 2010). Whereas a study involving characterization of undefined or unknown metabolites from plants is known as "untargeted" metabolomics and utilize LC-MS and GS-MS analysis techniques. Untargeted metabolomics may serve to be very useful for identification and characterization of bioactive metabolites and for evaluation of herbal drugs (Patti et al., 2012). The development of bioinformatics tools, metabolomics resources, and databases has been one of the significant promoters for the increasing metabolomics utilization in functional genomics as well as metabolic modeling.

Identification of metabolites from *R. serpentina* has been extensively carried out during the last several decades. Siddiqui et al. (1985a,b) reported the isolation and structure analysis of two alkaloids sandwicoline and sandwicolidine from *R. serpentina* roots. Later, Siddiqui et al. (1987a) isolated an alkaloid ajmalicine from *R. serpentina* roots and established its structure on the basis of spectroscopic and chemical methods. Further investigation by authors led to the identification of two more compounds, ajmalicidine (Siddiqui et al., 1987b) and rescinnamidine (Siddiqui et al., 1987c). Ruyter et al. (1988) isolated a glucoalkaloid, acetylrauglucine from cell suspension cultures of *R. serpentina* and carried out its structure determination using NMR spectroscopy. This alkaloid is present in very small amounts in *Rauwolfia* cultures.

Three MIAs of sarpagine group namely 19(*S*),20(*R*)-dihydroperaksine, 10-hydroxy-19(*S*),20(*R*)-dihydroperaksine, and 19(*S*),20(*R*)-dihydroperaksine-17-al along with 16 other already known alkaloids were isolated from *R. serpentina* hairy root cultures, and their structure elucidation was done by 1D and 2D NMR analyses (Sheludko et al., 2002).

Itoh et al. (2005) reported the identification and isolation of seven indole alkaloids from the dried roots of *R. serpentina*. These alkaloids included N_b-methylajmaline, N_b-methylisoajmaline, 3-hydroxysarpagine, yohimbinic acid, isorauhimbinic acid, an iridoid glucoside, 7-epiloganin, and a new sucrose derivative, 6′-*O*-(3,4,5-trimethoxybenzoyl) glomeratose A, along with 20 known compounds. The structures of these alkaloids were determined and the inhibitory activity of selected alkaloids was studied on topoisomerase I and II. Their cytotoxicity was also evaluated against human promyelocytic leukemia (HL-60) cell lines.

The application of a new mass spectroscopic technique, direct analysis in real time (DART-MS), has been studied in the analysis of *R. serpentina* hairy roots (Madhusudanan et al., 2008). The analysis of intact hairy roots was done by holding them in the space between the mass spectrophotometer and the DART source for measurements. The characterization of vomilenine and reserpine, the nitrogen-containing compounds, was done almost instantaneously from hairy roots using this technique. DART is combined to a time-of-flight (TOF) mass analyzer which provides selectivity and precise composition of elements through accurate mass measurement. This technique has been proved to be extremely useful for detection of metabolites directly without requirement of sample preparation through solvent extraction.

Recently, SerpentinaDB has been generated which is a database of plant-derived molecules of *R. serpentina* (Pathania et al., 2015). It is a structured compilation of 147 *R. serpentina* plant-derived molecules (PDMs) which comprises of alkaloids, phenols, fatty acids, phytosterols, glycosides, iridoid glucosides, and anhydronium bases. PDMs are well recognized to be an abundant source of diversified scaffolds that might serve as a base for rational drug designing and development (Lee and Schneider, 2001). SerpentinaDB is an extensive database which includes plant part source, chemical name and classification, IUPAC names, SMILES (Simplified Molecular-Input Line-Entry System) notations, and three-dimensional chemical structures for all mentioned PDMs with associated references. It also provides many physicochemical descriptors of these molecules which indicate their druglike properties. The availability of repertoire of PDMs such as SerpentinaDB can be of significant advantage to drug industry as well as academia. It will serve as a comprehensive reservoir in exploring prospection for therapeutic molecules from *R. serpentina*. Compilation of such type of datasets is of immense importance for in-silico drug designing that accelerates the process of screening of novel molecules with druglike properties from natural repertoire in terms of their biological activity and toxicity.

Metabolomics approaches using LC-MS−based techniques are very useful in evaluating the bioactive metabolites of medicinal plants. Simultaneous determination of bioactive MIAs (Ajmaline, Yohimbine, Ajmalicine, Serpentine, Reserpine) in ethanolic extract of seven *Rauwolfia* species has been carried out using ultrahigh-performance liquid chromatography coupled with hybrid triple quadrupole−linear ion trap mass spectrometry (UHPLC-QqQLIT-MS/MS) method (Kumar et al., 2016a). The quantification of MIAs in *R. serpentina* and *Rauwolfia verticillata* roots using high-performance thin layer chromatography (HPTLC), high-performance liquid chromatography (HPLC), and gas chromatography−mass spectrometry (GC-MS) has also been reported earlier (Klyushnichenko et al., 1995; Srivastava et al., 2006; Hong et al., 2013; Bindu et al., 2014). Different analytical techniques like high-performance liquid chromatography quadrupole time-of-flight tandem mass spectrometry (HPLC-Q-TOFMS) (Hong et al., 2010; Bindu et al., 2014) and direct analysis in real-time mass spectrometry (DART-MS) have been employed for identification and rapid fingerprinting of MIAs in roots and leaves of *R. serpentina* and other *Rauwolfia* species (Kumar et al., 2015). Characterization and distribution of reserpine using multiple-stage fragmentation analysis by UHPLC-MS (ultrahigh-performance liquid chromatography coupled mass spectrometry) has also been studied (Kumar et al., 2016b).

Identification of MIAs in ethanolic root extracts of *Rauwolfia* species (*R. serpentina*, *Rauwolfia hookeri*, *R. micrantha*, *R. verticillata*, *R. tetraphylla* and *R. vomitoria*) and their structural characterization has been reported using liquid chromatography coupled with quadrupole time-of-flight mass spectrometry (HPLC-ESI-QToF-MS/MS) (Kumar et al., 2016c). A total of 47 bioactive metabolites were identified and characterization was done based on their molecular formulas, accurate mass measurement, and MS/MS analysis. Ajmaline, serpentine, ajmalicine, yohimbine, and reserpine were clearly identified on the basis of comparison with authentic standards, while other metabolites were tentatively identified characterized from *Rauwolfia* roots.

A new ultrahigh-performance liquid chromatography—photo diode array—mass spectrometry (UHPLC-QToF-MS) method has been developed by Sagi et al. (2016) for the simultaneous detection of seven alkaloids namely, ajmaline, ajmalicine, yohimbine, corynanthine, serpentinine, serpentine, and reserpine from *R. serpentina* roots. Additionally, certain commercially available dietary supplements containing *Rauwolfia* roots were also analyzed for total alkaloids content. This analytical method can be employed for rapid detection of MIAs from *R. serpentina* as well as other *Rauwolfia* species.

Recently, 21-*O*-methylisoajmaline, a novel ajmaline-type alkaloid has been isolated from *R. serpentina* roots along with 21 previously known compounds in other plant species (Rukachaisirikul et al., 2017). These compounds include a mixture of β-sitosterol and stigmasterol, reserpine, reserpinine, tetrahydroalstonine, yohimbine, venoterpine, 6′-*O*-(3,4,5-trimethoxybenzoyl) glomeratose A, 3-*epi*-α-yohimbine, methyl 3,4,5-trimethoxy-*trans*-cinnamate, isoajmaline, a mixture of β-sitosterol 3-*O*-β-D-glucopyranoside and stigmasterol 3-*O*-β-D-glucopyranoside, 7-deoxyloganic acid, ajmaline, rescidine, suaveoline, tetraphyllicine, sarpagine, loganic acid, and 3-hydroxysarpagine. The structures of these compounds were elucidated analysis of spectroscopic data and comparison with available literature data. Some of these compounds have been identified in *Rauwolfia* for the first time. The vasorelaxant and anticholinesterase activity of certain isolated metabolites was also evaluated. The NMR data of sarpagine and rescidine have been reported for the first time in this study.

Gas chromatography coupled with mass spectrophotometry and Fourier transform infrared spectrometry (GC-MS/FT-IR) technique has been used for determination of bioactive constituents in roots of another *Rauwolfia* species, *R. vomitoria* (Okereke et al., 2017). Thirty phytochemical constituents were identified by comparing the peak values of chromatograms of unknown compounds with entries in National Institute Standard and Technology database. The functional groups possessed by these compounds were also identified. The different functional groups found to be present as confirmed through FTIR spectrum included alkyl halides, phenols, alcohols, secondary and tertiary alcohols, aldehydes, ketones, aromatic ethers, aliphatic nitro compounds, aromatic and aromatic organophosphorus compounds, carboxylic acid derivatives, saturated ketones, alkanes, and alkenes.

Metabolomic profiling of *R. serpentina* along with few other plant species against oxidative stress using LC-MS has been carried out (Afsheen et al., 2018). The major alkaloids detected in metabolomics profiling of *R. serpentina* roots were ajmaline, yohimbine, ajmalacine, and serpentine. Phenolic fingerprinting of another *Rauwolfia* species, *R. vomitoria*, has been carried out using HPLC recently (Oboh et al., 2019). The major phenolic compounds detected were gallic acid, catechin, chlorogenic acid, caffeic acid, rutin, kaempferol, luteolin, and apigenin.

8.6 Conclusion

The endangered status of *R. serpentina* and presence of pharmacologically important secondary metabolites like reserpine, ajmaline, ajmalicine, and serpentine necessitates the planning of conservation strategies for this important medicinal herb. The genetically stable superior accessions having high alkaloids content needs to be identified which can be further used for mass propagation to provide high quality raw material. There have been very limited efforts toward investigation of genetic diversity so far in this medicinal plant species. Studying the genetic diversity of available plant germplasm is a prerequisite for any breeding program. The knowledge of genetic diversity and structure among *R. serpentina* populations is essential for development of appropriate conservation and breeding strategies. Molecular marker techniques and forensic identification using DNA barcoding has also been carried out in *R. serpentina*. The information on genetic markers would be helpful in the development of DNA diagnostics for the authentication of quality plant material.

The knowledge of complete biosynthetic pathway and corresponding genes would be helpful in understanding molecular basis of MIAs biosynthesis in *R. serpentina* as well as planning genetic improvement strategies for enhancing alkaloids content. Majority of the genes and enzymes of the MIA biosynthesis pathway have been characterized and heterologously expressed but there have been limited efforts yet toward enhancing secondary metabolites production. Identification of key/regulatory genes of the biosynthetic route can have potential implications in molecular breeding and metabolic engineering after further functional validation using approaches like enzyme assays and gene silencing. Gene silencing approaches would be greatly helpful for identification and characterization of genes involved in MIA biosynthesis in this prominent medicinal plant. Further, the identification and analysis of TFs, transporters, and other regulatory elements associated with key genes would also be useful in designing a suitable genetic intervention strategy.

Mostly, all the enzymes of the biosynthetic route in *R. serpentina* have been characterized. The SBE, vinorine hydroxylase (VH), and norajmaline N-methyltransferase (NAMT) are the only enzymes which remain as the major candidates for expression studies in order to express heterologously the complete ajmaline biosynthetic pathway. Despite the near-complete discerning of the biosynthetic route in *R. serpentina*, existence of further bypasses to established routes as well as additional enzymes acting on intermediates and generating side products still cannot be excluded and yet to be identified.

Advances in high-throughput mass spectrometry—based techniques such as GC-MS and LC-MS have proved to be very helpful in separation and identification of several metabolites. Such analysis has been a valuable tool for identifying important bioactive molecules from *R. serpentina*. A large number of metabolites have been identified and reported from *R. serpentina* so far; however, comprehensive whole metabolome profiling studies are lacking. Such studies are urgently needed for better understanding of the metabolic framework of *R. serpentina* and its interaction with other biomolecules.

A significant number of individual omics datasets have been generated for *R. serpentina*; however, there is a lack of availability of integrated omics datasets. Individual datasets are insufficient to completely understand the complex biological frameworks. In order to completely understand a biological system/process, it is essential to understand the behavior of individual components of the system as well as the complex interactions or relationships between different components. Therefore, integration of information drawn from different omics-based studies is required for mechanistic studies on plant metabolism.

References

Afsheen, N., ur-Rehman, K., Jahan, N., Ijaz, M., Manzoor, A., Khan, K.M., Hina, S., 2018. Cardioprotective and metabolomic profiling of selected medicinal plants against oxidative stress. Oxid. Med. Cell. Longev. 2018, 9819360.

Agrawal, P., Mishra, S., 2013. Physiological, biochemical and modern biotechnological approach to improvement of *Rauvolfia serpentina*. J. Pharm. Biol. Sci. 6, 73—78.

Andreev, I.O., Spiridonova, K.V., Solovyan, V.T., Kunakh, V.A., 2005. Variability of ribosomal RNA genes in *Rauwolfia* species: parallelism between tissue culture-induced rearrangements and interspecies polymorphism. Cell Biol. Int. 29, 21—27.

Asada, K., Salim, V., Masada-Atsumi, S., Edmunds, E., Nagatoshi, M., Terasaka, K., Mizukami, H., De Luca, V., 2013. A 7-deoxyloganetic acid glucosyltransferase contributes a key step in secologanin biosynthesis in *Madagascar periwinkle*. Plant Cell 25, 4123—4134.

Aukerman, M.J., Sakai, H., 2003. Regulation of flowering time and floral organ identity by a microRNA and its APETALA2-like target genes. Plant Cell 15, 2730—2741.

Banerjee, M., Modi, P., 2010. A novel protocol for micropropagation of *Rauvolfia serpentina*: in low concentration of growth regulators with sucrose and phenolic acid. Int. J. Plant Sci. 5, 93–97.

Baran, R., Reindl, W., Northen, T.R., 2009. Mass spectrometry based metabolomics and enzymatic assays for functional genomics. Curr. Opin. Microbiol. 12, 547–552.

Barleben, L., Ma, X., Koepke, J., Peng, G., Michel, H., Stöckigt, J., 2005. Biochim. Biophys. Acta 1747, 89–92.

Bayer, A., Ma, X., Stöckigt, J., 2004. Acetyltransfer in natural product biosynthesis - functional cloning and molecular analysis of vinorine synthase. Bioorg. Med. Chem. 12, 2787–2795.

Benjamin, B.D., Roja, G., Heble, M.R., 1993. *Agrobacterium rhizogenes* mediated transformation of *Rauvolfia serpentina*: regeneration and alkaloid synthesis. Plant Cell Tissue Organ Cult. 35, 253–257.

Bernhardt, P., McCoy, E., O'Connor, S.E., 2007. Rapid identification of enzyme variants for reengineered alkaloid biosynthesis in periwinkle. Chem. Biol. 14, 888–897.

Bindu, S., Rameshkumar, K.B., Kumar, B., Singh, A., Anilkumar, C., 2014. Distribution of reserpine in *Rauvolfia* species from India–HPTLC and LC-MS studies. Ind. Crop. Prod. 62, 430–443.

Bino, R.J., Hall, R.D., Fiehn, O., Kopka, J., Saito, K., Draper, J., Nikolau, B.J., Mendes, P., Roessner-Tunali, U., Beale, M.H., Trethewey, R.N., Lange, B.M., Wurtele, E.S., Sumner, L.W., 2004. Potential of metabolomics as a functional genomics tool. Trends Plant Sci. 9, 418–425.

Bracher, D., Kutchan, T.M., 1992. Strictosidine synthase from *Rauvolfia serpentina*: analysis of a gene involved in indole alkaloid biosynthesis. Arch. Biochem. Biophys. 294, 717–723.

Brijesh, K.S., 2011. *Rauwolfia*: cultivation and collection. Biotech Articles. Web site. http://www.biotecharticles.com/Agriculture-Article/Rauwolfia-Cultivationand-Collection 892.html. (Accessed 23 May 2011).

Brugada, J., Brugada, P., Brugada, R., 2003. The ajmaline challenge in Brugada syndrome: a useful tool or misleading information? Eur. Heart J. 24, 1085–1086.

Bulgakov, V.P., 2008. Functions of *rol* genes in plant secondary metabolism. Biotechnol. Adv. 26, 318–324.

Cazares-Flores, P., Levac, D., De Luca, V., 2016. *Rauvolfia serpentina N*-methyltransferases involved in ajmaline and N_{β}–methylajmaline biosynthesis belong to a gene family derived from γ-tocopherol *C*-methyltransferase. Plant. J. 87, 335–342.

Chang, M.C., Keasling, J.D., 2006. Production of isoprenoid pharmaceuticals by engineered microbes. Nat. Chem. Biol. 2, 674–681.

Corbin, C., Lafontaine, F., Sepúlveda, L.J., Carqueijeiro, I., Courtois, M., Lanoue, A., de Bernonville, T.D., Besseau, S., Glévarec, G., Papon, N., Atehortúa, L., Giglioli-Guivarc'h, N., Clastre, M., St-Pierre, B., Oudin, A., Courdavault, V., 2017. Virus-induced gene silencing in *Rauwolfia* species. Protoplasma 254, 1813–1818.

Dang, T.T.T., Franke, J., Carqueijeiro, I.S.T., Langley, C., Courdavault, V., O'Connor, S.E., 2018. Sarpagan bridge enzyme has substrate-controlled cyclization and aromatization modes. Nat. Chem. Biol. 14, 760–763.

Dang, T.T.T., Franke, J., Tatsis, E., O'Connor, S.E., 2017. Dual catalytic activity of a cytochrome P450 controls bifurcation at a metabolic branch point of alkaloid biosynthesis in *Rauwolfia serpentina*. Angew. Chem. Int. Ed. 56, 9440–9444.

Dassonneville, L., Bonjean, K., Pauw-Gillet, M.C.D., Colson, P., Houssier, C., Quetin-Leclercq, J., Angenot, L., Bailly, C., 1999. Stimulation of topoisomerase II-mediated

DNA cleavage by three DNAintercalating plant alkaloids: cryptolepine, matadine, and serpentine. Biochem 38, 7719−7726.

Deshmukh, R., Sonah, H., Patil, G., Chen, W., Prince, S., Mutava, R., Vuong, T., Valliyodan, B., Nguyen, H.T., 2014. Integrating omic approaches for abiotic stress tolerance in soybean. Front. Plant Sci. 5, 244.

Dey, A., De, J.N., 2010. *Rauvolfia serpentina* (L). Benth. ex Kurz. − a review. Asian J. Plant Sci. 9, 285−298.

Dey, A, De, J.N., 2011. Ethnobotanical aspects of *Rauvolfia serpentina* (L). Benth. ex Kurz. In India, Nepal and Bangladesh. J. Med. Plant Res. 5, 144−150.

Dey, A., De, J.N., 2012. Anti-snake venom botanicals used by the ethnic groups of Purulia District, West Bengal, India. J. Herbs Spices Med. Plants 18, 152−165.

Dhanapal, A.P., Govindaraj, M., 2015. Unlimited thirst for genome sequencing, data interpretation, and database usage in genomic era: the road towards fast-track crop plant improvement. Genet. Res. Int. 2015, 684321.

Di Girolamo, F., Lante, I., Muraca, M., Putignani, L., 2013. The role of mass spectrometry in the "Omics" era. Curr. Org. Chem. 17, 2891−2905.

Dobbels, B., De Cleen, D., Ector, J., 2016. Ventricular arrhythmia during ajmaline challenge for the Brugada syndrome. EP Eur. 18, 1501−1506.

Dogru, E., Warzecha, H., Seibel, F., Haebel, S., Lottspeich, F., Stöckigt, J., 2000. Eur. J. Biochem. 267, 1397.

Dudley, E., Yousef, M., Wang, Y., Griffiths, W., 2010. Targeted metabolomics and mass spectrometry. Adv. Protein Chem. Struct. Biol. 80, 45−83.

Ellenhorn, M.J., Barceloux, D.G., 1988. Medical Toxicology. Elsevier Science Publishing Company, Inc, New York, NY, pp. 644−659.

Ellis, D.I., Dunn, W.B., Griffin, J.L., Allwood, J.W., Goodacre, R., 2007. Metabolic fingerprinting as a diagnostic tool. Pharmacogenomics 8, 1243−1266.

Eurlings, M.C.M., Lens, B.A.S.F., Pakusza, C., Peelen, T., Wieringa, J.J., Gravendeel, B., 2013. Forensic identification of Indian snakeroot (*Rauvolfia serpentina* Benth. ex Kurz) using DNA barcoding. J. Forensic Sci. 58, 822−830.

Facchini, P.J., Bohlmann, J., Covello, P.S., De Luca, V., Mahadevan, R., Page, J.E., Ro, D.K., Sensen, C.W., Storms, R., Martin, V.J., 2012. Synthetic biosystems for the production of high-value plant metabolites. Trends Biotechnol. 30, 127−131.

Faisal, M., Alatar, A.A., Ahmed, N., Anis, M., Hegazy, A.K., 2012a. An efficient and reproducible method for in vitro clonal multiplication of *Rauvolfia tetraphyll*a L. and evaluation of genetic stability using DNA-based markers. Appl. Biochem. Biotechnol. 168, 1739−1752.

Faisal, M., Alatar, A.A., Ahmed, N., Anis, M., Hegazy, A.K., 2012b. Assessment of genetic fidelity in *Rauvolfia serpentina* plantlets grown froms ynthetic (encapsulated) seeds following in vitro storage at 4°C. Molecules 17, 5050−5061.

Faisal, M., Alatar, A.A., Hegazy, A.K., 2013. Molecular and biochemical characterization in *Rauvolfia tetraphylla* plantlets grown from synthetic seeds following in vitro cold storage. Appl. Biochem. Biotechnol. 169, 408−417.

Furstenberg-Hagg, J., Zagrobelny, M., Bak, S., 2013. Plant defense against insect herbivores. Int. J. Mol. Sci. 14, 10242−10297.

Gahlaut, A., Dahiya, M., Gothwal, A., Kulharia, M., Chhillar, A.K., Hooda, V., Dabur, R., 2013. Proteomics and metabolomics: mapping biochemical regulations. Drug Invent. Today 5, 321−326.

Gantait, S., Kundu, S., Yeasmin, L., Ali, M.N., 2017. Impact of differential levels of sodium alginate, calcium chloride and basal media on germination frequency of genetically true artificial seeds of *Rauvolfia serpentina* (L.) Benth. Ex Kurz. J. Appl. Res. Med. Arom. Plant. 4, 75–81.

Gao, S., von Shumann, G., Stöckigt, J., 2002. A newly-detected reductase from *Rauvolfia* closes a gap in the biosynthesis of the antiarrhythmic alkaloid ajmaline. Planta Med. 68, 906–911.

Geissler, M., Burghard, M., Volk, J., Staniek, A., Warzecha, H., 2016. A novel cinnamyl alcohol dehydrogenase (CAD)-like reductase contributes to the structural diversity of monoterpenoid indole alkaloids in Rauvolfia. Planta 243, 813–824.

Gerasimenko, I., Ma, X., Sheludko, Y., Mentele, R., Lottspeich, F., Stöckigt, J., 2004. Purification and partial amino acid sequences of the enzyme vinorine synthase involved in a crucial step of ajmaline biosynthesis. Bioorg. Med. Chem. 12, 2781–2786.

Gerasimenko, I., Sheludko, Y., Ma, X., Stöckigt, J., 2002. Heterologous expression of a *Rauvolfia* cDNA encoding strictosidine glucosidase, a biosynthetic key to over 2000 monoterpenoid indole alkaloids. Eur. J. Biochem. 269, 2204–2213.

Geu-Flores, F., Sherden, N.H., Courdavault, V., Burlat, V., Glenn, W.S., Wu, C., Nims, E., Cui, Y., O'Connor, S.E., 2012. An alternative route tocyclic terpenes by reductive cyclization in iridoid biosynthesis. Nature 492, 138–142.

Gilman, A.F., Rall, W.T., Nies, A.D., Taylor, P., 1990. Goodman and Gilman's: the Pharmacologic Basis of Therapeutics, eighth ed. Pergamon Press, New York, New York, p. 795.

Goel, M.K., Goel, S., Banerjee, S., Shanker, K., Kukreja, A.K., 2010. *Agrobacterium rhizogenes* mediated transformed roots of *Rauwolfia serpentina* for reserpine biosynthesis. Med. Arom. Plant Sci. Biotechnol. 4, 8–14.

Goldberg, M.R., Robertson, D., 1983. Yohimbine: a pharmacological probe for study of the α2-adrenoceptor. Pharmacol. Rev. 35, 143–180.

Gongora-Castillo, E., Childs, K.L., Fedewa, G., Hamilton, J.P., Liscombe, D.K., Magallanes-Lundback, M., Mandadi, K.K., Nims, E., Runguphan, W., Vaillancourt, B., Varbanova-Herde, M., DellaPenna, D., McKnight, T.D., O'Connor, S., Buell, C.R., 2012. Development of transcriptomic resources for interrogating the biosynthesis of monoterpene indole alkaloids in medicinal plant species. PLoS One 7, e52506.

Hall, R., Beale, M., Fiehn, O., Hardy, N., Sumner, L., Bino, R., 2002. Plant metabolomics: the missing link in functional genomics strategies. Plant Cell 14, 1437–1440.

Han, R., Rai, A., Nakamura, M., Suzuki, H., Takahashi, H., Yamazaki, M., Saito, K., 2016. De novo deep transcriptome analysis of medicinal plants for gene discovery in biosynthesis of plant natural products. Methods Enzymol. 576, 19–45.

Hansen, B.O., Vaid, N., Musialak-Lange, M., Janowski, M., Mutwil, M., 2014. Elucidating gene function and function evolution through comparison of co-expression networks of plants. Front. Plant Sci. 5, 394.

Hao, D.C., Xiao, P.G., 2015. Genomics and evolution in traditional medicinal plants: road to a healthier life. Evol. Bioinform. 11, 197–212. Online.

Hong, B., Cheng, W., Wu, J., Zhao, C., 2010. Screening and identification of many of the compounds present in *Rauvolfia verticillata* by use of highpressure LC and quadrupole TOF MS. Chromatographia 72, 841–847.

Hong, B., Li, W.J., Song, A.H., Zhao, C.J., 2013. Determination of indole alkaloids and highly volatile compounds in *Rauvolfia* verticillata by HPLC-UV and GC-MS. J. Chromatogr. Sci. 51, 929–930.

Howes, L.G., Louis, W.J., 1990. *Rauvolfia* alkaloids (Reserpine), pharmacology of antihypertensive therapeutics. Handb. Exp. Pharmacol. 93, 263–285.

Hu, J., Rampitsch, C., Bykova, N.V., 2015. Advances in plant proteomics toward improvement of crop productivity and stress resistancex. Front. Plant Sci. 6, 209.

Ikeda, H., Esaki, N., Nakai, S., Hashimoto, K., Uesato, S., Soda, K., Fujita, T., 1990. Acyclic monoterpene primary alcohol: $NADP^+$ oxidoreductase of *Rauwolfia serpentina* cells: the key enzyme in biosynthesis of monoterpene alcohols. J. Biochem. 109, 341–347.

Itoh, A., Kumashiro, T., Yamaguchi, M., Nagakura, N., Mizushina, Y., Nishi, T., Tanahashi, T., 2005. Indole alkaloids and other constituents of *Rauwolfia serpentina*. J. Nat. Prod. 68, 848–852.

Jain, M., 2012. Next-generation sequencing technologies for gene expression profiling in plants. Brief. Funct. Genom. 11, 63–70.

Jerie, P., 2007. Milestones of cardiovascular therapy, IV: reserpine (in Czech). Cas. Lek. Cesk. 146, 573–577.

Klyushnichenko, V.E., Yakimov, S.A., Tuzova, T.P., Syagailo, Y.V., Kuzovkina, I.N., Wulfson, A.N., Miroshnikov, A.I., 1995. Determination of indole alkaloids from *R. serpentina* and *R. vomitoria* by high-performance liquid chromatography and high-performance thin-layer chromatography. J. Chromatogr. A 704, 357–362.

Koche, D., Shirsat, R., Imran, S., Bhadange, D.G., 2010. Phytochemical screening of eight traditionally used ethnomedicinal plants from Akola district (MS). India. Int. J. Pharm. Bio. Sci. 1, 253–256.

Koepke, J., Ma, X., Fritzsch, G., Michel, H., Stöckigt, J., 2005. Crystallization and preliminary X-ray analysis of strictosidine synthase and its complex with the substrate tryptamine. Acta Cryst. D61, 690–693.

Kolh, M.W., Draper, M.D., Keller, F., 1954. Alkaloids of *Rauvolfia serpentina* Benth III. Rescinnamine, a new hypotensive and sedative principle. J. Am. Chem. Soc. 76, 2843.

Kozlova, R.Y., Vinter, V.G., Fattakhov, S.G., Reznik, V.S., Konovalov, A.I., 2005. Melamine salt of bis(methylol)phosphinic acid (melaphene) as a regulator of *Rauwolfia serpentina* specialized metabolism. Dokl. Biochem. Biophys. 401, 136–138.

Kumar, S., Bajpai, V., Singh, A., Bindu, S., Srivastava, M., Rameshkumar, K.B., Kumar, B., 2015. Rapid fingerprinting of *Rauwolfia* species using direct analysis in real time mass spectrometry combined with principal component analysis for their discrimination. Anal. Methods 7, 6021–6026.

Kumar, S., Singh, A., Bajpai, V., Srivastava, M., Singh, B.P., Ojha, S., Kumar, B., 2016a. Simultaneous determination of bioactive monoterpene indole alkaloids in ethanolic extract of seven *Rauvolfia* species using UHPLC with hybrid triple quadrupole linear ion trap mass spectrometry. Phytochem. Anal. 27, 296–303.

Kumar, S., Singh, A., Bajpai, V., Kumar, B., 2016b. Identification, characterization and distribution of monoterpene indole alkaloids in *Rauwolfia* species by Orbitrap Velos pro mass spectrometer. J. Pharmaceut. Biomed. Anal. 118, 183–194.

Kumar, S., Singh, A., Bajpai, V., Srivastava, M., Singh, B.P., Kumar, B., 2016c. Structural characterization of monoterpene indole alkaloids in ethanolic extracts of *Rauwolfia* species by liquid chromatography with quadrupole time-of-flight mass spectrometry. J. Pharm. Anal. 6, 363–373.

Kutchan, T.M., 1989. Expression of enzymatically active cloned strictosidine synthase from the higher plant *Rauvolfia serpentina* in *Escherichia coli*. FEBS Lett. 1, 127–130.

Kutchan, T.M., Hampp, N., Lottspeich, F., Beyreuther, K., Zenk, M.H., 1988. The cDNA clone for strictosidine synthase from *Rauvolfia serpentina*: DNA sequence determination and expression in *Escherichia coli*. FEBS Lett. 257, 40–44.

Lee, M.L., Schneider, G., 2001. Scaffold architecture and pharmacophoric properties of natural products and trade drugs: application in the design of natural product-based combinatorial libraries. J. Comb. Chem. 3, 284–289.

Lehman, E., Haber, J., Lesser, S.R., 1957. The use of reserpine in autistic children. J. Nerv. Ment. Dis. 125, 351–356.

Lopez-Casado, G., Covey, P.A., Bedinger, P.A., Mueller, L.A., Thannhauser, T.W., Zhang, S., Fei, Z., Giovannoni, J.J., Rose, J.K., 2012. Enabling proteomic studies with RNA-Seq: the proteome of tomato pollen as a test case. Proteomics 12, 761–774.

Luijendijk, T.J.C., Stevens, L.H., Verpoorte, R., 1998. Purification and characterization of strictosidine β-dglucosidase from *Catharanthus roseus* cell suspension cultures. Plant Physiol. Biochem. 36, 419–425.

Ma, X., Koepke, J., Fritzsch, G., Diem, R., Kutchan, T.M., Michel, H., Stöckigt, J., 2004. Crystallization and preliminary X-ray crystallographic analysis of strictosidine synthase from *Rauvolfia*: the first member of a novel enzyme family. Biochim. Biophys. Acta 1702, 121–124.

Ma, X., Koepke, J., Panjikar, S., Fritzsch, G., Stöckigt, J., 2005. Crystal structure of vinorine synthase, the first representative of the BAHD superfamily. J. Biol. Chem. 280, 13576–13583.

Ma, X., Panjikar, S., Koepke, J., Loris, E., Stöckigt, J., 2006. The structure of *Rauvolfia serpentina* strictosidine synthase is a novel six-bladed β-propeller fold in plant proteins. Plant Cell 18, 907–920.

Madhusudanan, K.P., Banerjee, S., Khanuja, S.P.S., Chattopadhyay, S.K., 2008. Analysis of hairy root culture of *Rauvolfia serpentina* using direct analysis in real time mass spectrometric technique. Biomed. Chromatogr. 22, 596–600.

Maresh, J.J., Giddings, L.A., Friedrich, A., Loris, E.A., Panjikar, S., Trout, B.L., Stöckigt, J., Peters, B., O'Connor, S.E., 2008. Strictosidine synthase: mechanism of a Pictet-Spengler catalyzing enzyme. J. Am. Chem. Soc. 130, 710–723.

Marillonnet, S., Giritch, A., Gils, M., Kandzia, R., Klimyuk, V., Gleba, Y., 2004. In planta engineering of viral RNA replicons: efficient assembly by recombination of DNA modules delivered by *Agrobacterium*. Proc. Natl. Acad. Sci. USA 101, 6852–6857.

Marillonnet, S., Thoeringer, C., Kandzia, R., Klimyuk, V., Gleba, Y., 2005. Systemic *Agrobacterium tumefaciens*-mediated transfection of viral replicons for efficient transient expression in plants. Nat. Biotechnol. 23, 718–723.

Michael, T.P., Jackson, S., 2013. The first 50 plant genomes. Plant Genom. 6, 1–7.

Miettinen, K., Dong, L., Navrot, N., Schneider, T., Burlat, V., Pollier, J., Woittiez, L., van der Krol, S., Lugan, R., Ilc, T., Verpoorte, R., Oksman-Caldentey, K.M., Martinoia, E., Bouwmeester, H., Goossens, A., Memelink, J., Werck-Reichhart, D., 2014. The seco-iridoid pathway from *Catharanthus roseus*. Nat. Commun. 5, 3606.

Mitsuda, N., Ohme-Takagi, M., 2009. Functional analysis of transcription factors in *Arabidopsis*. Plant Cell Physiol. 50, 1232–1248.

Moore, B.D., Andrew, R.L., Kulheim, C., Foley, W.J., 2014. Explaining intraspecific diversity in plant secondary metabolites in an ecological context. New Phytol. 201, 733–750.

Morales, A., 2000. Yohimbine in erectile dysfunction: the facts. Int. J. Impot. Res. 12, 70–74.

Moreno, C., Lazar, J., Jacob, H.J., Kwitek, A.E., 2008. Comparative genomics for detecting human disease genes. Adv. Genet. 60, 655–697.

Moreno-Risueno, M.A., Busch, W., Benfey, P.N., 2010. Omics meet networks – using systems approaches to infer regulatory networks in plants. Curr. Opin. Plant Biol. 13, 126–131.
Mukherjee, E., Gantait, S., Kundu, S., Sarkar, S., Bhattacharyya, S., 2019. Biotechnological interventions on the genus *Rauvolfia*: recent trends and imminent prospects. Appl. Microbiol. Biotechnol. 103, 7325–7354.
Muranaka, T., Saito, K., 2013. Phytochemical genomics on the way. Plant Cell Physiol. 54, 645–646.
Nair, V.D., Raj, R.P.D., Panneerselvam, R., Gopi, R., 2014. Assessment of diversity among populations of *Rauvolfia serpentina* Benth. Ex. Kurtz. from Southern Western Ghats of India, based on chemical profiling, horticultural traits and RAPD analysis. Fitoterapia 92, 46–60.
Namjoshi, O.A., Cook, J.M., 2016. Sarpagine and related alkaloids. Alkaloids Chem. Biol. 76, 63–169.
Nammi, S., Boini, K.M., Koppula, S., Sreemantula, S., 2005. Reserpine-induced central effects: pharmacological evidence for the lack of central effects of reserpine methiodide. Can. J. Physiol. Pharmacol. 83, 509–515.
Nascimento, G.G.F., Lacatelli, J., Freitas, P.C., Silva, G.L., 2000. Antibacterial activity of plant extracts and phytochemicals on antibiotic-resistant bacteria. Braz. J. Microbiol. 31, 886–891.
Ncube, N.S., Afolayan, A.J., Okoh, A.I., 2008. Assessment techniques of antimicrobial products of natural compounds of plant origin: current methods and future trends. Afr. J. Biotechnol. 7, 1797–1806.
Niraimathi, K., Karunanithi, M., Brindha, P., 2012. Phytochemical and in-vitro screening of aerial parts of *Cleome viscosa* Linn. extracts (Capparidaceae). Int. J. Pharm. Pharmaceut. Sci. 4, 27–30.
O'Connor, S.E., Maresh, J.J., 2006. Chemistry and biology of monoterpene indole alkaloid biosynthesis. Nat. Prod. Rep. 23, 532–547.
Oboh, G., Adebayo, A.A., Ademosun, A.O., 2019. HPLC phenolic fingerprinting, antioxidant and anti-phosphodiesterase-5 properties of *Rauwolfia vomitoria* extract. J. Basic Clin. Physiol. Pharmacol. 30, 20190059.
Ojha, J., Mishra, U., 1985. Dhanvantari Nighantuh, with Hindi Translation and Commentary, first ed. Deptt. of Dravyaguna, Institute of Medical Sciences, BHU, Varanasi, p. 204.
Okereke, S.C., Ijeh, I.I., Arunsi, U.O., 2017. Determination of bioactive constituents of *Rauwolfia vomitoria* Afzel (Asofeyeje) roots using gas chromatography-mass spectrometry (GC-MS) and Fourier transform infrared spectrometry (FT-IR). Afr. J. Pharm. Pharmacol. 11, 25–31.
Okwu, D.E., Okwu, M.E., 2004. Chemical composition of *Spondias mombin* linn plant parts. J. Sustain. Agri. Env. 6, 140–147.
Pandey, P., Kaur, R., Singh, S., Chattopadhyay, S.K., Srivastava, S.K., Banerjee, S., 2014. Long-term stability in biomass and production of terpene indole alkaloids by hairy root culture of *Rauvolfia serpentina* and cost approximation to endorse commercial realism. Biotechnol. Lett. 36, 1523–1528.
Panjikar, S., Stöckigt, J., O'Connor, S.E., Warzecha, H., 2012. The impact of structural biology on alkaloid biosynthesis research. Nat. Prod. Rep. 29, 1176–1200.
Pathania, S., Acharya, V., 2016. Computational analysis of "-omics" data to identify transcription factors regulating secondary metabolism in *Rauvolfia serpentina*. Plant Mol. Biol. Rep. 34, 283–302.

Pathania, S., Bagler, G., Ahuja, P.S., 2016. Differential network analysis reveals evolutionary complexity in secondary metabolism of *Rauvolfia serpentina* over *Catharanthus roseus*. Front. Plant. Sci. 7, 1229.

Pathania, S., Ramakrishnan, S.M., Randhawa, V., Bagler, G., 2015. SerpentinaDB: a database of plant-derived molecules of *Rauvolfia serpentina*. BMC Compl. Altern. Med. 15, 262.

Patti, G.J., Yanes, O., Siuzdak, G., 2012. Innovation: metabolomics: the apogee of the omics trilogy. Nat. Rev. Mol. Cell. Biol. 13, 263–269.

Paul, M., Breithardt, G., Haverkamp, W., Eckardt, L., 2003. The ajmaline challenge in Brugada syndrome: diagnostic impact, safety, and recommended protocol. Eur. Heart J. 24, 1104–1112.

Polz, L., Schübel, H., Stoekigt, J., 1986. Characterization of 2β(*R*)-17-0-acetylajmalan: acetylesterase - a specific enzyme involved in the biosynthesis of the *Rauwolfia* alkaloid ajmaline. Z. Naturforsch. 42, 333–342.

Prakash, P., Rajakani, R., Gupta, V., 2015. Transcriptome-wide identification of *Rauvolfia serpentina* micro RNAs and prediction of their potential targets. Comput. Biol. Chem. 61, 62–74.

Pullaiah, J., 2002. Med. Plants India, vol. 2. Regency Publ, New Delhi, pp. 441–443.

Quanbeck, S.M., Brachova, L., Campbell, A.A., Guan, X., Perera, A., He, K., Rhee, S.Y., Bais, P., Dickerson, J.A., Dixon, P., Wohlgemuth, G., Fiehn, O., Barkan, L., Lange, I., Lange, B.M., Lee, I., Cortes, D., Salazar, C., Shuman, J., Shulaev, V., Huhman, D.V., Sumner, L.W., Roth, M.R., Welti, R., Ilarslan, H., Wurtele, E.S., Nikolau, B.J., 2012. Metabolomics as a hypothesis-generating functional genomics tool for the annotation of *Arabidopsis thaliana* genes of "Unknown Function". Front. Plant Sci. 3, 15.

Rai, A., Kamochi, H., Suzuki, H., Nakamura, M., Takahashi, H., Hatada, T., Saito, K., Yamazaki, M., 2017. De novo transcriptome assembly and characterization of nine tissues of *Lonicera japonica* to identify potential candidate genes involved in chlorogenic acid, luteolosides, and secoiridoid biosynthesis pathways. J. Nat. Med. 71, 1–15.

Rai, A., Umashankar, S., Rai, M., Kiat, L.B., Bing, J.A., Swarup, S., 2016. Coordinate regulation of metabolite glycosylation and stress hormone biosynthesis by TT8 in *Arabidopsis*. Plant Physiol. 171, 2499–2515.

Rana, S.K., Sehrawat, A.R., Chowdhury, V.R., 2015. Assessment of clonal fidelity in micropropagated plantlets of *Rauwolfia serpentina* Benth. ex. Kurz. Med. Plant 7, 258–263.

Ray, S., Majumdar, A., Bandyopadhyay, M., Jha, S., 2014. Genetic transformation of sarpagandha (*Rauvolfia serpentina*) with *Agrobacterium rhizogenes* for identification of high alkaloid yielding lines. Acta Physiol. Plant 36, 1599–1605.

Rhee, K., 2013. Minding the gaps: metabolomics mends functional genomics. EMBO Rep. 14, 949–950.

Rohela, G.K., Bylla, P., Kota, S., Abbagani, S., Chithakari, R., Reuben, T.C., 2013. In vitro plantlet regeneration from leaf and stem calluses of *Rauwolfia tetraphylla* (*R. canescens*) and confirmation of genetic fidelity of plantlets using the ISSR-PCR method. J. Herbs Spices Med. Plants 19, 66–75.

Rohela, G.K., Jogam, P., Bylla, P., Reuben, C., 2019. Indirect regeneration and assessment of genetic fidelity of acclimated plantlets by SCOT, ISSR, and RAPD markers in *Rauwolfia tetraphylla* L.: an endangered medicinal plant. BioMed. Res. Int. 2019, 3698742.

Rukachaisirikul, T., Chokchaisiri, S., Suebsakwong, P., Suksamrarn, A., Tocharus, C., 2017. A new ajmaline-type alkaloid from the roots of *Rauvolfia serpentina*. Nat. Prod. Commun. 12, 495–498.

Ruppert, M., Ma, X., Stöckigt, J., 2005b. Alkaloid biosynthesis in *Rauvolfia* – cDNA cloning of major enzymes of the ajmaline pathway. Curr. Org. Chem. 9, 1431–1444.

Ruppert, M., Panjikar, S., Barleben, L., Stöckigt, J., 2006. Heterologous expression, purification, crystallization and preliminary X-ray analysis of raucaffricine glucosidase, a plant enzyme specifically involved in *Rauvolfia* alkaloid biosynthesis. Acta Cryst. F62, 257–260.

Ruppert, M., Woll, J., Giritch, A., Genady, E., Ma, X., Stöckigt, J., 2005a. Functional expression of an ajmaline pathway-specific esterase from *Rauvolfia* in a novel plant-virus expression system. Planta 222, 888–898.

Ruyter, C.M., Schübel, H., Stöckigt, J., 1988. Novel glucoalkaloids from *Rauwolfia* cell cultures - acetylrauglucine and related glucosides. Z. Naturforsch. C. J. Biosci. 43, 479–484.

Sagi, S., Avula, B., Wang, Y.H., Khan, I.A., 2016. Quantification and characterization of alkaloids from roots of *Rauwolfia serpentina* using ultra-high performance liquid chromatography-photo diode array-mass spectrometry. Anal. Bioanal. Chem. 408, 177–190.

Saito, K., 2013. Phytochemical genomics – a new trend. Curr. Opin. Plant Biol. 16, 373–380.

Salim, V., Wiens, B., Masada-Atsumi, S., Yu, F., De Luca, V., 2014. 7-Deoxyloganetic acid synthase catalyzes a key 3 step oxidation to form 7-deoxyloganetic acid in *Catharanthus roseus* iridoid biosynthesis. PhytoChem 101, 23–31.

Salim, V., Yu, F., Altarejos, J., De Luca, V., 2013. Virus-induced gene silencing identifies *Catharanthus roseus* 7-deoxyloganic acid-7-hydroxylase, a step in iridoid and monoterpene indole alkaloid biosynthesis. Plant J. 76, 754–765.

Santos, P., Herrmann, A.P., Benvenutti, R., Noetzold, G., Giongo, F., Gama, C.S., 2017. Anxiolytic properties of N-acetylcysteine in mice. Behav. Brain Res. 317, 461–469.

Saravanan, S., Sarvesan, R., Vinod, M.S., 2011. Identification of DNA elements involved in somaclonal variants of *Rauvolfia serpentina* (L.) arising from indirect organogenesis as evaluated by ISSR analysis. Indian J. Sci. Technol. 4, 1241–1245.

Schmidt, D., Stöckigt, J., 1995. Enzymatic formation of the sarpagan-bridge: a key step in the biosynthesis of sarpagine- and ajmaline-type alkaloids. Planta Med. 61, 254–258.

Schwab, R., Palatnik, J.F., Riester, M., Schommer, C., Schmid, M., Weigel, D., 2005. Specific effects of microRNAs on the plant transcriptome. Dev. Cell 8, 517–527.

Senapati, S.K., Lahere, N., Tiwary, B.N., 2014. Improved in vitro clonal propagation of *Rauwolfia serpentina* L. Benth – an endangered medicinal plant. Plant Biosyst. 148, 885–888.

Sheludko, Y., Gerasimenko, I., Kolshorn, H., Stöckigt, J., 2002. New alkaloids of the sarpagine group from *Rauvolfia serpentina* hairy root culture. J. Nat. Prod. 65, 1006–1010.

Siddiqui, S., Ahmad, S.S., Haider, S.I., 1987a. A new alkaloid ajmalimine from the roots of *Rauwolfia serpentina*. Planta Med. 53, 288–289.

Siddiqui, S., Ahmad, S.S., Haider, S.I., Siddiqui, B.S., 1985a. Isolation and structure of a new alkaloid from the roots of *Rauwolfia Serpentina* Benth. Heterocycles 3, 617–622.

Siddiqui, S., Ahmad, S.S., Haider, S.I., Siddiqui, B.S., 1987b. Ajmalicidine an alkaloid from *Rauwolfia serpentina*. Phytochemistry 26, 875–877.

Siddiqui, S., Haider, S.I., Ahmad, S.S., 1987c. A new alkaloid from the roots of *Rauwolfia serpentina*. J. Nat. Prod. 50, 238–240.

Siddiqui, S., Haider, S.I., Ahmad, S.S., Siddiqui, B.S., 1985b. Isolation and structure of a new alkaloid from *Rauwolfia serpentina* Benth. Tetrahedron 41, 4577–4580.

Silja, V.P., Varma, K.S., Mohanan, K.V., 2008. Ethnomedical plant knowledge of the Mullu kuruma tribe of Wayanad district of Kerala. Ind. J. Trad. Knowl. 7, 604–612.

Singh, P.K., Kumar, V., Tiwari, R.K., Sharma, A., Rao, C.V., Singh, R.H., 2010. Medico-ethnobotany of 'chatara' block of district Sonebhadra, Uttar Pradesh, India. Adv. Biol. Res. 4, 65–80.

Srivastava, A., Tripathi, A.K., Pandey, R., Verma, R.K., Gupta, M.M., 2006. Quantitative determination of reserpine, ajmaline, and ajmalicine in *Rauvolfia serpentina* by reversed-phase high-performance liquid chromatography. J. Chromatogr. Sci. 44, 557–560.

Srivastava, M., Sharma, S., Misra, P., 2016. Elicitation based enhancement of secondary metabolites in *Rauwolfia serpentina* and *Solanum khasianum* hairy root cultures. Phcog. Mag. 12, S315–S320.

Steuer, R., 2007. Computational approaches to the topology, stability and dynamics of metabolic networks. Phytochemistry 68, 2139–2151.

Stöckigt, J., Pfitzner, A., Keller, P.J., 1983. Enzymatic formation of ajmaline. Tetrahedron Lett. 24, 2485–2486.

Stockigt, J., Zenk, M.H., 1977. Strictosidine (Isovincoside): the key intermediate in the biosynthesis of monoterpenoid indole alkaloids. J. Chem. Soc. Chem. Commun. 912–914.

Stöckigt, J., Zenk, M.H., 1995. Biosynthesis in *Rauvolfia serpentina* - modern aspects of an old medicinal plant. In: Cordell, G.A. (Ed.), The Alkaloids. Chemistry and Pharmacology. Academic Press, San Diego, pp. 115–172.

Sudha, C.G., Reddy, B.O., Ravishankar, G.A., Seeni, S., 2003. Production of ajmalicine and ajmaline in hairy root cultures of *Rauvolfia micrantha* Hook f., a rare and endemic medicinal plant. Biotechnol. Lett. 25, 631–636.

Sun, L., Chen, Y., Rajendran, C., Mueller, U., Panjikar, S., Wang, M., Mindnich, R., Rosenthal, C., Penning, T.M., Stöckigt, J., 2012. Crystal structure of perakine reductase, founding member of a novel aldo-keto reductase (AKR) subfamily that undergoes unique conformational changes during NADPH binding. J. Biol. Chem. 30 (287), 11213–11221.

Sun, L., Ruppert, M., Sheludko, Y., Warzecha, H., Zhao, Y., Stöckigt, J., 2008. Purification, cloning, functional expression and characterization of perakine reductase: the first example from the AKR enzyme family, extending the alkaloidal network of the plant *Rauvolfia*. Plant Mol. Biol. 67, 455–467.

Sweetlove, L.J., Fell, D., Fernie, A.R., 2008. Getting to grips with the plant metabolic network. Biochem. J. 409, 27–41.

Thakar, V.J., 2010. Historical development of basic concepts of Ayurveda from Veda up to Samhita. Ayu 31, 400–402.

Tomar, N., De, R.K., 2013. Comparing methods for metabolic network analysis and an application to metabolic engineering. Gene 521, 1–14.

Treimer, J.F., Zenk, M.H., 1979. Purification and properties of strictosidine synthase, the key enzyme in indole alkaloid formation. Eur. J. Biochem. 101, 225–233.

Tyler, V.E., Brady, L.R., Robbers, J.E., 1988. Pharmacognosy, ninth ed. Lea & Febiger, Philadelphia, PA, pp. 222–225.

Ulrich-Merzenich, G., Zeitler, H., Jobst, D., Panek, D., Vetter, H., Wagner, H., 2007. Application of the 'omic' technologies in phytomedicine. Phytomedicine 14, 70–82.

Unamba, C.I., Nag, A., Sharma, R.K., 2015. Next generation sequencing technologies: the doorway to the unexplored genomics of non-model plants. Front. Plant Sci. 6, 1074.

Vakil, R.J., 1949. A clinical trial of *Rauwolfia serpentina* in essential hypertension. Br. Heart J. 10, 350–355.

Vakil, R.J., 1955. *Rauwolfia serpentina* in the treatment of high blood pressure: a review of the literature. Circulation 12, 220–229.

Varchi, G., Battaglia, A., Samori, C., Baldelli, E., Danieli, B., Fontana, G., Guerrini, A., Bombardelli, E., 2005. Synthesis of deserpidine from reserpine. J. Nat. Prod. 68, 1629–1631.

von Schumann, G., Gao, S., Stöckigt, J., 2002. Vomilenine reductase - a novel enzyme catalyzing a crucial step in the biosynthesis of the therapeutically applied antiarrhythmic alkaloid ajmaline. Bioorg. Med. Chem. 10, 1913–1918.

Wang, Z., Gerstein, M., Snyder, M., 2009. RNA-Seq: a revolutionary tool for transcriptomics. Nat. Rev. Genet. 10, 57–63.

Warzecha, H., Obitz, P., Stöckigt, J., 1999. Purification, partial amino acid sequence and structure of the product of raucaffricine-*O*-β-D-glucosidase from plant cell cultures of *Rauwolfia serpentina*. Phytochemistry 50, 1099–1109.

Weiss, R.F., Fintelmann, V., 2000. Herbal. Med., second ed., vols. 229–230, pp. 387–416 Thieme, Stuttgart.

Weng, J.K., 2014. The evolutionary paths towards complexity: a metabolic perspective. New. Phytol. 201, 1141–1149.

Wiens, B., Luca, V.D., 2016. Molecular and biochemical characterization of a benzenoid/phenylpropanoid *meta/para-O*-methyltransferase from *Rauwolfia serpentina* roots. Phytochemistry 132, 5–15.

Williams, K., Kubelik, A.R., Rafalski, J.A., Tingey, S.V., 1990. DNA polymorphisms amplified by arbitrary primers are useful as genetic markers. Nucleic Acids Res. 18, 1631–1635.

Wink, M., Roberts, M.W., 1998. Alkaloids: Biochemistry, Ecology, and Medicinal Applications. Plenum Press, New York, ISBN 0-306-45465-3.

Woodson, R.E., Youngken, H.W., Schlittler, E., Schneider, J.A., 1957. *Rauvolfia*, Pharmacognosy, Chemistry, and Pharmacology, first ed. Little, Brown and Company, Toronto, Canada.

Wurtzel, E.T., Kutchan, T.M., 2016. Plant metabolism, the diverse chemistry set of the future. Science 353, 1232–1236.

Xiao, M., Zhang, Y., Chen, X., Lee, E.J., Barber, C.J., Chakrabarty, R., Desgagné-Penix, I., Haslam, T.M., Kim, Y.B., Liu, E., MacNevin, G., Masada-Atsumi, S., Reed, D.W., Stout, J.M., Zerbe, P., Zhang, Y., Bohlmann, J., Covello, P.S., De Luca, V., Page, J.E., Ro, D.K., Martin, V.J., Facchini, P.J., Sensen, C.W., 2013. Transcriptome analysis based on next-generation sequencing of non-model plants producing specialized metabolites of biotechnological interest. J. Biotechnol. 166, 122–134.

Yang, C.Q., Fang, X., Wu, X.M., Mao, Y.B., Wang, L.J., Chen, X.Y., 2012. Transcriptional regulation of plant secondary metabolism. J. Integr. Plant Biol. 54, 703–712.

Yang, L., Hill, M., Wang, M., Panjikar, S., Stöckigt, J., 2009. Structural basis and enzymatic mechanism of the biosynthesis of C9- from C10-monoterpenoid indole alkaloids. Angew. Chem. Int. Ed. Engl. 48, 5211–5213.

Zietkiewics, E., Rafalski, A., Labuda, D., 1994. Genome fingerprinting by simple sequence repeat (SSR) - anchored polymerase chain reaction amplification. Genomics 20, 176–183.

CHAPTER

Rhodiola imbricata

9

Archit Pundir[1], Anaida Kad[1], Hemant Sood[2]

[1]*University Institute of Engineering and Technology (UIET), Panjab University, Chandigarh, India;* [2]*Department of Biotechnology and Bioinformatics, Jaypee University of Information Technology, Waknaghat, Solan, Himachal Pradesh, India*

9.1 Introduction

Rhodiola imbricata Edgew. is an herbaceous, dioecious perennial plant, belonging to Crassulaceae family that consists of over 1400 species. These species are distributed in 33 genera including *Rhodiola* that is distributed worldwide especially in the regions of Northern Hemisphere and South Africa (Gupta et al., 2007). The word *Rhodiola* is made up of two words, "rhodon" and "iola" where the initial is a Greek work meaning rose owing to rose-scented roots of the plant, and the latter is a Latin word meaning diminutive. The *Rhodiola* genera consists of about 130 species (Lei et al., 2003), of which many have been abundantly used traditionally for curing chronic illness and weakness in the regions of Tibet and Western Himalayas belt for over 1000 (Rohloff, 2002). According to GBIF, the alpine habitats of India support the growth of 22 *Rhodiola* species, the most abundant of these being *Rhodiola tibetica, Rhodiola heterodonta, Rhodiola imbricata, Rhodiola quadrifida, Rhodiola sinuata, and Rhodiola wallichiana* (Chaurasia and Gurmet, 2006).

There are a number of names used for referring *R. imbricata* in common language such as Rose root (due to the rose-like fragrance of the fresh-cut rootstock), Golden root, Arctic root, Shrolo (as commonly called by the locals of Ladakh region), Solo (by localites of Rohtang–Manali region). It is also called stone crop or Himalayan stone crop in India because of its growth along the stony crevices and rocky slopes of high-altitude terrain of Himalayas (Ballabh and Chaurasia, 2007).

9.1.1 Classification and morphology

The botanical classification of *R. imbricata* has been illustrated in Table 9.1. The morphology of *Rhodiola* plants generally comprises an erect, succulent herb which reaches up to the height of 10–35 cm, with a thick subcylindrical rhizome, golden outside and pink inside which is sparsely branched and 2–2.5 cm long. Whereas the densely arranged leaves are generally 1.3–3 cm long, oblanceolate to narrowly elliptic, and nearly entire, sessile, glabrous with acute tip and round base, the plant

Himalayan Medicinal Plants. https://doi.org/10.1016/B978-0-12-823151-7.00014-3

Table 9.1 Botanical classification of *Rhodiola imbricata* is as follows.

Kingdom	Plantae
Phylum	Magnoliophyta
Class	Magnoliopsida
Order	Rosales
Family	Crassulaceae
Genus	*Rhodiola*
Species	*imbricata Edgew*

has a massive rose-scented rootstock. The flowers exist in the form of congested clusters of pale yellow color, surrounded by an involucre of leaves. The petals are angular-oblanceolate, with stamens distinctly longer than the petals, filaments of 5–8 mm, anthers of distinct dark purplish red color, carpels of 3–5 mm with 9–10 ovules. There are 4–5 fruits per plant having a number of seeds, but the season of flowering and fruiting is limited to summer season of the area, thus, limiting the availability of this pharmaceutical important plant to the months of July to September (Singh et al., 1996; Ballabh and Chaurasia, 2007).

9.1.2 Taxonomy

In third-world countries or underdeveloped countries, around 80% of inhabitants depend mostly on traditional medicine for the needs related to healthcare. A major portion of healthcare sector comprises the utilization of plant extracts or the bioactive compounds formed by the plants for general treatments. Quality control profile and standardization for accurate recognition of the concerned species, whether it is in fresh, dried, or powdered state, is one of the basic requirements of herbal medicines (Tayade, 2015). In preparation and administration of herbal medicine, real threat is the flawed changeover and species misclassification (Tayade, 2015). Most of the herbs which are mistaken for one another are the herbs having extremely comparable appearance to the inexperienced eye. The flawed classification of species and the mistaken substitution of herbs have also given rise to serious adverse effects (Tayade, 2015). Therefore, the taxonomic and botanical classification and recognition for the accurate species from its natural habitat of *R. imbricata* mentioned earlier is very important.

The current taxonomic status of the genus *Rhodiola* is quite complex due to the generally similar morphology (Brown et al., 2002; Liu et al., 2013). According to GBIF (2020), the genus *Rhodiola* comprises 175 accepted species, while the Plant List includes 192 scientific plant names of species rank for the genus *Rhodiola*. Of these, 98 are accepted species names, whereas the status of 12 is still doubtful.

9.1.3 Indigenous uses

In Tibet, Mongolia, and the upper Himalayan regions, *Rhodiola* species have been used as traditional medicines for over 1000 years for the treatment of undying weakness and illness due to infections (Rohloff, 2002). *R. imbricata* is not only an important traditional medicinal plant but also widely used as food crop and is distributed in the Trans-Himalayan cold desert regions. This edible plant is generally consumed in the form of a local Ladakhi delicacy called "Tantur," which is prepared by boiling the young shoots of the plant and mixing it with yoghurt.

Roots of *R. imbricata* are used for treatment of cold, cough, lung problems, fever, pulmonary complaints, and loss of energy in Tibetan and Amchi system of traditional medicine (Ballabh and Chaurasia, 2007). The plant has also proven its worth owing to the medicinal properties exploited in order to increase work productivity, physical endurance, longevity, and to treat asthma, fatigue, impotence, hemorrhage, and gastrointestinal ailments.

The plant has been extensively used for traditional medicines by the people of Leh and Ladakh. It also finds its mention in Tibetan and Chinese medicines, but due to easy availability of *Rhodiola rosea* and *Rhodiola crenulata,* the potential of *R. imbricata* is yet to be explored by the large pharmaceutical companies.

9.2 Geographical distribution

R. imbricata Edgew was initially believed to have its origin in the Himalayas and the mountainous regions of South West China. But the present scenario witnesses its distribution in mountainous as well as coastal habitats (Brown et al., 2002). The plant is native to the whole of the Northern hemisphere (Singh et al., 1996). In India, *R. imbricata* grows in the Trans-Himalayan cold desert, high Arctic latitudes, and mountain regions of Eurasia, primarily on the rocky slopes, wet places, and higher passes at high altitudes (12,000–18,380 ft above mean sea level) (Khanum et al., 2005). It is commonly found in Indus and Leh valley of Indian Trans-Himalayas (Singh et al., 1996; Ballabh and Chaurasia, 2007). The traces of vegetation of *R. imbricata* are also reported in Chang La, Pensi La, and Kumaon.

As per the data of Indian portal of Biodiversity, this plant has been spotted at three places namely, Rohtang Pass, Himachal Pradesh; Khardung La Pass, Jammu Kashmir; and Birje Ganj pass, Uttrakhand.

9.3 Biochemical composition

Analysis of different *Rhodiola* species revealed six groups of active principles in their chemical compositions (Khanum et al., 2005). These chemicals, generally secondary metabolites, form the essence of medical potential/importance of this plant (Table 9.2).

Table 9.2 Chemical composition.

1	Phenylpropanoids	The name Rosavin includes these three: Rosin, Rosavin, Rosarin
2	Triterpenes	β-sitosterol, Daucosterol
3	Phenol acids	Hydroxycinnamic, Chlorogenic, and Gallic acids
4	Flavonoids	Acetylrhodalgin, Rhodionin, Rhodiosin, Rhodiolin, and Tricin
5	Phenylethanol derivatives	Tyrosol, Rhodioloside which includes Salidroside and Rhodosin
6	Monoterpenes	Rosiridol

- The presence of Triandrine, p-coumaric alcohol, and its glucosides (Vimalin), p-cumaric acid, caffeic acid, β-sitosterol, Daucosterol, and Salidroside (in trace amounts) has also been detected in callus tissues cultures(Tayade, 2015).
- Hydrodistillation air-dried root of *Rhodiola* gives 0.05%–1% of essential oil.

Initially in the 1970s, the compound responsible for unique pharmacological properties of *Rhodiola* genus was believed to be Salidroside. According to the Russian Pharmacopeia (1989), the raw material of *R. rosea* should contain 0.8% salidroside (Furmanowa et al., 1998). However, further studies revealed that not only salidroside but also rosin derivatives are important bioactive compounds (Wagner et al., 1994).

The information regarding the fatty acid profile (Table 9.3), amino acid composition (Table 9.4), and mineral content (Table 9.5) of *R. imbricata* root can play a major role for the pharmacological properties, bioactivity, and provide new perception in the physiological adaptation aspects of the plant in stressful and difficult terrain of the Trans-Himalaya region (Tayade et al., 2017).

9.4 Pharmacological properties

The analysis of extensive work done on *R. imbricata* in the recent years has shown that in many in vitro and in vivo studies on cells and animals, the secondary metabolites of the plant influence a number of physiological functions and have shown strong biological activity. These include the regulation of neurotransmitter levels, central nervous system activity, and cardiovascular function. It is used to stimulate the nervous system, improve work performance, eliminate fatigue, reduce depression, and prevent diseases at high altitude including altitude sickness. Most of these effects have been attributed to components such as salidrosides (rhodiolosidos), rosavinas, colofonia, and p-tyrosol. Numerous pharmacological studies on *R. imbricata* have shown that this plant has adaptogenic and protective properties against stress (Darbinyan et al., 2000; Spasov et al., 2000; De Bock et al., 2004;

Table 9.3 Fatty acid content in root extract of *Rhodiola imbricata* (Tayade et al., 2017).

S. no.	Type of fatty acid	IUPAC name	Content in root (mg/g)
1.	Saturated fatty acid (SFA)		52.07 ± 1.28
	Capric acid (C10:0)	Decanoic acid	16.2 ± 0.41
	Behenic acid (C22:0)	Docosanoic acid	4.6 ± 0.12
	Palmitic acid (C16:0)	Hexadecanoic acid (9Z)	7.6 ± 0.19
	Caproic acid (C6:0)	Hexanoic acid	8.8 ± 0.22
	Lignoceric acid (C24:0)	Tetracosanoic acid	5.0 ± 0.13
2.	Unsaturated fatty acid (UFA)		47.22 ± 1.2
	Monounsaturated fatty acid (MUFA)		12.38 ± 0.31
	Oleic acid (C18:1 n9c)	(9Z)-Octadec-9-enoic acid	10.0 ± 0.25
	Poly unsaturated fatty acid (PUFA)		35.54 ± 0.89
	Arachidonic acid (C20:4 n6)	(5Z,8Z,11Z,14Z)-Eicosatetraenoic acid	6.8 ± 0.17
	cis-13,16-Docosadienoic acid (C22:2)	cis-13,16-Docosadienoic acid	4.9 ± 0.12
	Linoleic acid (C18:2n6c)	cis, cis-9,12-Octadecadienoic acid	12.2 ± 0.31
	Linolelaidic acid (C18:2 n6t)	(9E,12E)-Octadeca-9,12-dienoic acid	5.0 ± 0.14

Olsson et al., 2009) and acts as antioxidant (Mao et al., 2007; Chen et al., 2009; Schriner et al., 2009; Calcabrini et al., 2010), antitumor (Wójcik et al., 2008; Hu et al., 2010; Sun et al., 2012), antidepressant (van Diermen et al., 2009), protective activities for neurological wounds (Zhang et al., 2007; Yu et al., 2008), cardioprotective, antiinflammatory, and healing dermal wounds. It has also shown to have antiaging, immunostimulatory, radioprotective, and anticancer properties (Gupta et al., 2008). Some people have used *Rhodiola* sp. to treat diabetes, tuberculosis, aging, and liver damage. It also improves hearing, strengthens the central nervous system, and strengthens immunity. All these reports validate their use in the traditional medicine system. In addition, in the traditional medicine system, Amchi and Tibetan, the roots of *R. imbricata* are used against lung problems, colds, cough, and fever, loss of energy, and lung discomfort (Ballabh and Chaurasia, 2007).

Therefore, *Rhodiola* preparations can be applied therapeutically to humans to prevent or treat disorders such as neurodegenerative diseases, fatigue, hypoxia, cerebral ischemia, diabetes, cancer, and many others.

Table 9.4 Amino acid content in root extract of *Rhodiola imbricata* (Tayade et al., 2017).

S. no.	Amino acid	Abbreviation	Content (μg/g)
1.	L-2-amino-n-butyric acid	Abu	ND
2.	L-Alanine	Ala	1142.33 ± 11.02
3.	L-Arginine Arg	Arg	214.67 ± 7.09
4.	L-Aspartic acid	Asp	434.67 ± 8.74
5.	L-Cystine	Cys	239.33 ± 8.39
6.	L-Cystine HCl	Cys HCl	1136.33 ± 11.72
7.	L-Glutamic acid	Glu	320.67 ± 7.77
8.	L-Glycine	Gly	1640.67 ± 11.85
9.	L-Histidine	His	1434.33 ± 10.02
10.	L-Isoleucine	Ile	91.33 ± 7.77
11.	L-Leucine	Leu	928.67 ± 10.79
12.	L-Lysine	Lys	1329.33 ± 11.55
13.	L-Methionine	Met	736.67 ± 8.02
14.	L-Nor Leucine	Nor Leu	1038.67 ± 10.21
15.	L-Ornithine	Orn	ND
16.	L-Phenylalanine	Phe	855.33 ± 9.02
17.	L-Proline	Pro	1263.67 ± 10.50
18.	L-Serine	Ser	839.67 ± 10.97
19.	L-Threonine	Thr	1015.67 ± 8.02
20	L-Tryptophan	Trp	ND
21	L-Valine	Val	BDL

BDL, *below detection limit;* ND, *not detectable.*

Table 9.5 Mineral content of *Rhodiola imbricata* root (Tayade et al., 2017).

S. no.	Minerals	Symbols	Content (mg/kg)
1.	Calcium	Ca	11034.17 ± 332.04
2.	Chromium	Cr	7.27 ± 0.32
3.	Cobalt	Co	2.98 ± 0.06
4.	Copper	Cu	3.49 ± 0.12
5.	Iron	Fe	1441.17 ± 27.98
6.	Magnesium	Mg	581.99 ± 17.40
7.	Manganese	Mn	75.78 ± 2.21
8.	Molybdenum	Mo	2.65 ± 0.05
9.	Nickel	Ni	4.89 ± 0.21
10.	Phosphorous	P	376.72 ± 11.87
11.	Potassium	K	2143.25 ± 65.37
12.	Sodium	Na	109.75 ± 3.32
13.	Zinc	Zn	16.27 ± 0.54

9.4.1 Adaptogenic and antifatigue activity

Adaptogens are substances that enable the standardization of physiologic responses to various stresses, increase the stress tolerance of the body, and enhance work performance (Darbinyan et al., 2000; Grace et al., 2009). *Rhodiola* extracts have great utility in treating esthetic conditions which develop after intense physical or intellectual strain, including a decline in work performance, sleep difficulties, irritability, poor appetite, high blood pressure, headaches, and fatigue. Thus, it is effective in preventing oxidative stress following exhaustive exercise.

9.4.2 Elongation of life span and antiaging activity

An animal study found that *Rhodiola* extract could *inhibit the death of thymic T cells*, which is important as thymus function decreases with age. It reversed D-galactose–induced aging effects in neural and immune systems, improved motor activity, increased memory latency time, and enhanced lymphocyte mitogenesis and interleukin-2 production (Jafari et al., 2007; Mao et al., 2010a,b).

9.4.3 Antioxidant properties

Singlet oxygen scavenging, H_2O_2 scavenging, hypochlorite scavenging, ferric reducing, ferrous chelating, and protein thiol protection activities were noted (Chen et al., 2008). Salidroside reduced hydrogen peroxide–induced intracellular Reactive Oxygen Species production in human erythrocytes. Salidroside also increased cell survival and prevented human erythrocytes from undergoing eryptosis or erythroptosis mediated by H_2O_2 (Qian et al., 2012).

9.4.4 Antidepressant

It was noticed that orally administering salidroside for 2 weeks increased olfactory bulbectomy–induced hyperactivity in an open-field test and reduced immobility time in a forced swimming test. Reduction in TNF-a and IL-1b levels in the hippocampus was also noted. Salidroside also increased glucocorticoid receptor and brain-derived neurotrophic factor expression in the hippocampus of rats. In addition, salidroside attenuated corticotrophin-releasing hormone expression in the hypothalamus and the levels of serum corticosterone (Yang et al., 2014).

9.4.5 Skin treatments

Salidroside inhibited UVB-induced hyperpigmentation in brown guinea pig skin by reducing the number of DOPA-positive melanocytes in the basal layer of the epidermis and reducing tyrosinase activity and melanin synthesis in melanocytes (Peng et al., 2013). *R. rosea* extract, salidroside, and tyrosol may be effective skin-whitening agents.

9.4.6 Analysis of heavy metal

Studies have shown that the concentration of heavy metal in aqueous extracts of *R. imbricata*, except chromium, was less than the maximum permissible ranges proposed by the World Health Organization.

9.4.7 Radioprotective efficacy

In a study conducted on protection, lethal gamma irradiation (10 Gy) induced mortality in Swiss albino strain "A" mice. Preirradiation administration of extracts produced 83% to more than 90% survival, beyond the 30 days of observation period. The studies suggested that *R. imbricata* is a suitable radioprotector of herbal origin (Goel et al., 2006).

Hydroalcoholic rhizome extract of the plant showed antihemolytic capacity by preventing radiation-induced membrane degeneration of human erythrocytes. The study showed that *R. imbricata* rhizome renders in vitro and in vivo radioprotection via multifarious mechanisms (Arora et al., 2005).

9.4.8 Adjuvant activity

Antiinflammatory or immunosuppressive effect of *R. imbricata* rhizome was tested in adjuvant-induced arthritis model. The study suggested that the plant rhizome has adjuvant/immunopotentiating activity in terms of cell-mediated as well as humoral immune response (Mishra et al., 2010).

9.4.9 Mechanism and action of poststress caused by hypothermia induced by cold, hypoxia, and restraint (C-H-R) stress

Doses of extract administered 30 min prior to induced C-H-R stress and hypothermia induction decreased or maintained tissue glycogen and enzyme activities, viz., PFK, CS, G6-PD, and HK, in liver blood and liver, on attaining T(rec) 23°C and recovery. The results suggest that *R. imbricata* extract treatment in rats shifted anaerobic metabolism to aerobic, during C-H-R exposure and poststress recovery (Gupta et al., 2009).

Other physiological effects of *Rhodiola* and Salidroside include the following (Chiang et al., 2015):

- Antiinflammatory
- Protection against neuron damage
- Liver protection
- Reduction of oxidative stress in cardiovascular diseases
- Used in diabetes mellitus
- Obesity
- Antiviral
- Used against lung cancer

Salidroside protects human erythrocytes by its antioxidant activity and caspase-3 inhibition in a dose-dependent fashion (Qian et al., 2012). It also protects hematopoietic stem cells from oxidative stress by activating *PARP1*, a DNA repair enzyme actively involved in cell apoptosis (Li et al., 2014). Results of various clinical trials have revealed that Salidroside possesses various functions such as anticold, antifatigue, antianoxic, antivirus, antimicrowave radiation, and antitumor. It also possesses various medicinal properties such as preventing illness associated with old age, strengthening attention spans, delaying senility, and improving work efficiency. Due to its environmental acclimation activity, it plays important roles in healthcare, military, sports, and aerospace.

In the phytochemical study, the content of salidroside, rosavin, and its derivatives in *Rhodiola* plants depends on the morphological parts of the plant as if extracted from roots or rhizomes. In addition, it also depends on the age and sex of the plant, where the male rhizomes of the plants accumulated higher amounts of salidroside in *R. rosea* than their female counterparts (Platikanov and Evstatieva, 2008; Weglarz et al., 2008). The location and timing of collection also influence the salidroside content of the plants (Bykov et al., 1999). To date, document data related to the chemical profile of *Rhodiola* plants have shown that samples taken from natural sources have a higher salidroside and rosin content than in the cultivated fields.

9.5 Cultivation and propagation of *Rhodiola imbricata*

As the highest medical importance has been associated with secondary metabolites present in the roots of the plant, according to the culture experiments performed in southern Finland, *Rhodiola* spp. can be efficiently grown using organic plant culture methods (Tayade, 2015). For the natural stratification of winter, the seeds should be planted in the fall to produce seedlings. For approximately 1 year before transplant, these seedlings should be kept in pots due to the slow growth of the plants during the first 2–3 years. After 4 years of planting, the first root yield is harvested. The weight of the root and the yield of the roots of the plants depend to a great extent on the age (Fig. 9.1).

R. imbricata plant is propagated by two methods, namely sowing seeds and the cutting of rootstocks. In the general practice performed in tissue culture labs, 60%–65% of the germination of the seed is done in the fields, while in the case of the plantation of rootstocks, the scientists achieve a very high percentage of survival ranging between 85% and 90%. Therefore, better rates of dispersion have been achieved through the division of rootstocks. The plants of 3–5 years are considered ideal for the plantation of appropriate rootstocks (Sharma, 2016).

For micropropagation of *R. imbricata*, generally the proportionate concentration and combinations of Auxin, Cytokinin, and Gibberellic acid (like Indole-3-butyric acid, Kinetin, 6-Benzylaminopurine, GA_3, and Thidiazuron) are used. The most successful of these being MS media supplemented with BAP (1 mg/L) + IBA (2 mg/L) for multiple shoot culture (Pundir et al., 2019) (Figs. 9.2–9.4), while BAP (2 mg/L) + IBA (4 mg/L) was found good for root and shoot induction and growth (Sharma, 2016).

FIGURE 9.1

Rhodiola imbricata from Trans-Himalayan cold desert of Ladakh region, India (Tayade, 2015).

FIGURE 9.2

Micropropagated plants of *Rhodiola imbricata* growing in MS media supplemented with BAP (1 mg/L) + IBA (2 mg/L) (Pundir et al., 2019).

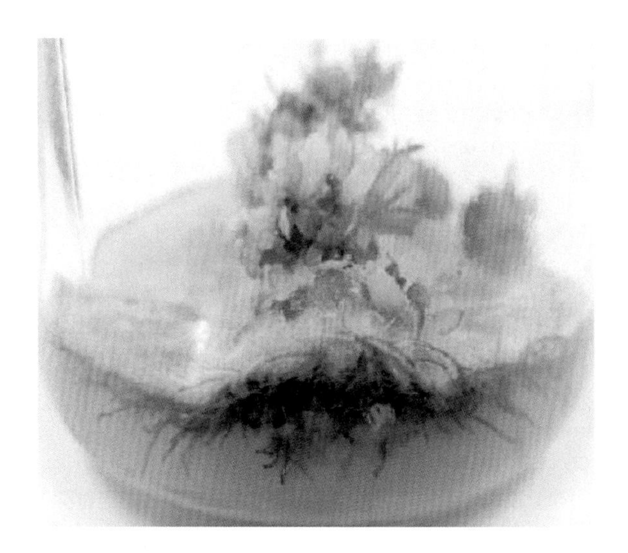

FIGURE 9.3

Roots induced in micropropagated plant growing in MS media supplemented with BAP (2 mg/L) + IBA (4 mg/L) (Sharma, 2016).

FIGURE 9.4

Callus of *Rhodiola imbricata* liquid MS media with BAP (1 mg/L) + IBA (2 mg/L) (Pundir et al., 2019).

Tasheva and Kosturkova (2010), developed efficient schemes for the regeneration and micropropagation of *R. rosea.* Zhao et al. (2012), developed a modified temporary immersion bioreactor with forced ventilation that reduces the rate of hyperhydration and improves the quality of outbreaks and multiplication rates in *R. crenulata.* The similar approach was used for *R. imbricata* where plant was micropropagated in liquid suspension culture containing MS media supplemented with BAP (1 mg/L) + IBA (2 mg/L) (Pundir et al., 2019). The idea behind using liquid suspension lies in short growth time and ease in multiplication of plant.

Zych et al. (2005), showed a successful encapsulation of differentiated callus and axillary buds in calcium alginate beads in *Rhodiola kirilowii.* These capsules were able to grow in shoots and seedlings in solid basal MS medium. This is yet to be explored in *R. imbricata* species.

Another significantly important approach used to seize the medical viability of plant in an efficient and immediate basis is using its callus culture for either directly attaining potential secondary metabolite or using callus as substitute of explants for micropropagation. For this purpose, MS media supplemented with TDZ (1 mg/L) is found most appropriate (Pundir et al., 2019) Fig. 9.5.

In recent years, a number of approaches have been used in order to increase content of pharmaceutically important secondary metabolites. Kapoor et al. (2018), studied the potential effect of light quality on biomass accumulation and production of industrially important secondary metabolites in callus cultures of *R. imbricata.* The results indicated blue light as promising light source for the enhanced production of flavonoids, secondary metabolite salidroside, and phenolics content in callus cultures.

FIGURE 9.5

Callus of plant growing in MS media supplemented with TDZ (1 mg/L) (Pundir et al., 2019).

In order to enhance the secondary metabolite production in *R. imbricata* cultures, various elicitors were tested. Jasmonic acid enhanced salidroside production, ascorbic acid content, total flavonoid content, and total phenolic in Callus Aggregate Suspension cultures. DPPH-scavenging activity and total antioxidant capacity were also enhanced upon Jasmonic acid treatment (Kapoor et al., 2019). Besides, treatment of shoot cultures for ultraviolet (UV) light for 30 min was found as a promising physical elicitor for providing growth along with optimized medium in *R. imbricata* shoot cultures (Pundir et al., 2019) Figs. 9.2 and 9.3.

9.6 Genetic diversity

In a study conducted by Gupta (2012), techniques like RAPD and ISSR markers were used to characterize and compare the genetic diversity in three collected populations of *R. imbricata*. The genetic closeness among the Khardung La and Chang La plants can be interpreted by the high rate of commonness in their individuality. The genetic similarity among these individuals is probably linked with their resemblance in their genomic and amplified region. AMOVA used for RAPD, ISSR, and RAPD + ISSR-combined markers were used for testing genetic variation. Results tabulated below showed significant ($P < .001$) genetic variation within population than among populations.

S. no.	Markers	Variation within population (%)	Variation among population (%)
1.	RAPD	56	44
2.	ISSR	78	22
3.	RAPD + ISSR	71	29

Similar reports have been made in ISSR studies of populations of *R. crenulata* (Lei et al., 2006), *Rhodiola chrysanthemifolia*, and *R. alsia* (Xia et al., 2005, 2007) which may be because of isolation of populations. Yan et al. (1999), reported *Rhodiola sachalinensis* had high genetic diversity within population at high altitude than that of growing at lower altitude.

As the plant grows in the Trans-Himalayan region of Ladakh, which is situated at more than 3000 m above mean sea level, there are numerous factors which can lead to partitioning of total genetic variation of a plant species which is different from the general pattern. Factors such as the low temperature, high UV radiations, the insufficient content of oxygen, short vegetation period (approximately 120 days) make the chance of seedling recruitment difficult and rare. Seed dispersal pattern and germination of *Rhodiola* at Ladakh and other Trans-Himalayan regions have irregular dispersal pattern which can be related to the effect of wind and other stressful habitat conditions (Gupta, 2012).

9.7 Omics

Secondary metabolites are often associated with having enormous curative properties. In *R. imbricata*, Salidroside—a Phenylethanol derivative and Rosavin—Phenylpropanoid—has been extensively studied for its various medicinal properties. The biosynthesis of salideroside and rosavin involves numerous important genes which have been illustrated below. The scientific community is making considerable conscious efforts in either stimulating or enhancing the functionality of these genes through number of bimolecular studies which have been highlighted in Figs. 9.6 and 9.7 defining the biosynthetic pathway of the metabolites.

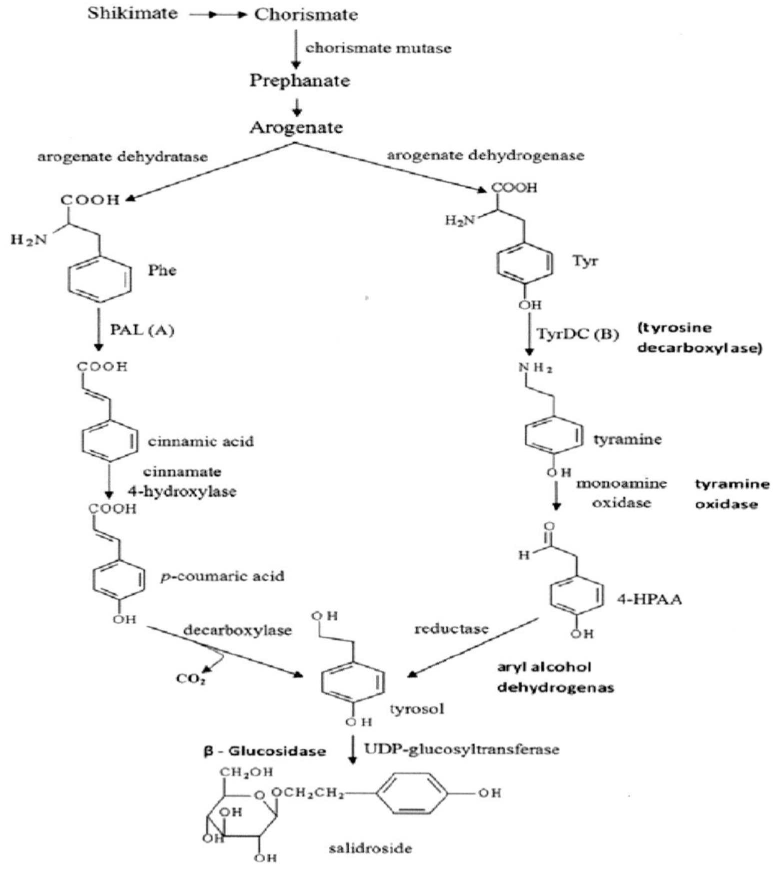

FIGURE 9.6

Propounded pathway responsible for biosynthesis of Salidroside (Ma et al., 2008).

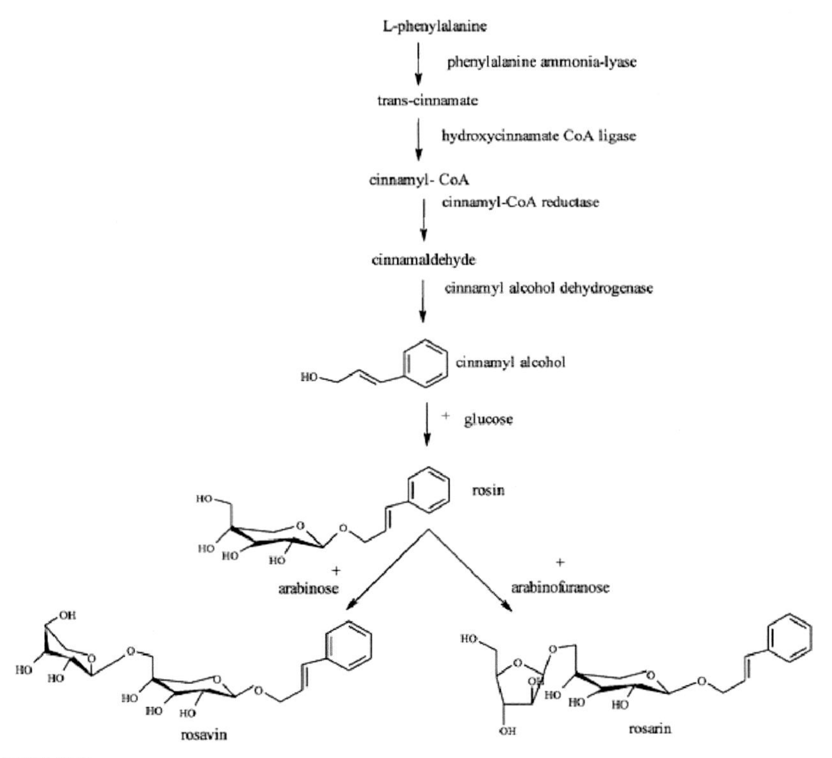

FIGURE 9.7

Pathway responsible for the biosynthesis of Rosin, Rosavin, and Rosarin (Grech-Baran et al., 2015).

9.7.1 Salidroside

• Tyramine precursor of tyrosol governed by **tyrosol glucosyltransferase (*TGase*)**	Xu et al. (1998), Yu et al. (2011), Zhang et al. (2011)
• Increased biomass and salidroside accumulation were achieved by medium supplementation with tyrosol than that in *Rhodiola sachalinensis*	Zhou et al. (2007)
• Increased salidroside by overexpression of the ***UGT73B* gene** in *R. sachalinensis*	Ma et al. (2007)
• **Uridine diphosphate (UDP)-glucosyltransferase cDNA (*UGT72B14*)** expression resulted in high salidroside production in vitro and in vivo in *R. sachalinensis*	Yu et al. (2011)
• **TyrDC gene** isolated from *Rhodiola rosea* had expression with the accumulation of salidroside	György et al. (2009)
• **RcTyrDC** enhanced the tyramine, tyrosol, and salidroside biosynthesis	Lan et al. (2013)

• **UDP-glucosyltransferase** • Anabolism reaction—transfers a glucosyl to tyrosol (the aglycon) and forms salidroside • Catabolism reaction—breakdown salidroside to glucose and tyrosol	Shi et al. (2007)

9.7.2 Rosavin

- Phenylalanine is a derivative of the shikimic—chorismic acid pathway and **Phenylalanine ammonia lyase** is the enzyme that directs the carbon atoms toward the phenylpropanoid metabolites biosynthesis.
- Both salidroside as well as the cinnamyl alcohol glycosides are the products of phenylpropanoid metabolism which is derived from phenylalanine.
- Cinnamyl-CoA ester is synthesized from cinnamic acid by **hydroxycinnamate: CoA ligase**.
- **Cinnamyl-CoA reductase** reduces cinnamyl-CoA ester to cinnamaldehyde.
- **Cinnamyl alcohol dehydrogenase** further reduces the cinnamaldehyde to cinnamyl alcohol.
- Rosin, which is the simplest glycoside of rose root, is formed by one glucose transfer.
- Rosavin is formed from rosin by the addition of an arabinose, and rosarin is formed from rosin by the addition of an arabinofuranose (Grech-Baran et al., 2015).

9.8 Conclusion and future prospects

The recent years observed the peaked interests of scientific community and the governmental agencies in conserving the traditional pharmacopoeia of the Trans-Himalayan regions. The myriad of assorted botanical flora of the region has been extensively researched with respect to their prophylactic and therapeutic potential, specially plants like *R. imbricata*. The species has been explored for its hidden treasures in solving multiple human diseases, for which it has also been referred as "magical concoction" or "Sanjeevani" by a number of renowned scientists over and over again. *R.a imbricata* has unprecedented capabilities in curing mankind of diseases such as Alzheimer's, cancer, renal cysts, edema of limb, burns, etc., along with the compounds offering immunostimulant, antifatigue, antidepressant, antihypoxia, antioxidant, antiradioactive, and hepatoprotective properties. Researchers are not only striving toward achieving best for pharmaceutical sector but also ensuring use of well-equipped tissue culture labs and high-end technologies for sustainable use of the vegetation in order to conserve the endangered species. *Rhodiola* holds strong foundation in food and lifestyle of local communities as an easy answer to all health ailments and a part of many delicacies. Hence, in order to avoid its overexploitation or overharvesting by the farmers, the need of the hour is to ascertain effective in situ conservation of the plant. Moreover, immediate aid and assistance through collaborative projects between indigenous people and associated experts of the national and international agencies for attaining choicest results are also desired. All in all, balance is the key for achieving best out of this plant while corroborating its presence in the wild and natural ecosystem in the future to come.

References

Arora, R., Chawla, R., Sagar, R., Prasad, J., Singh, S., Kumar, R., Sharma, A., Singh, S., Sharma, R.K., 2005. Evaluation of radioprotective activities of *Rhodiola imbricata* Edgew - a high altitude plant. Mol. Cell. Biochem. 273 (1–2), 209–223.

Ballabh, B., Chaurasia, O.P., 2007. Traditional medicinal plants of cold desert Ladakh-used in treatment of cold, cough and fever. J. Ethnopharmacol. 112 (2), 341–349.

Brown, R., Gerbarg, P., Ramazanov, Z., 2002. *Rhodiola rosea*: a phytomedicinal overview. HerbalGram 56, 40–52.

Bykov, V.A., Zapesochnaya, G.G., Kurkin, V.A., 1999. Traditional and biotechnological aspects of obtaining medicinal preparations from *Rhodiola rosea* L. (a review). Pharm. Chem. J. 33, 29–40. https://doi.org/10.1007/BF02508414.

Calcabrini, C., De Bellis, R., Mancini, U., Cucchiarini, L., Potenza, L., De Sanctis, R., Patrone, V., Scesa, C., Dachà, M., 2010. *Rhodiola rosea* ability to enrich cellular antioxidant defences of cultured human keratinocytes. Arch. Dermatol. Res. 302 (3), 191–200.

Chaurasia, O.P., Gurmet, A., 2006. Checklist on medicinal and aromatic plants of tran-Himalayan cold deserts. Herbal Drugs: A Twenty First Century Perspective, 1st ed. JPB, pp. 10–20. 19.

Chen, T.S., Liou, S.Y., Chang, Y.L., 2008. Antioxidant evaluation of three adaptogen extracts. Am. J. Chin. Med. 36 (6), 1209–1217.

Chen, Q.G., Zeng, Y.S., Qu, Z.Q., Tang, J.Y., Qin, Y.J., Chung, P., Wong, R., Hägg, U., 2009. The effects of *Rhodiola rosea* extract on 5-HT level, cell proliferation and quantity of neurons at cerebral hippocampus of depressive rats. Phytomedicine 16 (9), 830–838.

Chiang, H.M., Chen, H.C., Wu, C.S., Wu, P.Y., Wen, K.C., 2015. *Rhodiola* plants: chemistry and biological activity. J. Food Drug Anal. 23 (3), 359–369.

Darbinyan, V., Kteyan, A., Panossian, A., Gabrielian, E., Wikman, G., Wagner, H., 2000. *Rhodiola rosea* in stress induced fatigue - a double blind cross-over study of a standardized extract SHR-5 with a repeated low-dose regimen on the mental performance of healthy physicians during night duty. Phytomedicine 7 (5), 365–371.

De Bock, K., Eijnde, B.O., Ramaekers, M., Hespel, P., 2004. Acute *Rhodiola rosea* intake can improve endurance exercise performance. Int. J. Sport Nutr. Exerc. Metabol. 14 (3), 298–307.

Furmanowa, M., Skopińska-Rozewska, E., Rogala, E., Hartwich, M., 1998. *Rhodiola rosea* in vitro culture - phytochemical analysis and antioxidant action. Acta Soc. Bot. Pol. 67 (1), 69–73.

Goel, H.C., Bala, M., Prasad, J., Singh, S., Agrawala, P.K., Swahney, R.C., 2006. Radioprotection by *Rhodiola imbricata* in mice against whole-body lethal irradiation. J. Med. Food 9 (2), 154–160.

Grace, M.H., Yousef, G.G., Kurmukov, A.G., Raskin, I., Lila, M.A., 2009. Phytochemical characterization of an adaptogenic preparation from *Rhodiola heterodonta*. Nat. Prod. Commun. 4 (8), 1053–1058.

Grech-Baran, M., Sykłowska-Baranek, K., Pietrosiuk, A., 2015. Biotechnological approaches to enhance salidroside, rosin and its derivatives production in selected *Rhodiola* spp. in vitro cultures. Phytochem. Rev. 14, 657–674.

Gupta, S., 2012. Genetic diversity among natural populations of *Rhodiola imbricata* Edgew. from trans- Himalayan cold arid desert using random amplified polymorphic DNA (RAPD) and inter simple sequence repeat (ISSR) markers. J. Med. Plants Res. 6 (3), 405–415.

Gupta, A., Kumar, R., Upadhyay, N.K., Pal, K., Kumar, R., Sawhney, R.C., 2007. Effects of *Rhodiola imbricata* on dermal wound healing. Planta Med. 73 (8), 774–777.

Gupta, V., Saggu, S., Tulsawani, R.K., Sawhney, R.C., Kumar, R., 2008. A dose dependent adaptogenic and safety evaluation of *Rhodiola imbricata* Edgew, a high altitude rhizome. Food Chem. Toxicol. 46 (5), 1645–1652.

Gupta, V., Lahiri, S.S., Sultana, S., Kumar, R., 2009. Mechanism of action of *Rhodiola imbricata* Edgew during exposure to cold, hypoxia and restraint (C-H-R) stress induced hypothermia and post stress recovery in rats. Food Chem. Toxicol. 47 (6), 1239–1245.

György, Z., Jaakola, L., Neubauer, P., Hohtola, A., 2009. Isolation and genotype-dependent, organ-specific expression analysis of a *Rhodiola rosea* cDNA encoding tyrosine decarboxylase. J. Plant Physiol. 166 (14), 1581–1586.

Hu, X., Lin, S., Yu, D., Qiu, S., Zhang, X., Mei, R., 2010. A preliminary study: the anti-proliferation effect of salidroside on different human cancer cell lines. Cell Biol. Toxicol. 26 (6), 499–507.

Jafari, M., Felgner, J.S., Bussel, I.I., Hutchili, T., Khodayari, B., Rose, M.R., Vince-Cruz, C., Mueller, L.D., 2007. *Rhodiola*: a promising anti-aging Chinese herb. Rejuvenation Res. 10 (4), 587–602.

Kapoor, S., Raghuvanshi, R., Bhardwaj, P., Sood, H., Saxena, S., Chaurasia, O.P., 2018. Influence of light quality on growth, secondary metabolites production and antioxidant activity in callus culture of *Rhodiola imbricata* Edgew. J. Photochem. Photobiol. B Biol. 183, 258–265.

Kapoor, S., Sharma, A., Bhardwaj, P., Sood, H., Saxena, S., Chaurasia, O.P., 2019. Enhanced production of phenolic compounds in compact callus aggregate suspension cultures of *Rhodiola imbricata* Edgew. Appl. Biochem. Biotechnol. 187 (3), 817–837.

Khanum, F., Bawa, A.S., Singh, B., 2005. *Rhodiola rosea*: a versatile adaptogen. Compr. Rev. Food Sci. Food Saf. 4 (3), 55–62.

Lan, X., Chang, K., Zeng, L., Liu, X., Qiu, F., Zheng, W., Quan, H., Liao, Z., Chen, M., Huang, W., Liu, W., Wang, Q., 2013. Engineering salidroside biosynthetic pathway in hairy root cultures of *Rhodiola crenulata* based on metabolic characterization of tyrosine decarboxylase. PLoS One 8 (10), e75459.

Lei, Y., Nan, P., Tsering, T., Bai, Z., Tian, C., Zhong, Y., 2003. Chemical composition of the essential oils of two *Rhodiola* species from Tibet. Z. Naturforsch. C Biosci. 58 (3–4), 161–164.

Lei, Y., Gao, H., Tsering, T., Shi, S., Zhong, Y., 2006. Determination of genetic variation in *Rhodiola crenulata* from the Hengduan Mountains region, China using inter-simple sequence repeats. Genet. Mol. Biol. 29 (2), 339–344.

Li, X., Erden, O., Li, L., Ye, Q., Wilson, A., Du, W., 2014. Binding to WGR domain by salidroside activates parp1 and protects hematopoietic stem cells from oxidative stress. Antioxidants Redox Signal. 20 (12), 1853–1865.

Liu, Z., Liu, Y., Liu, C., Song, Z., Li, Q., Zha, Q., Lu, C., Wang, C., Ning, Z., Zhang, Y., Tian, C., Lu, A., 2013. The chemotaxonomic classification of *Rhodiola* plants and its correlation with morphological characteristics and genetic taxonomy. Chem. Cent. J. 7 (1), 118.

Ma, L.Q., Liu, B.Y., Gao, D.Y., Pang, X.B., Lü, S.Y., Yu, H.S., Wang, H., Yan, F., Li, Z.Q., Li, Y.F., Ye, H.C., 2007. Molecular cloning and overexpression of a novel UDP-glucosyltransferase elevating salidroside levels in *Rhodiola sachalinensis*. Plant Cell Rep. 26 (7), 989–999.

Ma, L.Q., Gao, D.Y., Wang, Y.N., Wang, H.H., Zhang, J.X., Pang, X.B., Hu, T.S., Lü, S.Y., Li, G.F., Ye, H.C., Li, Y.F., Wang, H., 2008. Effects of overexpression of endogenous phenylalanine ammonia-lyase (PALrs1) on accumulation of salidroside in *Rhodiola sachalinensis*. Plant Biol. 10 (3), 323–333.

Mao, Y., Li, Y., Yao, N., 2007. Simultaneous determination of salidroside and tyrosol in extracts of *Rhodiola L.* by microwave assisted extraction and high-performance liquid chromatography. J. Pharmaceut. Biomed. Anal. 45 (3), 510–515.

Mao, G.X., Deng, H.B., Yuan, L.G., Li, D.D., Li, Y.Y.Y., Wang, Z., 2010a. Protective role of salidroside against aging in a mouse model induced by D-galactose. Biomed. Environ. Sci. 23 (2), 161–166.

Mao, G.X., Wang, Y., Qiu, Q., Deng, H.B., Yuan, L.G., Li, R.G., Song, D.Q., Li, Y., yang, Y., Li, D.D., Wang, Z., 2010b. Salidroside protects human fibroblast cells from premature senescence induced by H_2O_2 partly through modulating oxidative status. Mech. Ageing Dev. 131 (11–12), 723–731.

Mishra, K.P., Chanda, S., Shukla, K., Ganju, L., 2010. Adjuvant effect of aqueous extract of *Rhodiola imbricata* rhizome on the immune responses to tetanus toxoid and ovalbumin in rats. Immunopharmacol. Immunotoxicol. 32 (1), 141–146.

Olsson, E.M.G., Von Schéele, B., Panossian, A.G., 2009. A randomised, double-blind, placebo-controlled, parallel-group study of the standardised extract SHR-5 of the roots of *Rhodiola rosea* in the treatment of subjects with stress-related fatigue. Planta Med. 75 (2), 105–112.

Peng, L.H., Liu, S., Xu, S.Y., Chen, L., Shan, Y.H., Wei, W., Liang, W.Q., Gao, J.Q., 2013. Inhibitory effects of salidroside and paeonol on tyrosinase activity and melanin synthesis in mouse B16F10 melanoma cells and ultraviolet B-induced pigmentation in Guinea pig skin. Phytomedicine 20 (12), 1082–1087.

Platikanov, S., Evstatieva, L., 2008. Introduction of wild golden root (*Rhodiola rosea* L.) as a potential economic crop in Bulgaria. Econ. Bot. 62 (4), 621–627.

Pundir, A., Kad, A., Sood, H., 2019. Elicitation of salidroside under tissue culture conditions in the trans- Himalayan plant *Rhodiola imbricata*. WSEAS Trans. Biol. Biomed. 16, 75–89.

Qian, E.W., Ge, D.T., Kong, S.K., 2012. Salidroside protects human erythrocytes against hydrogen peroxide-induced apoptosis. J. Nat. Prod. 75 (4), 531–537.

Rohloff, J., 2002. Volatiles from rhizomes of *Rhodiola rosea* L. Phytochemistry 59 (6), 655–661.

Schriner, S.E., Avanesian, A., Liu, Y., Luesch, H., Jafari, M., 2009. Protection of human cultured cells against oxidative stress by *Rhodiola rosea* without activation of antioxidant defenses. Free Radic. Biol. Med. 47 (5), 577–584.

Sharma, S., 2016. Effect of temperature on in vitro organogenesis of *Rhodiola imbricata* Edgew. – a medicinal herb. World J. Pharm. Pharmaceut. Sci. 5 (12), 1228–1243.

Shi, L.L., Wang, L., Zhang, Y.X., Liu, Y.J., 2007. Approaches to biosynthesis of salidroside and its key metabolic enzymes. For. Stud. 9 (4), 295–299.

Singh, B., Chaurasia, O.P., Jadhav, K.L., 1996. An ethnobotanical study of Indus Valley (Ladakh). J. Econ. Taxon. Bot. 12, 92–101.

Spasov, A.A., Wikman, G.K., Mandrikov, V.B., Mironova, I.A., Neumoin, V.V., 2000. A double-blind, placebo-controlled pilot study of the stimulating and adaptogenic effect of *Rhodiola rosea* SHR-5 extract on the fatigue of students caused by stress during an examination period with a repeated low-dose regimen. Phytomedicine 7 (2), 85–89.

Sun, L., Isaak, C.K., Zhou, Y., Petkau, J.C., Karmin, O., Liu, Y., Siow, Y.L., 2012. Salidroside and tyrosol from *Rhodiola* protect H9c2 cells from ischemia/reperfusion-induced apoptosis. Life Sci. 91 (5–6), 151–158.

Tasheva, K., Kosturkova, G., 2010. Bulgarian golden root in vitro cultures for micropropagation and reintroduction. Cent. Eur. J. Biol. 5 (6), 853–863.

Tayade, A.B., 2015. Phytochemical Characterization and Pharmacological Evaluation of *Rhodiola imbricata* Edgew. Root from Trans- Himalayan Cold Desert Region of Ladakh, India (Ph.D. thesis) Jaypee Univ. Inf. Technol. Solan, India.

Tayade, A.B., Dhar, P., Kumar, J., Sharma, M., Chaurasia, O.P., Srivastava, R.B., 2017. Trans-Himalayan *Rhodiola imbricata* Edgew. root: a novel source of dietary amino acids, fatty acids and minerals. J. Food Sci. Technol. 54 (2), 359–367.

van Diermen, D., Marston, A., Bravo, J., Reist, M., Carrupt, P.A., Hostettmann, K., 2009. Monoamine oxidase inhibition by *Rhodiola rosea* L. roots. J. Ethnopharmacol. 122 (2), 397–401.

Wagner, H., Nörr, H., Winterhoff, H., 1994. Plant adaptogens. Phytomedicine 1 (1), 63–76.

Weglarz, Z., Przybył, J.L., Geszprych, A., 2008. Roseroot (*Rhodiola rosea* L.): effect of internal and external factors on accumulation of biologically active compounds. In: Bioact. Mol. Med. Plants. Springer Berlin Heidelberg, pp. 297–315.

Wójcik, R., Siwicki, A.K., Sommer, E., Wasiutyński, A., Furmanowa, M., Malinowski, M., Mazurkiewicz, M., Skopńska-Rózewska, E., 2008. The effect of *Rhodiola quadrifida* extracts on cellular immunity in mice and rats. Pol. J. Vet. Sci. 11 (2), 105–111.

Xia, T., Chen, S., Chen, S., Ge, X., 2005. Genetic variation within and among populations of *Rhodiola alsia* (Crassulaceae) native to the Tibetan plateau as detected by ISSR markers. Biochem. Genet. 43 (3–4), 87–101.

Xia, T., Chen, S., Chen, S., Zhang, D., Zhang, D., Gao, Q., Ge, X., 2007. ISSR analysis of genetic diversity of the Qinghai-Tibet Plateau endemic *Rhodiola chrysanthemifolia* (Crassulaceae). Biochem. Systemat. Ecol. 35 (4), 209–214.

Xu, J.F., Su, Z.G., Feng, P.S., 1998. Activity of tyrosol glucosyltransferase and improved salidroside production through biotransformation of tyrosol in *Rhodiola sachalinensis* cell cultures. J. Biotechnol. 61 (1), 69–73.

Yan, T., Yan, X., Zu, Y., 1999. A primary discussion on the adaptive mechanism at different altitudinal levels of *Rhodiola sachalinensis* population. Bull. Bot. Res. 19, 201–206.

Yang, S.J., Yu, H.Y., Kang, D.Y., Ma, Z.Q., Qu, R., Fu, Q., Ma, S.P., 2014. Antidepressant-like effects of salidroside on olfactory bulbectomy-induced pro-inflammatory cytokine production and hyperactivity of HPA axis in rats. Pharmacol. Biochem. Behav. 124, 451–457.

Yu, S., Liu, M., Gu, X., Ding, F., 2008. Neuroprotective effects of salidroside in the PC12 cell model exposed to hypoglycemia and serum limitation. Cell. Mol. Neurobiol. 28 (8), 1067–1078.

Yu, H.S., Ma, L.Q., Zhang, J.X., Shi, G.L., Hu, Y.H., Wang, Y.N., 2011. Characterization of glycosyltransferases responsible for salidroside biosynthesis in *Rhodiola sachalinensis*. Phytochemistry 72 (9), 862–870.

Zhang, L., Yu, H., Sun, Y., Lin, X., Chen, B., Tan, C., Cao, G., Wang, Z., 2007. Protective effects of salidroside on hydrogen peroxide-induced apoptosis in SH-SY5Y human neuroblastoma cells. Eur. J. Pharmacol. 564 (1–3), 18–25.

Zhang, J.X., Ma, L.Q., Yu, H.S., Zhang, H., Wang, H.T., Qin, Y.F., Shi, G.L., Wang, Y.N., 2011. A tyrosine decarboxylase catalyzes the initial reaction of the salidroside biosynthesis pathway in *Rhodiola sachalinensis*. Plant Cell Rep. 30 (8), 1443–1453.

Zhao, Y., Sun, W., Wang, Y., Saxena, P.K., Liu, C.Z., 2012. Improved mass multiplication of *Rhodiola crenulata* shoots using temporary immersion bioreactor with forced ventilation. Appl. Biochem. Biotechnol. 166 (6), 1480–1490.

Zhou, X., Wu, Y., Wang, X., Liu, B., Xu, H., 2007. Salidroside production by hairy roots of *Rhodiola sachalinensis* obtained after transformation with *Agrobacterium rhizogenes*. Biol. Pharm. Bull. 30 (3), 439–442.

Zych, M., Furmanowa, M., Krajewska-Patan, A., Lowicka, A., Dreger, M., Mendlewska, S., 2005. Micropropagation of *Rhodiola kirilowii* plants using encapsulated axillary buds and callus. Acta Biologica Cracoviensia Series Botanica, 2nd ed. 47, pp. 83–87.

CHAPTER

Saussurea lappa

10

Ira Vashisht

Crop Genetics & Informatics Group, School of Computational and Integrative Sciences (SCIS), Jawaharlal Nehru University, New Delhi, India

10.1 Introduction

Saussurea lappa (syn. *Saussurea costus*) C.B. Clarke is an endangered medicinal plant, best known for its antiinflammatory, anticancerous, antiulcerogenic, and hepatoprotective properties (Butola and Samant, 2010; Gautam and Asrani, 2018; Zahara et al., 2014). Traditionally known as "kuth" or "kushtha," this high-altitude perennial herb forms an integral part of Ayurvedic, Unani, Siddha, and Tibetan (Amchi) medicine (Butola and Samant, 2010). Mention of *S. lappa* goes back to writings from Mesopotamia where evidence of medicinal herbs has been documented by Indus Valley people (Shah, 2019). R. Campbell Thompson was successful in deciphering the multiple names of plants, drugs, and minerals used in Assyrian botany (Mesopotamia). A clear mention of *S. lappa* import from India has been observed for treatment of jaundice (Thompson, 1949). Its ancient role in *panduroga* (jaundice) has been shown to echo in India as well (Dwivedi, 1963). It is a member of the large Asteraceae or Compositae family, comprising of approximately 410 globally distributed species which are mostly native to cold and temperate regions of Asia, Europe, and North America. Highest diversity has been observed to inhabit alpine regions of Central Asia and Himalayas (Butola and Samant, 2010).

S. lappa has become endangered due to overexploitation of its natural habitat for diverse medicinal and commercial purposes. It is one of main ingredients in closely 71 drug formulations documented in *The Handbook of Traditional Tibetan Drugs* (Tsarong, 1986), and in several popular polyherbal drug formulations like *Chandra Kalka, Ashtamangal Ghrita*, and *Sharkaradi Kalka* (Chamara et al., 2018; Singh, 2019). Bitter roots having a sweet and strong aromatic odor are the most valued part of *S. lappa* plants which are harvested for extraction of bioactive constituents. ***Costus oil*** is the main compound extracted from roots which is constituted by a number of secondary metabolites which have been prevalently used in medicine and perfumery (Table 10.1) (Butola and Samant, 2010). Most of the research in *S. lappa* is focused on investigation of bioactive constituents and corresponding therapeutic applications. A few scattered reports exist on attempts at large-scale cultivation and in vitro propagation of *S. lappa* for alleviating the endangered status of the

Himalayan Medicinal Plants. https://doi.org/10.1016/B978-0-12-823151-7.00012-X

Table 10.1 Constituents of *Saussurea lappa* essential oil or costus oil (Zahara et al., 2014; Abdelwahab et al., 2019; Maurer and Grieder, 1997; Dhillon et al., 1987).

S. No.	Compound	S. No.	Compound
1	Dehydrocostus lactone	20	(−)-α-costol
2	Costunolide	21	(+)-γ-costol
3	8-cedren-13-ol	22	(−)-elema-1,3,11 (13)-trien-12-ol
4	α-curcumene	23	(−)-α-costal
5	β-costol	24	(+)-γ-costal
6	δ-elemene	25	(−)-Elema-1,3,11(13)-trien-12-al
7	α-selinene	26	(−)-(E)-trans-bergamota-2,12-dien-14-al
8	β-selinene	27	(−)-ar-curcumene
9	α-costol	28	(−)-caryophyllene oxide
10	4-terpinol	29	12-methoxy dihydrodehydro costus lactone
11	Elemol	30	Eudesma-5,11(13)-dien-8,12-olide
12	α-ionone	31	Phenanthrenone
13	β-elemene	32	9,12-octadecadienoic acid (Z,Z)
14	(−)-γ-elemene	33	Cyclohexane
15	p-cymene	34	Germacra-1(10),4,11(13)-trien-12-oic acid,6à-hydroxy-,ç-lactone, (E,E)
16	2-β-pinene	35	Androstan-17-one, 3-ethyl-3-hydroxy-, (5à)
17	(−)-α-selinene	36	Bicyclo[10.1.0]tridec-1-ene
18	(+)-selina-4,11-diene	37	Naphthalene
19	(−)-α-trans-bergamotene	38	4a,8-dimethyl-2-(prop-1-en-2-yl)-1,2,3,4,4a,5,6,7-octahydronaphthalene

plant. Low germination potential has limited the application of breeding approaches in this herb; however, promising OMICS strategies are being explored for understanding molecular networks underlying biosynthesis of industrially valuable secondary metabolites. This chapter provides an elaborate account of botanical, biochemical, and therapeutic characteristics of *S. lappa* along with a discussion on challenges and efforts pertaining to conservation status and techniques which can open avenues for yield improvement and engineering of valuable secondary metabolites.

10.2 Botanical identification and classification

First botanical nomenclature for *S. lappa* was provided by Ainsile (Ainsile, 1813) who attempted to identify *koostum,* which was known in various vernaculars as koot (Sanskrit), kust (Arabian), kostum (Tamil), and Sippuday (Malayalam) among many others. Later, Royle (1839) discovered roots of this plant in Indian "bazar" and went on to call it "koot" exported from Cashmere where it thrives on mountains in areas surrounding Cashmere. Dr. Hugh Falconer encountered the plant in Drass, Kashmir, during his expedition in 1834–38 and corresponded with Royale regarding the medicinal plant, while naming it as a new genus, *Costia* (Shah, 2019). Later in 1845, Falconer went on to publish its botanical name as *Aucklandia costus* Falc. which was well accepted for quite a while. Further, Decne changed its genus to *Saussurea* and named it as *S. lappa* (Decne) Sch. Bip 1846 for the first time. Finally, in 1964, its name was accepted as *Saussurea costus* (Falc) Lipch. However, International Rules of Botanical Nomenclature have retained its multiple names as *Aplotaxis lappa* (Decne) (1843), *A. costus* Falc. (1845), *Aucklandia lappa* Decne (1875), and *Theodorea costus* Kuntze (1891) (The Plant List, 2013). The current widely accepted nomenclature is *S. lappa* ((Decne.) C. B. Clarke.) as described by the famous botanist Charles Baron Clarke (1876) (Waly, 2009).

10.3 Botany of *Saussurea lappa*

S. lappa is a tall and erect perennial herb having a stout and upright stem which is 1–2 m high (Pandey et al., 2007) (Fig. 10.1). Leaves are lobate, auricled at base, and

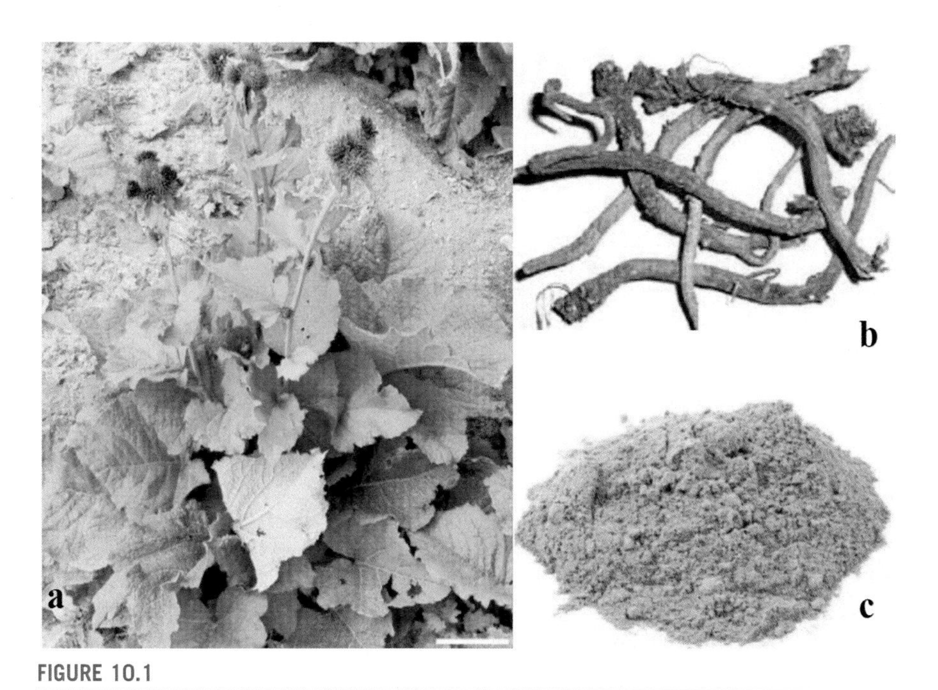

FIGURE 10.1

(a) *Saussurea lappa* plant in its natural habitat (Ladakh region). Scale bar = 15 cm (Warghat et al., 2016). (b, c) Dried roots and root powder of *S. lappa* (Amara et al., 2017).

irregularly toothed which are scaberulous above, while glabrate from beneath. Basal leaves are very long having winged stalks, whereas upper leaves are smaller, shortly petioled or subsessile having small lobes which almost clasp the stem (Pandey et al., 2007; Amara et al., 2017). The plant contains both cauline and radical leaves where former are small, irregularly toothed, and pubescent, while latter are relatively quite large and triangular having a long-winged petiole (Shah, 2019). While in the field, another plant, *Arctium lappa* has often been mistaken for *S. lappa* owing to exhibition of similar large leaves (Shah, 2019). The plant is headed by bluish-purple to black stalkless flowers which are clustered together in either terminal or axil part of leaves. It has many ovate-lanceolate and pointed involucral bracts and receptacle bristles are quite long (Pandey et al., 2007). Anther tails are fimbriate and corolla is about 2 cm long. Achnes are small, curved, and compressed with narrowed tip and pappus hair are brown and feathery (Pandey et al., 2007; Shah, 2019), while fruits are cupped, compressed, and curved (Pandey et al., 2007). Most valuable part of the plant is the root which is 15–30 cm thick and 40–60 cm long (Hajra et al., 1995) having a camphoraceous aroma (Shah, 2019) (Fig. 10.1). Secondary roots are often tubular and usually wrinkled and ridged (Shah, 2019). Roots are diced into small pieces and dried which appear muddy gray to creamy in color and possess a bitter taste. These dried roots of *S. lappa* are used as the crude drug available in the market (Fig. 10.1).

10.4 Origin and geographical distribution

S. lappa is a prominent member of the extensive Himalayan flora, found at elevation of 2700–4000 m amsl (Kaul, 1997; Nandkarni, 1954). Globally, the herb inhabits the cool arctic and temperate regions of Asia, North America, and Europe (Shah. 2006; Hajra et al., 1995). It has been an indigenous constituent of traditional medicine in India, China, and Pakistan (Shah, 2019). In India, *S. lappa* has a natural habitat in Jammu and Kashmir, Himachal Pradesh, and Uttarakhand (Kaul, 1997; Nandkarni 1954). Kashmir and adjoining areas harbor the majority of its native habitat where it can be found in Jhelum, Sonmarga, Drass, Kistwar, Zanskar valley (Ladakh), Chenab valley, and Kishenganga (Shah, 2019). Further, to quench the ever-increasing industrial demands, it is widely cultivated in Kashmir, Uttar Pradesh, and Tamil Nadu (Shah, 1982; Kamalpreet et al., 2019). In Pakistan, *S. lappa* can be observed on the moist, open hill slopes of the Himalayan region in Leepa, Neelam, and Kaghan valleys, while a scattered presence can be seen in valleys of Siran and Gurez along with Poonch and Bagh districts (Shah, 2019).

10.5 Biochemical/analytical properties

Roots are the primary source of the tremendous therapeutic potential of *S. lappa*. Studies on the biochemical properties of the herb date back to 1950s which have led to discovery of various active ingredients like terpenes, sesquiterpenoids, alkaloids, flavonoids, lignins, steroids, glycosides, and anthraquinones (Wei et al., 2014; Singh et al., 2017). Hydrodistillation of *S. lappa* root essential oil showed higher sesquiterpenoid content (79.80%) relative to monoterpenoids (13.25%) (Liu et al., 2012).

Principal constituents of *S. lappa* essential oil were dehydrocostus lactone and costunolide (Singh et al., 2017) (Table 10.1). Costus oil/*S. lappa* essential oil is the commercially valued essential oil extracted from *S. lappa* roots which is constituted by multiple bioactive compounds. The proportion of these constituents generally varies among plants which might be attributed to factors like phenophases, ecotypes, chemotype, genotype, and environmental variations like temperature, relative humidity, photoperiod, and irradiance (Marotti et al., 1994).

An account of various categories of chemical constituents isolated from *S. lappa* is provided in the sections below.

10.5.1 Terpenes

Monoterpenes isolated from roots include Phellandrene, Thymol, Anethole, Estragole, Citronellyl propionate, α-Pinene, β-Pinene, α-Thujene, Camphor, Camphene, Sabinene, Myrcene, Limonene, p-Cymene, γ-Terpinene, 1,8 Cineol, Menthone, α-Terpinolene, Citronellal, Terpinen-4-ol, Linalool, Cryptone, α-Terpineol, and Ocimene (Chang and Kim, 2008; Gwari et al., 2013).

S. lappa is an abundant source of sesquiterpenes which are roughly categorized into three groups based on the carbocyclic skeleton: Guaiane, Eudesmane, and Germacrane (Singh et al., 2017). These have been known to biosynthesize sequentially; however, due to instability of germacrene, half of the total sesquiterpenes are observed in form of guaianes and 40% turn up as eudesmanes, while germacrene constitutes the remaining fraction (Singh et al., 2017). Guaianes reportedly derived from *S. lappa* include Dehydrocostus lactone (Govindan and Bhattacharaya, 1977), Zaluzanin C, Isozaluzanin, 11β, 13-Dihydro-3-epizaluzanin C (Chhabra et al., 1998; Kalsi et al., 1983), Lappalone (Sun et al., 2003), Cynaropicrin (Cho et al., 1998), Saussureamine B, 12-Methoxy-dihydrodehyrocostus lactone (Dhillon et al., 1987), Saussureamine C (Yoshikawa et al., 1993), Dihydroglucoaluzanin C, Mokko lactone, 11,13, Saussurealdehyde, Isodehydrocostuslactone (Kalsi et al., 1983), 11,13-Epoxydehydrocostus lactone, Isodehydrocostuslactone-15-aldehyde (Kumar et al., 1995), 11,13-Epoxyisozaluzanin C (Chhabra et al., 1997), 4β-Methoxy-dehydrocostus lactone, 11, 13-Epoxy-3-ketodehyrocostus lactone(Chhabra et al., 1998), Lappadilactone, 15-Hydroxydehydrocostus lactone, etc. Eudesmanes type sesquiterpenoids include 13-Sulfodihydrosantamarine (Yin et al., 2005), Saussureal, Saussureamine D (Yoshikawa et al., 1993), Saussureamine E (Yoshikawa et al., 1993), 11β, 13-Dihydroreyosin, 13-Sulfodihydroreyosin, Reynosin (Cho et al., 1998), 1β, 6α-Dihydroxycostic acid ethyl ester (Sun et al., 2003), β-Costic acid (Govindan and Bhattacharaya, 1977), α-Cyclocostunolide, Isoalantolactone, Alantolactone, β-Cyclocostunolide (Govindan and Bhattacharaya, 1977), Hydroxyendesin-11(13)-en-12-al, Magnolialide, 4α-Hydroxy-4β-Methyldihydrocostol, Isocostic acid, 4β-α-Costol, Santamarine (Cho et al., 1998), Colartin and Arbusculin

A, etc. Some Germacrane type sesquiterpene lactones isolated from *S. lappa* are Dihydrocostunolide, Saussureamine A (Yoshikawa et al., 1993), Costunolide (Kim et al., 1999), Costunolide 15-o-β-D-glucopyranoside and 12-Methoxy dihydrocostunolide. α-Amyrin (Yang et al., 1997a,b), 3-β-Acetoxy-9(11)-baccharene, and α-Amyrin eicosanoate are a few triterpenes which have been isolated from *S. lappa* (Robinson et al., 2010).

10.5.2 Flavonoids

Multiple flavonoids isolated from *S. lappa* roots include Luteolin-7-O-β-D-glucoside, Apigenin-7-O-β-D-glucoside (Alaagib and Ayoub, 2015), and Rutin which have one glucoside substituent. Flavonoids with large substituents like three glucosides are relatively quite rare. Following acylated flavonoids have been reported from *S. lappa* roots by Rao et al. (2007): Kaempferol 3-O-β-D-glucopyranosyl-(1 → 4)-α-L-rhamnopyranosyl-(1 → 6)-β-D-galactopyranoside 7-O-(6‴ -O-acetyl-β-Dgluco-pyranosyl-(1 → 3)-[α-L-rhamnopyranosyl-(1 → 2)]-β-D-glucopyranoside, 3′[(3R)-3-Acetoxy-5,5-dimethylcyclopent-1-en-1-yl]-4′-Omethylscutellarein 7-O-(β-O-6‴-O-acetylglucopyranosyl-(1 3)-[α-L-rhamnopyranosyl -(1 → 2)]-β-D-glucopyranoside and Kaempferol 3-O-β-D-glucopyranosyl-(1 → 2)-β-D-(6 α′-O-caffeoyl) galactopyranoside 7-O-(β-D-6‴-O-acetyl-β-D-glucopyranosyl-(1 → 3)-[β-L-rhamnopyranosyl-(1 → 2)]-β-D-glucopyranoside, and Kaempferol 3-O-α-L-(2α′, 3α’-(E)-di-p-coumaroyl) rhamnoside 7-O-(6‴-O-acetyl-β-D glucopyranosyl-(1 → 3)-[α-Lrhamnopyranosyl-(1 → 2)]-β-D-glucopyranoside. Rao et al. (2007) reported four new flavonoids (KSR1-4) from ethanolic extract of powdered *S. lappa* roots.

10.5.3 Other constituents

A few anthraquinone compounds have also been isolated like Aloeemodin-8-O-β-D-glucopyranoside, Chrysophanol, and Rhein-8-O-β-dglucopyranoside which are involved in inhibition of protein tyrosine phosphatase (PTP-1B) enzyme. Phytosterols like 3-Epilappasterol, Lappasterol, β-Sitosterol, Pregnenolone, Daucosterol, and Lappalanasterol have also been reported. (E)-9-Isopropyl-6-methyl-5,9-decadiene-2-one, a terpenoid C14-ketone, α-amyrin starate, β-amyrin, and lupeol palmitates have been isolated from leaves (Pai, 1977; Bruno and Gunther, 1997). Amino acids like Saussureamines A–E, lignan glycoside (−) massoniresinol-4″-O-β-D-glucoside, Guaianolides like iso-zaluzanin-C, and isodehydrocostus lactone are some other compounds reported from *S. lappa*.

10.6 Therapeutic attributes of *Saussurea lappa*

S. lappa is well known for its multiple therapeutic applications which establish its suitability as a major ingredient in several drug formulations. Several reports have investigated the medicinal properties of *S. lappa*—derived bioactive compounds as documented below.

10.6.1 Anticancerous/antitumor properties

Costunolide, the main sesquiterpene lactone isolated from *S. lappa* root, was analyzed for its potential bioactivity in inducing apoptosis in Human Leukemia cells (*HL-60*). Measurement of Reactive Oxygen Species (ROS) and estimation of mitochondrial membrane potentials validated the apoptosis inducing activity of costunolide. It potentially causes mitochondrial permeability transition and release of cytochrome C into the cytosol. Costunolide treatment causes release of N-acetylcysteine which is responsible for causing a block in the mitochondrial alteration, production of ROS thereby leading to apoptotic death. The metabolite thus causes stimulation of ROS-mediated transition of mitochondrial permeability and the resultant Cytochrome C release (Lee et al., 2001). Anticarcinogenesis activity of costunolide was also evident from reporter gene assay which is triggered by tumor-endorsing compound phorbol ester 12-O-tetradecanoylphorbol-13-acetate. This compound amplifies the activity of nitric oxide synthase which has been found to be repressed by costunolide (Fukuda et al., 2001). Another study by Choi et al. (2009) reported the effect of the sesquiterpene lactone on telomerase activity via analysis on *MCF-7, MDA-MB-231*, thereby confirming the hindering activity. Costunolide has been observed to possess cytolytic activity and it performs via mechanistic means to inhibit granule exocytosis and represses amplification in tyrosine phosphorylation in a dose-dependent manner (Taniguchi et al., 1995). Costunolide causes inhibition of Vascular Endothelial Growth Factor (VEGF)-induced chemotaxis of human umbilical vein endothelial cells and causes selective inhibition of endothelial cell proliferation activated by VEGF. Thus, by blocking signaling of angiogenic factor pathway, it is potentially active in inhibition of angiogenesis (Jeong et al., 2002).

Hexane extract of *S. lappa* comprises of ***dehydrocostus lactone*** and was found to be effective in induction of apoptosis in human autonomous androgen prostrate cancer *DU145* cell lines for inhibition of cell growth (Kim et al., 1991). Hung et al. (2010) have also tested activity of dehydrocostus lactone against noncancer cell lines like $NCL\text{-}H_{460}$, $NCL\text{-}H_{520}$, and A_{549}. Flow cytometry studies have shown that dehydrocostus lactone actively facilitates arrest of cell cycle at G2/M stage, leading to inhibition of cell proliferation (Choi and Kim, 2010). It also affects cell cycle distribution, cell viability, and expression of ATP-binding cassette transporter in sarcoma cell lines. Furthermore, it led to activity of apoptosis indicators like caspase-3, caspases 3/7, and PARP cleavage (Kretschmer et al., 2012).

Cynaropicrin derived from *S. lappa* was found to possess immunomodulatory effects coupled with nitric oxide production. It depicted repression against *Eol-1, Jurkat T*, and U_{937} cell lines in a dose-dependent manner. Coupling of cynaropicrin with N-acetyl-L-cysteine or L-cysteine, ROS scavengers' rottlerin is capable of alleviating cynaropicrin-mediated cytotoxicity. Cynaropicrin was observed to have higher cytotoxic potential against leukocyte-derived cancer cells as compared to fibroblasts (Cho et al., 2004). Ethanolic extract of *S. lappa* has also depicted apoptotic activity against gastric cancer in a dose- and time-dependent manner (Ko et al., 2005). Thus, main compounds involved in anticancerous activity include costunolide, dehydrocostus lactone, and cynaropicrin.

10.6.2 Antibacterial properties

Various solvent extracts (ethanolic, methanolic, petroleum ether, and aqueous) have established the antibacterial activity of *S. lappa* against a diversity of resistant pathogens. Yang et al. (1998) reported the activity of ethanolic extract against five clinical strains of *Helicobacter pylori*. *S. lappa* extract has been found to be active against hepatitis B surface antigen (HbSAg) and other correlated antigens (Chen et al., 1995). Bioactive constituents of *S. lappa* have also been found to inhibit binding and transfer of R plasmids in pathogenic microorganism, *Shigella flexneri* (Li et al., 2010), and was also found to be effective against *Bacillus thuringiensis, Aspergillus, Pseudomonas aeruginosa, Staphylococcus aureus, Klebsiella pneumonia, Candida albicans, Proteus vulgaris, Escherichia coli*, and *Cornybacterium* in a concentration-dependent manner (Irshad et al., 2012, Thara and Zuhra, 2012). Ethanolic extract showed inhibitory activity against multidrug-resistant organisms *P. aeruginosa, S. aureus, E. coli*, and *K. pneumonia* (Hasson et al., 2013). Among multiple organic extracts, chloroform extract presented maximum antibacterial potential (Alaagib and Ayoub 2015).

10.6.3 Antiinflammatory activity

Sesquiterpenes are the major ingredients which possess activity for stabilization of endosomal release and cause prevention of cell proliferation via monitoring of nitric oxide and TNF-α levels in macrophage cells of mice (Damre et al., 2003). Methanolic extract exhibited >50% inhibition on induction of cytokine-induced neutrophil chemotactic factors (Lee et al., 1995). Ethanolic extract showed antiinflammatory activity via peritonitis and carrageenan induced edema in animal models (Gokhale et al., 2002). ***Costunolide*** was found to hinder mRNA and protein expression of interleukin-1b (Kang et al., 2004). ***Dehydrocostus lactone*** was found to possess inhibitory activity against oxidative osteoblast damage (Choi et al., 2009). It guides inactivation of nuclear transcription factor (NF-KB), causes inhibition of iNOS gene expression, and reduces generation of TNF-α and nitric oxide−induced through LPS (Lee et al., 1995; Jin et al., 2000). Similarly, ***saussureamines*** A and B also cause effective inhibition of NO caused by NF-κB activation and LPS (Matsuda et al., 2003). Among sesquiterpene lactones, ***cynaropicrin*** was found to be most active against inhibition of TNF-α (Cho et al., 1998).

10.6.4 Hepatoprotective properties

Acetone extract and costunolide were found to possess choleretic effect which could inhibit ulcers in mice (Yamahara et al., 1985). ***Costunolide*** and ***dehydrocostus***

lactone were found to be active against Human Hematome *Hep3B* cells and HBsAg, thus proving their potential in development of HBV drugs (Chen et al., 1995). Use of aqueous methanolic extract in a dose-dependent manner caused absence of parenchymal congestion, improved architectural detail, and decreased apoptotic cells and cellular swelling in hepatic damage repair (Yaeesh et al., 2010). Shao et al. (2005) reported choleretic effect and enhanced bile flow in rats. Examination of *S. lappa* extract depicted induction of gall bladder contraction in dogs as well (Liu et al., 2008).

10.6.5 Antiulcer and cholagogic properties

S. lappa is a major constituent of the popular antiulcer formulation, UL-409, which might be attributed to inflection of defensive factors via improved gastric cytoprotection (Mitra et al., 1996; Venkataranganna et al., 1998). In chronic superficial gastritis patients, decoction perfusion of *S. lappa* into patient's stomach resulted in increased endogenous motilin release and fastened gastric emptying (Chen et al., 1994). Herbal formulation of *S. lappa* was tested for antiulcer activity in Wistar rats which caused reduction in gastric ulceration induced by aspirin and alcohol (Mitra et al., 1996). Sutar et al. (2011) reported potential of the ethyl acetate extract against duodenal and gastric ulceration in rats. An apparent protective activity against acute damage to gastric mucosa of rats has also been observed (Wang, 2004). Saussureamines A, B, and C have been found to be active in repair of gastric damage inflicted by ethanol and hydrochloric acid, while saussureamine A is active in inhibition of stress-induced gastric ulcers in mice (Yoshikawa et al., 1993). In addition to saussureamines, dehydrocostus lactone and costunolide have also been known in treatment of gastric ulcers in mice (Matsuda et al., 2000).

10.6.6 Immunomodulatory properties

High doses of *S. lappa* extract have been shown to be immunomodulatory in cellular and humoral arms of immune system (Pandey, 2012). ***Dehydrocostus lactone*** and ***costunolide*** have also been observed to be active as inhibitors of cytotoxic T lymphocyte (CTL) activity. Costunolide prevents increase in tyrosine phosphorylation thereby inhibiting the killing potential of CTLs. Guaianolide moiety was also observed to exhibit substantial inhibitory activity toward CTLs and initiation of intercellular adhesion molecule-1 (Taniguchi et al., 1995; Yuuya et al., 1999).

10.6.7 Cardiovascular properties

S. lappa extract has been observed to cause lowering of blood pressure and prevention of blood coagulation while causing reduction in triglycerides and cholesterol in blood (Upadhyay et al., 1996). Aqueous decoction of *S. lappa* fortifies the fibrin content of blood (Yu, 1986) and costus oil has been reported to depict hypoglycemic effect (Gupta and Ghatak, 1967; Wang, 1997). ***Costunolide*** and ***dehydrocostus lactone*** present in the volatile oil are involved in inhibition of ADP-induced platelet coagulation (Hou et al., 2008). *S. lappa* extract is a source of betulinic acid, betulinic acid methyl ester, dehydrocostus lactone, mokko lactone, and anthraquinones which are reported to be active against obesity and hypertension associated with Type-II diabetes (Li et al., 2006; Choi et al., 2009, 2012).

10.6.8 Bronchitis

Alkaloid fraction isolated from *S. lappa* is nontoxic and exhibits noticeable spasmolytic effect on the tracheal and smooth (intestinal) muscle of pig lungs (Dutta et al., 1960). Tincture *Saussurea* petroleum ether extract and tincture *Saussurea* were evaluated for activity in bronchitis. Since tincture in petroleum ether extract has been reported to induce bronchoconstriction in guinea pigs, tincture *Saussurea* holds potential for development of drugs for asthma and chronic bronchitis (Sastry and Dutta, 1961).

10.6.9 Anticonvulsant properties

S. lappa petroleum ether extract has also been known to be effective against picrotoxin and pentylenetrazole-induced convulsions in mice by causing elevation of seizure threshold via GABAergic receptors (Ambavade et al., 2009). Alcoholic extract of *S. lappa* also reportedly exhibits significant activity against epilepsy (Gupta Pushpraj et al., 2009).

10.6.10 Antiparasitic properties

S. lappa extract has been known to be effective against nematodal infections via oral administration in rabbits infected with *Clonorchis sinensis* (Rhee et al., 1985). It also significantly reduces the percentage of fecal eggs of nematodes in children naturally infected with worms (Akhtar and Riffat, 1991).

10.6.11 Antihyperlipidemic properties

S. lappa aqueous extract shows significant hypolipidemic activity as tested through administration in rabbits (Upadhyay et al., 1996). Ethanolic extract is also involved in causing reduction in triglyceride levels coupled with an increase in HDL-C level in serum as well as tissues (Anbu et al., 2011).

10.6.12 Antidiarrheal properties

Methanolic extract of *S. lappa* was observed to exhibit substantial antidiarrheal activity in a dose-dependent manner. In fact, it was reported to have effects similar to the popular drug, loperamide in causing reduction of diarrhea stool (Hemamalini et al., 2011). Methanolic extract is also effective in diarrhea induced by castor oil in rats (Negi et al., 2013).

10.6.13 Angiogenesis activity

Costunolide has been known to suppress endothelial cell proliferation. Chemotaxis induced by VEGF of endothelia was observed to be suppressed effectively by *S. lappa*. Similarly, in vivo methods have also reported inhibition of VEGF-stimulated neovascularization in mouse cornea (Jeong et al., 2002; Thara and Zuhra, 2012; Saleem et al., 2013).

10.6.14 Spasmolytic activity

S. lappa is also capable of relaxing contractions induced by carbachol owing to activity of sesquiterpene lactones. These compounds have been recognized for their role in stimulation of *Sgc* which induces extrusion of K^+ ions leading to reduction of intrinsic calcium ions via activation of cyclic GMP and PKG pathways, thus relaxing the smooth muscles (Hsu et al., 2009).

10.6.15 Antimycobacterial activity

Investigation of in vitro antimycobacterial activity was investigated where ***costunolide*** as well as ***dehydrocostus lactone*** in whole oil and fractions depicted activity against *Mycobacterium tuberculosis H37Rv* strain. Both compounds reportedly exhibit synergistic activity as the mixture proved to be more effective as compared to pure compounds (Luna-Herrera et al., 2007).

10.6.16 Synthesis of nanoparticles

Recently, *S. lappa* has been successfully used in synthesis of silver nanoparticles. Nanoparticles offer a favorable option for usage in air/water treatment, optics, catalysis, mirrors, photography, medicine, drug delivery, electronics, clothing, food packaging, and electronics (Prabhu and Poulose, 2012). Nanoparticles have significant usage in nanobiotechnology as they enhance biomedical, optical, environmental, catalytic, and electrical properties, their applications and performance (Stark et al., 2015; Tonga et al., 2014). Silver nanoparticles due to their optical properties have been extremely successful in scientific applications like nanophotonics, sensors, photo-thermal therapy, medicine, and biological activity (El-Nour et al., 2010). Silver nanoparticles are the most widely used nanoparticles in biotechnology and can be effectively synthesized using either conventional but hazardous chemical and physical methods or green chemistry/biological methods which are natural, single-step, and eco-friendly. Plant extracts are preferable over other biological methods for uniform and controlled synthesis under natural conditions (El-Nour et al., 2010). Secondary metabolites procured from medicinal plants can be utilized for synthesis and capping of nanoparticles (Abdul Majeed Almashhedy and Al-Kawaz 2016). Mahapatra et al. (2018) proposed a rapid fabrication method for synthesis of silver nanoparticles (AgNPs) using root extract of *S. lappa* (RESL). An interesting flower-like morphology was observed for these nanoparticles and had a mean diameter of 4000−5000 nm. Surface Plasmon Resonance revealed the optical absorption band peak at 440 nm. In the biofabrication process, diffractogram planes depicted silver as the chief constituent. Presence of root phytoconstituents gave the appearance of crystalline peaks. Capping phenomenon for these bio-nanoparticles was validated using UV, X-ray diffraction, and scanning electron microscopy (SEM). Riaz et al., (2018) also reported the biogenic synthesis of AgNPs using aqueous and methanol extracts of *S. lappa* roots. Characterization of these nanoparticles was performed using UV−visible spectroscopy, SEM, as well as FT-IR. Antimicrobial activity of the synthesized nanoparticles was also observed against *E. coli* (11.0 mm) and *P. aeruginosa* (9.0 mm), where aqueous extract exhibited better potency. Recently, two more research groups reported synthesis of AgNPs using aqueous root extracts of *S. lappa*. Groach et al. (2019) synthesized spherical nanoparticles ranging in size from 7.13 to 24.0 nm, while (Ashwini and Kumar, 2019) synthesized cube-shaped nanoparticles having a size range of 500 nm−2 μm. Antibacterial activity was successfully observed against *E. coli* and *Bacillus cereus*, thereby proving their antimicrobial potential (Ashwini and Kumar, 2019; Groach et al., 2019). These studies have successfully proven the potential of *S. lappa* extract in synthesis of nanoparticles which can be promising candidates for usage in biomedicine.

10.7 Conservation status

Medicinal plants are under a constant threat owing to unregulated harvesting of their natural habitat for domestic as well as industrial purposes. Well-planned and effectively utilized strategies aimed at conservation of these plants remain the sole hope for preventing their extinction. These become especially relevant for popular herbs like *S. lappa* which have a well-recognized and documented potential in curative and preventive medicine. Butola and Samant attempted to study the distribution, diversity, habitat preference, endemism, nativity, status, as well as indigenous uses of *Saussurea* species in the Indian Himalayan Region (Butola and Samant, 2010). They recognized *S. lappa* as the most commercially viable species of *Saussurea* genus. It was first listed in Appendix II of CITES (Convention on International Trade in Endangered species of Wild Fauna and Flora) on July 01, 1975, and 10 years later, it was uplisted to Appendix I (Bano et al., 2018). According to IUCN, *S. lappa* has been categorized as a critically endangered species. Export of *S. lappa* has been prohibited due to inclusion in category of red list species according to Appendix I of CITES, 2003. Jammu and Kashmir which hosts its prime natural habitat has enforced a special Act called "The Kuth Act, 1978" for regulating the trade of *S. lappa* (Jain, 2001). It has also been included in the negative list of exports as imposed by Ministry of Commerce, Government of India along with listing in the "Schedule VI" of the Wildlife Protection Act, India. Trade of this valuable herb has been strictly prohibited under Foreign Trade Development Act 1992. Although the species is cultivated in regions like Lahaul valley, low and fluctuating market price limits the practice to a few scattered villages (Kuniyal et al., 2005).

Following in vitro propagation and ex situ strategies have been employed for conservation of the species.

10.7.1 In vitro propagation of *Saussurea lappa*

Genetic conservation is particularly important for endangered plant species like *S. lappa* owing to fear of their extinction and loss of valuable metabolites. Although germplasm storage in form of seeds appears to be the most lucrative option for conservation, it is not much feasible in case of plants which do not produce seeds or ones with low seed viability as in case of *S. lappa* (30%) (Wealth of India, 1972). The traditional methods intended for their maintenance are risky, laborious, and expensive. Tissue culture is the most suitable alternative for large-scale conservation in vegetative state which offers an option of production on demand as well. Arora and Bhojwani (1989) initiated the first attempt on in vitro propagation of *S. lappa* using different explants derived from aseptic plants. Among different seedling explants, maximum shoot regeneration was observed in case of leaves followed by cotyledons. However, roots and hypocotyl appeared to lack morphogenetic potential. Shoot multiplication of 3.5-folds was observed every 3 weeks on MS (Murashige and Skoog) media containing benzylaminopurine and gibberellin. Roots were observed with 90% efficiency on MS having 0.5 μM naphthaleneacetic acid. Interestingly, shoot cultures preserved in the dark at 5°C for 12 months, even without any intervening subculture showed 100% viability which presented a promising method for long-term storage. It was observed that shoots stored at cold temperature exhibited higher rates of multiplication in culture room conditions relative to untreated shoots. Later, role of TDZ (thiadiazuron) in achieving direct organogenesis was achieved by culturing shoot tips of 2-week-old seedlings on MS + TDZ (0.45 μM) media (Johnson et al., 1997). Among N6-benzyladenine-(BA) and TDZ containing media, latter was found to be more effective for induction of callus-free multiple shoots. Liquid medium was observed as a better culture media as compared to agar-solidified media. Shoots developed roots on MS + Napththalene-acetic acid media (NAA, 1.07 μM). The micropropagated plantlets were transferred to soil and 90% plants showed survival after the process of hardening. Verma et al. (2012) reported callus induction using root explants using 2.4-D and BAP during an initiative intended for in vitro propagation of 23 overexploited medicinal plants in India. Later, Warghat et al. (2016) optimized the plant regeneration under in vitro as well as in vivo conditions. On evaluation of multiple combinations of auxins as well cytokinins, 2, 4-dichlorophenoxyacetic acid (2, 4-D) (3 mg/L) and kinetin (Kin) (5 mg/L) were found to be most effective for callus induction in all tested explants. Explants used in the study included leaf, cotyledonary leaf, hypocotyl, epicotyl, stem, and root. Root and stem explants exhibited earlier response and better callus frequencies. Maximum number and length of shoots and roots were derived on medium having kinetin (2 mg/L) and indole-3-butyric-acid (2 mg/L) in the callus derived from root explant. Regenerated plants were transferred to a potting mixture (sand:soil:perlite (1:1:1)) for hardening followed by multiplication. 100% survival rate was observed for plants in greenhouse conditions, while well-developed healthy plants showed a survival rate of 80% in open conditions of herbal garden. Another study tested the callus induction potential of seedling, root, lamina, and petiole (Zaib-Un-Nisa et al., 2019). Callus appeared in shoot and seedling explants after 3 days, while lamina explant took the maximum time of 15–20 days. Recently, Sharma et al. (2019) developed an efficient protocol for in vitro multiplication of in vitro–grown seedling-derived shoot tip explants for generation of genetically uniform plants. MS medium supplemented with TDZ (1.14 μM) and NAA (2.68 μM) proved to be an optimal media combination for highest average shoot regeneration potential (73.33%) where maximum average number of shoots (11.4) and average shoot length (4.17 cm) was achieved. Maximum rooting of microshoots (77.78%) was observed on MS + indole butyric acid (2.46 μM) with average root number of 6.0 and average length of 3.07 cm. In vitro propagation via direct organogenesis as well as via callus induction have been established in *S. lappa*. However, future studies need to be planned for obtaining high content of secondary metabolites via tissue culture for establishing a supply system for medicinal purposes.

10.7.2 Ex situ strategies

1. *S. lappa* rhizomes were collected from its natural habitat and cut into small pieces having two to three active buds and were placed a day later in experimental plots (Sher et al., 2010) to evaluate their growth routine. However, the strategy was not very successful as the planted rhizomes had a very poor sprouting efficiency (Sher et al., 2010).
2. A comprehensive analysis was undertaken for assessing seed germination, seedling analysis, and survival percentage of *S. lappa* in higher and lower Himalayan altitudes in Uttarkashi (Parmar et al., 2012). A significant increase in root length was observed in polyhouse conditions at higher altitude. Overall, highest percentage of seed germination potential and survival percentage was noticed for high altitudinal places. Plant morphology at higher altitudes is indicative of adaptability measures employed for thermoregulation. Woolly hairs of these plants are densest at low temperature which prevent frost damage and UV damage. The study thus emphasized the relevance of natural habitat conditions for better growth and survival (Parmar et al., 2012).
3. Few reports have attempted to increase longevity of seed viability in *S. lappa*. Sharma et al. (2014) analyzed the physiological status of seeds procured from Lahaul for ambient long-term storage (66 months). The study demonstrated that seed viability and germination could be enhanced via chilling and GA_3 pretreatments. The seed viability was observed to maintain completely for at least 18 months and further, even beyond 30 months, 82% viability retention

was evident. Sustenance of reasonably high viability status, germination potential, and seedling vigor projects the suitability of this strategy for germplasm conservation via long-term storage.

10.7.3 Government initiatives

The Industrial Policy of 2001 launched in Uttarakhand specifically recognizes the massive potential of Herbal and Medicinal plant sector. This sector has remained mostly unexploited owing to unplanned and loosely coordinated strategies for cultivation coupled with a lack of integrated arrangements for commercial cultivation, processing as well as marketing (CITES, 2011). The State Government has prioritized *S. lappa* along with 26 medicinal and aromatic plants for encouraging cultivation. Since 2006, registration of farmers has also begun for taking up cultivation practices of *S lappa*. Moreover, strict regulations imposed by Indian government on trade and enforcement in India coupled with rising cultivation in countries like China have provided some respite to wild collection practices. Several stakeholders across Uttarakhand and Himachal Pradesh believe that despite the miniature distribution and population of *S. lappa*, some collection would still be possible periodically. However, this can be considered an opportunistic activity only which could only be utilized for local/domestic purposes and upscaling at a commercial level would not be possible. This can be attributed to fact that the primary natural habitat, i.e., Jammu and Kashmir itself seems incapable of supporting industrial scale of exploitation and trade. Absence of statistical data on wild population and the existing ban on wild collection needs review by the government for encouraging cultivation practices (CITES 2011) and strenuous efforts need to be developed for encouraging large-scale cultivation in India.

10.7.4 Challenges

The main reasons underlying low cultivation response of farmers include low cost of imports, cumbersome cultivation practices, persistent uncertainty in demand coupled with major time exhausting procedures required for procuring necessary cultivation certificates (CITES 2011). The restrictions which imply to wild collection are common to cultivation as well which makes the process of latter quite confusing and hard to attain. Great amount of uncertainty surrounds the attainment of permits and is compounded by a significant lack of transparency regarding proper rules and regulations along with an ambiguity in decision-making. The purpose of listing in CITES can draw benefits in conservation only when linked to suitable awareness and capacity building initiatives. Promotion of cultivation in consultation of CITES Management and Scientific authorities can help in removal of pressure from wild while ensuring that wild populations are not being impacted due to trade. Coupling awareness and information-sharing among local and national authorities can be seen as a major hope in promoting conservation practices of *S. lappa*.

10.8 Trade

Endangered status and low cultivation of *S. lappa* is unable to quench its massive industrial demand. It is cultivated in forest area in conditions similar to natural habitat. For maintaining export to Arabia and Red sea ports, roots are transported to medicine and perfumery industries in Calcutta and Bombay (Kamalpreet et al., 2019). France has been the prime importer for *S. lappa* and China stood as the largest exporter with a capacity of 1024 tonnes from 1983 to 2009 while India occupied the second position with export of 266 tonnes during the mentioned period (Kamalpreet et al., 2019). China still holds its strong position in trade of *S. lappa*. The product is marketed under trade name of costus root and costus root oil and is easily available in markets of Delhi, Calcutta, Amritsar, Mumbai, and Haridwar. However, myriad medicinal applicability of *S. lappa* has forced its illegal trade. Illegal extraction has been reported from Tilel and Gurez areas of Jammu and Kashmir where the product reportedly gets smuggled out in potato trucks (CITES 2011). In Lahaul and Spiti region of Himachal Pradesh, it is smuggled out of adjoining Sartha/Bani/Banderwah areas of Banderwah, Kathua, and Doda districts of Jammu and Kashmir state (CITES 2011).

10.9 Omics advancements in *Saussurea lappa*

Focal points of research for this extensively studied member of Asteraceae have always been limited to exploring biochemistry of the secondary metabolites and corresponding medical significance. However, an understanding of molecular pathways involved in biosynthesis and regulation of bioactive compounds would better equip breeders as well as genetic engineers for planning genetic advancement studies in *S. lappa*. Research progress in high throughput sequencing strategies have provided opportunities for elucidation of complex networks functional in synthesis of industrially valuable metabolites in endangered herbs.

Researchers have only recently started exploring OMICS tools in *S. lappa* for characterization of biosynthetic and regulatory pathways of secondary metabolism. Bains et al. (2019) utilized Illumina HiSeq 2000 platform for de novo transcriptome sequencing of leaf tissue leading to elucidation of genes involved in sesquiterpenoids and flavonoids biosynthesis. Sesquiterpene lactones are reportedly the predominant subclass of terpenoids present in approximately 8% of Asteraceae family including *S. lappa* (Nguyen et al., 2010). The essential oil derived from *S. lappa* is fairly rich in sesquiterpenoids (79.80% v/v) and monoterpenoids (13.25% v/v) (Liu et al., 2012). The proposed pathway elucidated the steps involved in biosynthesis of sesquiterpene lactones (STL), costunolides which are the main ingredients of pharmaceutical formulations. Almost all the genes reported to be involved in biosynthesis of STLs were observed in the transcriptome dataset (Fig. 10.2). Key enzymes noted in costunolide biosynthesis include germacrene A

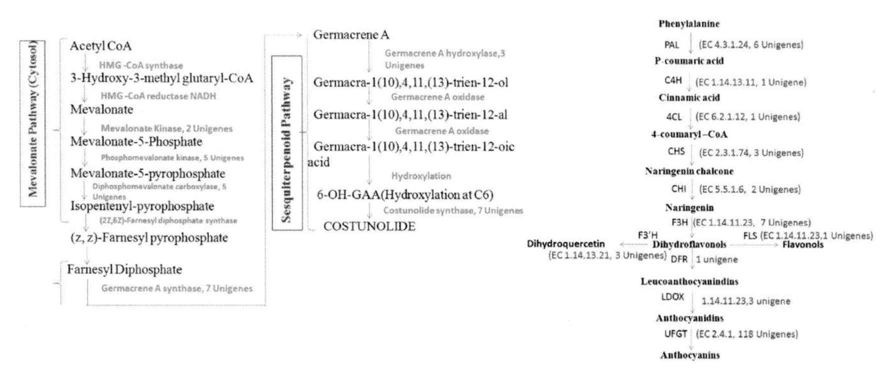

FIGURE 10.2

Proposed pathway for Costunolide and Anthocyanin Biosynthesis in *Saussurea lappa* (Bains et al., 2019).

synthase (GAS), germacrene A hydroxylase and costunolide synthase (COS). The first committed step for costunolide biosynthesis is catalyzed by GAS by directing conversion of FPP into germacrene A, which has been known as the source for deriving costunolide in most of the other plants as well (de Kraker et al., 1998, 2001a, 2001b). Further, as no downstream nonmevalonate pathway genes were obtained in the analysis, it was speculated that FPP derived from mevalonate pathway only contributes toward costunolide biosynthesis in *S. lappa* (Bains et al., 2019). The authors also elucidated the steps involved in biosynthesis of flavonoids and a total of 70 transcripts were observed as the constituents of biosynthetic pathway (Fig. 10.2). Most of the unigenes were found to code for UDP-glucose flavonoid 3-O glucosyltransferase (UFGT) which catalyzes production of anthocyanins which led to the postulated anthocyanin pathway in *S. lappa* (Bains et al., 2019). Relative expression analysis of costunolide biosynthesis genes among leaf and root tissues showed predominance in leaf thus proposing that the biosynthesis of the metabolite occurs in the leaf which is transported to the root tissue for long-term storage (Bains et al., 2019). These roots are the main plant material harvested for medicinal formulations. Several potential SSRs were also predicted in various genes involved in multiple metabolic pathways (Bains et al., 2019). These SSRs can be further examined for exploring the genetic diversity among different genotypes and can also allow analysis of variation between varying metabolite content in *S. lappa* plants.

Thakur et al. (2020) analyzed the crucial aspect of transcriptional regulation governing costunolide biosynthesis via a comparative transcriptome analysis of root and leaf tissues. Transcription factors were mined using Plant TFDB and TF gene coexpression networks were created. A total of 30,070 TFs were annotated representing 58 families comprising WRKY, bHLH, MYB, C2H2, NAC, and ERF TFs. Differential Gene Expression analysis was undertaken to identify the tissue-specific expression of costunolide biosynthetic genes. In this study, predominant expression was observed in case of root as compared to leaf which indicates toward the possibility that later steps of costunolide biogenesis like oxidation and hydroxylation might occur in roots as observed previously in *Tanacetum cinerariifolium* (Ramirez et al., 2013). Genes of mevalonate pathway were found to mostly coexpress with bHLH, MYB related and ERF transcription factor proteins, thus indicating their involvement in biosynthesis. The study also isolated the promoter region of *costunolide synthase gene (SlCOS1),* involved in catalysis of the final key step of costunolide biosynthesis. Analysis of corresponding cis-regulatory elements reported the presence of MYB-binding domains. Further studies are warranted to understand the exact mechanism of regulation exercised by MYB transcription factors in costunolide biosynthesis.

10.10 Conclusion and future perspectives

S. lappa has been effectively used in various indigenous medicinal systems globally. The bioactive constituents extracted from the plant are a source of anticancerous, antiinflammatory, hepatoprotective, antiulcerogenic, and antimicrobial properties. Although significant experimental evidence has proven its therapeutic importance, optimal clinical usage necessitates further research for establishing dosage forms and adverse side effects. Lack of cognizance among local farmers and uncoordinated government initiatives hinder large-scale cultivation and conservation of this endangered herb. Since field plants are the only source of industrially valuable roots, prioritized efforts need to be applied for conserving this bioresource in its natural habitat. In vitro propagation has been successfully established in *S. lappa* for optimizing growth conditions; however, further efforts need to be pushed for enhancing yield and metabolite content. Further, exploration of intricate molecular networks underlying biosynthesis and regulation of secondary metabolism would facilitate functional genomics and metabolic engineering approaches in this ancient herb.

References

Abdelwahab, S.I., Taha, M.M.E., Alhazmi, H.A., Ahsan, W., Rehman, Z.U., Bratty, M.A., Makeen, H., 2019. Phytochemical profiling of costus (*Saussurea lappa* Clarke) root essential oil, and its antimicrobial and toxicological effects. Trop. J. Pharmaceut. Res. 18, 2155–2160.

Abdul Majeed Almashhedy, L., Al-Kawaz, H.S., 2016. Green synthesis of silver nanoparticles using *Actinidia deliciosa* extracts. Res. J. Pharmaceut. Biol. Chem. Sci. 7, 2212.

Ainsile, W. Materia Medica of Hindustan, Madras, Government Gazette of Jammu & Kashmir, 4 May Chief Minister letter no.1109, 1813.

Akhtar, M.S., Riffat, S., 1991. Field trail of *Saussurea lappa* roots against nematodes and Nigella sativa seeds against cestodes in children. J. Pakistan Med. Assoc. 4, pp185–187.

Alaagib, R.M.O., Ayoub, S.M.H., 2015. On the chemical composition and antibacterial activity of *Saussurea lappa* (Asteraceae). Pharm. Innovation J. 4 (2), pp73–76.

Amara, U., Khan, A., Laraib, S., Wali, R., Sarwar, U., Ain, Q.T., Shakeel, S., 2017. Conservation status and therapeutic potential of *Saussurea lappa*: an overview. Am. J. Plant Sci. 8 (3), 602–614.

Ambavade, S.D., Mhetre, N.A., Muthal, A.P., Bodhankar, S.L., 2009. Pharmacological evaluation of anticonvulsant activity of root extract of *Saussurea lappa* in mice. Eur. J. Integr. Med. 1 (3), 131–137.

Anbu, J., Anjana, A., Purushothaman, K., Sumithra, M., Suganya, S., Bathula, N.K., Modak, S., 2011. Evaluation of antihyperlipidemic activity of ethanolic extract of *Saussurae lappa* in rats. Int. J. Pharm. Bio Sci. 2 (4).

Arora, R., Bhojwani, S.S., 1989. In vitro propagation and low temperature storage of *Saussurea lappa* CB Clarke – An endangered, medicinal plant. Plant Cell Rep. 8 (1), 44–47.

Ashwini, S.B., Kumar, J., 2019. Biogenic Synthesis, Antibacterial and Antioxidant Studies of Prepared Silver Nano Particles Using Root Extract of *Saussurea lappa*.

Bains, S., Thakur, V., Kaur, J., Singh, K., Kaur, R., 2019. Elucidating genes involved in sesquiterpenoid and flavonoid biosynthetic pathways in *Saussurea lappa* by de novo leaf transcriptome analysis. Genomics 111 (6), 1474–1482.

Bano, H., Bhat, J.I., Siddique, M.A.A., Noor, F., Bhat, M.A., Mir, S.A., 2018. Seed germination studies of *Saussurea Costus* Clarke, a step towards conservation of a critically endangered medicinal plant species of north western Himalaya. Plant Arch. 18 (1), 963–968.

Bruno, M., Gunther, O., 1997. (E)-9-isopropyl-6-methyl-5,9-decadiene-2-one, a terpenoid C18-ketone with a novel skeleton. J. Chem. Soc. 353–354.

Butola, J.S., Samant, S.S., 2010. Saussurea species in Indian Himalayan region: diversity, distribution and indigenous uses. Int. J. Plant Biol. 1 (1), e9.

Chamara, A.M.R., Kuganesan, A., Dolawatta, K.D., Amarathunga, I.M., Wickramasinghe, W.Y.H., Madushani, Y.M.P.K., Thiripuranathar, G., 2018. Evaluation of bioactivities of two polyherbal formulations found in sri lankan ayurvedic treatments. Int. J. Pharmaceut. Sci. Res. 9 (5), 2073–2079.

Chang, K.M., Kim, G.H., 2008. Comparison of volatile aroma components from *Saussurea lappa* CB Clarke root oils. Prev. Nutr. Food Sci. 13 (2), 128–133.

Chen, H.C., Chou, C.K., Lee, S.D., Wang, J.C., Yeh, S.F., 1995. Active compounds from *Saussurea lappa* Clarks that suppress hepatitis B virus surface antigen gene expression in human hepatoma cells. Antivir. Res. 27 (1–2), 99–109.

Chen, S.F., Li, Y.Q., He, F.Y., 1994. Effect of *Saussurea lappa* on gastric functions. Chin. J. Integr. Tradit. West. Med. 14 (7), 406–408.

Chhabra, B.R., Ahuja, N.M., Bhullar, M.K., Kalsi, P.S., 1998. Some C-3 oxygenated guaianolides from *Saussurea lappa*. Fitoterapia 69, 274–275.

Chhabra, B.R., Gupta, S., Dhillon, R.S., Kalsi, P.S., 1997. Minor sesquiterpene lactones from *Saussurea lappa* roots. Fitoterapia 68, 470–471.

Cho, J.Y., Kim, A.R., Jung, J.H., Chun, T., Rhee, M.H., Yoo, E.S., 2004. Cytotoxic and proapoptotic activities of cynaropicrin, a sesquiterpene lactone, on the viability of leukocyte cancer cell lines. Eur. J. Pharmacol. 492 (2–3), 85–94.

Cho, J.Y., Park, J., Yoo, E.S., Baik, K.U., Jung, J.H., Lee, J., Park, M.H., 1998. Inhibitory effect of sesquiterpene lactones from *Saussurea lappa* on tumor necrosis factor-α production in murine macrophage-like cells. Planta Med. 64 (07), 594–597.

Choi, E.J., Kim, G.H., 2010. Evaluation of anticancer activity of dehydrocostuslactone in vitro. Mol. Med. Rep. 3 (1), 185–188.

Choi, H.G., Lee, D.S., Li, B., Choi, Y.H., Lee, S.H., Kim, Y.C., 2012. Santamarin, a sesquiterpene lactone isolated from *Saussurea lappa*, represses LPS-induced inflammatory responses via expression of heme oxygenase-1 in murine macrophage cells. Int. Immunopharm. 13 (3), 271–279.

Choi, J.Y., Na, M., Hyun Hwang, I., Ho Lee, S., Young Bae, E., Yeon Kim, B., Seog Ahn, J., 2009. Isolation of betulinic acid, its methyl ester and guaiane sesquiterpenoids with protein tyrosine phosphatase 1B inhibitory activity from the roots of *Saussurea lappa* CB Clarke. Molecules 14 (1), 266–272.

Damre, A.A., Damre, A.S., Saraf, M.N., 2003. Evaluation of sesquiterpene lactone fraction of *Saussurea lappa* on transudative, exudative and proliferative phases of inflammation. Phytother Res. 17 (7), 722–725.

de Kraker, J.W., Franssen, M.C., Dalm, M.C., de Groot, A., Bouwmeester, H.J., 2001. Biosynthesis of germacrene A carboxylic acid in chicory roots. Demonstration of a cytochrome P450 (+)-germacrene A hydroxylase and NADP+-dependent sesquiterpenoid dehydrogenase (s) involved in sesquiterpene lactone biosynthesis. Plant Physiol. 125 (4), 1930–1940.

de Kraker, J.W., Franssen, M.C., de Groot, A., König, W.A., Bouwmeester, H.J., 1998. (+)-Germacrene A biosynthesis: the committed step in the biosynthesis of bitter sesquiterpene lactones in chicory. Plant Physiol. 117 (4), 1381–1392.

Dhillon, R.S., Kalsi, P.S., Singh, W.P., Gautam, V.K., Chhabra, B.R., 1987. Guaianolide from *Saussurea lappa*. Phytochemistry 26, 41209–41210.

Dutta, N.K., Sastry, M.S., Tamhane, R.G., 1960. Pharmacological actions of an alkaloidal fraction isolated from *Saussurea lappa*. Indian J. Pharm. 22, 6–7.

Dwivedi, K.P., 1963. Kushth, Dhanwantri, Vijayagarh, Aligarh (special ed.).

El-Nour, K.M.A., Eftaiha, A.A., Al-Warthan, A., Ammar, R.A., 2010. Synthesis and applications of silver nanoparticles. Arab. J. Chem. 3 (3), 135–140.

Fukuda, K., Akao, S., Ohno, Y., Yamashita, K., Fujiwara, H., 2001. Inhibition by costunolide of phorbol ester-induced transcriptional activation of inducible nitric oxide synthase gene in a human monocyte cell line THP-1. Cancer Lett. 164 (1), 7–13.

Gautam, H., Asrani, R., 2018. Phytochemical and pharmacological review of an ethno medicinal plant: *Saussurea lappa*. Vet. Res. 6 (01), 01–09.

Gokhale, A.B., Damre, A.S., Kulkarni, K.R., Saraf, M.N., 2002. Preliminary evaluation of anti-inflammatory and anti-arthritic activity of *S. lappa, A. speciosa* and *A. aspera*. Phytomedicine 9 (5), 433–437.

Govindan, S.V., Bhattacharaya, S.C., 1977. Alantolides and cyclocostunolides from *Saussurea lappa*. Indian J. Chem. 15, 956.

Groach, R., Yadav, K., Sharma, J., Singh, N., 2019. Biosynthesis and characterization of silver nanoparticles using root extract of *Saussurea lappa* (Decne.) Clarke and their antibacterial activity. J. Environ. Biol. 40 (5), 1060–1066.

Gupta, O.P., Ghatak, B.J., 1967. Pharmacological investigation on *Saussurea lappa* (Clarke). Indian J. Med. Res. 55, 1078–1083.

Gupta Pushpraj, S., Jadhav, S.S., Ghaisas, M.M., Deshpande, A.D., 2009. Anticonvulsant activity of *Saussurea lappa*. Pharmacologyonline 3, 809–814.

Gwari, G., Bhandari, U., Andola, H.C., Lohani, H., Chauhan, N., 2013. Volatile constituents of *Saussurea costus* roots cultivated in Uttarakhand Himalayas, India. Pharmacogn. Res. 5 (3), 179.

Hajra, P.K., Rao, R.R., Singh, D.K., Uniyal, B.P., 1995. Flora of India, vols. 12 & 13. Asteraceae. Botanical Survey of India, Calcutta.

Hasson, S.S.A., Al-Balushi, M.S., Al-Busaidi, J., Othman, M.S., Said, E.A., Habal, O., Sallam, T.A., Aljabri, A.A., AhmedIdris, M., 2013. Evaluation of anti–resistant activity of *Auklandia (Saussurea lappa)* root against some human pathogens. Asian Pac. J. Trop. Biomed. 3 (7), 557−562.

Hemamalini, K., Vasireddy, U., Nagarjun, G.A., Harinath, K., Vamshi, G., Vishnu, E., 2011. Anti-diarrhoeal activity of leaf extracts *Anogessius accuminata*. Int. J. Pharmaceut. Res. Dev. 3 (6), 55−57.

Hou, P.F., Chen, W.X., Zhao, X.H., S,U, S., Liu, H., Lu, Y., Duan, J., 2008. Analysis of chemical composition of the essential oil from *Radix aucklandiae* by GC-MS and the effect on platelet aggregation. Zhongguo Shi Yan Fang Ji Xue Za Zhi 14 (7), 26−30.

Hsu, Y.L., Wu, L.Y., Kuo, P.L., 2009. Dehydrocostuslactone, a medicinal plant-derived sesquiterpene lactone, induces apoptosis coupled to endoplasmic reticulum stress in liver cancer cells. J. Pharmacol. Exp. Therapeut. 329 (2), 808−819.

Hung, J.Y., Hsu, Y.L., Ni, W.C., Tsai, Y.M., Yang, C.J., Kuo, P.L., Huang, M.S., 2010. Oxidative and endoplasmic reticulum stress signaling are involved in dehydrocostuslactone-mediated apoptosis in human non-small cell lung cancer cells. Lung Cancer 68 (3), 355−365.

Irshad, S., Mahmood, M., Perveen, F., 2012. In vitro antibacterial activities of three medicinal plants using agar well diffusion method. Res. J. Biol. 2 (1), 1−8.

Jain, P., 2001. CITES and India. TRAFFIC India. WWF India & Ministry of Environment and Forests, New Delhi, India.

Jeong, S.J., Itokawa, T., Shibuya, M., Kuwano, M., Ono, M., Higuchi, R., Miyamoto, T., 2002. Costunolide, a sesquiterpene lactone from *Saussurea lappa*, inhibits the VEGFR KDR/Flk-1 signaling pathway. Cancer Lett. 187 (1−2), 129−133.

Jin, M., Lee, H.J., Ryu, J.H., Chung, K.S., 2000. Inhibition of LPS-induced NO production and NF-κB activation by a sesquiterpene from *Saussurea lappa*. Arch. Pharm. Res. 23 (1), 54−58.

Johnson, T.S., Narayan, S.B., Narayana, D.B.A., 1997. Rapid in vitro propagation of *Saussurea lappa*, an endangered medicinal plant, through multiple shoot cultures. In Vitro Cell. Dev. Biol. Plant 33 (2), 128−130.

Kalsi, S., Sharma, S., Kaur, G., 1983. Isodehydrocostus lactone and isozaluzanin C, two guaianolides from *Saussurea lappa*. Phytochemistry 22 (9), 1993−1995.

Kamalpreet, L.K., Singh, A., Kaur, J., Kaur, N., 2019. A brief review of remedial uses of *Saussurea lappa*. J. Pharmacogn. Phytochem. 8 (3), 4423−4430.

Kang, J.S., Yoon, Y.D., Lee, K.H., Park, S.K., Kim, H.M., 2004. Costunolide inhibits interleukin-1β expression by down-regulation of AP-1 and MAPK activity in LPS-stimulated RAW 264.7 cells. Biochem. Biophys. Res. Commun. 313 (1), 171−177.

Kaul, M.K., 1997. Medicinal Plants of Kashmir and Ladakh. Indus Publishing Co., New Delhi, p. 144.

Kim, R.M., Jeon, S.E., Choi, Y., 1991. EuiBangRuChui., vol. 6. Yeokang Publications, Seoul, p. 385.

Kim, J.S., Chi, H.J., Chang, S.Y., Ha, K.W., Kang, S.S., 1999. Isolation and quantitative determination of costunolide from *Saussurea* root. Korean J. Pharmacogn. 30 (1), 48−53.

Ko, S.G., Kim, H.P., Jin, D.H., Bae, H.S., Kim, S.H., Park, C.H., Lee, J.W., 2005. Saussurea lappa induces G2-growth arrest and apoptosis in AGS gastric cancer cells. Cancer Lett. 220 (1), 11−19.

Kretschmer, N., Rinner, B., Stuendl, N., Kaltenegger, H., Wolf, E., Kunert, O., Boechzelt, H., Leithner, A., Bauer, R., Lohberger, B., 2012. Effect of costunolide and dehydrocostus lactone on cell cycle, apoptosis, and ABC transporter expression in human soft tissue sarcoma cells. Planta Med. 78 (16), 1749−1756.

Kumar, S., Ahuja, N.M., Juawanda, G.S., Chhabra, B.R., 1995. New guaianolides from *Saussurea lappa* roots. Fitoterapia 66, 287–288.

Kuniyal, C.P., Rawat, Y.S., Oinam, S.S., Kuniyal, J.C., Vishvakarma, S.C., 2005. Kuth (*Saussurea lappa*) cultivation in the cold desert environment of the Lahaul valley, northwestern Himalaya, India: arising threats and need to revive socio-economic values. Biodivers. Conserv. 14 (5), 1035–1045.

Lee, G.I., Ha, J.Y., Min, K.R., Nakagawa, H., Tsurufuji, S., Chang, I.M., Kim, Y., 1995. Inhibitory effects of oriental herbal medicines on IL-8 induction in lipopolysaccharide-activated rat macrophages. Planta Med. 61 (01), 26–30.

Lee, M.G., Lee, K.T., Chi, S.G., PARK, J.H., 2001. Constunolide induces apoptosis by ROS-mediated mitochondrial permeability transition and cytochrome C release. Biol. Pharm. Bull. 24 (3), 303–306.

Li, S., An, T.Y., Li, J., Shen, Q., Lou, F.C., Hu, L., 2006. PTP1B inhibitors from *Saussurea lappa*. J. Asian Nat. Prod. Res. 8 (3), 281–286.

Li, Y., Gong, T., Yang, Q., Miao, Z., Cheng, G., Hou, Q., 2010. In vivo study of the inhibition of R plasmid transfer by conjugation in rats taking extracts of *Radix aucklandiae* and *Polygonatum cyrtonema*. Hua. Zhongguo Bingyuan Shengwuxue Zazhi/J. Pathog. Biol. 5 (2), 108–110.

Liu, J.J., Zheng, C.Q., Zhou, Z., 2008. The influence of herbal *Lysimachiare* and *Radix Aucklandiae* on the movement of dog gall bladder and its plasma CCK value. Sichuan Zhong Yi 26 (4), 31–32.

Liu, Z.L., He, Q., Chu, S.S., Wang, C.F., Du, S.S., Deng, Z.W., 2012. Essential oil composition and larvicidal activity of *Saussurea lappa* roots against the mosquito *Aedes albopictus* (Diptera: Culicidae). Parasitol. Res. 110 (6), 2125–2130.

Luna-Herrera, J., Costa, M.C., Gonzalez, H.G., Rodrigues, A.I., Castilho, P.C., 2007. Synergistic antimycobacterial activities of sesquiterpene lactones from *Laurus* spp. J. Antimicrob. Chemother. 59 (3), 548–552.

Mahapatra, D.K., Tijare, L.K., Gundimeda, V., Mahajan, N.M., 2018. Rapid biosynthesis of silver nanoparticles of flower-like morphology from the root extract of *Saussurea lappa*. Res. Rev. J. Pharmacol. 5 (1), 20–24.

Marotti, M., Piccaglia, R., Giovanelli, E., Deans, S.G., Eaglesham, E., 1994. Effects of planting time and mineral fertilization on peppermint (*Mentha x piperita* L.) essential oil composition and its biological activity. Flavour Fragrance J. 9 (3), 125–129.

Matsuda, H., Kageura, T., Inoue, Y., Morikawa, T., Yoshikawa, M., 2000. Absolute stereo structures and syntheses of saussureamines A, B, C, D and E, amino acid–sesquiterpene conjugates with gastroprotective effect, from the roots of *Saussurea lappa*. Tetrahedron 56 (39), 7763–7777.

Matsuda, H., Toguchida, I., Ninomiya, K., Kageura, T., Morikawa, T., Yoshikawa, M., 2003. Effects of sesquiterpenes and amino acid sesquiterpene conjugates from the roots of *Saussurea lappa* on inducible nitric oxide synthase and heat shock protein in lipopolysaccharide-activated macrophages. Bioorg. Med. Chem. 11 (5), 709–715.

Maurer, B., Grieder, A., 1977. Sesquiterpenoids from costus root oil (*Saussurea lappa* Clarke). Helv. Chim. Acta 60 (7), 2177–2190.

Mitra, S.K., Gopumadhavan, S., Hemavathi, T.S., Muralidhar, T.S., Venkataranganna, M.V., 1996. Protective effect of UL-409, a herbal formulation against physical and chemical factor induced gastric and duodenal ulcers in experimental animals. J. Ethnopharmacol. 52 (3), 165–169.

Nandkarni, A.K., 1954. Indian Materia Medica. Popular Book Depot, Mumbai, p. 1108.

Negi, J.S., Bisht, V.K., Bhandari, A.K., Bhatt, V.P., Sati, M.K., Mohanty, J.P., Sundriyal, R.C., 2013. Antidiarrheal activity of methanol extract and major essential oil contents of *Saussurea lappa* Clarke. Afr. J. Pharm. Pharmacol. 7 (8), 474–477.

Nguyen, D.T., Göpfert, J.C., Ikezawa, N., MacNevin, G., Kathiresan, M., Conrad, J., Spring, O., Ro, D.K., 2010. Biochemical conservation and evolution of germacrene A oxidase in Asteraceae. J. Biol. Chem. 285 (22), 16588–16598.

Pai, P.P., 1977. Isolation of alpha-amyrin stearate, beta-amyrin and lupeol palmitates from the costus leaves. Curr. Sci. 46, 261–262.

Pandey, M.M., Rastogi, S., Rawat, A.K.S., 2007. Evaluation of pharmacognostical characters and comparative morphoanatomical study of *Saussurea costus* (Falc.) Lipchitz and *Arctium lappa* L. roots. Nat. Prod. Sci. 13 (4), 304–310.

Pandey, R.S., 2012. *Saussurea lappa* extract modulates cell mediated and humoral immune response in mice. Der Pharm. Lett. 4 (6), 1868–1873.

Parmar, M.P., Negi, S.L., Ramola, S., 2012. Seeds germination and seedlings analysis of *Saussurea costus* Royle ex Benth. High and low altitudinal villages of district Uttarkashi (Uttarakhand). IOSR J. Pharm. 2, 25–30.

Prabhu, S., Poulose, E.K., 2012. Silver nanoparticles: mechanism of antimicrobial action, synthesis, medical applications, and toxicity effects. Int. Nano Lett. 2 (1), 32.

Ramirez, A.M., Saillard, N., Yang, T., Franssen, M.C., Bouwmeester, H.J., Jongsma, M.A., 2013. Biosynthesis of sesquiterpene lactones in pyrethrum (*Tanacetum cinerariifolium*). PLoS One 8 (5).

Rao, K.S., Babu, G.V., Ramnareddy, Y.V., 2007. Acylated flavone glycosides from the roots of *Saussurea lappa* and their antifungal activity. Molecules 12 (3), 328–344.

Rhee, J.K., Baek, B.K., Ahn, B.Z., 1985. Structural investigation on the effects of the herbs on *Clonorchis sinensis* in rabbits. Am. J. Chin. Med. 13 (01n04), 119–125.

Riaz, M., Altaf, M., Faisal, A., Shekheli, M.A., Miana, G.A., Khan, M.Q., Shah, M.A., Ilyas, S.Z., Khan, A.A., 2018. Biogenic synthesis of AgNPs with *Saussurea lappa* CB Clarke and studies on their biochemical properties. J. Nanosci. Nanotechnol. 18 (12), 8392–8398.

Robinson, A., Yashvanth, B.K., Rao, J.M., Madhavendra, S.S., 2010. Isolation of α-amyrin eicosanoate, a triterpenoid from the roots of *Saussurea lappa* Clarke-Differential solubility as an aid. J. Pharmaceut. Sci. Technol. 2, 207–212.

Royle, J.F., 1839. Illustrations of the Botany and Other Branches of the Natural History of the Himalayan Mountain and Flora of Cashmere. Allen & Co (Kostus 360).

Saleem, T.M., Lokanath, N., Prasanthi, A., Madhavi, M., Mallika, G., Vishnu, M.N., 2013. Aqueous extract of *Saussurea lappa* root ameliorate oxidative myocardial injury induced by isoproterenol in rats. J. Adv. Pharmceut. Technol. Res. 4 (2), 94.

Sastry, M.S., Dutta, N.K., 1961. A method for preparing tincture *saussurea*. Indian J. Pharm. 23, 247–249.

Shah, N.C., 1982. Herbal folk medicines in northern India. J. Ethnopharmacol. 6 (3), 293–301.

Shah, N.C., 2019. Kuṣṭha, *Saussurea costus* (*Saussurea lappa*): Its Unexplored History from the Atharvaveda.

Shah, R., 2006. Nature's Medicinal Plants of Uttaranchal: Herbs, Grasses and Ferns. Gyanodaya Prakashan, Kanpur.

Shao, Y., Huang, F., Wang, Q., Chengguie, D., 2005. Anti-inflammatory and cholagogic effects of aucklandiae. Jiangsu Yao Xue Yu Lin Chuang Yan Jiu 13 (4), 5–6.

Sharma, R.K., Sharma, S., Sharma, S.S., 2014. The long-term ambient storage-induced alterations in seed viability, germination and seedling vigour of *Saussurea costus*, a critically endangered medicinal herb of North West Himalaya. J. Appl. Res. Med. Aromat. Plant 1 (3), 92–97.

Sharma, S., Sharma, R., Sharma, P., Thakur, K., Dutt, B., 2019. Direct shoot organogenesis from seedling derived shoot tip explants of endangered medicinal plant *Saussurea costus* (Falc.) Lipsch. Proc. Natl. Acad. Sci. India B Biol. Sci. 89 (2), 755–764.

Sher, H., Hussain, F., Sher, H., 2010. Ex-situ management study of some high value medicinal plant species in Swat, Pakistan. Ethnobot. Res. Appl. 8, 017–024.

Singh, R., Chahal, K.K., Singla, N., 2017. Chemical composition and pharmacological activities of *Saussurea lappa*: a review. J. Pharmacogn. Phytochem. 6 (4), 1298–1308.

Singh, V.P., 2019. A review on pharmacodynamics of Ashtamangal ghrita and its uses in mental and physical growth in children. J. Pharmacogn. Phytochem. 8 (3), 3809–3812.

Stark, W.J., Stoessel, P.R., Wohlleben, W., Hafner, A.J.C.S.R., 2015. Industrial applications of nanoparticles. Chem. Soc. Rev. 44 (16), 5793–5805.

Sun, C.M., Syu, W.J., Don, M.J., Lu, J.J., Lee, G.H., 2003. Cytotoxic sesquiterpene lactones from the root of *Saussurea lappa*. J. Nat. Prod. 66 (9), 1175–1180.

Sutar, N., Garai, R., Sharma, U.S., Singh, N., Roy, S.D., 2011. Antiulcerogenic activity of *Saussurea lappa* root. Int. J. Pharm. Life Sci. 2 (1), 516–520.

Taniguchi, M., Kataoka, T., Suzuki, H., Uramoto, M., Ando, M., Arao, K., Magae, J., Nishimura, T., Otake, N., Nagai, K., 1995. Costunolide and dehydrocostus lactone as inhibitors of killing function of cytotoxic T lymphocytes. Biosci. Biotechnol. Biochem. 59 (11), 2064–2067.

Thakur, V., Bains, S., Pathania, S., Sharma, S., Kaur, R., Singh, K., 2020. Comparative transcriptomics reveals candidate transcription factors involved in costunolide biosynthesis in medicinal plant-*Saussurea lappa*. Int. J. Biol. Macromol. 150, 52–67.

Thara, K.M., Zuhra, K.F., 2012. Comprehensive In-vitro pharmacological activities of different extracts of *Saussurea lappa*. Eur. J. Exp. Biol. 2 (2), 417–420.

Thompson, R.C., 1949. Dictionary of Assyrian Botany. British Academy.

Tonga, G.Y., Saha, K., Rotello, V.M., 2014. 25th anniversary article: interfacing nanoparticles and biology: new strategies for biomedicine. Adv. Mater. 26 (3), 359–370.

Tsarong, T.J. (Ed.), 1986. Handbook of Traditional Tibetan Drugs: Their Nomenclature, Composition, Use, and Dosage. Tibetan Medical Publications.

Upadhyay, O.P., Singh, R.M., Dutta, K., 1996. Studies on antidiabetic medicinal plants used in Indian folk-lore. Aryavaidyan 9 (3), 159–167.

Venkataranganna, M.V., Gopumadhavan, S., Sundaram, R., Mitra, S.K., 1998. Evaluation of possible mechanism of anti-ulcerogenic activity of UL-409, a herbal preparation. J. Ethnopharmacol. 63 (3), 187–192.

Verma, P., Mathur, A.K., Jain, S.P., Mathur, A., 2012. In vitro conservation of twenty-three overexploited medicinal plants belonging to the Indian sub continent. Sci. World J. 2012.

Waly, N.M., 2009. Verifying the scientific name of Costus [*Saussurea lappa* ((Decne.) CB Clarke.)–Asteraceae]. Science 21 (2).

Wang, B.X., 1997. Pharmacology of Modern Chinese Medicine. Tianjin Science and Technology Press, Tianjing, China, p. 436.

Wang, X.Y., 2004. Influence of *Saussurea lappa* on acute damage of gastric mucosa of rats. Zhong Yi Yan Jiu 17 (2), 21–22.

Warghat, A.R., Bajpai, P.K., Rewang, S., Kapoor, S., Kumar, J., Chaurasia, O.P., Srivastava, R.B., 2016. In vitro callus induction and plantlet regeneration of *Saussurea lappa* (Clarke.) from Ladakh region of India. Proc. Natl. Acad. Sci. India B Biol. Sci. 86 (3), 651–660.

Wealth of India, 1972. A Dictionary of Indian Raw Materials and Industrial Products, vol. 9. CSIR Publication, New Delhi, pp. 240–243.

Wei, H., Yan, L.H., Feng, W.H., Ma, G.X., Peng, Y., Wang, Z.M., Xiao, P.G., 2014. Research progress on active ingredients and pharmacologic properties of *Saussurea lappa*. Studies 43, 48.

Yaeesh, S., Jamal, Q., Shah, A.J., Gilani, A.H., 2010. Antihepatotoxic activity of *Saussurea lappa* extract on D-galactosamine and lipopolysaccharide-induced hepatitis in mice. Phytother Res. 24 (S2), S229–S232.

Yamahara, J., Kobayashi, M., Miki, K., Kozuka, M., Sawada, T., Fujimura, H., 1985. Cholagogic and antiulcer effect of *Saussureae radix* and its active components. Chem. Pharmaceut. Bull. 33 (3), 1285–1288.

Yang, H., Xie, J., Sun, H., 1997. A new baccharane-type triterpenoid isolated from the roots of *Saussurea lappa* CB Clarke. Acta Bot. Sin. 39 (7), 667–669.

Yang, H., Xie, J., Sun, H., 1998. Research progress on the medicinal plant–*Saussurea lappa*. Nat. Prod. Res. Dev. 10 (2), 90–98.

Yin, H.Q., Fu, H.W., Hua, H.M., Qi, X.L., Li, W., Sha, Y., Pei, Y.H., 2005. Two new sesquiterpene lactones with the sulfonic acid group from *Saussurea lappa*. Chem. Pharmaceut. Bull. 53 (7), 841–842.

Yoshikawa, M., Hatakeyama, S., Inoue, Y., Yamahara, J., 1993. Saussureamines A, B, C, D, and E, new anti-ulcer principles from Chinese *Saussureae radix*. Chem. Pharmaceut. Bull. 41 (1), 214–216.

Yu, Z.J., 1986. Observation of action of 21 types of Chinese traditional medicine on in vitro solution of fibrous protein. Zhong Xi Yi Jie He Za Zhi 6 (8), 484.

Yuuya, S., Hagiwara, H., Suzuki, T., Ando, M., Yamada, A., Suda, K., Kataoka, T., Nagai, K., 1999. Guaianolides as immunomodulators. Synthesis and biological activities of dehydrocostus lactone, mokko lactone, eremanthin, and their derivatives. J. Nat. Prod. 62 (1), 22–30.

Zahara, K., Tabassum, S., Sabir, S., Arshad, M., Qureshi, R., Amjad, M.S., Chaudhari, S.K., 2014. A review of therapeutic potential of *Saussurea lappa*-an endangered plant from Himalaya. Asian Pac. J. Trop. Med. 7, S60–S69.

Zaib-Un-Nisa, S.J., Anwar, M., Sajad, S.H.S., Farooq, G., Ali, H., 2019. 60. Micropropagation through apical shoot explants and morphogenic potential of different explants of *Saussurea lappa*: an endangered medicinal plant. Pure Appl. Biol. 8 (1), 585–592.

CHAPTER

Stevia rebaudiana

11

Anita Kumari[1], Varun Kumar[1], Nikhil Malhotra[2]

[1]*Department of Ornamental Plants and Agricultural Biotechnology, Agricultural Research Organization, Volcani Center, Rishon LeZion, Israel;* [2]*ICAR-National Bureau of Plant Genetic Resources Regional Station, Shimla, Himachal Pradesh, India*

11.1 Introduction

The annual *Stevia rebaudiana* (Bertoni), a member of family Asteraceae, is a well-known plant to produce the low-calorie natural sweetener steviol that reportedly is around 300 times sweeter than the saccharose. The plant *Stevia* constitutes a plethora of primary and secondary metabolites; however, among them stevioside and rebaudioside A are the chief sweetening components that have market potential. Besides serving as a sweetener, the steviol glycoside also exerts biological activities that include antioxidant, antimicrobial, and antifungal actions. This plant discovery is credited to Dr. Moises Santiago Bertoni who discovered it in Paraguay in 1888; however, the scientific name is on the name of chemist Dr. Rebaudi from Paraguay (Kienle, 2016).

Owing to sweet taste, the herb is also known as candy leaf, honey leaf, and sweet leaf (Carakostas et al., 2008). The plant is commercially cultivated in a number of countries such as Brazil, Central America, China, India, Korea, Paraguay, and Thailand (Mizutani and Tanaka, 2002; Kim et al., 2002; Jaroslav et al., 2006). The *Stevia* cultivars reportedly differ in their nutritional value and this trait could be strategically utilized for the crop development programs aiming toward enhanced productivity and sustainability under different agro-climatic zones.

Stevia was introduced to Northern India by Council of Scientific Industrial Research Institute of Himalayan Bioresource Technology (CSIR-IHBT) during 2000 (Megeji et al., 2005). Over the years, several attempts to enhance the *Stevia* productivity in north-western Himalayan region were made. Kumar et al. (2014) concluded that under the mid-hill conditions of north-western Himalayas, the pinching of *Stevia* plants after 40 days of transplanting enhanced the dry leaf biomass over control that led to higher economical returns. In another report, planting of *Stevia* in broad bed and furrow (BBF) resulted in higher leaf yield over flat bed and camber bed. Moreover, harvesting *Stevia* once at 50% flower bud also proves to be beneficial over twice harvesting. In the same study, the application of 50:60: 50 Kg NPK/ha and planting *Stevia* in BBF in high rainfall area was found to be

Himalayan Medicinal Plants. https://doi.org/10.1016/B978-0-12-823151-7.00005-2

advantageous (Kumar et al., 2012a,b). Kumar et al. (2013) studied the effect of different shade levels (no shade, 25%, 50%, and 70% shade) on the growth of *Stevia* and revealed that reduction in plant growth, development, and total steviol glycosides was observed with increase in shade percentage level. The improved cultivars/variety released by CSIR-IHBT, India, includes Him Stevia, Madhuguna, and Madhuguni (Yadav et al., 2013).

In general, *Stevia* is a short-day plant that grows to the height of around 65–80 cm and flowers during January to March (Fig. 11.1). The flower is white with pale purple throat and arranged in the small corymbs (Goettemoeller and Ching, 1999; Singh and Rao, 2005). Sandy soil and sunny warm location are favorable for the growth of plant (Goettemoeller and Ching, 1999; Singh and Rao, 2005). The sweetener compounds stevioside and rebaudioside A are highest in the leaves, consequently leaf tissue is actively exploited for isolation of steviol. Moreover, the leaf of *Stevia* is also reported to constitute the mixture of eight diterpene glycosides namely isoStevial, stevioside, rebaudiosides (A, B, C, D, E, F), steviolbioside, and dulcoside A (Rajasekaran et al., 2008; Goyal et al., 2010). Though the *Stevia* leaves are sweet tasting, there are reports on the bitter aftertaste that is attributed to the presence of essential oils, tannins, and flavonoids (Phillips, 1987). The climate and

FIGURE 11.1

A mature *Stevia rebaudiana* plant.

agronomical practices are reported to fairly impact the quality and quantity of stevioside and low day conditions are suggested to be more appropriate. As far as harvesting time for *Stevia* leaves is concerned, it is determined by the land type, cultivars of the *Stevia*, and the growing season. In general, it is advisable to do first harvest after 4 months of plantations and subsequently after 3 months. Drying temperature also plays critical role in determining the quality of *Stevia* products, and drying at the higher temperature is reported to negate the medicinal and commercial value of products.

With awareness among masses regarding development of chronic diseases upon consumption of harmful sugars derived either from synthetic ingredients or from chemical synthesis, the inclination toward safer natural sweetener such as *Stevia* is observed. The toxicological studies also revealed that the consumption of *Stevia* sweetener did not lead to allergic, mutagenic, or carcinogenic effects (Pol et al., 2007). Further, the sweetener exhibits zero glycemic index and no caloric values (Raymond, 2010; Atteh et al., 2008; Kroyer, 2009; Puri et al., 2011a,b).

The stevioside constitutes 4%–20% dry matter of leaves; however, it is reported to vary among cultivars and propagation techniques (Brandle and Rosa, 1992; Geuns, 2000). Although the rebaudioside A content is lower than that of stevioside, it significantly contributes toward sweet pleasant taste (Sahelin and Gates, 1999, Mitchell, 2008). The *Stevia* stem also encompasses lower concentration of stevioside; however, to reduce the processing cost, they are removed during the extraction process (Brandle and Rosa, 1992).

The chloroplast is considered as a site for the synthesis of steviol glycosides precursor and the tissue encompassing lesser chlorophyll contain less or trace of steviol glycosides (Brandle and Rosa, 1992; Singh and Rao, 2005; De Oliveira et al., 2011). The stevioside concentration in the leaves is reportedly highest prior to onset of flowering indicating that for commercial use, the leaves should be harvested before flowering (Brandle and Rosa, 1992). The *Stevia* is available in market as green or white depigmented powder and as an extracted solution (Mishra et al., 2010; Brandle and Rosa, 1992; Abou-Arab et al., 2010).

The other remarkable attributes of *Stevia* include high content of protein, dietary fibers, mineral, and essential amino acids (Abou-Arab et al., 2010). Moreover, the *Stevia* leaves have gained the interest of scientific investigations due to antimicrobial (Jayaraman et al., 2008), antiviral (Kedik et al., 2009), antifungal (Silva et al., 2008), antihypertensive (Chan et al., 1998; Lee et al., 2001; Hsieh et al., 2003), antihyperglycemic (Jeppesen et al., 2002; Benford et al., 2006), antitumor (Jayaraman et al., 2008), antiinflammatory, antidiarrheal, diuretic, antihuman rotavirus activities (Das et al., 1992; Takahashi et al., 2001), anti-HIV (Takahashi et al., 1998), hepatoprotective (Mohan and Robert, 2009), and immunomodulatory properties (Jaroslav et al., 2006; Chatsudthipong and Muanprasat, 2009). So keeping in the view of increasing demand, the growing information regarding *Stevia*, its products, medicinal value, and crop improvement is highly desirable.

11.1.1 Details of general cultivation practices adopted for *Stevia* farming

Stevia plants require well-drained nutrient-rich soil, semihumid climate, and the optimal temperature is around 23.8°C. The red and loamy soil with a pH of six to seven enriched with farmyard manure (approximately 25 tons/hectare) is amiable for the commercial cultivation of plant. The extreme low temperature, i.e., below 5°C, and saline soils are found to be unfavorable for the plant growth (Jaitak et al., 2011). Low seed germination rate is limiting factor for its commercial cultivation and thus, clonal propagation, air layering, budding, and cutting are preferred methods. For induction of roots from the cut plants, 75–100 ppm GA_3 was found to be effective. The microcutting technique also offers the way to clone selected genotypes and achieve mass production (Osman et al., 2013). Manual weeding is preferred (Walia et al., 2011) and the leaf yield and stevioside accumulation were found to be governed by environmental and agronomic practices (Pal et al., 2015). So the scope of improvement of yield, biomass, and secondary metabolite profile could be attained through intense scrutiny of growing locations and cultivation management techniques especially nutrient management.

11.2 Medical significance of *Stevia*

Zero calorie sweetener *Stevia* has gained popularity as a substitute to synthetic white sugar, and the purified extract rebiana encompasses 95% and above steviol glycoside (Ashwell, 2015). Joint Food and Agriculture Organization/WHO Expert Committee on Food Additives and Codex are the regulatory bodies that approve high-purity *Stevia* extracts for usage in food and beverages industry (Ashwell, 2015). In albino rats, the aqueous extract of *Stevia* is effective for caloric management and weight control by decreasing the levels of random blood glucose, fasting glucose, and glycosylated hemoglobin indicating its suitability for neutraceutical therapy (Ahmad and Ahmad, 2018). *Stevia* intake is also reported to lower the energy intakes that eventually contributed toward reduction and prevention of obesity (Nettleton et al., 2019). The presence of myriads of biochemical components in *Stevia* imparts to the medicinal properties of plant. Table 11.1 briefly summarizes the medicinal properties and the underlying mechanism.

Thus, not only the sugar levels but the *Stevia* consumption also lead to lowering other health-related issues such as caries, obesity, and diabetes (Savita et al., 2004). The high molecular weight hydrophilic stevioside that is not absorbed in the intestine and the digestive enzymes from animals and humans have also been demonstrated to be ineffective to degrade stevioside (Wingard et al., 1980; Hutapea et al., 1997; Koyama et al., 2003).

Table 11.1 Medicinal properties and mode of action of *Stevia*.

S. No.	Medicinal properties	Mechanism	References
1.	Hypotensive effect	Inhibition of the Ca(2^+) influx	Liu et al. (2003), Savita et al. (2004), Onakpoya and Heneghan (2014)
2.	Hypoglycemic effect	Peroxisome proliferator −activated receptor-γ −dependent mechanism	Assaei et al. (2016), Gregersen et al. (2004), Jeppesen et al. (2003), Holvoet et al. (2015)
3.	Antiinflammatory and immunomodulatory effects	Inhibition of the secretion of proinflammatory cytokines TNF-α, interleukin (IL)-6, IL-1 by downregulating NF-κB	Boonkaewwan et al. (2006), Mizushina et al. (2005), Ajagannanavar et al. (2014)
4.	Anticancerous activity	G1-phase cell cycle arrest, activation of mitochondrial apoptotic pathway	Gupta et al. (2017), Deshmukh and Kedari (2014), Mann et al. (2014), Vaško et al. (2014), Khaybullin et al. (2014)
5.	Antioxidant activity	Free radical scavenging	Bender et al. (2015)

Atteh et al. (2011) reported that the steviol glycosides are broken down to steviol and glucose by the intestinal flora. While the released glucose is not absorbed into the bloodstream, the steviol is transformed to glucuronide in the liver and finally released in urine and feces. The acceptable daily intake of steviol glycoside is generally expressed as steviol equivalents and is 4−5 mg/kg body weight per day (Chatsudthipong and Muanprasat, 2009; Awney et al., 2011).

The few members of Asteraceae family have been shown to have allergenic effect; however, the incidences of allergy upon consumption of highly purified *Stevia* extracts need the scientific evidence to support such claims (Urban et al., 2015). Samuel et al. (2018) analyzed the effects of highly purified steviol glycosides on toxicity and carcinogenicity on rodents and they did not observe adverse effects.

Currently, *Stevia* is widely used to sweeten the products such as soft drink, beverages, ice cream, confectionary, and bakery. Although this potent sweetener possesses several health-promoting properties, the research in the area of determining the interaction among *Stevia* metabolites with other food components along with determining the acceptable daily intake of food additive is highly desirable. Further, to draw conclusions regarding the observed medicinal effect of *Stevia* extract and the stevioside, the purity of compound is a crucial factor and the steviol pharmacology needs attention (Chen et al., 2005; Ferreira et al., 2006; Wong et al., 2006).

11.3 Breeding attempts for *Stevia* improvement

The general objectives of *Stevia* selection and the breeding program are to target the increased production of steviol glycoside along with raising the multistress-resistant plants (Tavarini et al., 2018). For crop improvement programs, the germplasm collection and conservations are essential in terms of providing source of superior gene pool for broadening the genetic diversity. However, the success of *Stevia* breeding is determined by numerous factors such as choice of parents, crosses, raising of adequate population, and selection procedure. Prior to breeding attempts, the total glycoside concentration in leaf tissue was reportedly around 2%–10% on dry-weight basis which has improved to up to 20% via breeding and selection efforts (Huang et al., 1995). The *Stevia* selection procedure for higher glycoside content is more effective and economically cheaper in late-growing season due to higher genetic variability (Yao et al., 1999). The extensive *Stevia* cross breeding/selection programs have been undertaken by several countries such as China, India, Japan, Korea, Russia, and Taiwan. These attempts led to release of several improved high-yielding cultivars such as Suweon 2, Suweon 11 (released by Korea), Yunri, Yunbing, and Zongping from China, and Madhuguna and Madhuguni from India (Yadav et al., 2013). Fig. 11.2 summarizes the desirable traits that need to be targeted for the *Stevia* improvement programs.

Since the landrace *Stevia* did not possess the optimal concentration of steviol glycoside required for the product development, the development and release of superior cultivars such as RSIT-94-1306 and RSIT 94-751 are considered landmark (Sys et al., 1998; Marsolais et al., 1998). The development of RSIT 95-166-13 is

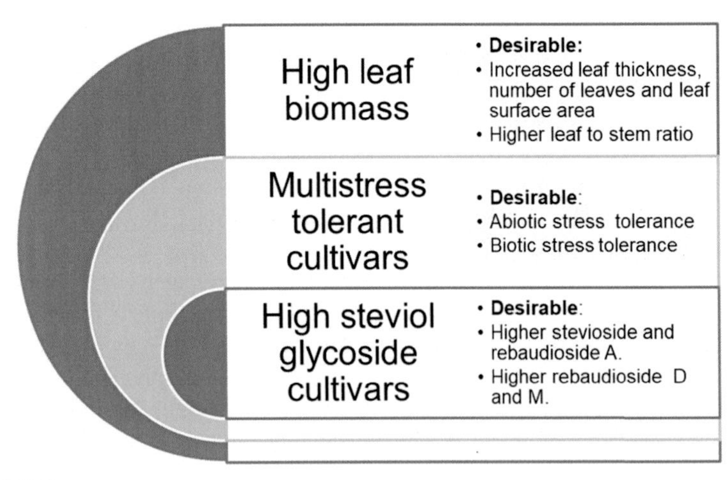

FIGURE 11.2

Desirable traits to be targeted for *Stevia* breeding programs.

Based on Tavarini, S., Passera, B., Angelini, L.G., 2018. Crop and Steviol Glycoside Improvement in Stevia by Breeding. pp. 1–31.

also one of the landmarks in *Stevia* breeding program as it possesses unique combination of high rebaudioside-C:stevioside ratio that distinguished it from its parents as well as from the other *Stevia* cultivars (Tavarini et al., 2018).

Since the observed heterogeneity is higher in *Stevia* owed to its self-incompatible trait, the recurrent selection is the effective way to enhance the foliage yield and total glycoside content. The development of synthetic cultivar via intercrossing clones or sibbed lines through recurrent selection is highly desirable to obtain rebaudioside-A and steviol glycoside—rich plants. Due to the higher natural out-crossing and absence of efficient pollination control system in *Stevia*, the composites and synthetics hold importance as they capture the available heterosis. The development of AC Black Bird (the synthetic cultivar) that exhibits high level of total glycosides (up to 14%) and 9.1:1 ratio of rebaudioside-A to stevioside and ATCC Accession No. PTA-444 with at least 2.56 times more rebaudioside-A than stevioside and capable of being cultivated by seed propagation is a big achievement in this field (Brandle, 2001).

Another alternative to the time- and labor-consuming conventional breeding procedure is to broaden the variability by the means of mutagenesis. Broadly, there are two types of mutagenic agents namely physical and chemical mutagenic agents that are reportedly used in the breeding programs. The commonly used mutagenic agents include X-rays, γ-rays, ethyl methanesulfonate, diethylsulfate, N-methyl-N-nitrosourea, N-ethyl-N-nitrosourea, N-methyl-N-nitrosourea, and N-ethyl-N-nitrosourea.

In *Stevia*, Nurhidayah et al. (2014) conducted radiosensitivity test with an objective to identify the effective dose for in vitro mutagenesis. For this, the shoot tips were irradiated with gamma radiation at 0, 10, 20, 30, 40, 60, and 80 Gy. Among all, 60 and 80 Gy were proved to be lethal for the shoot tip survival, while the effective doses were found to be 10, 20, 30, and 40 Gy. An experiment conducted by Pande and Khetmalas (2011) on the gamma-irradiated seeds (doses of gamma radiation are 5, 10, 15, 20, 25, and 30 kiloradians (kR)) of *Stevia* revealed that the low dose of gamma rays 5—15 kR is suitable for callus induction, proliferation, and for somatic embryogenesis, while the higher dose is reported to cause necrosis in explants.

Stevia breeding through induction of polyploidy could also be a promising approach to enhance the steviol glycosides. In general, polyploid plants are desirable as they are large and healthy (Zhang et al., 2018). The polyploidy arises spontaneously as a consequence of cell division failures or due to fusion of unreduced gametes (Comai, 2005). In breeding programs, colchicine is a widely used chemical for induction of polyploidy (Blakeslee and Avery, 1937). Zhang et al. (2018) reported that the colchicine treatment (0.05% for 48 h or 0.1% for 24 h) effectively induced polyploidy in germinating seeds of *Stevia* as evidenced by flow cytometry analysis. The tetraploid plants produced in the study were also shown to accumulate the higher amount of stevioside and rebaudioside A, more glands, higher chlorophyll content, and are more vigorous as in comparison to diploid plants indicating the suitability of polyploidy breeding for commercial production.

Reports on induction of polyploidy exploiting various tissues such as seeds (Ghonema et al., 2015; Yadav et al., 2013) and buds (Ghonema et al., 2015; Chavan et al., 2014; Hegde et al., 2015; Mahadi, 2012) by either involving imbibing or spraying with different concentrations of colchicine are also available. In an effort to develop polyploid strain of *S. rebaudiana*, Valois (1992) performed a study in which seeds were soaked in varying concentration of colchicine (0.001%–0.5% for 18 h). The results revealed that the lowest colchicine level is apt for inducing polyploidy.

In another set of study, in order to achieve the higher biomass, the seeds of *S. rebaudiana* (Bertoni) were subjected to colchicine treatment at varying concentrations (0.01%, 0.05%, 0.1%, 0.2%, 0.4%, and 0.6%) for 24 h. The results revealed that although the survival rate of seedling above the concentration of 0.2% is below 50%, the maximum number of tetraploids was obtained with 0.6% colchicine treatment, while mixoploidy was observed at 0.2% colchicine. The striking feature of autotetraploid included increased leaf size, thickness, chlorophyll content, and reduction in internodal length (Yadav et al., 2013).

To sum up, the breeding program of *Stevia* should target the genetic and environmental variation along with the interaction between the two variations. As considerable genetic and phenotypic variations in terms of plant size, flowering period, steviol yield, and varying composition of glycosides are observed, these variations could be the key for development of desirable cultivars. However, the extensive research in this direction is still required to generate *Stevia* cultivars that exhibit resiliency toward environmental changes.

11.4 Biochemical profile analysis of *Stevia*

For unveiling the medicinal properties of *Stevia*, dissecting the biochemical profile of *Stevia* is utmost crucial (Allam et al., 2001). The proximate and mineral content analyses of dried *Stevia* leaves revealed that besides sweetener, it is also a good source of crude fiber, carbohydrate, calcium, phosphorus, sodium, potassium, iron, magnesium, and zinc (Gasmalla et al., 2014). The other diterpene glycosides isolated from *Stevia* leaf are rebaudioside A, B, C, D, E, and dulcoside A. This makes *Stevia* a popular choice for food, cosmetic, and pharma industry. Extensive prominent analytical techniques were exploited to assess the distribution of glycosides such as thin layer chromatography (Tanaka, 1982), droplet countercurrent chromatography (Kinghorn et al., 1982), overpressured layer chromatography (Fullas et al., 1991), and capillary electrophoresis (Liu and Li, 1995). The advanced exploited techniques also involve silica gel and high-performance liquid chromatography (HPLC) (Nikolova-Damyanova et al., 1994), hydrophobic chromatography (Vanek et al., 2001), HPLC-ESIMS (Choi et al., 2002), highly sensitive reversed phase (RP-HPLC) (Minne et al., 2004), and LC-MS-ESI (Rajasekaran et al., 2008).

For human consumption of stevioside, the enzymatic extraction appears to be more appropriate alternative. Report on optimization of enzymatic extraction of stevioside from the leaf tissue is available, wherein the hemicellulose was observed to give the highest stevioside yield followed by cellulose and pectinase (Puri et al., 2011a,b). However, more research effort in this field is required.

The root and leaf extracts of *S. rebaudiana* were also shown to constitute the naturally occurring polysaccharides, i.e., fructooligosaccharides that indicates toward the possibility of utilization of *Stevia* explants extracts as a dietary supplement. The leaves of *Stevia* also constitute palmitic acid, stearic acid, palmitoleic, oleic, linoleic, and linolenic acids (Tadhani and Subhash, 2006). The antioxidant activity of aqueous extract of *S. rebaudiana* leaves and calli is also established. Wherein the folic acid is found to be predominant vitamin in leaf extract followed by vitamin C, while for callus, it is *vice-a-versa*. Among phenolics, pyrogallol was found to prevail in both leaf extract and callus (Kim et al., 2011).

It was also established by a study that glycol-aqueous extract of *Stevia* possesses the highest amount of phenols (twofold higher) and flavonoids over aqueous and ethanolic extracts. While the highest protein content (226.83 mg/g of *Stevia* leaves) was found in aqueous followed by glycol-aqueous (374.67 mg/g of *Stevia* leaves) suggesting the suitability of glycol-aqueous and aqueous extracts as an excellent source of biologically active protein and peptides (Gaweł-Beben et al., 2015). With advancement in cutting edge technologies in the area of chemical and metabolic profiling, it is anticipated that in near future, our current understanding of biosynthetic routes and spatial distribution of stevioside and related diterpenoids could be elucidated.

11.5 Details on steviol glycoside biosynthesis pathway and the associated genes

Steviols are prominently found in high concentration in the leaf tissue, so their probable role as an insect repellent or regulator of gibberellic acid could not be negated (Smith and Van-Stadin, 1992). In *Stevia*, the divergence of steviol glycoside and gibberellin pathways is at kaurene which reportedly is converted to steviol via mevalonate pathway followed by glycosylation and rhaminosylation to form the chief sweetener. The general route of the formation of steviols involved precursor synthesis in chloroplast followed by precursor transportation to endoplasmic reticulum to golgi apparatus and finally to the vacuoles.

Around 16 enzymes are reportedly involved in steviol glycoside biosynthesis of which the 7 are involved in catalyzing various steps of methylerythritol-4-phosphate pathway (Wanke et al., 2001) (Fig. 11.3). These seven enzymes include deoxyxylulose phosphate synthase (DXS), deoxyxylulose phosphate reductase, 4-diphosphocytidyl-2-C-methyl-d-erythritol synthase, 4-diphosphocytidyl-2-C-methyl-d-erythritol kinase (CMK), 4-diphosphocytidyl-2-Cmethyl-d-erythritol 2,4-cyclodiphosphate synthase, 1-hydroxy2-methyl-2(E)-butenyl-4-diphosphate synthase (HDS), and 1-hydroxy-2-methyl-2(E)-butenyl-4-diphosphate reductase (HDR). The next four steps are

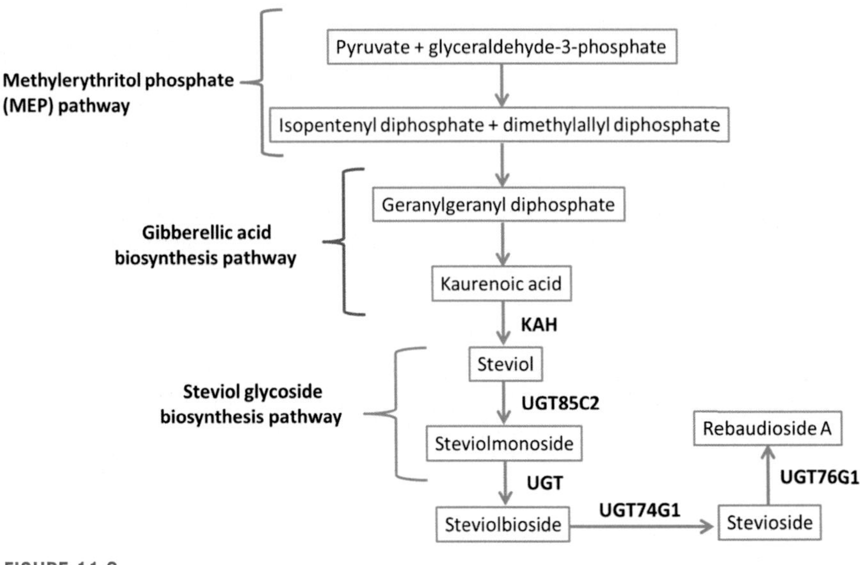

FIGURE 11.3

General biosynthetic pathway of steviol glycoside biosynthesis. Enzyme abbreviations are KAH, kaurenoic acid hydroxylase and (UGTs) uridine diphosphate (UDP)-dependent glycosyltransferases.

Adapted from Guleria, P., Yadav, S.K., 2013a. Insights into steviol glycoside biosynthesis pathway enzymes through structural homology modeling. Am. J. Biochem. Mol. Biol. 3, 1–19; Guleria, P., Yadav, S.K., 2013b. Agrobacterium mediated transient gene silencing (AMTS) in Stevia rebaudiana: insights into steviol glycoside biosynthesis pathway. PLoS One 8 (9).

catalyzed by geranylgeranyl diphosphate synthase (GGDPS), copalyl diphosphate synthase (CDPS), kaurene synthase (KS), and kaurene oxidase (KO) that mediate the formation of kaurenoic acid from geranylgeranyl diphosphate.

As mentioned above, the steviol glycoside and gibberellin pathways diverge at kaurene where two different endoplasmic reticulum membrane—located cytochrome P450 monooxygenases (CYPs) act to convert ent-kaurenoic acid into either steviol by kaurenoic acid hydroxylase (KAH) or gibberellic acid by kaurenoic acid oxidase. The final phase involves the glycosylation of steviol in cytosol by UDP-glycosyltransferases (UGTs) such as UGT85C2, UGT74G1, and UGT76G1 to form various types of steviol glycosides. Most of the enzymes involved in the biosynthesis of SGs in *Stevia* have been identified. Therefore, in order to increase the production of SGs, current researches should focus on metabolic engineering of these biosynthetic pathways. Silencing three major UGT genes through *Agrobacterium*-mediated gene transformation of *S. rebaudiana* was found to increase SGs production (Guleria and Yadav, 2013a,b).

Understanding the regulation of genes involved in steviol glycoside biosynthesis, their cloning, and characterization is crucial for the optimization of SGs production in the *Stevia* leaves. Kumar et al. (2012a,b) reported that the SGs biosynthesis pathway genes were found to be responsive to GA_3 treatment and exhibited upregulation of SrMCT, SrCMK, SrMDS, and SrUGT74G1, while methyl jasmonate (MeJA) and kinetin treatment were found to downregulate the expression of most of the genes of the pathway. The authors also cloned seven cDNA sequences to full length employing RACE-PCR techniques.

Understanding a set of total transcripts under a particular condition which is more precisely termed as transcriptome is anticipated to provide a way to identify the transcript abundance during SGs production. Chen et al. (2014) performed an RNA-seq analysis of three *Stevia* genotypes that differed in their rebaudioside and SGs content. The analysis generated around 191,590,282 high-quality reads and a total of 80,160 unigenes were annotated. The gene sequences of all the enzymes that are associated with SGs biosynthesis were found in the data. Additionally, the data mining identified 10, 070 SSRs and 44,510 SNPs. The involvement of UGT in the production of SGs is known; thus, the data are also mined to identify UGTs related to SGs biosynthesis. The data generated in this study provided a genomic resource for the species.

In another set of study, Kim et al. (2015) integrated metabolomics and transcriptomic analysis to understand the diterpenoid metabolism in *Stevia*. They performed the RNA-seq analysis of trichomes and leaf without trichomes. The analysis revealed that SGs accumulation occurs in leaf cells, while the trichomes accumulates labdane-type diterpenoids such as oxomanoyl oxide and agatholic acid. The analysis also revealed that 20% of assembled unigenes were differentially expressed in trichomes, while less than 12% genes showed expression specificity in the leaves without trichomes. Most of the SGs biosynthesis pathway genes showed predominant expression in leaf trichomes at the site of SG accumulation, while the copalyl diphosphate synthase (SrCPS2) and KS1-like (SrKSL), homologs of SrCPS and SrKS1, were specifically expressed in trichomes. The study also highlighted the significance of multiomic study to provide useful information for specialized diterpenoids that exhibit preferential accumulation in different tissues of *Stevia* leaves.

Singh et al. (2017) conducted a study to understand the molecular mechanism of SGs biosynthesis by using global transcriptome analysis approach in leaf tissues during vegetative, budding, and flowering phases. The study revealed the advantages of high-throughput genomics to accelerate our understanding of developmental phase transitions involved in SGs biosynthesis. The results of study revealed that there is homeostatic balance among primary and secondary metabolism. Additionally, the study also identified putative candidates, for example, CYPs (124) and UGTs (45) that assist the modification and diversification of SGs in *Stevia*. Moreover, in the same study, the protein–protein interactome network analysis revealed that the interaction among various proteins was significantly higher in vegetative phase as compared to bud and flowering phase providing the hint of putative interacting protein hub toward regulation of SGs accumulation. This study opens the avenue for genetic engineering, molecular breeding program, and varietal improvement program in *Stevia*.

The study concerning the identification of stable reference genes in *S. rebaudiana* is also highly relevant for expression analysis of genes involved in SGs biosynthetic pathway. Lucho et al. (2018) examined the stability of seven reference genes (18S ribosomal RNA, Actin, Aquaporin, Calmodulin, Eukaryote elongation factor 1-a, Malate dehydrogenase, and Ubiquitin) under the effect of three stress-related elicitors (MeJA, salicylic acid (SA), and spermidine (SPD)) in *Stevia*. The outcome of the study indicated that Actin and Ubiquitin showed the stable expression under all the experimental conditions and thus can be used in future studies to eliminate erroneous result in studies concerning gene expression analysis.

In another study, Lucho et al. (2018) examined the effect of elicitor MeJA, SPD, SA, and paclobutrazol (PBZ) on SGs contents and transcription levels of SGs biosynthetic pathway genes. The elicitors (100 μM) were applied to plants grown in hydroponic system for different time interval (24, 48, 72, and 96 h). Among the elicitors, MeJA and SPD resulted in upregulation of SGs biosynthetic genes, while PBZ downregulated expression of most of the genes. Different patterns were observed for SA. The study concluded that overall HDR, GGDPS, CDPS, KS, KO, and KAH are elicitor-responsive genes and can be effectively regulated at least at the transcriptional level providing the opportunities to explore elicitor-responsive genes regulation in the SGs biosynthesis pathway. So in near future, multiomic analysis is anticipated to shed light on the gene regulation. The integration of multiomic tool and metabolic engineering is also expected to contribute toward enhanced yield of desired compounds without compromising on agronomic traits.

11.6 Approaches to improve steviol glycosides in *Stevia rebaudiana*

The advanced biotechnological techniques tools such as transcriptomics, metabolomics, and proteomics offer the novel horizons in the metabolic engineering of steviol glycoside biosynthesis pathway in *Stevia*. The other techniques that were reported to be exploited are random amplified polymorphic DNA (RAPD), intersimple sequence repeat (ISSR), amplified fragment length polymorphism, simple sequence repeat (SSR), high-performance thin-layer chromatography, and next-generation sequencing technology. Since these techniques are rapid, widely applicable, and cost-effective, they are useful to understand the underlying mechanism of secondary metabolism in *Stevia*.

The use of molecular markers makes significant contribution toward the management and utilization of crop genetic diversity. The report on the utilization of molecular markers as RAPD (Chester et al., 2013) and ISSR (Heikal et al., 2008) has been used in previous studies to analyze the diversity in small collection of individuals. For the construction of first *S. rebaudiana* genetic map in 1999, the RAPD technique was used (Yao et al., 1999). However, the lack of repeatability is the major issue associated with these markers. The private consortium PureCircle/Coca-Cola/

KeyGene has released the genome sequence of *S. rebaudiana* in 2017. Among the markers, SSR has wider applicability owed to high polymorphism, easy scoring, neutrality, and Mendelian inheritance. The pioneer studies in the area of development of SSR marker were by Kaur et al. (2015) and Bhandawat et al. (2015). They reported a set of 52 (Kaur et al., 2015) and 17 (Bhandawat et al., 2015) SSR markers via screening of 5548 *Stevia* ESTs sequences from leaf tissues retrieved from the NCBI. The SSR markers that were developed are exploited to classify 40 genotypes from selection or 12 local genotypes from Northern India. Recently, in an effort to assess the applicability of EST-SSRs markers in landrace genotypes, and to identify the genetic diversity and population structure and link between genetic variability and its structure and phenotypic SG variability, a major experimental study was conducted (Cosson et al., 2019). The study concluded the development of 18 EST-SSR from around 150,258 EST from The Compositae Genome Project of UC Davis. Additionally, they genotyped 145 *Stevia* individuals issued from 31 cultivars and 31 landrace of various origins worldwide. The landraces were reported to reveal twice the allelic richness (Cosson et al., 2019).

Recently, in a comprehensive study, the level of genetic and biochemical diversity was assessed for the *Stevia* genotypes collected from different locations of the world using HPLC analysis and RAPD markers. Additionally, the influence of the genotypes on the content of steviol glycosides, antioxidants, phenols, flavonoids, and tannins was analyzed using phytochemical assays. The study revealed that the genotypes from Morocco, Poland, Egypt, and Nigeria can be defined as samples of higher quality compared to other genotypes analyzed in terms of the amount of steviol glycosides. For the selection criterion, rebaudioside A/stevioside ratio of the genotypes of Australia, China, India, and Pakistan was shown to be valuable. Further, the genotypes from Morocco, Egypt, Poland, Nigeria, China, and India were suggested to be apt for *Stevia* breeding programs as they are genetically different (Dyduch-Siemińska et al., 2020). Thus, the biotechnological tools could not only contribute substantially to improve steviol glycoside production in *Stevia* but will also prove worthy to improve overall agronomy, biochemical traits, morphological, and physiological characteristics. The manipulation of steviol glycoside pathway in heterologous host such as *Escherichia coli* and *Saccharomyces cerevisiae* in near future could prove to be the stepping stone in the biotechnological production of steviol glycoside. The *E. coli* is shown to serve as a promising host for heterologous production of SGs.

Kong et al. (2015) identified a novel kaurenoic acid 13α-hydroxylase (KAH KAHn2) from *S. rebaudiana* based on the mining of RNA-seq data that characterized the activity of KAH_ACD93722 with the activity of 13α-hydroxylase and a UDP-glucosyltransferase UGT91D2w with the activity of steviol-13-monoglucoside-1,2-glucosyltransferase in *E. coli*. In another report, the *ent-copalyl diphosphate synthase* (CPPS) and *ent-kaurene synthase* (KS) genes from *S. rebaudiana* were heterologously expressed in *E. coli*. A step forward with an aim to enhance ent-kaurene production in *E. coli*, six GGPPSs from various microorganism and eight strains of *E. coli* were compared. The study revealed that the *E. coli* strain MG1655 coexpressing synthetic

CPPS-KS module and GGPPS from *Rhodobacter sphaeroides* produced highest ent-kaurene content (41.1 mg/L). The production of *ent*-kaurene was further elevated up to 179.6 mg/L by overexpression of the three key enzymes for isoprenoid precursor, 1-deoxyxylulose-5-phosphate synthase (DXS), farnesyl diphosphate synthase, and isopentenyl diphosphate isomerase from *E. coli* (Kong et al., 2015).

The attempts to enhance the stevioside synthesis in micropropagated plants via elicitor treatments are also reported. In this attempt, the node explants with an axillary bud were cultured in woody plant medium supplemented with varying concentrations of elicitors that include alginate (ALG), casein hydrolysate, pectin, yeast extract (YE), MeJA, SA, or chitosan. The most favorable elicitor treatment for maximum stevioside production was found to be 0.5 g/L ALG or 2.0 g/L YE. The study paved the path for further studies targeting the deliberate enhancement of stevioside in bioreactor system or in the field conditions (Bayraktar et al., 2016).

The effect of various PGRs and agar concentrations was also tested for biomass and steviol glycoside production. For callus induction, MS medium supplied with 2,4-dichlorophenoxyacetic acid alone or in combination with α-naphthalene acetic acid or indole acetic acid with two different concentrations (3.5 and 7.0 g/L) was established. The result revealed that lower concentration of agar is more effective for callusing and root induction, while among PGRs, the combination of BA (3.0 mg/L) with Kin (3.0 mg/L) and 3.5 g/L agar increases dulcoside-A and stevioside content. This study holds significance as the result of this study has further implications for the selection of appropriate PGRs for reliable regeneration and enhancement of SGs production for commercial purposes (Aman et al., 2013).

Generally, for the propagation of *Stevia* plant, vegetative propagation is reliable technique; however, it is not the soundproof method for production of large number of consistent chemotypes. Thus, both ex vitro and in vitro attempts are needed to optimize the leaf biomass production and SGs content. In one of the recent attempts, wherein the tissue culture method was utilized to support SGs production under field conditions has been investigated. The nodal segment was employed to assess the effect of various PGRs (BAP and KIN) on shoot formation. The regenerated shoot was rooted and then after the period of 3 weeks of acclimatization under greenhouse, the regenerants and seed-derived seedlings were transferred to field. After 16 weeks in the field, the SGs content in the leaves was analyzed and it did not differ significantly in both the populations. The crux of this study was the plant growth stage that did not hamper the SG content in the leaf tissue (Yücesan et al., 2016).

The growth and SG content of *Stevia* shoot grown in roller bioreactor were also investigated. The analysis revealed shoot growth intensity and the SGs production was 1.5–2.0 times higher than tube-grown shoots. The study also concluded that enhanced SG production could be attributed to chlorenchyma cells differentiation and the formation of specific subcellular structure (Bondarev et al., 2002). The effect of endophytic bacterial community in *Stevia* leaves along with deciphering the correlation between the endophyte and SGs accumulation is also investigated. By exploiting the pyrosequencing of 16S rRNA gene, the bacterial communities were studied in the leaf sample of *Stevia* at different growth stages with and without fulvic acid (FA) treatment. The analysis revealed predominance of *Proteobacteria*, by *Actinobacteria*, *Bacteroides*, *Firmicutes*, *Gemmatimonadetes*, and *Acidobacteria* regardless of growth stage and FA treatment. *Agrobacterium* and *Erwinia* dominated in seedling stage and then declined along with growth stages, i.e., vegetative and flowering stage, while *Sphingomonas* and *Methylobacterium* dominated in mature leaves of *Stevia*. The positive correlation for *Sphingomonas* and *Methylobacterium* while negative for Erwinia, *Agrobacterium*, and *Bacillus* for stevioside content was observed. This report is highly significant as it sheds light on the effect of growth regulators on the plant leave–associated endophytes (Yu et al., 2015). The field of induction of hairy root culture was also explored by the researchers to enhance the stevioside content. The work of Yamazaki et al. (1991) pointed that no stevioside was observed in the hairy roots of *Stevia*. This could be due to the involvement of plastids for stevioside formation and thus it could not be synthesized in the roots. Several other researchers attempted to optimize the hairy root culture protocols. Michalec-Warzecha et al. (2016) established a protocol that had 50% of transformation efficiency. The protocol involved infection of leaves and internodes with the strains LBA 9402 and ATCC 15384. However, they did not analyze the stevioside content in their samples.

Although the potential of hairy root cultures could not be exploited so far for the production of fair amount of stevioside, more attempts in future may succeed. The establishment of genetic transformation system for various cultivars could also provide the feasibility for transfer of desired traits in *Stevia*. The overexpression of target enzymes in transgenic approach is anticipated to bring a breakthrough in the production of stevioside content. The research concerning the precursor feeding, overexpression analysis, gene regulation studies, and pathway reconstitution in heterologous host are the other areas that could contribute significantly to achieve the goal of high stevioside production.

11.7 Conclusion and future prospects

Stevia popularity is anticipated to enhance more in the near future that could be attributed to the advancing knowledge of its sweetening as well as medicinal properties. However, several characteristic attributes such as the bitter aftertaste, low rebaudioside, and stevioside need to be improvised. Concentrated efforts of conventional plant breeding and biotechnological approaches are required to produce the cultivars/varieties having desirable agronomic traits. Improvement in our current understanding of the steviol glycoside biosynthesis pathway and its regulation with amalgamate of metabolic engineering is anticipated in near future and the production of steviol glycosides could be improved tremendously without compromising on quality.

References

Abou-Arab, A.E., Abou-Arab, A.A., Abu-Salem, M.F., 2010. Physico-chemical assessment of natural sweeteners steviosides produced from *Stevia rebaudiana* Bertoni plant. Afr. J. Food Sci. 4 (5), 269–281.

Ahmad, U., Ahmad, R.S., 2018. Anti diabetic property of aqueous extract of *Stevia rebaudiana* Bertoni leaves in streptozotocin-induced diabetes in albino rats. BMC Complement. Altern. Med. 18 (1), 179.

Ajagannanavar, S.L., Shamarao, S., Battur, H., Tikare, S., Al-Kheraif, A.A., Al Sayed, M.S.A.E., 2014. Effect of aqueous and alcoholic Stevia (*Stevia rebaudiana*) extracts against *Streptococcus* mutans and Lactobacillus acidophilus in comparison to chlorhexidine: an in vitro study. J. Int. Soc. Prev. Community Dent. 4 (2), S116.

Allam, A.I., Nassar, A., Besheti, S.Y., 2001. Nitrogen fertilizer requirements of *Stevia rebaudiana* under Egyptian conditions. Egypt. J. Agric. Res. 79, 1005–1018.

Aman, N., Hadi, F., Khalil, S.A., Zamir, R., Ahmad, N., 2013. Efficient regeneration for enhanced steviol glycosides production in *Stevia rebaudiana* (Bertoni). Comptes Rendus Biol. 336 (10), 486–492.

Ashwell, M., 2015. Stevia, nature's zero-calorie sustainable sweetener: a new player in the fight against obesity. Nutr. Today 50 (3), 129.

Assaei, R., Mokarram, P., Dastghaib, S., Darbandi, S., Darbandi, M., Zal, F., Akmali, M., Omrani, G.H.R., 2016. Hypoglycemic effect of aquatic extract of Stevia in pancreas of diabetic rats: PPARγ-dependent regulation or antioxidant potential. Avicenna J. Med. Biotechnol. 8 (2), 65.

Atteh, J., Onagbesan, O., Tona, K., Decuypere, E., Geuns, J., Buyse, J., 2008. Evaluation of supplementary Stevia (*Stevia rebaudiana* Bertoni) leaves and stevioside in broiler diets: effects on feed intake, nutrient metabolism, blood parameters and growth performance. J. Anim. Physiol. Anim. Nutr. 92 (6), 640–649.

Atteh, J., Onagbesan, O., Tona, K., Buyse, J., Decuypere, E., Geuns, J., 2011. Potential use of *Stevia rebaudiana* in animal feeds. Arch. Zootec. 60 (229), 133–136.

Awney, H.A., Massoud, M.I., El-Maghrabi, S., 2011. Long-term feeding effects of stevioside sweetener on some toxicological parameters of growing male rats. J. Appl. Toxicol. 31, 431–438.

Bayraktar, M., Naziri, E., Akgun, I.H., Karabey, F., Ilhan, E., Akyol, B., Bedir, E., Gurel, A., 2016. Elicitor induced stevioside production, in vitro shoot growth, and biomass accumulation in micropropagated *Stevia rebaudiana*. Plant Cell Tissue Organ Cult. 127 (2), 289–300.

Bender, C., Graziano, S., Zimmermann, B.F., 2015. Study of *Stevia rebaudiana* Bertoni antioxidant activities and cellular properties. Int. J. Food Sci. Nutr. 66 (5), 553–558.

Benford, D.J., DiNovi, M., Schlatter, J., 2006. "Safety evaluation of certain food additives: steviol glycosides"(PDF). In: WHO Food Additives Series (World Health Organization Joint FAO/WHO Expert Committee on Food Additives (JECFA), vol. 54, p. 140.

Bhandawat, A., Sharma, H., Nag, A., Singh, S., Ahuja, P.S., Sharma, R.K., 2015. Functionally relevant novel microsatellite markers for efficient genotyping in *Stevia rebaudiana* Bertoni. J. Genet. 94 (1), 75–81.

Blakeslee, A.F., Avery, A.G., 1937. Methods of inducing doubling of chromosomes in plants: by treatment with colchicine. J. Hered. 28 (12), 393–411.

Bondarev, N., Reshetnyak, O., Nosov, A., 2002. Features of development of *Stevia rebaudiana* shoots cultivated in the roller bioreactor and their production of steviol glycosides. Planta Med. 68 (08), 759–762.

Boonkaewwan, C., Toskulkao, C., Vongsakul, M., 2006. Anti-inflammatory and immunomodulatory activities of stevioside and its metabolite steviol on THP-1 cells. J. Agric. Food Chem. 54 (3), 785–789.

Brandle, J., 2001. *Stevia rebaudiana* with Altered Steviol Glycoside Composition. U.S. Patent no. US 6255557 B1.

Brandle, J.E., Rosa, N., 1992. Heritability for yield, leaf: stem ratio and stevioside content estimated from a landrace cultivar of *Stevia rebaudiana*. Can. J. Plant Sci. 72 (4), 1263–1266.

Carakostas, M.C., Curry, L.L., Boileau, A.C., Brusick, D.J., 2008. Overview: the history, technical function and safety of rebaudioside A, a naturally occurring steviol glycoside, for use in food and beverages. Food Chem. Toxicol. 46 (7), S1–S10.

Chan, P., Xu, D.Y., Liu, J.C., Chen, Y.J., Tomlinson, B., Huang, W.P., Cheng, J.T., 1998. The effect of stevioside on blood pressure and plasma catecholamines in spontaneously hypertensive rats. Life Sci. 63, 1679–1684.

Chatsudthipong, V., Muanprasat, C., 2009. Stevioside and related compounds: therapeutic benefits beyond sweetness. Pharmacol. Ther. 121 (1), 41–54.

Chavan, R.N., Mahadi, S., Ashok, T.H., Shashidhar, H.E., Vasundhara, M., 2014. Induction of genetic variability in *Stevia rebaudiana* Bertoni. Ecol. Environ. Conserv. 20, 1273–1281.

Chen, T.H., Chen, S.C., Chan, P., Chu, Y.L., Yang, H.Y., Cheng, J.T., 2005. Mechanism of the hypoglycemic effect of stevioside, a glycoside of *Stevia rebaudiana*. Planta Med. 71 (02), 108–113.

Chen, J., Hou, K., Qin, P., Liu, H., Yi, B., Yang, W., Wu, W., 2014. RNA-Seq for gene identification and transcript profiling of three *Stevia rebaudiana* genotypes. BMC Genom. 15 (1), 571.

Chester, K., Tamboli, E.T., Parveen, R., Ahmad, S., 2013. Genetic and metabolic diversity in *Stevia rebaudiana* using RAPD and HPTLC analysis. Pharm. Biol. 51 (6), 771–777.

Choi, Y.H., Kim, I., Yoon, K.D., Lee, S.J., Kim, C.Y., Yoo, K.P., Choi, Y.H., Kim, J., 2002. Supercritical fluid extraction and liquid chromatographic-electrospray mass spectrometric analysis of stevioside from *Stevia rebaudiana* leaves. Chromatographia 55 (9–10), 617–620.

Comai, L., 2005. The advantages and disadvantages of being polyploid. Nat. Rev. Genet. 6 (11), 836–846.

Cosson, P., Hastoy, C., Errazzu, L.E., Budeguer, C.J., Boutié, P., Rolin, D., Schurdi-Levraud, V., 2019. Genetic diversity and population structure of the sweet leaf herb, *Stevia rebaudiana* B., cultivated and landraces germplasm assessed by EST-SSRs genotyping and steviol glycosides phenotyping. BMC Plant Biol. 19 (1), 436.

Das, S., Das, A.K., Murphy, R.A., Punwani, I.C., Nasution, M.P., Kinghorn, A.D., 1992. Evaluation of the cariogenic potential of the intense natural sweeteners stevioside and rebaudioside A. Caries Res. 26, 363–366.

De Oliveira, A.J.B., Gonçalves, R.A.C., Chierrito, T.P.C., dos Santos, M.M., de Souza, L.M., Gorin, P.A.J., Sassaki, G.L., Iacomini, M., 2011. Structure and degree of polymerisation of fructooligosaccharides present in roots and leaves of *Stevia rebaudiana* (Bert.) Bertoni. Food Chem. 129 (2011), 305–311.

Deshmukh, S.R., Kedari, V.R., 2014. Isolation, purification and characterization of sweetners from *Stevia rebaudiana* (Bertoni) for their anticancerous activity against colon cancer. World J. Pharm. Pharmaceut. Sci. 3 (5), 1394–1410.

Dyduch-Siemińska, M., Najda, A., Gawroński, J., Balant, S., Świca, K., Żaba, A., 2020. Stevia rebaudiana bertoni, a source of high-potency natural sweetener—biochemical and genetic characterization. Molecules 25 (4), 767.

Ferreira, E.B., Neves, F.D.A.R., da Costa, M.A.D., do Prado, W.A., Ferri, L.D.A.F., Bazotte, R.B., 2006. Comparative effects of *Stevia rebaudiana* leaves and stevioside on glycaemia and hepatic gluconeogenesis. Planta Med. 72 (08), 691–696.

Fullas, F., Kim, J., Compadre, C.M., Kinghorn, A.D., 1991. Separation of natural product sweetening agents using overpressured layer chromatography. J. Chromatogr. A 464, 213–219.

Gasmalla, M.A.A., Yang, R., Amadou, I., Hua, X., 2014. Nutritional composition of *Stevia rebaudiana* Bertoni leaf: effect of drying method. Trop. J. Pharma. Res. 13 (1), 61–65.

Gaweł-Beben, K., Bujak, T., Nizioł-Łukaszewska, Z., Antosiewicz, B., Jakubczyk, A., Karaś, M., Rybczyńska, K., 2015. *Stevia rebaudiana* Bert. leaf extracts as a multifunctional source of natural antioxidants. Molecules 20 (4), 5468–5486.

Geuns, J.M., 2000. Safety of stevia and stevioside. Recent Res. Dev. Phytochem. 4, 75–88.

Ghonema, M., Khaled, A.E., Abdelsalam, N.R., Ibrahim, N.M., 2015. Physico-chemical properties of chromatin, proline content; and induction of polyploidy in *Stevia rebaudiana* (Bertoni). Alexandria Sci. Exch. J 36, 147–156.

Goettemoeller, J., Ching, A., 1999. Seed germination in *Stevia rebaudiana*. In: Perspectives on New Crops and New Uses. ASHS Press, Alexandria, VA, pp. 510–511.

Goyal, S.K., Samsher, G.R., Goyal, R.K., 2010. Stevia (*Stevia rebaudiana*) a bio-sweetener: a review. Int. J. Food Sci. Nutr. 61 (1), 1–10.

Gregersen, S., Jeppesen, P.B., Holst, J.J., Hermansen, K., 2004. Antihyperglycemic effects of stevioside in type 2 diabetic subjects. Metabolism 53 (1), 73–76.

Guleria, P., Yadav, S.K., 2013a. Insights into steviol glycoside biosynthesis pathway enzymes through Structural Homology Modeling. Am. J. Biochem. Mol. Biol. 3, 1–19.

Guleria, P., Yadav, S.K., 2013b. Agrobacterium mediated transient gene silencing (AMTS) in *Stevia rebaudiana*: insights into steviol glycoside biosynthesis pathway. PLoS One 8 (9), e74731.

Gupta, N., Bhadauria, R., Kitchlu, S., Gupta, A.P., Rai, A., Penna, S., Gudipati, T., 2017. Assessment of genetic and chemo-diversity among different Indian ecotypes of *Stevia rebaudiana* (Bertoni). Int. J. Curr. Res. Biosci. Plant Biol. 4 (11), 96–105. https://doi.org/10.20546/ijcrbp.2017.411.011.

Hegde, S.N., Rameshsing, C.N., Vasundhara, M., 2015. Characterization of *Stevia rebaudiana* Bertoni polyploids for growth and quality. Med. Plants-Int. J. Phytomed. Related Ind. 7 (3), 188–195.

Heikal, A.H., Badawy, O.M., Hafez, A.M., 2008. Genetic relationships among some Stevia (*Stevia rebaudiana* Bertoni) accessions based on ISSR analysis. Res. J. Cell Mol. Biol. 2 (1), 1–5.

Holvoet, P., Rull, A., García-Heredia, A., López-Sanromà, S., Geeraert, B., Joven, J., Camps, J., 2015. Stevia-derived compounds attenuate the toxic effects of ectopic lipid accumulation in the liver of obese mice: a transcriptomic and metabolomic study. Food Chem. Toxicol. 77, 22–33.

Hsieh, M.H., Chan, P., Sue, Y.M., Liu, J.C., Liang, T.H., Huang, T.Y., Tomlinson, B., Chow, M.S., Kao, P.F., Chen, Y.J., 2003. Efficacy and tolerability of oral stevioside in patients with mild essential hypertension: a two-year, randomized, placebo-controlled study. Clin. Ther. 25, 2797–2808.

Huang, Y.S., Guo, A.G., Qian, Y., Chen, L.Y., Gu, H.F., 1995. Studies on the variation of steviosides content and selection of type RA in *Stevia rebaudiana*. J. Plant Resour. Environ. 4 (3), 28–32.

Hutapea, A.M., Toskulkao, C., Buddhasukh, D., Wilairat, P., Glinsukon, T., 1997. Digestion of stevioside, a natural sweetener, by various digestive enzymes. J. Clin. Biochem. Nutr. 23 (3), 177–186.

Jaitak, V., Kaul, K., Kaul, V.K., Singh, V., Singh, B., 2011. *Stevia rebaudiana*-A natural substitute for sugar. In: Genetic Resources, Chromosome Engineering, and Crop Improvement, Medicinal Plants, vol. 6, pp. 885–910.

Jaroslav, P., Barbora, H., Tuulia, H., 2006. Characterization of *Stevia rebaudiana* by comprehensive two-dimensional liquid chromatography time-of-flight mass spectrometry. J. Chromatogr. A 1150, 85–92.

Jayaraman, S., Manoharan, M.S., Illanchezian, S., 2008. *In-vitro* antimicrobial and antitumor activities of *Stevia rebaudiana* (Asteraceae) leaf extracts. Trop. J. Pharma. Res. 7 (4), 1143–1149.

Jeppesen, P.B., Gregersen, S., Alstrupp, K.K., Hermansen, K., 2002. Stevioside induces antihyperglycaemic, insulinotropic and glucagonostatic effects *in vivo*: studies in the diabetic Goto-Kakizaki (GK) rats. Phytomedicine 9, 9–14.

Jeppesen, P.B., Gregersen, S., Rolfsen, S.E.D., Jepsen, M., Colombo, M., Agger, A., Xiao, J., Kruhøffer, M., Ørntoft, T., Hermansen, K., 2003. Antihyperglycemic and blood pressure-reducing effects of stevioside in the diabetic Goto-Kakizaki rat. Metabolism 52 (3), 372–378.

Kaur, R., Sharma, N., Raina, R., 2015. Identification and functional annotation of expressed sequence tags based SSR markers of *Stevia rebaudiana*. Turk. J. Agric. For. 39 (3), 439–450.

Kedik, S.A., Yartsev, E.I., Stanishevskaya, I.E., 2009. Antiviral activity of dried extract of Stevia. Pharm. Chem. J. 43 (4).

Khaybullin, R.N., Zhang, M., Fu, J., Liang, X., Li, T., Katritzky, A.R., Okunieff, P., Qi, X., 2014. Design and synthesis of isosteviol triazole conjugates for cancer therapy. Molecules 19 (11), 18676–18689.

Kienle, U., 2016. *Stevia rebaudiana*: a look into the future. J. Verbr. Lebensm. 11, 1–2. https://doi.org/10.1007/s00003-015-1011-3.

Kim, J., Choi, Y.H., Choi, Y.H., 2002. Use of stevioside and cultivation of *Stevia rebaudiana* in Korea. In: Kinghorn, A.D. (Ed.), Stevia, the Genus Stevia, Medicinal and Aromatic Plants—Industrial Profiles, vol. 19. Taylor and Francis, London, pp. 196–202.

Kim, I.S., Yang, M., Lee, O.H., Kang, S.N., 2011. The antioxidant activity and the bioactive compound content of *Stevia rebaudiana* water extracts. LWT Food Sci. Technol. 44 (5), 1328–1332.

Kim, M.J., Jin, J., Zheng, J., Wong, L., Chua, N.H., Jang, I.C., 2015. Comparative transcriptomics unravel biochemical specialization of leaf tissues of Stevia for diterpenoid production. Plant Physiol. 169 (4), 2462–2480.

Kinghorn, A.D., Nanayakkara, N.P.D., Soejarto, D.D., Medon, P.J., Kamath, S., 1982. Potential sweetening agents of plant origin: I. Purification of *Stevia rebaudiana* sweet constituents by droplet counter-current chromatography. J. Chromatogr. A 237 (3), 478–483.

Kong, M.K., Kang, H.J., Kim, J.H., Oh, S.H., Lee, P.C., 2015. Metabolic engineering of the *Stevia rebaudiana* ent-kaurene biosynthetic pathway in recombinant *Escherichia coli*. J. Biotechnol. 214, 95–102.

Koyama, E., Sakai, N., Ohori, Y., Kitazawa, K., Izawa, O., Kakegawa, K., Fujino, A., Ui, M., 2003. Absorption and metabolism of glycosidic sweeteners of Stevia mixture and their aglycone, steviol, in rats and humans. Food Chem. Toxicol. 41 (6), 875–883.

Kroyer, G., 2009. Stevioside and Stevia-sweetener in food: application, stability and interaction with food ingredients. J. Consum. Protect. Food Safety 1–5. https://doi.org/10.1007/s00003-010-0557-3.

Kumar, R., Sharma, S., Ramesh, K., Prasad, R., Pathania, V.L., Singh, B., Singh, R.D., 2012a. Effect of agro-techniques on the performance of natural sweetener plant–stevia (*Stevia rebaudiana*) under western Himalayan conditions. Indian J. Agron. 57 (1), 74–81.

Kumar, H., Kaul, K., Bajpai-Gupta, S., Kaul, V.K., Kumar, S., 2012b. A comprehensive analysis of fifteen genes of steviol glycosides biosynthesis pathway in *Stevia rebaudiana* (Bertoni). Gene 492 (1), 276–284.

Kumar, R., Sharma, S., Ramesh, K., Singh, B., 2013. Effects of shade regimes and planting geometry on growth, yield and quality of the natural sweetener plant stevia (*Stevia rebaudiana* Bertoni) in north-western Himalaya. Arch. Agron Soil Sci. 59 (7), 963–979.

Kumar, R., Sharma, S., Sharma, M., 2014. Growth and yield of natural-sweetener plant stevia as affected by pinching. Indian J. Plant Physiol. 19 (2), 119–126.

Lee, C.N., Wong, K., Liu, J., Chen, Y., Chen, J., Chan, P., 2001. Inhibitory effect of stevioside on calcium influx to produce anti-hypertension. Planta Med. 67, 796–799.

Liu, J., Li, S.F.Y., 1995. Separation and determination of Stevia sweeteners by capillary electrophoresis and high performance liquid chromatography. J. Liq. Chromatogr. Relat. Technol. 18 (9), 1703–1719.

Liu, J.C., Kao, P.K., Chan, P., Hsu, Y.H., Hou, C.C., Lien, G.S., Hsieh, M.H., Chen, Y.J., Cheng, J.T., 2003. Mechanism of the antihypertensive effect of stevioside in anesthetized dogs. Pharmacology 67 (1), 14–20.

Lucho, S.R., do Amaral, M.N., Benitez, L.C., Milech, C., Kleinowski, A.M., Bianchi, V.J., Braga, E.J.B., 2018. Validation of reference genes for RT-qPCR studies in *Stevia rebaudiana* in response to elicitor agents. Physiol. Mol. Biol. Plants 24 (5), 767–779.

Mahadi, S., 2012. Induction of Genetic Variability by Colchicine Treatment in *Stevia rebaudiana* Bertoni (M. Sc. thesis). University of Agriculture Sciences (GKVK), Bangalore, Karnataka.

Mann, T.S., Agnihotri, V.K., Kumar, D., Pal, P.K., Koundal, R., Kumar, A., Padwad, Y.S., 2014. *In vitro* cytotoxic activity guided essential oil composition of flowering twigs of *Stevia rebaudiana*. Nat. Prod. Commun. 9 (5), 1934578X1400900535.

Marsolais, A.A., Brandle, J., Sys, E.A., 1998. U.S. Patent Application No. 08/657,463.

Megeji, N.W., Kumar, J.K., Singh, V., Kaul, V.K., Ahuja, P.S., 2005. Introducing *Stevia rebaudiana*, a natural zero-calorie sweetener. Curr. Sci. 801–804.

Michalec-Warzecha, Ż., Pistelli, L., D'Angiolillo, F., Libik-Konieczny, M., 2016. Establishment of highly efficient *Agrobacterium rhizogenes*-mediated transformation for *Stevia rebaudiana* Bertoni explants. Acta Biol. Crac. Ser. Bot. 58 (1), 113–118.

Minne, V.J., Compernolle, F., Toppet, S., Geuns, J.M., 2004. Steviol quantification at the picomole level by high-performance liquid chromatography. J. Agric. Food Chem. 52 (9), 2445–2449.

Mishra, A.N., Bhadauria, S., Gaur, M.S., Pasricha, R., Kushwah, B.S., 2010. Synthesis of gold nanoparticles by leaves of zero-calorie sweetener herb (*Stevia rebaudiana*) and their nanoscopic characterization by spectroscopy and microscopy. Int. J. Green Nanotechnol. Phys. Chem. 1 (2), P118–P124.

Mitchell, H. (Ed.), 2008. Sweeteners and Sugar Alternatives in Food Technology. John Wiley & Sons. Blackwell Pub.

Mizushina, Y., Akihisa, T., Ukiya, M., Hamasaki, Y., Murakami, N.C., Kuriyama, I., Takeuchi, T., Sugawara, F., Yoshida, H., 2005. Structural analysis of isosteviol and related compounds as DNA polymerase and DNA topoisomerase inhibitors. Life Sci. 77, 2127–2140.

Mizutani, K., Tanaka, O., 2002. Stevia, the genus Stevia. Medicinal and aromatic plants—industrial profiles. In: Use of *Stevia rebaudiana* Sweeteners in Japan, vol. 19, pp. 178–195.

Mohan, K., Robert, J., 2009. Hepatoprotective effects of *Stevia rebaudiana* Bertoni leaf extract in CCl4-induced liver injury in albino rats. Med. Aromat. Plant Sci. Biotechnol. 3, 59–61.

Nettleton, J.E., Klancic, T., Schick, A., Choo, A.C., Shearer, J., Borgland, S.L., Chleilat, F., Mayengbam, S., Reimer, R.A., 2019. Low-dose Stevia (Rebaudioside A) consumption perturbs gut microbiota and the mesolimbic dopamine reward system. Nutrients 11 (6), 1248.

Nikolova-Damyanova, B., Bankova, V., Popov, S., 1994. Separation and quantitation of stevioside and rebaudioside A in plant extracts by normal-phase high performance liquid chromatography and thin-layer chromatography: a comparison. Phytochem. Anal. 5 (2), 81–85.

Nurhidayah, S., Norazlina, N., Rusli, I., 2014. Effect of acute gamma irradiation on in vitro growth of *Stevia rebaudiana* Bertoni. Innov. Plant Prod. Qual. 22, 214–217.

Onakpoya, Heneghan, C., 2014. Effect of the natural sweetener, steviol glycoside, on cardiovascular risk factors: a systematic review and meta-analysis of randomised clinical trials. Eur. J. Prev. Cardiol. 2, 1–9.

Osman, M., Samsudin, N.S., Faruq, G., Nezhadahmadi, A., 2013. Factors affecting microcuttings of Stevia using a mist-chamber propagation box. Sci. World J. 2013.

Pal, P.K., Kumar, R., Guleria, V., Mahajan, M., Prasad, R., Pathania, V., Gill, B.S., Singh, D., Chand, G., Singh, B., Singh, R.D., 2015. Crop-ecology and nutritional variability influence growth and secondary metabolites of *Stevia rebaudiana* Bertoni. BMC Plant Biol. 15 (1), 67.

Pande, S., Khetmalas, M., 2011. Biological effect of gamma irradiations on *in vitro* culture of *Stevia rebaudiana*. Indian J. Appl. Res. 1 (2), 11–20.

Phillips, K.C., 1987. Stevia: steps in developing a new sweetener. In: Grenby, T.H. (Ed.), Developments in Sweeteners. Elsevier, New York, pp. 1–5.

Pól, J., Ostrá, E.V., Karásek, P., Roth, M., Benešová, K., Kotlaříková, P., Čáslavský, J., 2007. Comparison of two different solvents employed for pressurised fluid extraction of stevioside from *Stevia rebaudiana*: methanol versus water. Anal. Bioanal. Chem. 388 (8), 1847–1857.

Puri, M., Sharma, D., Tiwari, A.K., 2011a. Downstream processing of stevioside and its potential applications. Biotechnol. Adv. 29, 781–791.

Puri, M., Kaur, A., Schwarz, W.H., Singh, S., Kennedy, J.F., 2011b. Molecular characterization and enzymatic hydrolysis of naringin extracted from kinnow peel waste. Int. J. Biol. Macromol. 48 (1), 58–62.

Rajasekaran, T., Ramakrishna, A., Udaya Sankar, K., Giridhar, P., Ravishankar, G.A., 2008. Analysis of predominant steviosides in *Stevia rebaudiana* Bertoni by liquid chromatography/electrospray ionization-mass spectrometry. Food Biotechnol. 22 (2), 179–188.

Raymond, K.W., 2010. General Organic and Biological Chemistry. John Wiley & Sons, USA, pp. 364–368 (516).

Sahelian, R., Gates, D., 1999. The Stevia Cookbook. Penguin., Avery, New York, USA, pp. 2–29 (181).

Samuel, P., Ayoob, K.T., Magnuson, B.A., Wölwer-Rieck, U., Jeppesen, P.B., Rogers, P.J., Rowland, I., Mathews, R., 2018. Stevia leaf to Stevia sweetener: exploring its science, benefits, and future potential. J. Nutr. 148 (7), 1186S–1205S.

Savita, S.M., Sheela, K., Sunanda, S., Shankar, A.G., Ramakrishna, P., 2004. *Stevia rebaudiana*–A functional component for food industry. J. Hum. Ecol. 15 (4), 261–264.

Silva, P.A., Oliveira, D.F., Prado, N.R., Carvalho, D.A., Carvalho, G.A., 2008. Evaluation of the antifungal activity by plant extracts against *Colletotrichum gloeosporioides* PENZ. Cienc. E Agrotecnol 32, 420–428.

Singh, S.D., Rao, G.P., 2005. Stevia: the herbal sugar of 21st century. Sugar Tech. 7 (1), 17–24.

Singh, G., Singh, G., Singh, P., Parmar, R., Paul, N., Vashist, R., Swarnkar, M.K., Kumar, A., Singh, S., Singh, A.K., Kumar, S., 2017. Molecular dissection of transcriptional reprogramming of steviol glycosides synthesis in leaf tissue during developmental phase transitions in *Stevia rebaudiana* Bert. Sci. Rep. 7 (1), 1–13.

Smith, J., Van-Stadin, H., 1992. Subcellular pathway of glycoside synthesis. South Afr. J. Sci. 88, 206.

Sys, E.A., Marsolais, A.A., Brandle, J., 1998. U.S. Patent Application No. 08/652,712.

Tadhani, M., Subhash, R., 2006. Preliminary studies on *Stevia rebaudiana* leaves: proximal composition, mineral analysis and phytochemical screening. J. Med. Sci. 6 (3), 321–326.

Takahashi, K., Iwata, Y., Mori, S., Shigeta, S., 1998. In-vitro anti-HIV activity of extract from *Stevia rebaudiana*. Antiviral Res. 37, A59.

Takahashi, K., Matsuda, M., Ohashi, K., Taniguchi, K., Nakagomi, O., Abe, Y., Mori, S., Sato, N., Okutani, K., Shigeta, S., 2001. Analysis of antirotavirus activity of extract from *Stevia rebaudiana*. Antiviral Res. 49, 15–24.

Tanaka, O., 1982. Steviol-glycosides: new natural sweeteners. Trends Anal. Chem. 1, 246–248.

Tavarini, S., Passera, B., Angelini, L.G., 2018. Crop and Steviol Glycoside Improvement in Stevia by Breeding, pp. 1–31.

Urban, J.D., Carakostas, M.C., Taylor, S.L., 2015. Steviol glycoside safety: are highly purified steviol glycoside sweeteners food allergens? Food Chem. Toxicol. 75, 71–78.

Valois, A.C.C., 1992. *Stevia rebaudiana* Bert: uma alternative econômica. Comunicado Te'cnico. Cenargen 13, 1–13.

Vaněk, T., Nepovı&m, A., Valı&ček, P., 2001. Determination of stevioside in plant material and fruit teas. J. Food Compos. Anal. 14 (4), 383–388.

Vaško, L., Vašková, J., Fejerčáková, A., Mojžišová, G., Poráčová, J., 2014. Comparison of some antioxidant properties of plant extracts from *Origanum vulgare*, *Salvia officinalis*, *Eleutherococcus senticosus* and *Stevia rebaudiana*. In Vitro Cell. Dev. Biol. Anim. 50 (7), 614–622.

Wanke, M., Skorupinska-Tudek, K., Swiezewska, E., 2001. Isoprenoid biosynthesis via 1-deoxy-D-xylulose 5-phosphate/2-C-methyl-D-erythritol 4-phosphate (DOXP/MEP) pathway. Acta Biochim. Pol. 48 (3), 663–672.

Walia, U.S., Walia, S.S., Kler, D.S., Singh, D., 2011. Science of Agronomy. Scientific Publishers.

Wingard, R.E., Brown, J.P., Enderlin, F.E., Dale, J.A., Hale, R.L., Seitz, C.T., 1980. Intestinal degradation and absorption of the glycosidic sweeteners stevioside and rebaudioside A. Experientia 36 (5), 519–520.

Wong, K.L., Lin, J.W., Liu, J.C., Yang, H.Y., Kao, P.F., Chen, C.H., Loh, S.H., Chiu, W.T., Cheng, T.H., Lin, J.G., Hong, H.J., 2006. Antiproliferative effect of isosteviol on angiotensin-II-treated rat aortic smooth muscle cells. Pharmacology 76 (4), 163–169.

Yadav, A.K., Singh, S., Yadav, S.C., Dhyani, D., Bhardwaj, G., Sharma, A., Singh, B., 2013. Induction and morpho-chemical characterization of Stevia rebaudiana colchiploids. Indian J. Agric. Sci. 83 (2), 159–165.

Yamazaki, T., Flores, H.E., Shimomura, K., Yoshihira, K., 1991. Examination of steviol glucosides production by hairy root and shoot cultures of Stevia rebaudiana. J. Nat. Prod. 54 (4), 986–992.

Yao, Y., Ban, M., Brandle, J., 1999. A genetic linkage map for *Stevia rebaudiana*. Genome 42 (4), 657–661.

Yu, X., Yang, J., Wang, E., Li, B., Yuan, H., 2015. Effects of growth stage and fulvic acid on the diversity and dynamics of endophytic bacterial community in *Stevia rebaudiana* Bertoni leaves. Front. Microbiol. 6, 867.

Yücesan, B., Büyükgöçmen, R., Mohammed, A., Sameeullah, M., Altuğ, C., Gürel, S., Gürel, E., 2016. An efficient regeneration system and steviol glycoside analysis of *Stevia rebaudiana* Bertoni, a source of natural high-intensity sweetener. In Vitro Cell. Dev. Biol. Plant 52 (3), 330–337.

Zhang, H., An, S., Hu, J., Lin, Z., Liu, X., Bao, H., Chen, R., 2018. Induction, identification and characterization of polyploidy in *Stevia rebaudiana* Bertoni. Plant Biotechnol. 35 (1), 81–86.

Further reading

Brandle, J.E., Gijzen, M., Starratt, A., 2000. *Stevia rebaudiana*, Its Biological, Chemical and Agricultural Properties (No. 00991).

Lucho, S.R., do Amaral, M.N., López-Orenes, A., Kleinowski, A.M., do Amarante, L., Ferrer, M.Á., Calderón, A.A., Braga, E.J.B., 2019. Plant growth regulators as potential elicitors to increase the contents of phenolic compounds and antioxidant capacity in Stevia plants. Sugar Tech. 21 (4), 696–702.

Wang, Y.H., Avula, B., Tang, W., Wang, M., Elsohly, M.A., Khan, I.A., 2015. Ultra-HPLC method for quality and adulterant assessment of steviol glycosides sweeteners–*Stevia rebaudiana* and Stevia products. Food Addit. Contam. 32 (5), 674–685.

CHAPTER

Swertia chirayita

12

Vijay Singh[1], Vikrant Jaryan[2], Vikas Sharma[3], Himanshu Sharma[4], Indu Sharma[2], Vikas Sharma[2]

[1]*Department of Botany, Mata Gujri College, Fatehgarh Sahib, Punjab, India;* [2]*Department of Botany, Sant Baba Bhag Singh University, Khiala, Jalandhar, Punjab, India;* [3]*Department of Molecular Biology and Genetic Engineering, Lovely Professional University, Jalandhar, Punjab, India;* [4]*Agri-Biotechnology Division, National Agri-Food Biotechnology Institute, Mohali, Punjab, India*

12.1 Introduction

Swertia chirayita Roxb. ex Fleming commonly known as Chirata is one of the most important medicinal plants distributed from temperate to the alpine region of the Indian Himalayas, ranging across the altitudinal gradient from 1200 to 3000 m above sea level (Clarke, 1885; Tandon et al., 2010; Pradhan and Badola, 2015). Sometimes it is also found up to 3400 m above sea level (Bora and Singh, 2016). It is an indigenous species of the Himalaya and distributed from Kashmir to Sikkim. Besides, this species is also found to grow in Khasi hills at an altitude from 1200 to 1500 m (Tandon et al., 2010). It is also found to grow in Nepal, Bhutan, Pakistan, and Afghanistan (Pradhan and Badola, 2015). This plant species belong to family Gentianaceae and is one of the important ingredients of Ayurvedic, Unani, Siddha, and modern medicines (Bora and Singh, 2016). It is a synonym of *Gentiana chirayita* Roxb. ex Flem.; *G. chirayita* Wall. It is an old traditional herb and its medicinal value was known from earlier times. It was known Kiratatikta in Sanskrit, while Nilavaembu in Tamil (Selvam, 2012). The plant is native of the Himalaya and highly threatened due to its high demand as a medicinal plant and its occurrence in narrow range of specialized habitat. In this chapter, we will discuss the occurrence, habitat, morphology, cytology, ecology, biochemical constituents and genetic diversity of the *S. chirayita* and its comparative account to related species of *Swertia*.

12.1.1 Morphology of plant

S. chirayita is an annual or biennial herb with 60−125 cm in height or sometimes reaches up to 1.5 m in height (Aleem and Kabir, 2018; Kumar and Van Staden, 2016; Selvam, 2012). Its stem is sometimes robust, branched, cylindrical below, four angled upward containing a large pith (Selvam, 2012). Stem is yellowish or purplish in color (Aleem and Kabir, 2018). Its leaves are opposite, sessile, broadly ovate or lanceolate, decussate five nerved with round base, acuminate at apex,

Himalayan Medicinal Plants. https://doi.org/10.1016/B978-0-12-823151-7.00016-7

glabrous, cordate or obtuse at base with margins entire, and three to seven prominent lateral veins. Plant flowers generally from July to August in rainy season. Its inflorescence has large leafy panicles of solitary axillary or axillary with three to five flowers arranged in terminal clusters. Flowers are generally yellowish-green outside and purple inside. They are tetramerous, drooping, or erect. Fruit is a capsule which is ovoid or ellipsoid. Seeds are numerous, minute, globose, and brownish.

12.1.2 Habitat and climatic conditions

S. chirayita generally grows in subtropical to temperate forests in open forest margins. It prefers cool and moist places with shady moist slopes and tall grasses (Selvam, 2012). Many studies confirm that the species performance varies among microhabitats (Shrestha and Jha, 2010; Pradhan and Badola, 2012). Present climate change is the biggest threat for the survival of this threatened species because of its narrow range of specialized habitat and has a greater risk of instability which can lead to the extinction of this species (Brys et al., 2005; Samant et al., 1996). Therefore, there is great demand to understand the climatic condition of these microhabitats to conserve narrow range of this specialized habitat species which can be helpful for planning its conservation and management (Hegland et al., 2001; Colling and Matthies, 2006; Kalliovirta et al., 2006).

This plant prefers sandy (light), loamy (medium), as well as clay (heavy) soil conditions. The plant flourishes well in acidic, neutral, as well as alkaline soils. The plant prefers semishade or woodland conditions or needs humid or damp soil. The Chirata plants can withstand temperatures as low as −15°C and still continue to grow well (Kumar et al., 2010).

12.2 Cytological studies of genus *Swertia*

Cytology is one of the most important fields in biology and deals with the study of the cells, especially their structure, function, and chemistry with a focus on the chromosome. The chromosomes are thread-like structures in which DNA molecule is packaged tightly and cemented with the help of histone proteins and carry the genetic information from parents to offspring through meiosis. But the genetic information is carried out through the generation by the process of meiosis only by chromosomes of germline cells. However, the somatic cells help only in the regeneration and maintenance within the generation by the process of mitosis. Therefore, meiosis is required to maintain a genetic balance over generations, although it produces variations via independent assortment and crossing over between nonsister chromatids of homologous chromosomes. Meiosis is best known for its reliability over sexual life cycles, as it maintains genome stability through many events (cellular and molecular), such as DNA/chromosome replication in G1 phase, chromosome pairing and recombination in prophase-1, chromosomal segregation in anaphase-I/II and cytokinesis. Further, the process of meiosis involves reductional and equational division and results in four haploid gametes (half genome compared to the parent cell). Homologous recombination through crossing over in the pachytene stage (prophase-I) is a dynamic process by which DNA sequences/strands are exchanged, hence induce structural aberrations and ultimately cause the evolution of gene/genome.

It is a widely accepted view that cytotaxonomy is one of the key determinants in the plant systematics as chromosome numbers significantly contribute to grouping plant species/taxa. The first ever report of the cytological work in the flowering plants (orchids) was given by Strasburger in 1882 (cf. Fedorov, 1969), while the first chromosomal report for Indian populations was made by Johnson in 1910 (cf. Darlington and Wylie, 1955). The determination of chromosome numbers coupled with detailed meiotic studies is always a valuable effort that provides a future platform for the researchers in applied sciences. The chromosomes through their numerical (euploid or aneuploid cytotype) and structural changes have played a key role in chromosomal evolution by producing reduced or doubled gametes and the phenomenon is quite common in Angiosperms (Stebbins, 1950; Ornduff et al., 1963; Heywood et al., 1977; Robinson et al., 1981).

In the family Gentianaceae, at world level, a number of prominent cytologists have explored the family, which includes Favarger (1952), Skalinska (1952), Löve (1953), etc. In Indian populations, the previous detailed chromosome work in the family was undertaken from the Western Himalayas (Mehra and Gill, 1968; Vasudevan, 1975), South India (Subramanyam and Kamble, 1966), and Kashmir Himalaya (Khoshoo and Tandon, 1963; Khoshoo et al., 1966; Koul and Gohil, 1973; Gohil et al., 1981; Jee et al., 1985, 1989). Recently, the meiotic study was undertaken in the three genera covering nine species of the family and revealed first ever chromosome counts in *Gentiana aprica* ($2n = 20$), *Gentiana argentea* var. *albescens* ($2n = 20$), and *Gentiana pygmaea* ($2n = 20$) and new/varied chromosome counts in *Swertia ciliata* ($2n = 26$) and *Swertia purpurascens* ($2n = 18$) at the global level (Bala et al., 2015). Recently, three species of *Swertia* viz. *S. ciliata*, *S. cordata*, and *S. petiolata* have been worked out cytologically by our group. The results of this investigation are given in Fig. 12.1 and discussed below:

***S. ciliata* (D. Don ex G. Don) B. L. Burtt.** Meiotic analysis in the species revealed a diploid cytotype with $2n = 26$ showing 13 bivalents at metaphase-I and 13:13 equal distribution of chromosomes at each pole in anaphase-I (Fig. 12.1; 1a,b). The present chromosome count of $2n = 26$ is in line with the previous reports from India. Besides, the species is also known to exist with $2n = 18$, 20, and 24 in Indian accessions (Table 12.1).

***S. cordata* Wall.** The species is reported to exist at diploid level with $2n = 26$ (Fig. 12.1; 2a), which is in line with the previous report from India (Table 12.1).

***S. petiolata* Royle.** Cytological study of the species revealed a diploid cytotype with $2n = 26$ showing 13:13 distribution of chromosomes at anaphase-I (Fig. 12.1; 3a). The species is known with a single chromosome count of $2n = 26$, which has been confirmed in the present study as well (Table 12.1).

FIGURE 12.1

1: *Swertia ciliata*, 1a: PMC at M-I showing 13 bivalents connections, 1b: PMC at A-I showing 13 chromosomes at one pole. 2: *Swertia cordata*; 2a: PMC at A-I showing 13:13 distribution of chromosomes at both poles. 3: *Swertia petiolata*; 3a: PMC at A-I showing 13:13 distribution of chromosomes at each pole.

Out of a total of 170 taxonomically known species, the chromosome numbers have been reported in 52 species with a range from 2n = 14–60, with 12 different chromosome numbers in between, which depict its polybasic nature (x = 7, 8, 9, 10, 11, 12, 13), of which x = 10, 12, and 13 are prominent ones. In India, 22 species are cytologically known so far and chromosome number ranges from 2n = 16–26, of which 2n = 26 being the most common, although x = 8, 9, and 10 are also reported in Indian accessions.

Table 12.1 Table showing names of taxons and chromosome numbers in cytologically worked out species at world level.

Name of the taxon	Chromosome number in cytologically worked out species			
	Outside India		India	
	n	2n	n	2n
1. *Swertia abyssinica* Hochst.	–/–	18+2B (Morton, 1993) 20 (Nemomisa, 1998)	–/–	–/–
2. *Swertia alata* (Royle ex D. Don) C.B. Clarke	13 (Khatoon and Ali, 1993)	–/–	13 (Vasudevan, 1975)	–/–
3. *Swertia albicaulis* var. *albicaulis* Douglas ex Kuntze	26 (Chambers et al., 1998)			
4. *Swertia angustifolia* Burkill	–/–	–/–	13 (Vasudevan, 1975)	26 (Mallikarjuna, 1985)
5. *Swertia beddomei* C.B. Clarke	–/–	–/–	–/–	26 (Mallikarjuna, 1985; Mallikarjuna et al., 1987)
6. *Swertia bifolia* Batalin	–/–	c.26 (Huang et al., 1996) 28 (He et al., 1999)		
7. *Swertia bimaculata* (Siebold and Zucc.) Hook. F.and Thomson ex C.B. Clarke	–/–	26 (Shigenobu 1982, 1983)	11 (Vasudevan, 1975) 13 (Sharma, 1970)	–/– –/–
8. *Swertia calycina* Franch.	–/–	20 (Auquier and Renard, 1975)		
9. *Swertia chinensis* Franch.	–/–	20 (Wada, 1966) 28 (Kawakami, 1930)		
10. *Swertia chirata* B. ham ex. Wall.	–/–	20, 24 (Wada, 1966)	26 (Khoshoo and Tandon 1963)	13 (Vasudevan, 1975)

Continued

Table 12.1 Table showing names of taxons and chromosome numbers in cytologically worked out species at world level.—*cont'd*

Name of the taxon	Chromosome number in cytologically worked out species			
	Outside India		India	
	n	2n	n	2n
11. *Swertia ciliata* (D. Don) B. L. Burtt	–/–	–/–	09 (Bala and Gupta, 2011) 10 (Vasudevan, 1975) 12 (Mehra and Gill 1968) 13 (Malik et al., 2011)	2 (Khoshoo and Tandon, 1963) 24 (Mehra and Gill, 1968)
12. *Swertia cordata* (Wall. ex G. Don) C.B. Clarke	–/–	–/–	13 (Vasudevan, 1975)	–/–
13. *Swertia corymbosa* Wight ex Griseb.	–/–	–/–	26 (Mallikarjuna, 1985)	
14. *Swertia crassiuscula* Gilg	–/–	20 (Thulin, 1970)	–/–	–/–
15. *Swertia cuspidata* (Maxim.) Kitag.	–/–	28 (Shigenobu, 1983)	–/–	–/–
16. *Swertia densifolia* (Grisebach) Kashyapa	–/–	–/–	13 (Mallikarjuna, 1985)	–/–
17. *Swertia diluta* (Turcz.) Benth. and Hook. f.	–/–	20 (Yuan and Küpfer, 1993)	–/–	–/–
18. *Swertia emarginata* Schrenk	–/–	16 (Ma et al., 1990)	–/–	–/–
19. *Swertia engleri* Gilg	10 (Nemomissa, 1998)	–/–	–/–	–/–
20. *Swertia fimbriata* (Hochst.) Cufod.	–/–	26 (Nemomissa, 1998)	–/–	–/–
21. *Swertia franchetiana* Harry Sm	–/–	20 (He et al., 1999)	–/–	–/–

22. *Swertia hickinii* Burkill	–/–	20 (He et al., 1999)	–/–	–/–
23. *Swertia iberica* Fisch. and C.A. Mey.	–/–	26 (Davlianidze, 1984; Gagnidze and Gviniaschvili, 1984; Gvinianidze and Avazneli, 1982)	–/– –/–	–/–
24. *Swertia japonica* (Schult.) Makino	–/–	20, 21 (Shigenobu, 1983)		
25. *Swertia kilimandscharica* Engl.	–/–	26 (Hedberg and Hedberg, 1977)	–/–	–/–
26. *Swertia komarovii* Pissjauk	–/–	26 (Krogulevich, 1978)	–/–	–/–
27. *Swertia lawii* Burkill	–/–	–/–	–/–	26 (Mallikarjuna, 1985)
28. *Swertia lugardiae*	10 (Nemomissa, 1998)	–/–	–/–	–/–
29. *Swertia lurida* Royle ex D. Don	–/–	–/–	13 (Vasudevan, 1975; Mehra and Vasudevan, 1972)	–/–
30. *Swertia macrosepala* Gilg	–/–	26 (Hedberg and Hedberg, 1977)	–/–	-/
31. *Swertia minor* (Grisebach) Knobl.	–/–	–/–	–/–	20 (Mallikarjuna, 1985)
32. *Swertia nervosa* (Wall. ex G. Don) C.B. Clarke	–/–	–/–	–/–	26 (Sharma, 1970)
33. *Swertia paniculata* Wall.	8 (Khatoon and Ali, 1993)	–/–	8 (Vasudevan, 1975)	
34.			(Bala and Gupta, 2011)	–/–
35. *Swertia perennis* L.	14 (Post, 1983)	28 (Love and Love, 1986; Dawe and Murray, 1979)	–/–	–/–
36. *Swertia petiolata* D. Don	–/–	–/–	13 (Jee et al., 1989)	–

Continued

Table 12.1 Table showing names of taxons and chromosome numbers in cytologically worked out species at world level.—*cont'd*

Name of the taxon	Chromosome number in cytologically worked out species			
	Outside India		India	
	n	2n	n	2n
37. *Swertia pseudochinensis* H. Hara	–/–	20 (Shigenobu, 1983)	–/–	–/–
38. *Swertia punctata* Baumg.	–/–	28 (Tam and Vladimirov, 2001)	–/–	–/–
39. *Swertia speciosa* Wall.	–/–	–/–	13 (Vasudevan, 1975)	–/–
40. *Swertia stenopetala* Pissjauk.	–/–	14 (Zhukova et al., 1973)	–/–	–/–
41. *Swertia subnivalis* T.C.E. Fr.	–/–	26 (Hedberg and Hedberg, 1977)	–/–	–/–
42. *Swertia swertopsis* Makino	–/–	52 (Shigenobu, 1982, 1983)	–/–	–/–
43. *Swertia tashiroi* Makino	–/–	60 (Shigenobu, 1982, 1983)	–/–	–/–
44. *Swertia tetragona* R. H. Miao	–/–	–/–	–/–	18 (Vasudevan, 1975; Khoshoo and Tandon, 1963)
45. *Swertia tetrandra* Hochst.	10 (Nemomissa, 1998)	–/–	–/–	–/–
46. *Swertia tetraptera* Maxim.	–/–	14 (He et al., 1999)	–/–	–/–
47. *Swertia thomsonii* C. B. Clarke	–/–	–/–	13 (Vasudevan, 1975; Khoshoo and Tandon, 1963)	–/–
48. *Swertia trichotoma* Wall.	–/–	–/–	26 (Mallikarjuna et al., 1987)	–/–
49. *Swertia tongluensis* Burkill	–/–	–/–	–/–	18 (Sharma and Sarkar, 1967–68)
50. *Swertia uniflora* Mildbr.	–/–	26 (Hedberg and Hedberg, 1977)	–/–	–/–
51. *Swertia veratroides* Maxim. ex Kom.	–/–	26 (Probatova, 2006)	–/–	–/–
Swertia wolfgangiana Grüning	–/–	28 (He et al., 1999)	–/–	–/–

The family Gentianaceae represents a lot of polyploidy and dysploidy in its species, as a wide range of base numbers exist, i.e., x = 5, 6, 7, 8, 9, 10, 11, 12, 13, 15, 17, 19, or higher. But it is dominated by x = 9 in most of the genera of the family. According to Favarger (1952), the original base number of the family is $x = 5$, and that the base number of $x = 11$ might be a result of polyploidy followed by fragmentation. The same explanation may be given to another base number x = 13 which may have originated from = 7 by polyploidization and fusion. The same idea was adopted by Skalinska (1952) while studying chromosome number in *Gentiana frigida* (2n = 24) and reveals that x = 11 (5 + 6) and x = 13 (6 + 7) have been evolved from x = 5, 6, and 7.

12.3 Genetic diversity studies

Evaluation of diversity in a plant is important for knowing its life dynamics and sustainability in future. Estimates of DNA polymorphism show the capability of adaptive behavior in changing climates and evolutionary history of that plant species. However, studies of DNA polymorphism in *S. chirayit*a are severely lacking and only two reports are available which used only limited accessions of this species because of its declining populations in nature. Joshi and Dhawan (2007a,b) analyzed 13 accessions of *S. chirayita* using ISSR markers and revealed high genetic diversity among analyzed samples. In a recent study, Kaur et al. (2019) explored the interspecific genetic diversity of five *Swertia* species including five accessions of *S. chirayita*. They used ISSR and RAPD markers and concluded that *S. chirayita* showed low diversity. There is urgent need of studies with more numbers of populations and accessions which should be analyzed using better marker systems such as Simple Sequence Repeat markers so that accurate estimates can be drawn which can help in designing conservation and management strategies in future.

12.3.1 Genes and metabolite association studies

Molecular data related to the biosynthesis of secondary metabolites of *S. chirayita* are lacking. However, few workers have tried to explore the pathways and relative contents of secondary metabolite and their association to regulatory genes and miRNAs (Vaidya et al., 2013; Fan et al., 2014; Kumar et al., 2014; Padhan et al., 2015, 2016; Koul et al., 2016; Liu et al., 2017; Pal et al., 2018; Padhan, 2018). Padhan et al. (2015) reported the expression profiling of swertiamarin, amarogentin, and mangiferin biosynthesis pathway genes and their correlation with the respective metabolites content in different tissues of *S. chirayita.* They observed that root tissues of greenhouse-grown plants contained the maximum amount of swertiamarin, whereas maximum accumulation of mangiferin in floral organs. Ten genes of the secoiridoids biosynthesis pathway and five genes of mangiferin biosynthesis identified and correlated to corresponding metabolite contents. Their results can be important in its genetic improvement works. Liu et al. (2017) generated transcriptome sequences from the root, leaf, stem, and flower tissues of *S. mussotii* to understand the secoiridoid biosynthesis pathway and identified 39 candidate transcripts encoding the key enzymes for secoiridoid biosynthesis. Their study revealed that the levels of three bioactive compounds, i.e., sweroside, swertiamarin, and gentiopicroside, were variable in different tissues. They found no significant correlation with the expression profiles of key genes and suggested complex biological behaviors in the coordination of metabolite biosynthesis and accumulation. Pal et al. (2018) generated comparative transcriptomes of *S. chirayita* to decipher the genes and other regulatory components related to secondary metabolites biosynthesis. They found that 19 genes from primary metabolism showed higher in silico expression indicating their involvement in regulating the central carbon pool. When validated by qRT-PCR, 10 genes showed similar expression pattern across both the methods. The authors identified differentially expressed transcription factors and ABC-type transporters putatively associated with secondary metabolism in *S. chirayita*.

12.4 Bioactivity and medicinal uses

This plant is used as expectorant, laxative, antispasmodic, antioxidant, antidiabetic, antipyretic, antitussive, stomachic, anthelmintic, and antidiarrhea. It has many medicinal properties such as antiinflammatory, hypoglycemic, hepatoprotective, antibacterial and wound healing (Laxmi et al., 2011; Tabassum et al., 2012; Mahmood et al., 2014; Alam et al., 2009). Different ailments cured by *S. chirayita* are described in Table 12.2. Besides, many medicinal compounds were isolated from natural herbs which have anticancer, antitumor, and anti-AIDS properties (Sultana and Ahmed, 2013).

Table 12.2 Ailments cured by using different procedure and parts of *Swertia chirayita.*

S. No.	Ailments	Parts used	Procedure
1	Malaria	Whole plant	Overnight dipped plants in water and juice extracted
2	Digestive organs	Whole plant	Decoction as tonic
3	Bronchial asthma, cough, cold, headache and fever, diarrhea, constipation, dyspepsia, burning of the body, and skin diseases	Whole plant	Paste of plant
4	Liver diseases and urinary disorders	Roots	Root juice
5	Joint pain	Roots	Roots crushed and paste rubbed over joints
6	Boils and scabies	Leaves	Leaves warmed and paste prepared with mustard oil applied over boils and scabies

12.4.1 Active principles

Swertia contain many bioactive compounds in the form of phenolics compounds, alkaloids, and other secondary metabolites with wide therapeutic importance and potential applications (Singh et al., 2012). The major classes of phytochemicals include many Xanthone compounds and other phytochemicals specifically swerchirin, swertiamarin, swertanone, swertenol, episwertinol, chiratenol, gammacer-16-en-3β-ol, 21-a-H-hop-22(29)-en-3β-ol, taraxerol, oleanolic acid, ursolic acid, ophelic acid, and 1,3,6,7-tetrahydroxyxanthone-C-2-β-D-glucoside (mangiferin) which have shown different therapeutic effects (Singh, 2005; Ghosal et al., 2006; Iqbal et al., 2006; Selvam, 2012; Singh et al., 2012).

12.5 Tissue culture studies in *Swertia chirayita*

Plant tissue culture methods have been used in conservation of medicinal plants for past many years. Over time, these methods have been developed and modified for exploiting medicinal plant potential and helping in establishment of numerous plant products—based industries like pharmaceutical, nutrition, and life style. IHR (Indo-Himalayan region) is well known for its biodiversity, flora, and fauna. *S. chirayita* is critically endangered and listed in Red Data book. There have been so many conservation measures taken to conserve the plant and to increase its natural populations. But these programs were not efficient in achieving the goal as natural propagation in this plant is not good because of seed dormancy, seed viability, and natural pathogen attack or excessive grazing in its natural habitats. Vegetative propagation programs are also highly discouraged because of unavailability of quality planting materials, difficult to establish root stocks; moreover, it cannot be propagated through cutting or grafting. Therefore, this plant has remained quite important among the plant biotechnologists and has been studied extensively for establishment of in vitro propagation protocols across the globe. *S. chirayita* is having high commercial potential because of its therapeutic properties and variety of bioactives. The various plant tissue culture strategies such as tissue and organ cultures, suspension cultures/callus cultures, somatic embryogenesis, and authentication of micropropagated plantlets and more recently in vitro co-culture, have been studied in this plant. Therefore, current study has been done to compile these studies like micropropagation, caulogenesis, rhizogenesis, callogenesis, somatic embryogenesis, and their advancements.

Seeds, nodal segments, axillary buds, and leaves as explants have been used from field-grown young juvenile plants, and explants from micropropagated plants have been tested for enhanced multiplication of *S. chirayita*. Nodal segments or nodal explants have been found more suitable for enhanced shoot multiplication and production of genetically true-to-type plants among all explants tested (Sharma et al., 2013, 2016). Almost all studies in this plant have reported the use of MS as basal tissue culture media; Gamborg B5 have also been reported in some studies with modifications for in vitro bud break and culture establishment. Optimum shoot development and maturation have been observed in MS medium (Sharma et al., 2013, 2016; Wang et al., 2009; Chaudhuri et al., 2007; Wawrosch et al., 1999). The type of carbon source and its concentration plays a very significant role in micropropagation of plants. Optimum caulogenesis has been reported when medium was fortified with sucrose as carbon source at a concentration of 2.5%−3% (Sharma, 2011). *In vitro propagation* of *S. chirayita* has been reported by many researchers using different explants, like shoot tips and axillary buds (Wawrosch et al., 1999; Joshi and Dhawan, 2007a,b; Ahuja et al., 2003), leaf segments (Chaudhuri et al., 2008; Wang et al., 2009), cultured root explant (Pant et al., 2010), and nodal segment (Sharma et al., 2013, 2015, 2016; Chaudhuri et al., 2007).

One of the safer approaches to decrease the somaclonal incidences and confirming genetic fidelity of micropropagated plantlets is to use axillary buds and nodal segments. The importance of PGRs (auxins and cytokinins) in breaking axillary bud dormancy, or shoot multiplication, as well as their effect alone or in combination for shoot proliferation, multiplication, and elongation has been well established in *S. chirayita*.

The effect of BAP with respect to induction and shoot enhancement in initial phase of culture establishment has been reported by many workers. Also the synergistic effect of BAP and IAA or additives like adenine sulfate in enhancing the shoot multiplication rate and reducing the use of other cytokinins has been obtained in *S. chirayita* (Sharma et al., 2013, 2016). Cytokinins BAP; Kinetin; 2-iP (alone or in combination with Kinetin or auxins IAA; IBA; NAA) have been used. BAP alone or in combination with other cytokinin or auxins in lower proportions has been found optimum in shoot multiplication from nodal segments and leaf explants (Sharma et al., 2016; Wang et al., 2009). One recent study on micropropagation of *Withania somnifera* involving the use of in vitro co-culture with PGPRs has resulted in shoot proliferation with faster rates, enhanced and vigorous rhizogenesis, and use of very reduced concentration of PGRs. Reduced use of phytohormones results in lesser incidences of basal callusing during shoot proliferation and thus reduces the chances of somaclonal variations and confirms the clonal identity (Sharma et al., 2015).The in vitro co-culture can also serve as a bioassay for screening of positive plant microbe interactions that can be further exploited for various other beneficial outputs. Addition of growth factors like polyamines, reduced nitrogen sources like adenines, and coconut milk has profound effect on in vitro caulogenesis (Sharma et al., 2013, 2016; Sharma, 2011). Indirect organogenesis via callus formation is also a good means of mass multiplication; somatic embryogenesis has been studied in *S. chirayita* using nodal segment, leaf explant, in vitro grown seedlings, and internode (Chaudhuri et al., 2009; Sharma, 2011). The superiority of 2,4-D in establishing the callus and its maturation has been reported by the workers. The suitability of leaf explant for inducing embryogenic callus over intermodal segments has also been reported. It has been suggested that it is better to take leaf explants from micropropagated plantlets as they bypass the additional steps of sterilization and give better response. Root emergence, development, and elongation are vital steps in every tissue culture study. Same has been extensively studied in Chirata also. Influence of phytohormones and media strength has morphogenic effects on the events of rooting. Auxins alone or in combination with other auxins in rooting the microshoots were effective in *S. chirayita* (Joshi and Dhawan, 2007a,b; Sharma et al., 2016; Wang et al., 2009). Direct organogenesis via leaf explant has also been reported in *S. chirayita* (Chaudhuri et al., 2008). One study has also reported the in vitro flowering in *S. chirayita* through tissue culture (Sharma et al., 2014).

12.6 Ecological status

As *S. chirayita* is a highly important medicinal plant and used as an ingredient in Ayurvedic, Unani, Siddha, and modern medicines, it has become vulnerable in its natural state in forest due to its overextractions. It is in great demand in pharmaceutical industries both nationally and internationally and, therefore, its overexploitation, unsustainable harvesting, overgrazing, and illegal trading has put this plant under critical endangered status (Kumar and Van Staden, 2016; Badola and Pal, 2002). Destruction of its microhabitat and overexploitation leads to drastic reduction in its natural population all over the Himalayan region (Bhatt et al., 2005). Its low seed viability, long gestation period, and low seed germination percentage are also the major causes for its dwindling natural population (Samant et al., 1998; Joshi and Dhawan, 2005). Exceeding developmental activities, building of new roads and infrastructure, unorganized urbanization, and increasing anthropogenic pressure all over the Himalayan region are the main driving forces for its habitat destruction. It can leads to the loss of biodiversity and extinction of species from its natural habitat. The extinction rate of plant species was estimated to 100–1000 times faster than the natural speed of extinction (Kumar and Van Staden, 2016). Recently, ministry of environment and forestry has banned the export of this plant along with many other Himalayan medicinal plants. Efforts were made for its in situ and ex situ conservations (Nishteswar, 2014). The main focus is to conserve the habitat of the *S. chirayita* by maintaining the open spaces with moist slopes around the forest and stopping the surrounding shrubs to encroach the habitat. The surrounding shrubs overtake the habitat of *S. chirayita* and modify the microhabitat resulting in the decrease in its population. Maintaining the microhabitat, allowing the sustainable harvest, and clearance of shrubs are helpful in preserving the microhabitat of this species as an excellent strategy for in situ conservation. Approaches have also started to conserve the species through in vitro tissue culture techniques. Mass multiplication has been carried out by tissue culture technique via direct shoot multiplication through leaf explants Chuadhury et al. (2008). Restoration of natural population can be carried out by planting the multiplied material through tissue culture. Cultivation is another approach for the ex situ conservation of natural population of *S. chirayita* and can save it from the overexploitation from natural habitat (Shukla et al., 2017). Species Distribution Modeling and Ecological Niche Modeling can be some excellent approaches for identification of the suitable niche (microclimatic region) for restoration of the natural population of *S. chirayita* in the Himalayan region (Gaikwad et al., 2011; Jaryan et al., 2013).

12.7 Conclusion

S. chirayita is an important medicinal plant which yields many important active principles. However, declining populations due to habitat destruction and human disturbances is the major concern. The plant has been explored to some extent including chromosomal, genetic, genomic, biochemical, tissue culture, and ecological levels but more research works are required to make its best use in a sustainable way. The different ecotypes if any present in natural populations should be explored and documented. Genetic diversity of available germplasm using molecular tools needs to be assessed to make insights into past and future population dynamics of the species. The elite accession having high contents of active principles present in the natural populations needs to be identified and multiplied either through conventional methods or using tissue culture technology. More insights into biosynthetic pathways of active principles of the plants are required so that desired genetic manipulations can be done in future to increase the yield. Hence, these are some major research fields where the future works can be focused for the conservation and utilization of this incredible plant species.

References

Ahuja, A., Koul, S., Kaul, B.L., Verma, N.K., Kaul, M.K., Raina, R.K., Qazi, G.N., 2003. Media Compositions for Faster Propagation of Swertia chirayita. WO 03/045132 A1.

Alam, K.D., Ali, M.S., Parvin, S., Mahjabeen, S., Akbar, M.A., Ahamed, R., 2009. In vitro antimicrobial activities of different fractions of *Swertia chirata* ethanolic extract. Pak. J. Biol. Sci. 12 (19), 1334–1337.

Aleem, A., Kabir, H., 2018. Review on *Swertia chirata* as traditional uses to its pyhtochemistry and phrmacological activity. J. Drug Deliv. Therapeut. 8 (5-s), 73–78.

Auquier, P., Renard, R., 1975. Nombres chromosomiques de quelques angiospermes du Rwanda, Burundi et Kivu (Zaïre). I. Bull. Jard. Bot. Belg. 45, 421–445.

Badola, H.K., Pal, M., 2002. Endangered medicinal plant species in Himachal Pradesh. Curr. Sci. 83, 797–798.

Bala, S., Gupta, R.C., 2011. IAPT/ IOPB chromosome data 12. In: Marhold, K. (Ed.), Taxon 60 (6), 1784–1786.

Bala, S., Malik, R.A., Gupta, R.C., 2015. New chromosome counts in some gentians from Western Himalayas. Caryologia 68 (2), 147–153.

Bhatt, A., Rawal, R.S., Dhar, U., 2005. Ecological features of a critically rare medicinal plant, *Swertia chirayita*, in Himalaya. Plant Species Biol. 21, 49–52.

Bora, M., Singh, V., 2016. Chirata (*Swertia chirayita* Roxb. Ex fleming) in alpine zone of Kumaun Himalaya: a study of the Khalia top medows. Int. J. Sci. Nat. 7 (2), 251–254.

Brys, R., Jacquemyn, H., Endels, P., Blust, G.D., Hermy, M., 2005. Effect of habitat deterioration on population dynamics and extinction risks in previously common perennial. Conserv. Biol. 19 (5), 1633–1643.

Chambers, K.L., Green, D., Potampa, S., McMahan, L., 1998. IOPB chromosome data 13. Newslett. Int. Organ. Pl. Biosyst. (Oslo) 29, 18–22.

Chaudhuri, R.K., Pal, A., Jha, T.B., 2008. Conservation of *Swertia chirata* through direct shoot multiplication from leaf explants. Plant Biotechnol. Rep. 2, 213–218.

Chaudhuri, R.K., Pal, A., Jha, T.B., 2007. Production of genetically uniform plants from the nodal explants of *Swertia chirayita* Buch.-Ham. ex Wall.- an endangered medicinal herb. In Vitro Cell Dev. Bio.-Pl 43 (5), 467–472.

Chaudhuri, R.K., Pal, A., Jha, T.B., 2009. Regeneration and characterization of Swertia chirata Buch.-Ham. ex Wall. plants from immature seed cultures. Sci. Horticul. 120 (1), 107–114.

Clarke, C.B., 1885. Gentianaceae. In: Hooker, J.D. (Ed.), The Flora of British India IV. L. Reeve and Co., Ltd., London, pp. 121–130.

Colling, G., Matthies, D., 2006. Effects of habitat deterioration on population dynamics and extinction risk of an endangered, long-lived perennial herb (*Scorzonera humilis*). J. Ecol. 94, 959–972.

Darlington, C.D., Wylie, A.P., 1955. Chromosome Atlas of Flowering Plants. Allen and Unwin, London.

Davlianidze, M.T., 1984. Investigatio cytogeographics speciorum nonnullarum altimontanarum e Caucaso. Not. Syst. Georg. Inst. Bot. Thbilissi. 40, 56–66.

Dawe, J.C., Murray, D.F., 1979. IOPB chromosome number reports LXIII. Taxon 28, 265–268.

Fan, G., Luo, W.Z., Luo, S.H., Li, Y., Meng, X.L., Zhou, X.D., et al., 2014. Metabolic discrimination of *Swertia mussotii* and *Swertia chirayita* known as "Zangyinchen" in traditional Tibetan medicine by 1H NMR-based metabolomics. J. Pharmaceut. Biomed. Anal. 98, 364–370.

Favarger, C., 1952. Contribution h I'Ctude caryologique et biologique des Gentianac Ces.11-. Bull. Soc. Bot. Suisse 62, 144–257.

Fedorov, A.A. (Ed.), 1969. Chromosome Numbers of Flowering Plants. Izdatelstvo "Nauk,", Leningrad.

Gagnidze, R.I., Gviniaschvili, T.N., 1984. Chromosome numbers of some high mountain species from Georgia. Bot. Zhurn. SSSR 69 (12), 1703–1704.

Gaikwad, J., Wilson, P.D., Ranganathan, S., 2011. Ecological niche modeling of customary medicinal plant species used by Australian Aborigines to identify species-rich and culturally valuable areas for conservation. Ecol. Model. 222 (18), 3437–3443.

Ghosal, Sharma, P.V., Chaudhuri, R.K., Bhattacharya, S.K., 2006. J. Pharma. Sci. 62, 926–930.

Gohil, R.N., Ashraf, M., Raina, R., 1981. Cytotaxonomical conspectus of the flora of Kashmir II. Chromosome numbers of 51 dicotyledonous species. Herba Hung. 20, 43–49.

Gvinianidze, Z.I., Avazneli, A.A., 1982. Khromosomnye chisla nekotorykh predstavitelej vysokogornykh floristicheskikh kompleksov Kavkaza. Soobkskc. Akad. Nauk Gruzinskoi SSR, Inst. Bot., Trudy, Ser. Geobot. 106 (3), 577–580.

He, T.Z., Wang, W., Xue, C.Y., 1999. A karyomorphological study on 5 species of Swertia (Gentianaceae). Acta Bot. Boreal.-Occid. Sin. 19 (3), 546–551.

Hedberg, I., Hedberg, O., 1977. Chromosome numbers of afroalpine and afromontane angiosperms. Bot. Not. 130, 1–24.

Hegland, S.J., Van Leeuwen, M., Oostermeijer, J.G.B., 2001. Population structure of *Salvia pratensis* in relation to vegetation and management of Dutch dry floodplain grasslands. J. Appl. Ecol. 38 (6), 1277–1289.

Heywood, V.H., Harborne, J.B., Turner, B.L. (Eds.), 1977. The Biology and Chemistry of the Compositae, vols. 1 and II. Academic Press, London.

Huang, R.F., Shen, S.D., Lu, X.F., 1996. Studies on the chromosome number and polyploidy for a number of plants in the north-east Qinghai-Xizang Plateau. Acta Bot. Boreal.-Occid. Sin. 16 (3), 310–318.

Iqbal, Z., Lateef, M., Khan, M.N., Jabbar, A., Akhtar, M.S., 2006. Anthelmintic activity of *Swertia chirata* against gastrointestinal nematodes of sheep. Fitoterapia 77 (6), 463–465.

Jaryan, V., Datta, A., Uniyal, S., Kumar, A., Gupta, R., Singh, R., 2013. Modelling potential distribution of Sapium sebiferum – an invasive tree species in western Himalaya. Curr. Sci. 105 (9), 1282–1288.

Jee, V., Dhar, U., Kachroo, P., 1989. Cytogeography of some endemic taxa of Kashmir Himalaya. Proc. Indian Natl. Sci. Acad. B 55, 177–184.

Jee, V., Dhar, U., Kachroo, P., 1985. Chromosomal conspectus of some alpine subalpine taxa of Kashmir Himalaya. Chromosome Inf. Serv. 39, 33–35.

Joshi, P., Dhawan, V., 2007a. Axillary multiplication of *Swertia chirayita* (Roxb. Ex Fleming) H. Karst., a critically endangered medicinal herb of temperate Himalayas. In Vitro Cell Dev. Bio.-Pl 43, 631–638.

Joshi, P., Dhawan, V., 2007b. Analysis of genetic diversity among *Swertia chirayita* genotypes. Biol. Plant. 51 (4), 764–768.

Joshi, P., Dhawan, V., 2005. *Swertia chirayita* – an overview. Curr. Sci. India 89 (4), 635–640.

Kalliovirta, M., Ryttari, T., Hiekinen, R.K., 2006. Population structure of a threatened plant, Pulsatilla patens, in boreal forests: Modelling relationships to overgrowth and site closure. Biodiversity Conserv. 15, 3095–3108.

Kaur, P., Pandey, D.K., Gupta, R.C., Dey, A., 2019. Assessment of genetic diversity among different population of five *Swertia* species by using molecular and phytochemical markers. Ind. Crop. Prod. 138 https://doi.org/10.1016/j.indcrop.2019.111569.

Kawakami, J., 1930. Chromosome numbers in Leguminosae. Bot. Mag. (Tokyo) 44, 319–328.

Khatoon, S., Ali, S.I., 1993. Chromosome Atlas of the Angiosperms of Pakistan. Department of Botany, University of Karachi, Karachi.

Khoshoo, T.N., Khushu, F.A.S., Khushu, C.L., 1966. Biosystematics of Indian Plants II. The problem of *Centaurium pulchellum* complex. Proc. Indian Acad. Sci. B 63 (3), 152–160.

Khoshoo, T.N., Tandon, S.R., 1963. Cytological, morphological and pollination studies on some Himalayan species of *Swertia*. Caryologia 16, 445–477.

Koul, S., Suri, K., Dutt, P., Sambyal, M., Ahuja, A., Kaul, M., 2016. Protocol for in vitro regeneration and marker glycoside assessment in *Swertia chirata* BuchHam. In: Protocols for In Vitro Cultures and Secondary Metabolite Analysis of Aromatic and Medicinal Plants, pp. 139–153.

Koul, A.K., Gohil, P.N., 1973. Cytotaxonomical conspectus of the flora of Kashmir. I. Chromosome numbers of some common plants. Phyton 15, 57–66.

Krogulevich, R.E., 1978. Kariologicheskij Analiz Vidov Flory Vostochnogo Sajana. V Flora Pribajkal'ja. 19–48. Nauka, Novosibirsk.

Kumar, K.P.S., Bowmik, D., Chiranjib, B., Chandira, M., 2010. *Swertia chirata*: a traditional herb and its medicinal uses. J. Chem. Pharm. Res. 2 (1), 262–266.

Kumar, V., Singh, S.K., Bandopadhyay, R., Sharma, M.M., Chandra, S., 2014. In vitro organogenesis secondary metabolite production and heavy metal analysis in *Swertia chirayita*. Cent. Eur. J. Biol. 9, 686–698.

Kumar, V., Van Staden, J., 2016. A review of *Swertia chirayita* (Gentianaceae) as a traditional medicinal plant. Front. Pharmacol. 6, 308.

Laxmi, A., Siddhartha, S., Archana, M., 2011. Antimicrobial screening of methanol and aqueous extracts of *Swertia chirata*. Int. J. Pharm. Pharm. Sci. 3 (4), 142–146.

Liu, Y., Wang, Y., Guo, F., Zhan, L., Mohr, T., Cheng, P., Huo, N., Gu, R., Danning Pei, D., Sun, J., Tang, L., Long, C., Huang, L., Gu, Y.Q., 2017. Deep sequencing and transcriptome analyses to identify genes involved in secoiridoid biosynthesis in the Tibetan medicinal plant *Swertia mussotii*. Sci. Rep. 7, 4308.

Löve, D., 1953. Cytotaxonomical remarks on the Gentianaceae. Hereditas 39, 225–235.

Love, A., Love, D., 1986. Chromosome number reports 93. Taxon 35, 897–899.

Ma, X.H., Ma, X.Q., Li, N., 1990. Chromosome observation of some drug plants in Xinjiang. Acta Bot. Boreal.-Occid. Sin. 10, 203–210.

Mahmood, S., Hussain, S., Tabassum, S., Malik, F., Riaz, H., 2014. Comparative phytochemical, hepatoprotective and antioxidant activities of various samples of *Swertia chirayita* collected from various cities of Pakistan. Pak. J. Pharm. Sci. 27 (6), 1975–1983.

Mallikarjuna, M.B., Sheriff, A., Krishnappa, D.G., 1987. Chromosome number reports 97. Taxon 36, 766–767.

Malik, R.A., Gupta, R.C., Kumari, S., 2011. In IAPT/IOPB Chromosome data 12. Taxon 60 (6), E47–49.

Mallikarjuna, M.B., 1985. Karyomorphological and Cytotaxonomic Studies in the Family Gentianaceae. Ph.D. thesis. Bangalore University.

Mehra, P.N., Gill, L.S., 1968. IOPB chromosome number reports XVI. Taxon 17, 199–204.

Mehra, P.N., Vasudevan, K.N., 1972. In: Löve A., (Ed.), IOPB chromosome number reports XXXVI. Taxon 21(2/3), 333–346.

Morton, J.K., 1993. Chromosome numbers and polyploidy in the flora of Cameroon Mountain. Opera Bot. 121, 159–172.

Nemomissa, S., 1998. A synopsis of Swertia (Gentianaceae) in east and northeast tropical Africa. Kew Bull. 53 (2), 419–436.

Nishteswar, K., 2014. Depleting medicinal plant resources: a threat for survival of Ayurveda. Ayu 35 (4), 349–350.

Ornduff, R., Raven, P.H., Kyhos, D.W., Krukeberg, A.R., 1963. Chromosome numbers in compositae III. Senecioneae. Am. J. Bot. 50, 131–139.

Padhan, J.K., 2018. Understanding Biosynthesis of Major Chemical Constituents in Swertia Chirayita (Thesis).

Padhan, J.K., Kumar, P., Sood, H., Chauhan, R., 2016. Prospecting NGS-transcriptomes to assess regulation of miRNA-mediated secondary metabolites biosynthesis in *Swertia chirayita*, a medicinal herb of the North-Western Himalayas. Med. Plants - Int. J. Phytomed. Relat. Ind. 8 (3), 219–228.

Padhan, J.K., Kumar, V., Sood, H., Singh, T.R., Chauhan, R.S., 2015. Contents of therapeutic metabolites in *Swertia chirayita* correlate with the expression profiles of multiple genes in corresponding biosynthesis pathways. Phytochemistry 116, 38–47.

Pal, T., Padhan, J.K., Kumar, P., Sood, H., Chauhan, R.S., 2018. Comparative transcriptomics uncovers differences in photoautotrophic versus photoheterotrophic modes of nutrition in relation to secondary metabolites biosynthesis in *Swertia chirayita*. Mol. Biol. Rep. 45, 77–98.

Pant, M., Bisht, P., Gusain, M.P., 2010. De novo shoot organogenesis from cultured root explants of *Swertia chirata* Buch.-Ham. Ex Wall.: an endangered medicinal plant. Nat. Sci. 8 (9), 244–252.

Post, D.M., 1983. In IOPB chromosome number reports. Taxon 32, 509.

Pradhan, B.K., Badola, H.K., 2015. *Swertia chirayta*, a threatened high-value medicinal herb: microhabitats and conservation challenges in Sikkim Himalaya, India. Mt. Res. Dev. 35 (4), 374–381.

Pradhan, B.K., Badola, H.K., 2012. Effects of microhabitat, light and temperature on seed germination of a critically endangered Himalayan medicinal herb, *Swertia chirayita*: conservation implications. Plant Biosyst. 146 (2), 345–351.

Probatova, 2006. Chromosome numbers of plants of the primorsky territory, the Amur river basin and Magadan region. Bot. Zhurn. (Mosc. Leningr.) 91 (3), 491–509.

Robinson, H., Powell, A.M., King, R.M., Weedin, J.F., 1981. Chromosome numbers in compositae, XII: Heliantheae. Smithson. Contrib. Bot. 52, 1–28.

Samant, S.S., Dhar, U., Palni, L.M.S., 1998. Medicinal Plants of Indian Himalayas: Diversity, Distribution and Potential Values. Himavikas Publication No. 13, Almora, Uttaranchal, India (GB Pant Institute of Himalayan Environment and Development).

Samant, S.S., Dhar, U., Rawal, R.S., 1996. Conservation of rare endangered plants: the context of Nanda Devi biosphere reserve. In: Ramakrishnan, P.S., Purohit, A.N., Saxena, K.G., Rao, K.S., Maikhuri, R.K. (Eds.), Conservation and Management of Biological Resources in Himalaya. Oxford & IBH, New Delhi, India, pp. 521–545.

Selvam, A.B.D., 2012. Pharmacognosy of Negative Listed Plants, pp. 215–230.

Sharma, A.K., 1970. Annual report, 1967-1968. Res. Bull. Univ. Calcutta Cytogenetics Lab. 2, 1–50.

Sharma, V., 2011. In Vitro Rapid Mass Multiplication and Molecular Validation of *Swertia chirayita*. A Ph.D. thesis submitted to HNB Garhwal University Srinagar, Garhwal.

Sharma, V., Barkha, K., Srivastava, N., Negi, Y., Dobriyal, A.K., Jadon, V.S., 2016. Assessment of genetic fidelity of *in vitro* raised plants in *Swertia chirayita* through ISSR, RAPD analysis and peroxidase profiling during organogenesis. Braz. Arch. Biol. Technol. 59, 1–11.

Sharma, V., Barkha, K., Srivastava, N., Negi, Y., Dobriyal, A.K., Jadon, V.S., 2015. Enhancement of in vitro growth of *Swertia chirayita* Roxb. Ex Fleming co-cultured with plant growth promoting rhizobacteria. Plant Cell Tissue Organ Cult. https://doi.org/10.1007/s11240-014-0696-9.

Sharma, V., Kamal, B., Srivastava, N., Dobriyal, A.K., Jadon, V., 2013. Effects of additives in shoot multiplication and genetic validation in *Swertia chirayita* revealed through RAPD analysis. Plant Tissue Cult. Biotech. 23 (1), 11–19.

Sharma, A., Sarkar, A.K. (Eds.) 1967–68. Chromosome number reports of plants. In: Annual Report of Cytogenetics Laboratory, Department of Botany University of Calcutta. Research Bulletin. 2, pp. 38–48.

Sharma, V., Srivastava, N., Kamal, B., Dobriyal, A.K., Jadon, V.S., 2014. In vitro Flower induction from shoots regenerated from cultured axillary buds of endangered medicinal herb *Swertia chirayita* H. Karst. Biotechnol. Res. Int. 2014 https://doi.org/10.1155/2014/264690.

Shigenobu, Y., 1982. Cytological relationship between *Swertia bimaculata* and *S. swertopsis*. J. Jpn. Bot. 57, 353–357.

Shigenobu, Y., 1983. Karyomorphological studies in some genera of Gentianaceae II. Gentiana and its allied four genera. Bull. Coll. Child Dev. Kochi Womens Univ. 7, 65–84.

Shrestha, B.B., Jha, P.K., 2010. Life history and population status of the endemic Himalayan *Aconitum* naviculare. Mt. Res. Dev. 30 (4), 353–364.
Shukla, J.K., Dhakal, P., Uniyal, R.C., Paul, N., Sahoo, D., 2017. Ex-situ cultivation at lower altitude and evaluation of Swertia chirayita, a critically endangered medicinal plant of Sikkim Himalayan region, India. South African J. Bot. 109, 138–145.
Singh, A.P., 2005. Promising phytochemicals from Indian medicinal plants. Ethnobot. Leafl. 2005 (1), 18.
Singh, R.L., Singh, P., Agarwal, A., 2012. Chemical constituents and bio-pharmacological activities of *Swertia chirata*: a review. Nat. Prod. Indian J. 8 (6), 238–247.
Skalinska, M., 1952. Cytologigal studies in *Gentiana*–species from the Tatra and Pieniny Mts. Bulletin Academie Polonaise des Sciences el des Lettres–B 119–136, 1951.
Stebbins, G.L., 1950. Variation and Evolution in Plants. Columbia University Press, New York, p. 643.
Subramanyam, K., Kamble, N.P., 1966. IOPB chromosome number reports VII. Taxon 15, 155–163.
Sultana, M.J., Ahmed, F.R.S., 2013. Phytochemical investigations of the medicinal plant *Swertia chirata* Ham. Biochem. Anal. Biochem. 2 (145), 2161-1009.
Tabassum, S., Mahmood, S., Hanif, J., Hina, M., Uzair, B., 2012. An overview of medicinal importance of *Swertia chirayita*. Int. J. Appl. 2 (1).
Tan, K., Vladimirov, V., 2001. Swertia punctata Baumg. (Gentianaceae) in Bulgaria. Bocconea 13, 461–466.
Tandon, P., Kumaria, S., Kayang, H., 2010. Conservation of medicinal and aromatic plants of northeast India. In: Ahmad, A., Siddiqi, T.O., Iqbal, M. (Eds.), Medicinal Plants in Changing Environment. Capital Publishing Company, New Delhi, India, pp. 203–212.
Thulin, M., 1970. Chromosome numbers of some vascular plants from East Africa. Bot. Notiser. 123 (8), 488–494.
Vaidya, H., Goyal, R.K., Cheema, S.K., 2013. Anti-diabetic Activity of swertiamarin is due to an active metabolite, gentianine, that upregulates PPAR-γ gene expression in 3T3-L1 cells. Phytother Res. 27, 624–627.
Vasudevan, K.N., 1975. Contribution to the cytotaxonomy and cytogeography of the flora of the Western Himalayas (with an attempt to compare it with the flora of Alps). Part II. Bericht der Schweizerischen Botanischen Gesellschaft 85, 210–252.
Wada, Z., 1966. Chromosome numbers in Gentianaceae. Chromosome Inf. Serv. 7, 28–30.
Wang, L., An, L., Hu, Y., Wei, L., Li, Y., 2009. Influence of phytohormonesand medium on the shoot regeneration from leaf of *Swertia chirata* Buch.-Ham. ex Wall. in vitro. Afr. J. Biotechnol. 18 (11), 2513–2517.
Wawrosch, C., Maskay, N., Kopp, B., 1999. Micropropagation of the threatened Nepalese medicinal plant *Swertia chirata* Buch.-Ham.ex Wall. Plant Cell Rep. 18, 997–1001.
Yuan, Y., Küpfer, P., 1993. Karyological studies of Gentianopsis Ma and some related genera of Gentianaceae from China. Cytologia 58, 115–123.
Zhukova, P.G., Petrovsky, V.V., Plieva, T.N., 1973. The chromosome numbers and taxonomy of some plant species from Siberia and Far East. (in Russian) Bot. zurn. (Moscow & Leningrad) 58, 1331–1342.

Further reading

Davis, A.P., Govaerts, R., Bridson, D.M., Ruhsam, M., Moat, J., Brummitt, N.A., 2009. A global assessment of distribution, diversity, endemism, and taxonomic effort in the Rubiaceae. Ann. Mo. Bot. Gard. 96, 68–78.

Jee, V., Dhar, U., Kachroo, P., 1983. Chromosome numbers of some alpine subalpine taxa of Kashmir Himalaya. Herba Hung. 22, 23–31.

Mosaleeyanon, K., Zobayed, S., Afreen, F., Kozai, T., 2005. Relationships between net photosynthetic rate and secondary metabolite contents in St. John's wort. Plant Sci. 169, 523–531.

Roy, S.C., Ghosh, S., Chatterjee, A., 1988. A cytological survey of eastern Himalayan plants. II. Cell Chromosome Res. 11, 93–97.

Suryawanshi, S., Asthana, R.K., Gupta, R.C., 2009. Assessment of systemic interaction between *Swertia chirata* extract and its bioactive constituents in rabbits. Phytother Res. Int. J. Devot. Pharmacol. Toxicol. Eval. Nat. Prod. Deriv. 23 (7), 1036–1038.

CHAPTER

Trillium govanianum

13

Vishal Kumar[1], Pradeep Singh[2], Pramod Kumar Singh[3], Mohammed Saba Rahim[5], Vikas Sharma[4], Joy Roy[5], Himanshu Sharma[5]

[1]*Govt. Senior Secondary School, Bhadwar, Kangra, Himachal Pradesh, India;* [2]*Department of Biotechnology, Guru Nanak Dev University, Amritsar, Punjab, India;* [3]*Department of Biosciences, Christian Eminent College, Indore, Madhya Pradesh, India;* [4]*Department of Botany, Sant Baba Bhag Singh University, Khiala, Jalandhar, Punjab, India;* [5]*Agri-Biotechnology Division, National Agri-Food Biotechnology Institute, Mohali, Punjab, India*

13.1 Introduction

Plants are major backbone to mankind in various ways. Since from early of civilization, people are dependent on natural resources and nature for various needs which we got from plant-based medicines/products as one of them. The Himalayan zone is harboring very rich diversity of flora region. In India, the Himalayas are present in an area of ~591,000 km^2 and lie between 27°50′ and 37°06′ N and 72°30′ and 97°25′ E. Because of the Himalayas, India is considered in among 10 most comprehensively forested areas in the world and it is covered by 18% of India's geographical area and subsequently forms more than 50% of the country's forest cover and 40% of the species endemic to the Indian subcontinent (Saxena et al., 2002). Different climatic conditions are mainly accountable for different types of diverse environments, which lead to higher biodiversity in the Himalayas. Rana and Samant (2009) reported that there are more than 18,000 plants species that reside in the Himalayas which include more than 1700 medicinally important plants (Samant et al., 1998). Plants present in temperate and alpine climate are facing more stress as compared to plants found in the subtropics, and due to this, they have unique medicinal and aromatic properties. Medicinal plants are mainly found in diverse habitats and habits, they may be annual, biennial, and others are perennial. Mankind has been using herbs and produce from different plants for healing ailments and improving well-being since antiquity.

Recent rises in cases of illness, side effects of allopathic medicines raises a concern on traditional medicines for various diseases and this enlightenment ultimately leads to the shift in the use of allopathic medicines to traditional medicines of plant origin. Several plants are used for the various medicinal purposes from immemorial times. Further plant-based medicines explained by various Unani, Ayurveda, and other medicine system impact a respectable position today, especially in the tribal regions of developing countries, where still health services are limited or

Himalayan Medicinal Plants. https://doi.org/10.1016/B978-0-12-823151-7.00006-4

not available for poor man. The popularity of traditional medicines which are more effective, safe, and inexpensive raised the concern among both developing and developed countries. Information of traditional medicine effectiveness has played a vital role in the discovery of novel products from plants as chemotherapeutic agents (Katewa et al., 2004). In addition to this, approximately 25% of pharmaceutical raw material comes directly from natural plant products (Schmidt et al., 2008). Continuous demands of traditional medicinal compound lead to the overexploitation of 1000 years conserved biodiversity. This leads to the creation of a situation where demand is huge and supply is lacking day by day. In most of the cases, demand is fulfilled by collecting material directly from the wild habitat which took very long time to revive and results into replenishment of that species. As already explained Himalayan region is one of the hotspots of biodiversity and also home for thousands of medicinal plant species. In the region of Himalayas, still 85% population relied on the traditional knowledge of medicinal plants (Farnsworth, 1988). Emphasis of the World Health Organization further puts pressure on this precious treasure of medicinal plants. Herbal industries are mainly dependent on its requirements on the medicinal plants of the Himalayan region (Dhar et al., 2000). Continuous widening of gaps between demand and supply leads to illegal harvesting which in result puts threat to many species at the brink of extinction (Vidyarthi et al., 2013). Among the medicinal plants present in Himalayan region, *Trillium govanianum* is one of the fast-emerging medicinal herbs, with many pharmacological activities.

13.1.1 *Trillium govanianum*

T. govanianum Wall ex D. Don, commonly known as Himalayan Trillium, Nag chhatri, or Teen patra, is a perennial herb endemic to the Himalayas (Samant et al., 1998; Kubota et al., 2006). It belongs to the plant family Melanthiaceae, having 181 species related to 17 genera of perennial herbs. Mostly distributed in the temperate region of Northern Hemisphere and considered for its traditional and modern medicinal properties (Zomlefer et al., 2001; Christenhusz and Byng, 2016), *T. govanianum* belongs to genus *Trillium*, one of the largest genus of Melanthiaceae, comprising of 50 species, 39 American species of Arcto-Tertiary origin, and 11 Asian species (http://www.theplantlist.org/1.1/browse/A/Melanthiaceae/). This genus is further divided into two subgenera, *Phyllantherum* Raf. comprising of sensile-flowered species (26 American species) and *Trillium* with pedicellate-flowered species (all Asian and remaining American species) (Ohara and Kawano, 2006; Schilling et al., 2019). Only two species, namely, i.e., *T. govanianum* and *Trillium tschonoskii* of genus *Trillium* are found in the Himalayan region.

T. govanianum has an important place in the traditional medicine system due to the presence of various therapeutically important compounds in its rhizomes (Zhan, 1994; Shah, 2006; Khan et al., 2016). Recently, a trend rises in the collection of *T. govanianum* from the forests attracts the attention of biodiversity conservators and researchers. The important part for which it is exploited is underground rhizome containing trillarin, a raw material for medicines mainly sex stimulants. In traditional medicines system, rhizome of *T. govanianum* is used for the cure of dysentery, boils, menstrual and sexual disorders, etc. The recent studies led to the identification of several pharmacological activities possessed by *T. govanianum* such as antiseptic, analgesic, antiinflammatory, antifungal, free radical scavenging, as well as cytotoxicity against prostate and cervical carcinoma cells (Ur Rahman et al., 2016).

Recently, important active components are explored and their corresponding pharmacological activities from *Trillium* are of utmost importance. It is specifically distributed in the Indian Himalayas, and collecting its rhizome for commercial activities has become common in the Indian Himalayas. However, in reality, the collection of species from natural habitats is unjustifiable. Understanding the phenomenon like socioecological dynamics of a species and making a strategy for its sustainable use is a challenging mission in the Himalayas. There should be potential and effective strategies for conservation of the species while retaining community incomes. This chapter provides a focus on overview of the biology, uses, and conservation approaches that can be followed for the sustainable utilization of *T. govanianum* in the Indian Himalayas.

13.2 Classification, origin, distribution, and cytotaxonomy

13.2.1 Classification and morphology

Kingdom: Plantae
Subkingdom: Tracheobionta
Superdivision: Spermatophyta
Division: Magnoliophyta
Class: Liliopsida
Subclass: Liliidae
Order: Liliales
Family: Melanthiaceae
Genus: Trillium L.
Species: *govanianum*

T. govanianum is a small 100–200 mm creeping stem rising from a short, tuberous rhizome and having one or three ovate, acute, stalked leaves. At the time of reproduction, a flower emerges at the shoot apex, surrounded by three leaves. Flowers are 2–3 cm long having six distinctly yellow stamens filament attached to base, a whorl each of petals and sepals, and a three-celled purplish brown ovary that produces multiple seeds. The fruit is a globular red berry and seeds are ovoid with pulpy lateral appendages (Fig. 13.1). Seeds of *Trillium* spp. are numerous and their dispersal is low, such that seeds typically remain close to the parent plant (Ohara and Kawano, 2005).

FIGURE 13.1

Trillium govanianum plant growing in wild habitat.

13.2.2 Distribution

This plant is mainly distributed in the range of 2500–4000 mts across the Himalayas (Vidyarthi et al., 2013). The plant is sciophyte having three leaves on purple red stem bearing trimerous flowers. *T. govanianum* is most commonly found under the canopies of mix temperate (*Abies pindrow*, *Betula utilis*, *Cedrus deodara, Juglans regia*, *Juniperus indica*, *Picea smithiana*, *Quercus* spp., *Rhododendro*n spp., and Salix spp.) and subalpine forests (*Rhododendron* spp.) with habitat of thick humus decomposing litter. Due to these specific habitat conditions, the species has patchy and limited distribution to specific pockets in the Himalayas (Chauhan et al., 2020). The species is more common in western regions of the Himalayas as compared to eastern region (Chauhan et al., 2018). The main areas of its distribution are as follows: in Jammu and Kashmir, the species grows in Fatehpur, Gulmarg, Gurez, Kanzalwan, Pahalgam, Poonch, Sonamarg in Uttarakhand, Gangotri, Govind Pashu Vihar, Harshil, Kedarnath, Munsiyari, Panchachuli, Pindari, Sunderdhunga, and Tungnath, while in areas of Himachal Pradesh, the species is commonly reported from Kullu, Kinnaur, Lahaul-Spiti, and Shimla.

13.2.3 Origin and cytotaxonomy

With a base number ($x = 5$), genus *Trillium* possesses different ploidy levels. Among the Asian species, except *Trillium camschatcense* (diploid, $2n = 10$) all the species are allopolyploids involving the hybridization of different genomes, whereas American species are diploid $2n = 10$, except for few reports of autotriploids (Darlington and Wylie, 1956; Darlington and Shaw, 1959). The karyological analysis of *T. govanianum* has revealed that it is a tetraploid, $2n = 20$, having 10 pairs (A–J) of unusual long chromosomes. Based on length and position of centromere, chromosome pairs are categorized as (i) metacentric–submetacentric: first two large pairs (A and B), (ii) submetacentric: two next large (C and D) and two small (I and J), and (iii) acrocentric: four moderate sized (E, F, G, and H) (Mehra and Sachdeva, 1975).

Based on cytological and morphological characteristics, the origin of *T. govanianum* is considered as a historic evolutionary event for survival and perpetuation against several cold and dry seasons during Pleistocene age in Asia. *T. govanianum* originates from the rare natural intergeneric hybridization of genus *Trillium* and *Daiswa* (Fukuda, 2001).

13.3 Biochemical analysis

The *genus Trillium* having around 31 species and few of them consists of rich sources of bioactive compounds which possess medicinal value with wide application in pharmaceuticals application (Gracie and Lamont, 2012). The major classes of phytochemicals include mainly steroids, glycosides, terpenoids, sterol, saponins, and flavonoids (Ismail et al., 2015). Various protocols have been reported for efficient extraction of metabolite from dried rhizomes of *T. govanianum* and characterized using various chromatographic techniques such as ^{1}H-NMR, ^{13}C-NMR, COSY, NOESY, HSQC, HMBC, FAB, HR-FAB, IR, and UV. Previously only four compounds, namely Govanoside A, Pennogenin, Borossoside E and Diosgenin were detected. Recently, Singh et al. (2020) also identified 24 steroidal saponins in *T. govanianum* by using UHPLC-QTOF-MS/MS (Fig. 13.2).

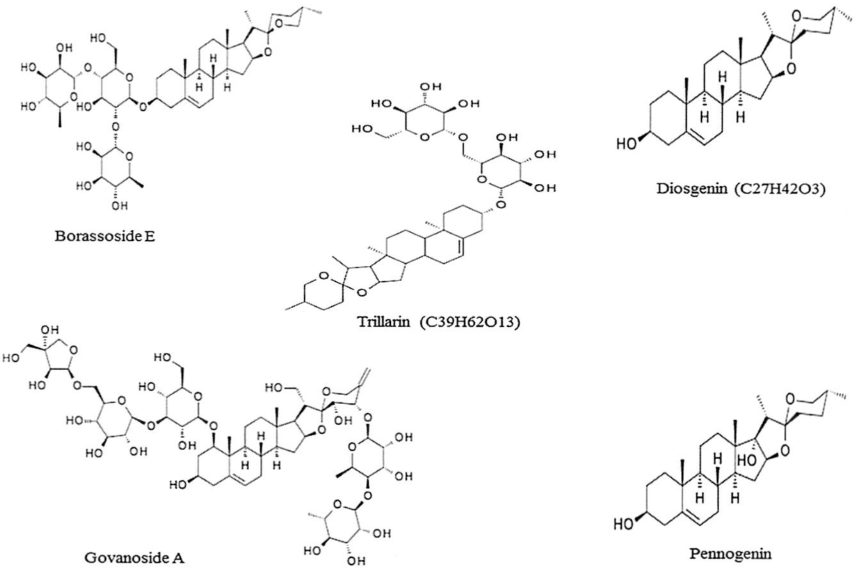

Borassoside E

Trillarin (C39H62O13)

Diosgenin (C27H42O3)

Govanoside A

Pennogenin

FIGURE 13.2

Biochemical structure of phytochemicals present in *Trillium govanianum*.

Based on Ur Rahman, S., Adhikari, A., Ismail, M., Shah, M.R., Khurram, M., Anis, I., Ali, F., 2017a. A new trihydroxylated fatty acid and phytoecdysteroids from rhizomes of Trillium govanianum. *Record Nat Prod 11, 323–327; Ur Rahman, S., Ismail, M., Khurram, M., Ullah, I., Rabbi, F., Iriti, M., 2017b. Bioactive steroids and saponins of the genus* Trillium. *Molecules 22(12), 2156.*

Table 13.1 List of bioactive compounds in the *Trillium species*.

S. no.	Source	Chemical compound	Class	References
1	*Trillium govanianum*	Govanoside A	Sterol saponin	Ur Rahman et al. (2015a,b)
2	*T. govanianum*	Borossoside E	Sterol saponin	Ur Rahman et al. (2015a,b)
3	*T. govanianum*	Pennogenin	Sterol saponin	Ur Rahman et al. (2015a,b)
4	*T. govanianum*	Diosgenin	Sterol saponin	Ur Rahman et al. (2015a,b)
5	*T. govanianum*	Govanoside A, Govanoside B, Protodioscin, Pregna-chacotrioside, Pennogenin tetraglycosides, Pennogenin diglycosides, Borassoside E, Borassoside D, Diosgenin, Pennogenin-[O-β-D-glucopyranosyl-S1 or its isomer, (1β, 23S,24S)-1-[O-β-D-glucopyranosyl (1 → 3)-O-(β-D-xylopyranosyl-(1 → 2)-O-α-L-rhamnopyranosyl]-23 hydroxyspirosta-5,25-dienyl-24-[O-β-D-6-deoxygulopyranoside] or its isomer, pirosta-5,25-dienyl-[O-β-D-glucopyranosyl-S3 or its isomer, Protodioscin, Pennogenin-[O-β-D-glucopyranosyl-S4 or its isomer, Diosgenin-[O-β-D-glucopyranosyl-S5 or its isomer, Pennogenin-[O-β-D-glucopyranosyl-S4 or its isomer, Pennogenin-[O-β-D-glucopyranosyl-S6, Gentrogenin 3-O-β-chacotrioside or its isomer, Pennogenin-[O-β-D-glucopyranosyl-	Sterol saponin	Singh et al. (2020)

Table 13.1 List of bioactive compounds in the *Trillium species.—cont'd*

S. no.	Source	Chemical compound	Class	References
		S7 isomer, Pennogenin-[O-β-D-glucopyranosyl or its isomer, Borassoside E isomer, Borassoside D, Borassoside D isomer		
6	*Trillium tschonoskii*	2,3-S-trans,10R,6E)-7,11-dimethyl-3-methylene-1, 6-dodecadien-10, 11-diol 10-O-b-D-glucopyranosyl-(1→4) -O-b-D-glucopyranosyl-(1→4)-Ob-D-glucopyranoside	Sesquiterpenoid glycoside	Chai et al. (2014)
7	*Trillium erectum*	25R)-17a-hydroxyspirost -5-en-3b-yl O-a-L-rhamnopyranosyl -(1→2)-b-D-glucopyranoside	Steroidal glycosides	Nohara et al. (1975)
8	*T. erectum*	(25R)-17a-hydroxyspirost -5-en-3b-yl O-a-L-rhamnopyranosyl -(1→4)-b-D-glucopyranoside	Steroidal glycosides	Mahato et al. (1981)
9	*T. erectum*	(25S)-17a, 27-dihydroxyspirost -5-en-3b-yl O-a-Lrhamnopyranosyl-(1→2)-b -D-glucopyranoside	Steroidal glycosides	Ono et al. (2007)
10	*T. erectum*	(25S)-Spirost-5-ene-3b,17a,27-triol	Steroidal glycosides	Yokosuka and Mimaki (2008)
11	*T. erectum*	(25S)-27-[(b-D-Glucopyranosyl)oxy] -17a-hydroxyspirost-5-en-3b-yl O-a-L-rhamnopyranosyl-(1→2) -b-D-glucopyranoside	Steroidal glycosides	Yokosuka and Mimaki (2008)
12	*T. tschonoskii*	7-b-hydroxy trillenogenin 1-O-b-D apiofuranosyl -(1 → 3)-a-rhamnopyranosyl-(1 → 2)-[b-D xylopyranosyl-(1 → 3)]-a-Larabinopyranoside	Steroidal glycosides	Wang et al. (2007)

Continued

Table 13.1 List of bioactive compounds in the *Trillium species.—cont'd*

S. no.	Source	Chemical compound	Class	References
13	*T. tschonoskii*	(23S,24S,25S)-spirost-5-en-1b,3b,21,23, 24-pentaol-1-O-b-D xylopyranosyl-(1→!3) -[O-a Lrhamnopyranosyl-(1→2)]-O-a-L-arabinopyranoside	Steroidal glycosides	Chai et al. (2014)

The other species of *Trillium* also have diverse bioactive compounds which have different medicinal properties listed in Table 13.1.

13.4 Medical significance

T. govanianum plants have been used in the traditional as well as in the modern systems of medicine due to the presence of bioactive compounds called phytochemicals. The species of *Trillium* found in the North part of America have properties like uterine stimulant, antimicrobial, antifungal, and antibacterial properties (Huang and Zou, 2011; Ono et al., 2007). The rhizome of *T. govanianum* is used for the treatment of various disorders like dysentery, backache, healing of wounds, inflammation, skin boils, and menstrual and sexual disorder (Mahmood et al., 2012; Rani et al., 2013; Sharma and Samant, 2014). There are several reports available on the uses of powdered form of *T. govanianum* plants as anthelmintic, antifungal, and antimicrobial properties (Lone et al., 2013). Some of the other medicinal properties of *T. govanianum* are listed in Table 13.2.

White amorphous powder known as Govanoside A ($C_{56}H_{88}O_{29}$) and Borassoside E ($C_{45}H_{72}O_{16}$) and steroidal saponins obtained from the dried rhizome of *T. govanianum* are used as the antifungal agents (Ur Rahman et al., 2016). Govanoside A shows its antifungal effect against *Candida albicans* and *Candida glabrata*

Table 13.2 Medicinal properties of *Trillium govanianum.*

S. no.	Diseases	References
1	Skin infection	Lone et al. (2013)
2	Antianalgesic	Ur Rahman et al. (2016)
3	Antiinflammatory	Ur Rahman et al. (2016)
4	Treating sexual disorder	Rani et al. (2013)
5	Antidiarrheal and antiseptic	Mahmood et al. (2012), Sharma and Samant (2014)
6	Antifertility	Huang and Zou (2011)

(Ur Rahman et al., 2016). Borassoside E contains three sugar units' residues and shows better antifungal activity as compared with Govanoside A which has five sugar units' residues. The antifungal activities of these two compounds are due to their polar nature which may help in the membrane permeability in the fungal membrane (Ur Rahman et al., 2016).

Khan et al. (2018) reported first time the cytotoxicity effect of methanolic extract of root of *T. govanianum* and its solid-phase extraction fraction against four human carcinoma cell lines: breast, liver, lung, and urinary bladder using MTT assay (Hayes et al., 2009). The methanolic extract showed differential response in different cell line and the highest level of cytotoxicity was reported in urinary bladder cell line (EJ138) but considerably active against breast (MCF-7), liver (HepG2), and lung (A549) cell line (Yokosuka and Mimaki, 2008). The cytotoxicity effect of *T. govanianum is* due to the presence of phytochemicals saponins and sapogenins (Nooter and Herweijer, 1991).

Ur-Rahman et al. (2015a,b) investigated the n-hexane fraction of *T. govanianum* using GC/MS analysis and revealed the presence of steroids, glycosides, saponins, saturated, and unsaturated fatty acids that act as biologically active compounds with antifungal, antibacterial, and anticancer activities (Ching, 2008; Qiong et al., 2011).

Sagar et al. (2017) in their study reported the presence of various endophytes and antibacterial activity of *T. govanianum* plants. They isolated the endophytes according to their occurrence in various seasons and different parts of the plants such as stem, leaves, and rhizomes. The isolated endophytes were screened as *Alternaria* sp., *Aspergillus nidulans*, *Aspergillus niger*, *Aspergillus wentii*, *Fusarium solani*, *Mucor plumbeus*, *Phoma* sp., *Pythium* sp., *Rhizopus nigricans*, *Rhizopus oryzaoe*, *Stachybotrys atra*, and Trichoderma *viride*. Antibacterial activity was performed using the different solvent extracts such as ethanol, acetone, and distilled water using well diffusion method. The screening showed the highest activity of methanolic extract against *Staphylococcus aureus*, *Escherichia coli*, and *Yersinia pestis* (the human pathogenic bacteria).

13.5 Molecular breeding and genetic mapping

The cultivation of medicinal plants is important as there is increased market of phytomedicines. Despite increasing worldwide demand of secondary metabolites like phenylproponoids, alkaloids, flavonoids, polyketides, isoprenoids, etc. There are not much efforts have been made to develop more content of naturally occurring therapeutic compounds. Due to polygenic inheritance of these traits, it is very difficult to analyze the clear-cut effect of minor genes and segregation pattern in tradition breeding. We can employ advanced OMICs tools like genomic, transcriptomics, proteomics, and metabolomics to improve the content of secondary metabolites that have medicinal value. By integration of these tools, we can understand about genetics of medicinally important secondary metabolites. Through population or reverse genetics, development of molecular marker facilitates the detection of those quantitative trait loci (QTL) that control these traits. Method and approaches for marker development, QTLs identification and marker-trait association well described in other crops (Bhandawat et al., 2015; Bhardwaj et al., 2014; Jayaswall et al., 2019; Rahim et al., 2018; Rahim, et al., 2020; Sharma et al., 2011, 2020b, 2020a, 2020d; Sharma, et al., 2020; Sharma et al., 2020c; Singh et al., 2015; Singh and Sharma, 2020). We can find out the desirable genotypes of medicinally important plants in view of augmented contents of herbal medicinal. There are few studies reported on genetic linkage maps of medicinal plants, such as stevia (Yao et al., 1999), poppy (Straka and Nothnagel, 2002), and periwinkle (Gupta et al., 2007). We can perform molecular mapping of *T. govanianum* to evaluate and detection of QTL governing the therapeutic traits, and through advanced breeding skill such as genomic selection (Rahim et al., 2020b), we can improve herbal medicinal value and improved variety of *T. govanianum*. QTL's detection and use of flanking marker have been successfully achieved in some forest crop plants (Kole, 2011; Kole and Abbott, 2008).

In *T. govanianum,* such efforts are lacking due to the limited genetic information and large genome sizes. However, recent reports involve solitary transcriptomic analysis and two studies related to development of SSR markers and diversity characterizations. The spatial transcriptomic study involving the sequencing of four different tissues has characterized the steroidal saponin biosynthesis pathway genes and their expression pattern including other key genes like CYPs and UGTs and transcription factors. This study has revealed that all the tissues are actively involved in steroidal biosynthesis with accumulation in rhizome (Singh et al., 2017). Firstly, Sharma et al. (2017) have developed 21 SSRs markers and characterized them in 20 *T. govanianum* accessions. Secondly, Dhyani et al. (2020) created functionally relevant genome-wide marker resource of 5337 SSR markers in *T. govanianum* and identified 105 polymorphic markers. They also reported that low genetic diversity also existed in the populations of *T. govanianum*, so there will be an urgent need of strict conservation plans for *T. govanianum* in the Indian Himalayas.

13.5.1 Future goals for *Trillium govanianum*

1. Conservation and evaluation of *T. govanianum* germplasms.
2. Development of in vitro and ex situ propagation protocols.
3. Assessment of genetic diversity, agronomic performance, and contents of the herbal medicines.
4. Construct of genome-wide linkage maps using SNPs marker.
5. To identify QTLs controlling the contents of the herbal medicines.
6. Marker-assisted selection or genomic selection approaches can be employ for the selection of high-quality genotypes or to develop improved species.

13.6 Molecular database studies

The recent rise in the illicit trafficking and overexploitation of this species raises concern and leads to find a solution for this with the help of ever-expanding data in public domain. In *T. govanianum*, only 5358 microsatellite markers, 66 nucleotides, 214 protein sequences, and 4 transcriptome (https://www.ncbi.nlm.nih.gov/search/all/?term=trillium%20govanianum) analyses provide various insights into conserving this endemic plant. These studies lead to the assessment of various synthesis pathways. Knowledge of key genes in steroidal saponins synthesis can be explored for commercial use.

13.7 Threat

Continuous overexploitation of rhizome of plants, specific habitat, and unique climate requirements poses threat of extinction to this species. *T. govanianum* vulnerability is mainly due to illegal trade, habitat destruction, predation, and change in climatic condition (Chauhan et al., 2019). Increase in collection for commercial purposes in nonsustainable way and further rise in price leads to illegal undocumented trade which puts pressure on this species. Next the need of specific habitats puts its revival at risk with habitat destruction and climate change. Studies also documented the threat from herbivores to both wild and domesticated animals. The reports of various levels of inbreeding depressions due to self-pollination and limited pollination lead to production of less viable seeds. Vegetative propagation is lacking or very little which further adds vulnerability.

13.8 Conclusions and future perspectives

T. govanianum has many important active ingredients and medicinal properties, due to which it is used in the treatment of numerous critical diseases. Due to illegal overexploitation, the species is at the verge of extinction in the Indian Himalayas. The atypical life cycle, narrow range of distribution, constrained habitation, overexploitation, and increased value in the market are the major threats to the survival of the species. In the past, there were very few reports in literature that showed limited research has been done in ex situ propagation, genomics, biochemical, transcriptomics, and metabolomics of the *T. govanianum.* So, there will be an urgent need of actions and policies for conservation of the species like in vitro and ex situ propagation protocols, creation of genomic resource, surveillance of harvesting, and trade practices for security of the sustainable usage of the species for future generations.

Acknowledgments

National Agri-Food Biotechnology Institute (NABI), Mohali Punjab, Department of Biotechnology, Govt. of India, is acknowledged for support. DeLCON (DBT-Electronic Library Consortium), Gurugram, India, is acknowledged for access to the e-resources.

References

Bhandawat, A., Sharma, H., Nag, A., Singh, S., Ahuja, P.S., Sharma, R.K., 2015. Functionally relevant novel microsatellite markers for efficient genotyping in *Stevia rebaudiana* Bertoni. J. Genet. 94 (1), 75–81.

Bhardwaj, P., Sharma, R.K., Kumar, R., Sharma, H., Ahuja, P.S., 2014. SSR marker based DNA fingerprinting and diversity assessment in superior tea germplasm cultivated in Western Himalaya. Proc. Indian Natn. Sci. Acad. 80 (1), 157–162.

Chai, J., Song, X., Wang, X., Mei, Q., Li, Z., Cui, J., et al., 2014. Two new compounds from the roots and rhizomes of *Trillium tschonoskii*. Phytochem. Lett. 10, 113–117.

Chauhan, H.K., Bisht, A.K., Bhatt, I.D., Bhatt, A., Gallacher, D., 2019. *Trillium*—toward sustainable utilization of a biologically distinct genus valued for traditional medicine. Bot. Rev. 85 (3), 252–272.

Chauhan, H.K., Bhatt, I.D., Bisht, A.K., 2020. Biology, uses and conservation of *Trillium govanianum*. In: Socio-Economic and Eco-Biological Dimensions in Resource Use and Conservation. Springer, Cham, pp. 235–247.

Chauhan, H.K., Bisht, A.K., Bhatt, I.D., Bhatt, A., Gallacher, D., Santo, A., 2018. Population change of *Trillium govanianum* (Melanthiaceae) amid altered indigenous harvesting practices in the Indian Himalayas. J. Ethnopharmacol. 213, 302–310.

Ching, T.H., 2008. New bioactive fatty acids. Asia Pac. J. Clin. Nutr. 17, 192–195.

Christenhusz, M.J., Byng, J.W., 2016. The number of known plants species in the world and its annual increase. Phytotaxa 261 (3), 201–221.

Darlington, C.D., Shaw, G.W., 1959. Parallel polymorphism in the heterochromatin of *Trillium* species. Heredity 13 (1), 89–121.

Darlington, C.D., Wylie, A.P., 1956. Chromosome Atlas of Flowering Plants.

Dhar, U., Rawal, R.S., Upreti, J., 2000. Setting priorities for conservation of medicinal plants—a case study in the Indian Himalaya. Biol. Conserv. 95 (1), 57–65.

Dhyani, P., Sharma, B., Singh, P., Masand, M., Seth, R., Sharma, R.K., 2020. Genome-wide discovery of microsatellite markers and, population genetic diversity inferences revealed high anthropogenic pressure on endemic populations of *Trillium govanianum*. Ind. Crop. Prod. 154, 112698.

Farnsworth, N.R., 1988. Screening plants for new medicines. Biodiversity 15 (3), 81–99.

Fukuda, I., 2001. The origin and evolution in *Trillium*. Cytologia 66 (3), 319–327.

Gracie, C., Lamont, E., 2012. Spring wildflowers of the Northeast: A natural history. Princeton University Press.

Gupta, S., Pandey-Rai, S., Srivastava, S., Naithani, S.C., Prasad, M., Kumar, S., 2007. Construction of genetic linkage map of the medicinal and ornamental plant *Catharanthus roseus*. J. Genet. 86 (3), 259–268.

Hayes, P.Y., Lehmann, R., Penman, K., Kitching, W., De Voss, J.J., 2009. Steroidal saponins from the roots of *Trillium erectum* (Beth root). Phytochemistry 70 (1), 105–113.

Huang, W., Zou, K., 2011. Cytotoxicity of a plant steroidal saponin on human lung cancer cells. Asian Pac. J. Cancer Prev. APJCP 12 (2), 513–517.

Ismail, M., Shah, M.R., Adhikari, A., Anis, I., Ahmad, M.S., Khurram, M., 2015. Govanoside A, a new steroidal saponin from rhizomes of *Trillium govanianum*. Steroids 104, 270–275.

Jayaswall, K., Sharma, H., Bhandawat, A., Sagar, R., Yadav, V.K., Sharma, V., Mahajan, V., Roy, J., Singh, M., 2019. Development of intron length polymorphic (ILP) markers in onion (*Allium cepa* L.), and their cross-species transferability in garlic (*A. sativum* L.) and wild relatives. Genet. Resour. Crop Evol. 66 (7), 1379–1388.

Katewa, S.S., Chaudhary, B.L., Jain, A., 2004. Folk herbal medicines from tribal area of Rajasthan, India. J. Ethnopharmacol. 92 (1), 41–46.

Khan, K.M., Nahar, L., Al-Groshi, A., Zavoianu, A.G., Evans, A., Dempster, N.M., Sarker, S.D., 2016. Cytotoxicity of the roots of *Trillium govanianum* against breast (MCF7), liver (HepG2), lung (A549) and urinary bladder (EJ138) carcinoma cells. Phytother Res. 30 (10), 1716–1720.

Khan, K.M., Nahar, L., Mannan, A., Ul-Haq, I., Arfan, M., Ali Khan, G., Sarker, S.D., 2018. Cytotoxicity, in vitro anti-leishmanial and fingerprint HPLC-photodiode array analysis of the roots of *Trillium govanianum*. Nat. Prod. Res. 32 (18), 2193–2201.

Kole, C., 2011. Molecular breeding in medicinally active plants: bitter melon as a model. Key note presentation. In: Second Annual Conference of the American Council for Medicinally Active Plants. Huntsville, AL, 17–20 January, pp. 35–36.

Kole, C., Abbott, A.G., 2008. Principles and practices of plant genomics. Science Publishers.

Kubota, S., Kameyama, Y., Ohara, M., 2006. A reconsideration of relationships among Japanese *Trillium* species based on karyology and AFLP data. Plant Syst. Evol. 261 (1), 129–137.

Lone, P.A., Bhardwaj, A.K., Bahar, F.A., 2013. Traditional knowledge on healing properties of plants in Bandipora district of Jammu and Kashmir, India. Int. J. Recent Sci. Res. 4 (11), 1755–1765.

Mahato, S.B., Sahu, N.P., Ganguly, A.N., 1981. Steroidal saponins from *Dioscorea floribunda:* structures of floribundasaponins A and B. Phytochemistry 20 (8), 1943–1946.

Mahmood, A., Mahmood, A., Malik, R.N., 2012. Indigenous knowledge of medicinal plants from Leepa valley, Azad Jammu and Kashmir, Pakistan. J. Ethnopharmacol. 143, 338–346.

Mehra, P.N., Sachdeva, S.K., 1975. Cytology of some W. Himalayan cyperaceae. Cytologia 40, 497–515.

Nohara, T., Miyahara, K., Kawasaki, T., 1975. Steroid saponins and sapogenins of underground parts of *Trillium kamtschaticum* Pall. II. Pennogenin-and kryptogenin 3-O-glycosides and related compounds. Chem. Pharm. Bull. 23 (4), 872–885.

Nooter, K., Herweijer, H., 1991. Multidrug resistance (mdr) genes in human cancer. Br. J. Canc. 63 (5), 663.

Ohara, M., Kawano, S., 2005. Life-history monographs of Japanese plants. 2: *Trillium camschatcense* Ker-Gawl.(Trilliaceae). Plant Species Biol. 20 (1), 75–82.

Ohara, M., Kawano, S., 2006. Life-history monographs of Japanese plants. 5: *Trillium tschonoskii* Maxim.(Trilliaceae). Plant Species Biol. 21 (1), 53–60.

Ono, M., Takamura, C., Sugita, F., Masuoka, C., Yoshimitsu, H., Ikeda, T., Nohara, T., 2007. Two new steroid glycosides and a new sesquiterpenoid glycoside from the underground parts of *Trillium kamtschaticum*. Chem. Pharmaceut. Bull. 55 (4), 551–556.

Qiong, M.X., Yan-Li, L., Xiao-Ran, L., Xia, Li., Shi-Lin, Y., 2011. Three new fatty acids from the roots of Boehmeria nivea and their antifungal activities. Nat. Prod. Res. 25, 640–647.

Rahim, M.S., Sharma, H., Parveen, A., Roy, J.K., 2018. Trait mapping approaches through association analysis in plants. In: Plant Genetics and Molecular Biology. Springer, Cham, pp. 83–108.

Rahim, M.S., Mishra, A., Katyal, M., et al., 2020a. Marker-trait association identified candidate starch biosynthesis pathway genes for starch and amylose–lipid complex gelatinization in wheat (*Triticum aestivum* L.). Euphytica 216 (127).

Rahim, M.S., Bhandawat, A., Rana, N., Sharma, H., Parveen, A., Kumar, P., Madhawan, A., Bisht, A., Sonah, H., Sharma R., T., Roy, J., 2020b. Genomic selection in cereal crops: methods and applications. In: Accelerated Plant Breeding, vol. 1. Springer, Cham, pp. 51–88.

Rana, M.S., Samant, S.S., 2009. Prioritization of habitats and communities for conservation in the Indian Himalayan region: a state-of-the-art approach from Manali wildlife sanctuary. Curr. Sci. 326–335.

Rani, S., Rana, J.C., Rana, P.K., 2013. Ethnomedicinal plants of Chamba district, Himachal Pradesh, India. J. Med. Plants Res. 7 (42), 3147–3157.

Sagar, A., Chauhan, V., Prakash, V., 2017. Studies on endophytes and antibacterial activity of *Trillium govanianum* Wall. ex D. Don. J. Drug Deliv. Therapeut. 7 (2), 5–10.

Samant, S.S., Dhar, U., Palni, L.M.S., 1998. Medicinal Plants of Indian Himalaya. Gyanodaya Prakashan.

Saxena, K.G., Rao, K.S., Sen, K.K., Maikhuri, R.K., Semwal, R.L., 2002. Integrated natural resource management: approaches and lessons from the Himalaya. Conserv. Ecol. 5 (2).

Schilling, E.E., Floden, A., Lampley, J., Patrick, T.S., Farmer, S.B., 2019. A new species of *Trillium* (Melanthiaceae) from central georgia and its phylogenetic position in subgenus Sessilium. Syst. Bot. 44 (1), 107–114.

Schmidt, B., Ribnicky, D.M., Poulev, A., Logendra, S., Cefalu, W.T., Raskin, I., 2008. A natural history of botanical therapeutics. Metabolism 57, S3–S9.

Shah, R., 2006. Nature's Medicinal Plant of Uttaranchal. Gyanodaya Prakashan, Nainital.

Sharma, M., Rahim, M.S., Kumar, P., et al., 2020. Large-scale identification and characterization of phenolic compounds and their marker–trait association in wheat. Euphytica 216 (127).

Sharma, P., Samant, S., 2014. Diversity, distribution and indigenous uses of medicinal plants in Parbati valley of Kullu district in Himachal Pradesh, Northwestern Himalaya. Asian J. Adv. Basic Sci. 2 (1), 77–98.

Sharma, H., Kumar, P., Singh, A., Aggarwal, K., Roy, J., Sharma, V., Rawat, S., 2020a. Development of polymorphic EST-SSR markers and their applicability in genetic diversity evaluation in *Rhododendron arboreum*. Mol. Biol. Rep. 1–11.

Sharma, H., Bhandawat, A., Kumar, P., Rahim, M.S., Parveen, A., Kumar, P., Roy, J., 2020b. Development and characterization of bZIP transcription factor based SSRs in wheat. Gene 144912.

Sharma, H., Bhandawat, A., Rahim, M.S., Kumar, P., Choudhoury, M.P., Roy, J., 2020c. Novel intron length polymorphic (ILP) markers from starch biosynthesis genes reveal genetic relationships in Indian wheat varieties and related species. Mol. Biol. Rep. 1–16.

Sharma, H., Bhandawat, A., Rawat, S., 2020d. Cross-transferability of SSR markers developed in Rhododendron species of Himalaya. Mol. Biol. Rep. 47, 6399–6406.

Sharma, H., Kumar, R., Sharma, V., Kumar, V., Bhardwaj, P., Ahuja, P.S., Sharma, R.K., 2011. Identification and cross-species transferability of 112 novel unigene-derived microsatellite markers in tea (*Camellia sinensis*). Am. J. Bot. 98 (6), e133–e138.

Sharma, V., Wani, M.S., Singh, V., Kaur, K., Gupta, R.C., 2017. Development and characterization of novel microsatellite markers in *Trillium govanianum*: a threatened plant species from North-Western Himalaya. 3 Biotech. 7 (3), 190.

Singh, P., Sharma, R.K., 2020. Development of informative genic SSR markers for genetic diversity analysis of *Picrorhiza kurroa*. J. Plant Biochem. Biotechnol. 29 (1), 144–148.

Singh, P., Sharma, H., Nag, A., Bhau, B.S., Sharma, R.K., 2015. Development and characterization of polymorphic microsatellites markers in endangered *Aquilaria malaccensis*. Conserv. Genet. Resour. 7 (1), 61–63.

Singh, P., Singh, G., Bhandawat, A., Singh, G., Parmar, R., Seth, R., Sharma, R.K., 2017. Spatial transcriptome analysis provides insights of key gene (s) involved in steroidal saponin biosynthesis in medicinally important herb *Trillium govanianum*. Sci. Rep. 7 (1), 1–12.

Singh, P.P., Bora, P.S., Suresh, P.S., Bhatt, V., Sharma, U., 2020. Qualitative and quantitative determination of steroidal saponins in *Trillium govanianum* by UHPLC-QTOF-MS/MS and UHPLC-ELSD. Phytochem. Anal. 31 (6), 861–873.

Straka, P., Nothnagel, T., 2002. A genetic map of *Papaver somniferum* L. based on molecular and morphological markers. J. Herbs Spices Med. Plants 9, 235–241.

Ur Rahman, S., Adhikari, A., Ismail, M., Shah, M.R., Khurram, M., Anis, I., Ali, F., 2017a. A new trihydroxylated fatty acid and phytoecdysteroids from rhizomes of *Trillium govanianum*. Record Nat. Prod. 11, 323–327.

Ur Rahman, S., Ismail, M., Khurram, M., Ullah, I., Rabbi, F., Iriti, M., 2017b. Bioactive steroids and saponins of the genus *Trillium*. Molecules 22 (12), 2156.

Ur Rahman, S., Ismail, M., khurram, M., Haq, I.U., 2015a. Pharmacognostic and ethnomedicinal studies on *Trillium govanianum*. Pak. J. Bot. 47, 187–192.

Ur Rahman, S., Ismail, M., Shah, M.R., Iriti, M., Shahid, M., 2015b. GC/MS analysis, free radical scavenging, anticancer and _-glucuronidase inhibitory activities of *Trillium govanianum* rhizome. Bangladesh J. Pharmacol. 10, 577–583.

Ur Rahman, S., Adhikari, A., Ismail, M., Raza Shah, M., Khurram, M., Shahid, M., Li, F., Haseeb, A., Akbar, F., Iriti, M., 2016. Beneficial effects of *Trillium govanianum* rhizomes in pain and inflammation. Molecules 21, 1095.

Vidyarthi, S., Samant, S.S., Sharma, P., 2013. Dwindling status of *Trillium govanianum* Wall. ex D. Don-A case study from Kullu district of Himachal Pradesh, India. J. Med. Plants Res. 7 (8), 392–397.

Wang, J., Zou, K., Zhang, Y., Liu, C., Wu, J., Zhou, Y., Zhang, Y., 2007. An 18-norspirostanol saponin with inhibitory action against COX-2 production from the underground part of *Trillium tschonoskii*. Chem. Pharmaceut. Bull. 55 (4), 679–681.

Yao, Y., Ban, M., Brandle, J., 1999. A genetic linkage map for *Stevia rebaudiana*. Genome 42 (4), 657–661.

Yokosuka, A., Mimaki, Y., 2008. Steroidal glycosides from the underground parts of *Trillium erectum* and their cytotoxic activity. Phytochemistry 69, 2724–2730.

Zhan, Y.H., 1994. Resources of Medicinal Plants in Shennongjia of China. Hubei Scientific and Technologic Press, Wuhan, China.

Zomlefer, W.B., Williams, N.H., Whitten, W.M., Judd, W.S., 2001. Generic circumscription and relationships in the tribe Melanthieae (Liliales, Melanthiaceae), with emphasis on *Zigadenus*: evidence from ITS and trnL-F sequence data. Am. J. Bot. 88 (9), 1657–1669.

CHAPTER

Valeriana jatamansi

14

Pushpender Bhardwaj[1], Shiv Rattan[2], Avilekh Naryal[1], Ashwani Bhardwaj[1], Ashish R. Warghat[2]

[1]*Defence Institute of High Altitude Research, Defence R & D Organization, Leh, Ladakh, India;* [2]*Biotechnology Division, CSIR-Institute of Himalayan Bioresource Technology, Palampur, Himachal Pradesh, India*

14.1 Introduction

There are around 250 species associated with the genus *Valeriana* representing the Valerianaceae family. As far as the global distribution is concerned, it inhabits temperate regions of the world. In India, around 16 species are present, among which 2 subspecies and 5 species of this genus are found in the high-altitude region of central Himalayas. Herbaceous *Valeriana jatamansi* is also known as Indian Valerian, Sugandhbala, and Tagar in Hindi and Sanskrit, respectively (Patan et al., 2018). In Himalayan territories, it is found at an altitude of 3000 m, whereas in Khasi and Jaintia Hills, it dwells between 1500 and 1800 m. However, geographically isolated temperate regions and altitudinal variation stipulate its genetic and morphological features directly affecting the accumulation of active ingredients, volatile, and nonvolatile components. It is being used in Indian medicine systems since ages particularly as a substitute of European *Valeriana officinalis* (Prakash, 1999). Studies regarding the genetic diversity of this species help to understand the available genetic abundance. The biochemical analysis of this plant has an economical value and leads to the selection of superior variety for the active ingredient production (Singh et al., 2015).

14.2 Origin and distribution

The *Valeriana* word was primarily used in the 19th and 20th centuries (Evans, 2008). Due to its medicinal properties, this plant is highly valued in the Indian Ayurveda system, the Unani system in ancient Greek and Arab, and in old Egypt and Rome. It is reported to be used in numerous Ayurvedic and modern medicines (Bhatt et al., 2012; Jugran et al., 2012). This plant grows wildly in all temperate and subtropical regions of the world except Australia (Jain, 1968; Polunin and Stainton, 1987). About 250 species of the genus *Valeriana* have been reported from Chile,

Himalayan Medicinal Plants. https://doi.org/10.1016/B978-0-12-823151-7.00013-1

Brazil, South Africa, and subtropical Asia, among which 16 species and 2 subspecies of the genus were found in India. Moreover, in India, five species inhabit an altitude range of 3000 m and 1500–1800 m in Kumaon and Garhwal regions of central Himalayas (Polunin and Stainton, 1987; Rao et al., 1997; Prakash 1999). Besides this many of the genera of this family are reported in the Mediterranean region (Valerianella, Fedia, and Centranthus). In North America, the species of this family are mostly reported in subalpine forests, moist paddocks, stream beds, and rarely beyond the tree line. The family is commonly divided into three tribes specifically, Triplostegieae, Patriniene, and Valerianaceae (Graebner, 1906). Earlier many authors have documented 14 genera in Valerianaceae (Weberling, 1970; Cronquist, 1988). Between these, tribe Triplostegieae has a single genus *Triplostegiea*, while two genera *Patrinia* and *Nardostachys* are sited in Patriniene, although recent studies have sited these species of the South American taxa in Valeriana, thus reducing the number of genera within Valerianaceae to eight (Borsini, 1944; Larsen, 1986; Eriksen, 1989).

14.3 Morphology

V. jatamansi is mostly known as Indian Valerian, Muskbala, Sugandhbala, or Tagar (Table 14.1). The species occurs in diverse geographic localities and retains inclusive morphological and genetic features. The species reproduces through sexual (seeds) and asexual (rhizome) means (Ankush et al., 2011). The plant prefers a temperate climate and mostly grows randomly in steep areas, moist, rocky, disturbed grassy slopes, and on stones with coarse sandy loam soil, respectively. In the Himalayan region, this species grows frequently under the canopy of *Qurecus leucotrichophora*, *Pinus roxburghii*, mixed forests, and grassy habitats. It is an aromatic, hairy dwarf, and rhizomatous perennial herb which reaches up to 50 cm in height, covered with horizontal descending fibers and pubescent stem and leaves.

Table 14.1 Vernacular names (Patan et al., 2018).

Language	Vernacular names
Hindi	Balchhari, Mansi, Mushkbala, Nihani, Smak, Sumaya, Tagar
Kannada	Jatale, Naatijatamaansi, Nandubatlu, Tagara
Malayalam	Takaram
Marathi	Thagarmool
Sanskrit	*Jatamansi*, Natah, Tagarah
Tamil	Shadamangie, Takaram
Telugu	Tagara
Ayurvedic name	Tagar
Unani name	Tagar
Trade name	Mushkbala, Tagar

FIGURE 14.1

Representation of plant morphology.

The rootstock is 6–10 cm thick and long fibrous roots tangled by irregular circular edges. The plant has numerous stems which are 15–45 cm long. Leaves are of two types, radical and cauline. Radical leaves (2.5–8 cm long and 1–3 cm in diameter) are cordate ovate, long stalked, sinuate, or toothed, while cauline leaves are few, small, entire, or lobulated (Patan et al., 2018; Sundaresan and Ilango 2018). Flowers are white or tinted with pink color, ensue in flat-topped corymbose clusters on erect, almost leafless peduncles (Fig. 14.1). Flowers are unisexual; male and female flowers seem on different plants. Corolla is with five lobes and funnel shaped. Fruits are crowned with a persistent pappus like a calyx. Flowering, fruiting, and seed ripening occur in March, April, and May. *Valeriana* can be propagated by both seeds or asexual through rhizome, preferably in the rainy season. A total of 24 species of Valerianaceae belonging to 4 different genera have been reported in India (Prakash, 1999).

14.4 Botanical classification

The botanical classification of the plants is given below:

Kingdom: Plantae
Division: Mangnoliophyta
Class: Mangnoliopsida
Order: Dipsacales

Family: Valerianaceae
Genus: *Valeriana*
Species: *jatamansi*
Botanical name: *Valeriana jatamansi*

14.5 Agronomy technique

Valeriana can be propagated by seeds or using portions of the rootstock, preferably in the rainy season. It is generally suitable to raise the crop by suckers, since crop raised by seeds takes extra time to get mature. Seeds can be poised in April—May and sown directly in the nursery. For raising the crop by rootsuckers, a single mother nursery can be retained. About 2.5—3 kg seeds are required to raise planting stock for 1 ha of land. No specific treatment to seed is essential. Though the rootstock is chosen as propagules, if the crop is to be raised through seeds, then the nursery is prepared separately in April—May. Seeds germinated within 15—20 days are pricked into polybags for further growth. The seedlings are ready for planting in about 3 months.

The traditional farming system has always been time-consuming and cannot produce adequate plant material to meet industrial demands. Modern agriculture practices viz., hydroponic and aeroponic cultivation provide a better solution to produce higher yield with no compromise to crop loss. The nutrient solution is the most important factor required for the improvement of crop production. Plants that are usually collected from wild for local and trade purposes somehow affect its natural habitat. It is fascinating to know that herbal medicine in India roots back to 2000 years. They traditionally used and relied heavily on these medicinal plants. Nevertheless, around 80% of the Indian population still uses herbal medicines. Therefore, herb production and sustainable utilization have become a growing concern with the ever-growing population. The gap between demand and supply of medicinal plants is estimated at 40,000 tonnes and expected to rise to 1,52,000 tonnes by 2025. Thus, ever-rising demands lead to a situation where species become rare and vulnerable. Therefore, plants cultivated under controlled conditions fulfill the growing demands and offer an opportunity for quality produce. Hydroponic is a water culture system and in aeroponic, nutrients are sprayed on the roots. Nearly about 90%—70% of the water requirement is preluded. Similarly, less fertilizer is needed to achieve good nutritive value (Linden and Stoner, 2013). Another advantage of this system is the reliable quality biomass production free from any contaminants (Hayden, 2006). A nutrient solution includes inorganic ions and each element has a physiological role in performing the life cycle of plants (Taiz and Zeiger, 1998). The nutrient solutions composition determines the electrical conductivity (EC) and osmotic potential of the solution. The pH of the nutrient solution indicates the relationship between the concentration of ions and ranges between 0 and 14. In a hydroponic and aeroponic system, plant growth is closely related to nutrient uptake and pH of the nutrient solution (Marschner, 1995). EC is an index that describes the total amount of salts in a solution.

14.6 Phytochemistry

Natural products are popular to have a crucial role in the search of new compounds which leads to drug development (Cragg et al., 1997; Newman et al., 2003; Newman and Cragg, 2007). Compounds classified as low polarities have always been an interest from *Valeriana* species and majorly included in two groups, i.e., essential oil sesquiterpenes and valepotriates. Like most medicinal plants, its roots and rhizome are rich in medicinally important components, i.e., valepotriates (Chopra et al., 1956). The roots of this plant are exploited for the commercial production of Valerian which is known and used for its antibacterial, leprosy, lewy body dementia, and antiprotozoal activity (Anonymous, 1976; Bagchi and Hooper, 2011). 13 new iridoids have been isolated from fractions of the ethyl acetate portion of 95% ethanol extract having intermediate polarity from the whole plant (Table 14.2) (Lin et al., 2010). Navarrete et al. (2006) have isolated five lignans from the roots of the plant. Chen et al. (2005) have reported the presence of 11-methoxyviburtinal, in acetone fraction of ethanol extract of roots of the plant. Valeriandoids A–C, together with three known analogues (chlorovaltrate, isovaltrate isovaleroyloxyhydrin, 1,5-dihydroxy-3,8-epoxyvalechlorine), have been isolated from methanolic extract of the roots of *V. jatamansi* (Xu et al., 2011a,b). Li et al. (2013) have isolated jatamanins N–P along with the seven known iridiods from 95% ethanolic extract of roots extract (Table 14.2). Glaser et al. (2015) have reported podophyllotoxin and 4′-demethylpodophyllotoxin in *V. jatamansi*. Additionally, jatamanvaltrates R–S and jatamanin Q were reported in the roots of the plant (Dong et al., 2015). Further, Quan et al. (2019) have isolated four new 3,8-epoxy iridoids from 95% EtOH extract of the roots and rhizomes. Further, an iridoid (valeridoid A) and five undescribed bis-iridoids (valeridoids B–F) have been isolated from the roots and rhizomes (Quan et al., 2020). Navarrete et al. (2006) have isolated five lignans from the roots of the plant (Table 14.2). In another study, jatadoids A and B were also reported in methanol extract of the root along with jatamanvaltrate H along with a sesquiterpene valeriananoid C (Xu et al., 2012). Li et al. (2013) have isolated jatamanins N–P along with the seven known iridoids, from 95% ethanolic extract of roots extract. 15 chlorinated valepotriates were identified in *V. jatamansi* (Table 14.2) (Lin et al., 2013).

Among the terpene class, Valerilactones A and B were reported in the roots of *V. jatamansi* along with bakkenolide-H and bakkenolide-B (Xu et al., 2011a,b). Valeriananoids D–E and clovane-2β-isovaleroxy-9α-ol were isolated from the roots of *V. jatamansi* Jones (Dong et al., 2015). Jatamansone (Valerenone),

Table 14.2 Chemical compound identified in *Valeriana jatamansi*.

Identified compound	References
Total 13 compounds, i.e., jatamanins A–M and new lignan (+)-9′-isovaleroxylariciresinol along with 4-hydroxy-8-methoxy-3-methyl-10-methylene-2,9-dioxatricyclo(4,3,1,03,7)-decane, patriscabrol, rupesin E, lariciresinol, pinsepiol, (+)-1-hydroxypinoresinol, (+)-acetoxypinoresinol, (+)-medioresinol, (+)-monomethylpinoresinol, (+)-syringaresinol, (+)-pinoresinol, (+)-5′-hydroxypinoresinol, (+)-cyclo-olivil, (−)-massoniresinol, 4,4′,9,7′-tetrahydroxy-3,3′-dimethoxy-7,9′-epoxylignan, and (−)-berchemol	Lin et al. (2010)
Isolated 5 lignans, i.e., massoniresinol-4′-O-β-D-glucoside, berchemol-4′-O-β-D-glucoside, pinoresinol-4,4′-di-O-β-D-glucoside, 8-hydroxypinoresinol-4′-O-β-D-glucoside, and pinoresinol-4-O-β-D-glucoside, from roots of the plant	Navarrete et al. (2006)
Iridoid (valeridoid A) and five undescribed bis-iridoids (valeridoids B–F)	Quan et al. (2020)
Valeriandoids A–C, together with three known analogues (chlorovaltrate, isovaltrate isovaleroyloxyhydrin, 1,5-dihydroxy-3,8-epoxyvalechlorine)	Xu et al. (2011a,b)
Isolated jatamanins N–P along with the seven known iridiods, i.e., volvatrate A, (3S,4R,5S,7S,8S,9S)-3,8-epoxyoctahydro-4,8-dimethylcyclopenta[c]pyran-7-ol, (3S,4S,5S,7S,8S,9S)-3,8-epoxy-7-hydroxy-4,8 dimethylperhydrocyclopenta[c]-pyran, jatamanin G, jatamanin A, hexahydro-6-hydroxy-7-(hydroxymethyl)-4-methylenecyclopenta[c]pyran-1(3H)-one and (4b,8b)-8-methoxy-3-methoxy-10-methylene-2,9-dioxatricyclo [4.3.1.03,7]decan-4-ol) 4–10, from 95% ethanolic extract of roots extract	Li et al. (2013)
15 chlorinated valepotriates, designated as chlorovaltrates A–O, together with 6 known analogues, (1S,3R,5R,7S,8S,9S)-3,8-epoxy-1,5-dihydroxyvalechlorine, volvaltrate B, chlorovaltrate, rupesin B, (1S,3R,5R,7S,8S,9S)-3,8-epoxy-1-*O*-ethyl-5-hydroxyvalechlorine, and (1R,3R,5R,7S,8S,9S)-3,8-epoxy-1-*O*-ethyl-5-hydroxyvalechlorine	Lin et al. (2013).
However, linarin, linarin-isovalerianate, linarin-2-O-methylbutyrate, 6-methylapigenin/hispidulin, hesperetin-7-O-β-rutinoside [2S(−) hesperidin], acacetin-7-O + rutinoside, 7-O-β-sophoroside and acacetin 7-O-(600-O-a-L-rhamnopyranosyl)-β-sophoroside, kaempferol 3-O-β-rutinoside, rutin, kaempferol 3-O-β-D-glucopyranoside, quercetin 3-O-β-D-glucopyranoside, kaempferol, quercetin, acacetin 7-O-β-D-glucopyranoside, apigenin 7-O-β-D-glucopyranoside, daucosterol, and trans-caffeic acid are also reported in roots and rhizomes of this plant	Glaser et al. (2015), Marder et al. (2003), Tang et al. (2003), Jugran et al. (2019).

hydroxyvalerenic acid, acetoxyvalerenic acid, valerenic acid, and isopatriniosid were also identified in *V. jatamansi* (Navarrete et al., 2006; Tan et al., 2016). Among the flavones glycosides, acacetin 7-O-b-sophoroside and acacetin 7-O-(600-O-a-Lrhamnopyranosyl)-b-sophoroside have been reported along with linarin, acacetin 7-O-b-D-glucopyranoside, apigenin 7-O-b-D-glucopyranoside, kaempferol 3-O-b-rutinoside, rutin, kaempferol 3-O-b-D-glucopyranoside, quercetin 3-O-b-D-glucopyranoside, kaempferol, quercetin, daucosterol, and trans-caffeic acid (Tang et al., 2003). However, Glaser et al. (2015), Marder et al. (2003), Tang et al. (2003) and Jugran et al. (2019) also reported compounds in roots and rhizomes of this plant (Table 14.2). Among the phenolic class, gallic acid, catechin, hydroxybenzoic acid, caffeic acid, chlorogenic acid, and p-coumaric acid were reported in the root and aerial part of *V. jatamansi* (Bhatt et al., 2012; Jugran et al., 2016).

Essential oil of *V. jatamansi* composition investigated patchouli alcohol, α-bulnesene, aguaiene, guaiol, seychellene, viridiflorol, 8-acetoxypatchouli alcohol, (E)-b-caryophyllene, α-patchoulene, D-cadinene, β-elemene, γ-patchoulene, kessane, bulnesol, and β-gurjunene as major components (Verma et al., 2011). GC-MS analysis of dried roots shows the presence of major compounds that include Isovaleric acid, 3-Methylvaleric acid, Valeric acid, α-Pinene, Camphene, β-Pinene, 1,8-Cineol, Camphor, α-Terpineol, Methyl thymol ether, Bornyl acetate, Copaene, β-Patchoulene, β-Elemene, α-Gurjunene, Caryophyllene, Calarene, α-Guaiene, α-Caryophyllene, Seychellene, α-Bulnesene, D-Cadinene, Selina-3,7(11)-dien, Viridiflorol, Guaiol, Patchoulol, and Valeranone, among which patchoulol, α-bulnesene, isovaleric acid, α-guaiene, and 3-methylvaleric acid were the major compounds (Liu et al., 2013). Isovaleric acid, α-santalene, β-gurjunene, ar-curcumene, xanthorrhizol, bornyl isovalerate maaliol, valtrate, didrovaltrate, patchouli alcohol, 3-methylvaleric acid, 8-acetoxypatchouli alcohol, and α-patchoulene, β-patchoulene, γ-patchoulene, α-santalene, viridiflorol, α-bulnesene, α-guaiene, bornyl acetate, 7-epi-α-selinene, and β-elemene, carotol, germacrene B, cis-β-farnesene, α-humulene, humulene epoxide-II, patchoulol, α-bulnesene, patchouli alcohol, seychellene, calarene-β-gurjunene, and α-santalene, bornyl acetate, α-guaiene, α-bulnesene/delta-guaiene, 7-epi-α-selinene, kessane, spathulenol, and viridiflorol were reported in the *V. jatamansi* (Sati et al., 2005; Verma et al., 2011; Agnihotri et al., 2011; Raina and Negi, 2015; Jugran et al., 2016). Further, baldrinal, prinsepiol-4-O-β-D-glucoside, coniferin, hexacosanic acid, trans-p-coumaric acid, β-sitosterol, behenic acid, nonadecyl alcohol, decursidin, decursitin B, decursitin A, 3′(S)-acetoxy-4′(R)-angeloyloxy-3′,4′-dihydroxanthyletin, dibutyl phthalate, cinnamic acid bornyl ester derivatives, bornyl caffeate, and villoside aglycone were also reported from the roots of *V. jatamansi* (Chen et al., 2005; Jugran et al., 2019). Singh et al. (2006) have standardized a high-performance thin-layer chromatographic method for the estimation of the marker compound valerenic acid which is a sesquiterpenoid with a limit of detection for valerenic acid as 80 ng and limit of quantification as 500 ng.

14.7 Conservation approach

14.7.1 In situ conservation

Day by day, *V. jatamansi* demand is increasing due to tremendous pharmaceutical importance and wide usage in various herbal formulations. Efforts on in situ and ex situ conservation have been taken through various approaches. Conventionally, *V. jatamansi* is propagated through seeds and roots. The plants obtained through the *in vitro* propagation system are not only true to type but also help to meet the increasing market demands. Moreover, tissue culture protocols are widely being used, with focused research on developing the propagation system of this medicinal herb through shoot regeneration using nodal explants. A different combination of 6-benzyl amino purine 1.5 μM, α-naphthalene acetic acid 0.5 μM, and gibberellic acid (GA_3) 0.1 μM was used for shoot initiation and proliferation (Purohit et al., 2015; Das et al., 2013; Chen et al., 2014). A high-efficiency propagation system has been developed from leaf and rhizomes explants using different plant growth regulators. The efficiency of callus induction was significantly found best in rhizome explant and 0.5 mg/L 2, 4-D was found suitable for callus induction in *V. jatamansi* (Das et al., 2013). Furthermore, various attempts for in situ conservation for *V. jatamansi* have been reported in the North-Eastern part of India by Shankar and Rawat (2013). Plantlets obtained through seeds failed to convert into mature plants, whereas the usage of rhizomes as propagule significantly showed good results (Raina et al., 2011; Shankar and Rawat, 2013).

14.7.2 Ex situ conservation

Ex situ conservation corresponds to conserve the plants using natural conditions and acclimatizing those plants in their native places using these sites as seed banks (Hamayun et al., 2006). Several researchers have reported the successful acclimatization of *V. jatamansi* in their native places. This herb exhibits tremendous flowering and growth but found poor germination percentage when seeds were used as explants. Mukherjee et al. (2009) revealed the survivability of 60%–90% plants using rhizome. The production yield of biomass was found significantly higher, but this can be varied from site to site and altitudinal variations. Further, Nawchoo et al. (2012) revealed the successful conservation of this endangered herb. Moreover, Raina et al. (2011) cultivated *V. jatamansi* in July and August in the Mid Hill area of Himachal Pradesh and found substantial growth of these plants and reported a significant increase in secondary metabolite (Valepotriate) after 2–3 years of plantation in the ex situ environment.

14.8 Molecular characterization

Comprising the molecular aspects of *V. jatamansi*, RAPD analysis of 13 populations showed 368 amplicons by using 45 oligo primers which lead to 79.61% polymorphism (Singh et al., 2015). The statistical analysis of the study by PCA and AMOVA revealed that variation was higher among the populations rather than within the population. In another study, authors have found two genotypically different groups using genetic and phytocomponents analysis and concluded that it does not just depend on genetic features but also on environmental factors (He et al., 2018). Recently, Joglekar and Barve (2019) developed a method for accurate identification of dry root samples. They have developed a protocol to extract genomic DNA from the fresh and dry roots of Indian Valerian which is a modification of the CTAB method. Jugran et al. (2013) reported a high level of genetic variability in the populations. 159 bands showed 125 polymorphic ISSR loci which were detected using 20 selected ISSR primers. Further, with a mean of 78.6%, the percentage of polymorphic loci ranged from 65.4% (Dwali, 2730 m asl) to 91.2% (Pithoragarh, 1872 m asl). Nei's genetic diversity index (He) ranged from 0.25 (Surkunda, 2775 m asl) to 0.37 (Pithoragarh, 1872 m asl), with a mean value of 0.31. Singh et al. (2014) performed a molecular analysis using simple sequence repeats (SSRs) to evaluate the polymorphism in 12 genotypes of *V. jatamansi*. In comparison to their percentage polymorphism, mononucleotide repeats (42.9%) showed the highest followed by di- (21.4%), tri- and hexa- (14.3%), and tetra- (7.14%) nucleotides.

14.9 Omics approach

For the understanding of the iridoid biosynthesis pathway in *V. jatamansi*, transcriptome sequences of leaves and tissues have been generated followed by a de novo sequence assembly. In total, 183,524,060 transcripts and 61,876 unigenes were obtained in 13.28 Gb reads. Among all the unigenes which were analyzed by public databases, 56,641 unigenes were annotated, while 5235 unigenes remained unannotated. Furthermore, 5195 unigenes were identified which contain SSR by MISA analysis. Overall, 24 unigenes were reported which are related to the iridoid biosynthesis. Further six genes of the MVA pathways, nine genes of the MEP pathways, and nine genes of the iridoid pathway were also reported. The qRT-PCR analysis revealed their expression of these genes in different tissues. The author suggests these genes to be a potential target for a biotechnological approach to engineer an improved pathway for iridoid compounds yield in *V. jatamansi* (Shuang and Chenshu, 2020).

14.10 Formulated products

Industrially important *V. jatamansi* root part is widely used in the preparation of various herbal formulations. In the present world, many products are available and formulated using the root extract containing major metabolites, i.e., valerenic acid, acetoxyvalerenic acid, and hydroxyl valerenic acid and oil. The industries include Vadik Herbs, Dr. Axez Gangaram Mohanlal, Pure and Natural that have commercialized the oil as the product, and Herbal Hills, Nature's Way, Swisslove, and Cureveda, respectively, produced and commercialized root powder. Beside these, companies such as Vita green, Vayam Ayurveda, and Bliss Wellness commercialized it in the form of the capsule (500 mg) which contains a combination of jatamansi and Ashwagandha.

14.11 Conclusion and future perspectives

In this chapter, the literature survey shows that *V. jatamansi* is a well-explored plant herb for medicinal purposes and commercial use. With ongoing advancements in research and new developing technologies, many formulations, new pharmaceutical drugs, and products have been developed for human welfare. The major sector for the commercial use of this plant is an essential oil. Drastically, the decline of the herb has been reported from the natural habitats, and it is mainly due to the increased demand in the perfume and other industries. Hence, this herb needs rigorous protection by developing new cultivation techniques and improved varieties to meet market demand. Hydroponics is the new and growing technique with commanding potential to meet the demand for the production of biomass yield, essential oil, and bioactive molecule. In addition to this, there is still a need to further explore this plant for its biologically active compounds and study the underlying pathways for their bioactive active compound production. Moreover, the breeding programs are also required for suffice biomass production.

References

Agnihotri, S., Wakode, S., Ali, M., 2011. Chemical composition, antimicrobial and topical anti-inflammatory activity of *Valeriana jatamansi* Jones. Essential oil. J. Essent. Oil Bear Plants 14, 417–422.

Ankush, K., Susheel, V., Puneet, S., 2011. Stylar movement in *Valeriana wallichii* DC.-a contrivance for reproductive assurance and species survival. Curr. Sci. 100 (8), 143–1144.

Anonymous, 1976. The Wealth of India: Raw Materials, vol. (x). Sp-D CSIR Publication, New Delhi, India, pp. 424–426.

Bagchi, P., Hooper, W., 2011. In: International Conference on Bioscience, Biochemistry and Bioinformatics. IASCIT Press, Singapore, p. 5.

Bhatt, I.D., Dauthal, P., Rawat, S., Gaira, K.S., Jugran, A., Rawal, R.S., Dhar, U., 2012. Characterization of essential oil composition, phenolic content, and antioxidant properties in wild and planted individuals of *Valeriana jatamansi* Jones. Sci. Hortic. 136, 61–68.

Borsini, O., 1944. Valerianaceae. In: Descole (Ed.), Genera et Species Plantarum Argentinarum, 2, pp. 275–372 tab. 132–160.

Chen, R., Zhang, M., Lü, J., Zhang, X., da Silva, J.A.T., Ma, G., 2014. Shoot organogenesis and somatic embryogenesis from leaf explants of *Valeriana jatamansi* Jones. Sci. Hortic. 165, 392–397.

Chen, Y.G., Yu, L.L., Huang, R., Lv, Y.P., Gui, S.H., 2005. 11-Methoxyviburtinal, a new iridoid from *Valeriana jatamansi*. Arch Pharm. Res. 28 (10), 1161–1163.

Chopra, R.N., Nayyar, S.O., Chopra, I.C., 1956. Glossary of Indian Medicinal Plants. CSIR Publication, New Delhi, p. 251.

Cragg, G.M., Newman, D.J., Snader, K.M., 1997. Natural products in drug discovery and development. J. Nat. Prod. 60, 52–60.

Cronquist, A., 1988. The Evolution and Classification of Flowering Plants, second ed. (New York).

Das, J., Mao, A.A., Handique, P.J., 2013. Callus-mediated organogenesis and effect of growth regulators on production of different valepotriates in Indian valerian (*Valeriana jatamansi* Jones). Acta Physiol. Plant. 35 (1), 55–63.

Dong, F.W., Yang, L., Wu, Z.K., Zi, C.T., Yang, D., Luo, H.R., Zhou, J., Hu, J.M., 2015. Iridoids and sesquiterpenoids from the roots of *Valeriana jatamansi* Jones. Fitoterapia 102, 27–34.

Eriksen, B., 1989. Notes on generic and infrageneric delimitation in the Valerianaceae. Nord. J. Bot. 9 (2), 179–187.

Evans, W.C., 2008. Trease and Evans Pharmacognosy, fifteenth ed. Elseveir, Noida, India.

Glaser, J., Schultheis, M., Moll, H., Hazra, B., Holzgrabe, U., 2015. Antileishmanial and cytotoxic compounds from *Valeriana wallichii* and identification of a novel nepetolactone derivative. Molecules 20 (4), 5740–5753.

Graebner, P., 1906. The genera of the natural family of the Valerianaceae. Bot. Jahrb. Syst. 37, 464–480.

Hamayun, M., Khan, S.A., Kim, H.Y., Na, C.I., Lee, I.J., 2006. Traditional knowledge and ex situ conservation of some threatened medicinal plants of Swat Kohistan, Pakistan. Int. J. Bot. 2 (2), 205–209.

Hayden, A., 2006. Aeroponic and hydroponic system for medicinal herb, rhizome and root crops. Hortic. Sci. 41 (3), 536–538.

He, X., Wang, S., Shi, J., Sun, Z., Lei, Z., Yin, Z., Qian, Z., Tang, H., Xie, H., 2018. Genotypic and environmental effects on the volatile chemotype of *Valeriana jatamansi* Jones. Front. Plant Sci. 9, 1003.

Jain, S.K., 1968. Medicinal Plants. National Book Trust, India, New Delhi, p. 154.

Joglekar, N.H., Barve, S.S., 2019. Efficient genomic DNA extraction and molecular analysis of medicinally rich *Valeriana jatamansi* dry roots. Int. J. Pharmaceut. Sci. Res. 10 (8), 3979–3983.

Jugran, A.K., Bahukhandi, A., Dhyani, P., Bhatt, I.D., Rawal, R.S., Nandi, S.K., 2016. Impact of altitudes and habitats on valerenic acid, total phenolics, flavonoids, tannins, and antioxidant activity of *Valeriana jatamansi*. J. Appl. Biochem. Biotechnol. 179, 911–926.

Jugran, A., Rawat, S., Dauthal, P., Mondal, S., Bhatt, I.D., Rawal, R.S., 2012. Association of ISSR markers with some biochemical traits of *Valeriana jatamansi* Jones. Ind. Crop. Prod. 44, 671–676.

Jugran, A.K., Bhatt, I.D., Rawal, R.S., Nandi, S.K., Pande, V., 2013. Patterns of morphological and genetic diversity of *Valeriana jatamansi* Jones in different habitats and altitudinal range of West Himalaya, India. Flora-Morphol. Distrib. Funct. Ecol. Plant 208 (1), 13–21.

Jugran, A.K., Rawat, S., Bhatt, I.D., Rawal, R.S., 2019. *Valeriana jatamansi*: an herbaceous plant with multiple medicinal uses. Phytother. Res. 33 (3), 482–503.

Larsen, B.B., 1986. A taxonomic revision of *Phyllactis* and *Valeriana* sect. Bracteata (Valerianaceae). Nord. J. Bot. 6 (4), 427–446.

Li, Y., Wu, Z., Li, H., Li, H., Li, R., 2013. Iridoids from the roots of *Valeriana jatamansi*. Helv. Chim. Acta 96 (3), 424–430. https://doi.org/10.1002/hlca.201100465.

Lin, S., Chen, T., Liu, X.H., Shen, Y.H., Li, H.L., Shan, L., Liu, R.H., Xu, X.K., Zhang, W.D., Wang, H., 2010. Iridoids and lignans from *Valeriana jatamansi*. J. Nat. Prod. 73 (4), 632–638.

Lin, S., Zhang, Z.X., Chen, T., Ye, J., Dai, W.X., Shan, L., Su, J., Shen, Y.H., Li, H.L., Liu, R.H., Xu, X.K., 2013. Characterization of chlorinated valepotriates from *Valeriana jatamansi*. Phytochemistry 85, 185–193.

Linden, J.C., Stoner, R.J., 2013. II Agricultural Homeopathic Elements for Biocontrol. Patent Publication number: 20130260993; Application date: 3.10.

Liu, X.C., Zhou, L., Liu, Z.L., 2013. Identification of insecticidal constituents from the essential oil of *Valeriana jatamansi Jones* against *Liposcelis bostrychophila* Badonnel. J. Chem., 853912

Marder, M., Viola, H., Wasowski, C., Fernandez, S., Medina, J.H., Paladini, A.C., 2003. 6-Methylapigenin and hesperidins: new *Valeriana* flavonoids with activity on the CNS. Pharmacol. Biochem. Behav. 75, 537–545.

Marschner, H., 1995. Mineral Nutrition of Higher Plants. Academic Press, New York, USA. ISBN 0-12-473542-8.

Mukherjee, D., Chakraborty, S., Roy, A., Mokthan, M.W., 2009. Differential approach of germplasm conservation of high value of medicinal plants in north eastern himalayan region. Int. J. Agric. Environ. Biotechnol. 2 (4), 332–340.

Navarrete, A., Avula, B., Choi, Y.W., 2006. Chemical fingerprinting of *Valeriana* species: simultaneous determination of valerenic acids, flavonoids, and phenylpropanoids using liquid chromatography with ultraviolet detection. J. AOAC Int. 89, 8–15.

Nawchoo, A.I., Rather, M.A., Ganie, H.A., Jan, R.T., 2012. Need for unprecedented impetus for monitoring and conservation of *Valeriana jatamansi*, a valuable medicinal plant of Kashmir Himalaya. Agric. Sci. Res. J. 2 (7), 369–373.

Newman, D.J., Cragg, G.M., 2007. Natural products as sources of new drugs over the last 25 years. J. Nat. Prod. 70, 461–477.

Newman, D.J., Cragg, G.M., Snader, K.M., 2003. Natural products as sources of new drugs over the period 1981–2002. J. Nat. Prod. 66, 1022–1037.

Patan, A., Alekhya, K., Aanandhi, V., Tharagesh, K., Anish, A., 2018. *Valeriana jatamansi*: an ethnobotanical review. Asian J. Pharmaceut. Clin. Res. 11 (4), 38–40.

Polunin, O., Stainton, A., 1987. Concise Flowers of the Himalaya. Oxford University Press, London, p. 255.

Prakash, V., 1999. Indian Valerianaceae - A Monograph on Medicinally Important Family. Scientific Publishers, India, pp. 1–2.

Purohit, S., Rawat, V., Jugran, A.K., Singh, R.V., Bhatt, I.D., Nandi, S.K., 2015. Micropropagation and genetic fidelity analysis in *Valeriana jatamansi Jones*. J. Appl. Res. Med. Aromatic Plants 2 (1), 15–20.

Quan, L.Q., Hegazy, A.M., Zhang, Z.J., Zhao, X.D., Li, H.M., Li, R.T., 2020. Iridoids and bis-iridoids from *Valeriana jatamansi* and their cytotoxicity against human glioma stem cells. Phytochemistry 175, 112372.

Quan, L.Q., Su, L.H., Qi, S.G., Xue, Y., Yang, T., Liu, D., Zhao, X.D., Li, R.T., Li, H.M., 2019. Bioactive 3, 8-epoxy iridoids from *Valeriana jatamansi*. Chem. Biodivers. 16 (5), e1800474.

Raina, A.P., Negi, K.S., 2015. Essential oil composition of *Valeriana jatamansi* Jones from Himalayan regions of India. Indian J. Pharmaceut. Sci. 77, 218–222.

Raina, R., Chand, R., Sharma, Y.P., 2011. Conservation strategies of some important medicinal plants. Int. J. Med. Aromatic Plants 1, 342–347.

Rao, P.S., Sethi, H., Randhawa, G.S., Badhwar, R.L., 1997. Aromatic plants of India. Indian J. Hortic. 34, 323–333.

Sati, S., Chanotiya, C.S., Mathela, C.S., 2005. Comparative investigations on the leaf and root oils of *Valeriana wallichii* DC from northwestern Himalaya. J. Essent. Oil Res. 17, 408–409.

Shankar, R., Rawat, M.S., 2013. Conservation and cultivation of threatened and high valued medicinal plants in north East India. Int. J. Biodivers. Conserv. 5, 584–591.

Shuang, Z.H.A.O., Chenshu, W.A.N.G., 2020. Deep sequencing and transcriptome analyses to identify genes involved in iridoid biosynthesis in the medicinal plant *Valeriana jatamansi Jones*. Not. Bot. Horti Agrobot. Cluj-Napoca 48 (1), 189–199.

Singh, N., Gupta, A.P., Singh, B., Kaul, V.K., 2006. Quantification of valerenic acid in *Valeriana jatamansi* and *Valeriana officinalis* by HPTLC. Chromatographia 63 (3–4), 209–213.

Singh, S.K., Katoch, R., Kapila, R.K., 2015. Genetic and biochemical diversity among *Valeriana jatamansi* populations from Himachal Pradesh. Sci. World J., 863913

Singh, V., Srivastava, S.R., Shitiz, K., Chauhan, R.S., 2014. Valepotriates content variation and genetic diversity in high-value medicinal herb, *Valeriana jatamansi*. Int. J. Genet. Eng. Biotechnol. 5, 179–184.

Sundaresan, N., Ilango, K., 2018. Review on *Valeriana* species-*Valeriana wallichii* and *Valeriana jatamansi*. J. Pharmaceut. Sci. Res. 10 (11), 2697–2701.

Taiz, L., Zeiger, E., 1998. Plant Physiology, second ed. Sinauer Associates, Inc. Publishers, Sunderland, Massachusetts.

Tan, Y.Z., Yong, Y., Dong, Y.H., Wang, R.J., Li, H.X., Zhang, H., Guo, D.L., Zhang, S.J., Dong, X.P., Xie, X.F., 2016. A new secoiridoid glycoside and a new sesquiterpenoid glycoside from *Valeriana jatamansi* with neuroprotective activity. Phytochem. Lett. 17, 177–180.

Tang, Y.P., Liu, X., Yu, B., 2003. Two new flavone glycosides from *Valeriana jatamansi*. J. Asian Nat. Prod. Res. 5 (4), 257–261.

Verma, R.S., Verma, R.K., Padalia, R.C., Chauhan, A., Singh, A., Singh, H.P., 2011. Chemical diversity in the essential oil of Indian Valerian (*Valeriana jatamansi Jones*). Chem. Biodivers. 8 (10), 1921–1929.

Weberling, F., 1970. Valerianaceae. In: Hegi (Ed.), Illustrierter flora von Mitteleuropa. 2 Aufl. 6, pp. 97–172.

Xu, J., Yang, B., Guo, Y., Jin, D.Q., Guo, P., Liu, C., Hou, W., Zhang, T., Gui, L., Sun, Z., 2011a. Neuroprotective bakkenolides from the roots of *Valeriana jatamansi*. Fitoterapia 82 (6), 849–853.

Xu, J., Zhao, P., Guo, Y., Xie, C., Jin, D.Q., Ma, Y., Hou, W., Zhang, T., 2011b. Iridoids from the roots of *Valeriana jatamansi* and their neuroprotective effects. Fitoterapia 82 (7), 1133–1136.

Xu, J., Li, Y., Guo, Y., Guo, P., Yamakuni, T., Ohizumi, Y., 2012. Isolation, structural elucidation, and neuroprotective effects of iridoids from *Valeriana jatamansi*. Biosci. Biotechnol. Biochem. 76 (7), 1401–1403.

CHAPTER

Withania somnifera

15

Indu Sharma[1], Rahul Kumar[2], Vikas Sharma[3], Baldev Singh[4], Pratap Kumar Pati[4], Ashutosh Sharma[2]

[1]*Department of Botany, Sant Baba Bhag Singh University, Khiala, Jalandhar, Punjab, India;* [2]*Faculty of Agricultural Sciences, DAV University, Jalandhar, Punjab, India;* [3]*Department of Molecular Biology and Genetic Engineering, Lovely Professional University, Jalandhar, Punjab, India;* [4]*Department of Biotechnology, Guru Nanak Dev University, Amritsar, Punjab, India*

15.1 Introduction

Withania somnifera (L.) Dunal is a medicinal plant of high repute and has been immensely utilized in African, Ayurvedic, and Unani systems of medicine (Mishra et al., 2000; Singh et al., 2015; Wadhwa et al., 2016). It is commonly known as Ashwagandha, Asgandh, Indian ginseng, Pevetti or Winter cherry, etc., and belongs to the family Solanaceae. Its generic name *Withania* is in the honor of the Sir Henry Thomas Maire Witham (an English paleobotanist), with an orthographic variation of the last alphabet "*m*" into "*n*" with the addition of a commemorative termination "*ia*," whereas the specific epithet "*somnifera*" is derived from two Latin words "*somnus*" (meaning sleep) and "*fero*" (meaning to bear) that are linked to its sleep-inducing properties (Mir et al., 2012). Its most common Indian name Ashwagandha has been derived from two words "*ashva*" (meaning horse) and "*gandha*" (meaning smelling) because of the resemblance of the smell of its roots to a sweating horse (Rajeswara Rao et al., 2012). It is one of the 36 plants that is under cultivation already, due to their high demand and require the crop improvement efforts to develop/provide the elite cultivars to the growers, in order to increase their farm income (Ved and Goraya, 2007). Due to its immense therapeutic potential, it has also appeared in the World Health Organization (WHO) monographs on selected medicinal plants (Mirjalili et al., 2009).

The medicinal use of plants/green pharmacy has been described in the alternative ancient Indian medicine science for life care, i.e., Ayurveda (Singh, 2018). Ashwagandha is considered as a nootropic herb and in Ayurvedic system of medicine, it is also mentioned as "*Medhya Rasayana*" (neuronutrient) and/or as "Queen of Ayurveda" because of its immense therapeutic potential (Ray and Ray, 2015; Singh, 2018; Wadhwa et al., 2016). In Ayurveda, the "*Medhya*" is related to intelligence and "*Rasayana*" means drugs (collectively "*Medhya Rasayana*"); therefore, it is used to promote various functions of brain, prevent and/or treat mental/mood

Himalayan Medicinal Plants. https://doi.org/10.1016/B978-0-12-823151-7.00007-6

disorders, psychiatric and psychosomatic diseases; and promote retention power, intellect, and memory. According to Acharya Charak, the Medhya Rasayana drugs possess powers of Grasping (*Grahan-shakti*), Retention (*Dharana-shakti*), Discrimination (*Vivek-shakti*), and Recollection (*Smriti-shakti*) (Chaudhari and Murthy, 2014; Ray and Ray, 2015).

Keeping in mind the immense therapeutic values of this medicinal plant, this chapter deals the bioactives present in *W. somnifera*, its phytochemistry, medicinal properties, major challenges in its cultivation, efforts made for its crop improvement, and the enhancement of bioactives in it using various methods including the biotechnological approaches. We have also tried to discuss the recent advances made in the OMICS applications for the improved understanding of its overall biology. Therefore, the aim of this chapter is to help the intermediate to advance level researchers in getting a holistic view of the challenges and opportunities involved in working on this plant.

15.2 Botany

W. somnifera is an important medicinal plant of the dicot family, Solanaceae. Despite some variation in the morphological characters of the plants in different cultivated and wild accessions (Khanna et al., 2014; Kumar et al., 2007), the plants are generally erect, evergreen, annual or perennial undershrub with branched, hairy-tomentose stem bearing broadly ovate, pubescent, and petiolate leaves. The leaves are usually entire along the margins (sometimes wavy), acute to obtuse from the apex, dorsiventral with a prominent midrib (with broad abaxial hump), and its lamina is bilaterally symmetrical with abundant epidermal trichomes on its abaxial side. Leaves on their axil bear umbellate cyme cluster of pale green flowers, which appear throughout the year. Flowers are sessile to subsessile, shortly pedicellate, pentamerous, hermaphrodite, yellow to dull green in color. Calyx is gamosepalous, campanulate, green, and sepals are five in number. Corolla is gamopetalous, campanulate, greenish yellow, lobes that are triangular and tomentose to the outer side. Anthers are usually five in number, subincluded, and anther filaments are elongated. Fruit (berry) varying from yellow-orange to deep red in color is enclosed in persistent inflated calyx, globose, containing numerous, small, subpyriform to reniform, minutely reticulate-foveolate, yellowish-brown, smooth discoid seeds.

The floral development in *W. somnifera* has been extensively studied to understand its reproductive behavior. Its flowers exhibit partial temporal dichogamy of protogynous type, in which stigma attains receptivity about 48 h prior to anther dehiscence and it remains exerted beyond the reach of staminal cone. Normally, the insect pollinators facilitate the cross pollination of receptive stigma. However, in case of nonreceipt of cross-pollen due to absence of insect pollinators, the autogamous fertilization is assured by upward movement of staminal cone by elongation of filaments, and keeps the staminal cone in close proximity to stigma. Therefore,

W. somnifera plants are believed to be of mixed mating type (both cross and self-pollinating), which could be the reason of the high reproductive success in the plant (Lattoo et al., 2007). However, there is a season variation in the fertility of pollens, which are more fertile in winters (85%–96%) in comparison to summers (65%–75%) (Singh, 2009).

15.3 Origin and distribution

W. somnifera is believed to be originated in Mediterranean region and Orientalis (Singh and Kumar, 1998), and it is widely distributed in dry arid soils of tropical and subtropical regions. The plant grows in wild condition in Southern Mediterranean region, Canary Islands, South and East Africa, Congo, Madagascar, Palestine Israel, Jordan, Egypt, Sudan Iran, Afghanistan, Baluchistan, India, and Pakistan (Gifri, 1992; Kumar et al., 2007; Rajeswara Rao et al., 2012; Singh and Kumar, 1998; Udayakumar et al., 2013a; Wood, 1908). In India, the plant grows well throughout the drier parts and in the subtropical and semitemperate regions, including the states of Maharashtra, Madhya Pradesh, Gujarat, Rajasthan, Uttar Pradesh, Haryana, and Punjab extending a height of 1700 m in Himalaya (mainly Himachal Pradesh and Jammu and Kashmir) (Deb, 1980; Gupta, 1964; Mirjalili et al., 2009; Nayar, 1964; Nigam, 1984; Verma et al., 2012). It is cultivated mostly in Mansa, Neemuch, and Jawad tehsils of Mandsaur district of Madhya Pradesh, some parts of Kota district of Rajasthan, Andhra Pradesh, and Uttar Pradesh (Kakaraparthi et al., 2013; Kumar et al., 2007, 2009a; Mir et al., 2012; Mirjalili et al., 2009; Nigam, 1984).

W. somnifera grows well in north-western Himalayan region of both India and Pakistan (Aslam et al., 2017; Bhatt and Negi, 2006; (Sharma et al., 2012a,b). It also grows extensively in the subtropical, low to mid hill areas of Himachal Pradesh, Uttarakhand, and Jammu and Kashmir (Bhatt and Negi, 2006; Dar et al., 2017; Dawa et al., 2013; Sharma et al., 2012a,b, 2013a). In a study, it was concluded that growing *W. somnifera* could be profitable in mid hills of western Himalayas as an intercrop in the local agro-forestry–based land ecosystems (Verma and Thakur, 2010).

15.4 Medical significance

Ashwagandha is known to possess a wide array of therapeutic properties against a number of ailments. It is known to possess antiinflammatory, antioxidant, antiaging, antineoplastic, antigenotoxic, immunomodulatory, anticancer, antitumor, antistress, arthritis-preventive, cardioprotective, prophylactic, somnogenic, neuroprotective, and nephroprotective potential (Bharti et al., 2016; Dutta et al., 2017; Gupta and Singh, 2014; Kaushik et al., 2017; Kim et al., 2020; Mohanty et al., 2004; Rani et al., 2004; Rasheed et al., 2020; Singh et al., 2015; Srivastava et al., 2018). The medicinal values of *W. somnifera* are attributed to the presence of a wide range of secondary metabolites present in it (discussed in Section 15.6). Being a potent aphrodisiac, adaptogen, and neuronutrient, Ashwagandha has been used to treat asthma, anxiety, depression, attention deficit hyperactivity disorder, diabetes, dermatitis, diarrhea, insect bite, uterine weakness, impotency, frequent miscarriage, emaciation, infertility, anemia, gout, osteoarthritis, rheumatoid arthritis, stress, cerebellar ataxia, high cholesterol, fibromyalgia, Parkinson's disease, cardiovascular diseases, neurodegenerative and some types of seizures, etc (Poojari et al., 2019; Shukla et al., 2011; Srivastava et al., 2018).

The tumor-preventing potential of *W. somnifera* has been reported by Singh et al. (1986). In adult male albino mice, lung adenomas were induced by urethane treatment (biweekly 125 mg/kg dose for 7 months) in all the animals. As a result, decline in body weight, leucopenia, increase in rate of mortality, and decline in lymphocyte percentage was observed. However, treatment of Ashwagandha (200 mg/kg daily dose) prevented urethane-induced tumors (lung adenomas), weight loss, declined mortality, enhanced lymphocyte percentage, and total leukocyte count through immunostimulant potential or stimulation in adaptogen. Senthilnathan et al. (2006) have also reported the antitumor effects of root extracts of *W. somnifera* through its immunomodulatory properties in the male Swiss albino mice. Administration of benzo(a)pyrene (carcinogen) caused lung cancer and consequent immune dysfunction in the albino mice. Further treatment with root extracts of Ashwagandha significantly altered the levels of immunoglobulins, immunocompetent cells, and immune complexes. This restoration of number of immune cells and their functions by Ashwagandha revealed its immunomodulatory potential in treatment of lung cancer.

Ashwagandha may also be effective against the snake bites. Machiah et al. (2006) reported that the external application of a glycoprotein purified from *W. somnifera* could inhibit the hyaluronidase activity of cobra and viper snake venoms. Besides its other roles mentioned above, it is also known for its fertility-enhancing properties. An investigation to analyze the efficacy of *W. somnifera* treatment on sperm apoptosis and intracellular concentrations of reactive oxygen species of spermatozoa and the levels of essential metal ions (iron, copper, gold, and zinc) in seminal plasma from the infertile men was carried out (Shukla et al., 2011). Therefore, Ashwagandha can also be used to improve the semen quality by managing oxidative stress, cell death, and also by enhancing the levels of essential metal ions in seminal plasma.

It has also been documented that the root extracts of Ashwagandha ameliorate the hypobaric hypoxia through its prophylactic and neuroprotective potential that can induce memory impairment. In a study, it was found that the Ashwagandha treatment prevented neurodegeneration and memory impairment which was induced by hypobaric hypoxia in male rats (Baitharu et al., 2013), because of the modulation of the levels of nitric oxide, corticosterone, and acetylcholine, the activity of acetylcholine esterase in hippocampal region of rats. The supplementation of sodium nitroprusside along with Ashwagandha root extracts resulted in enhanced concentrations of corticosterone and number of pyknotic cells during hypobaric hypoxia. Therefore, Ashwagandha treatments could ameliorate memory impairment and neurodegeneration induced by hypobaric hypoxia in hippocampus through nitric oxide—mediated alterations in levels of corticosterone.

The antioxidant and antiinflammatory effects of Ashwagandha root extracts has been reported by Gupta and Singh (2014) in collagen-induced arthritic rats. Because of its antioxidant and antiinflammatory properties, the treatment of Ashwagandha significantly suppressed the behavioral changes, radiological score, and showed arthritis-protective effects in rats. The cardiorespiratory potential of Ashwagandha root extracts and its positive effect on the quality of life of male/female athletic adults have been reported (Choudhary et al., 2015). Ashwagandha treatments induced cardiorespiratory endurance and improved various parameters of quality of life (as mentioned by WHO) such as psychological and physical health as well as social relationships and maximum oxygen consumption in athletes.

A bioactive component of *W. somnifera*, withanolide A, had been reported to delay age-associated physiological changes, extend life span, and improve health span of *Caenorhabditis elegans* (Akhoon et al., 2016). Furthermore, the neuroprotective potential, antiamyloidogenic properties of withanolide A has been reported via modulation of acetylcholine (neural mediators) and mitigation of α-synuclein aggregation in the nematode (*C. elegans*). Besides this, withanolide A mediates revealed antistress potential and prolonged the life span in *C. elegans* through stimulation of pathway of insulin or insulin-like growth factor. The findings of this study emphasized that the *W. somnifera* can be employed to treat Parkinson's or Alzheimer's disease. Kaushik et al. (2017) demonstrated that the water extracts of *W. somnifera* contain an active somnogenic (sleep-inducing) component, i.e., triethylene glycol and it can be explored further for the treatment of insomnia.

The root extracts of *W. somnifera* reduced glial activation, prevented the phosphorylation of NF-κB (nuclear factor kappa B), altered the expression of cytokines/chemokines, and also acted as autophagy inducer in reported on transgenic mouse model (expressing G93A *sod1* mutant) of Amyotrophic Lateral Sclerosis with TDP-43 proteinopathy (Dutta et al., 2018). Thus, the root extracts of *W. somnifera* may be used to treat Amyotrophic Lateral Sclerosis at its early stages.

In a study focused to analyze the antistress potential, mood-enhancing properties of Shoden (a standardized Ashwagandha root extract) and its safety profile of healthy adults suffering from mild stress, it was recorded that with the supplementation of Ashwagandha (Shoden), decrease in morning cortisol, self-report depression, anxiety, stress scale-21, and clinician-administered Hamilton Anxiety Rating Scale was observed, whereas the levels of testosterone were enhanced in the males (Lopresti et al., 2019). This study emphasized on the stress-relieving potential of *W. somnifera* through regulatory properties on the hypothalamus—pituitary—adrenal axis. The root extracts (ethanolic) of *W. somnifera* have also been reported to inhibit lipogenesis (fatty acid synthesis) in the human prostate cancer cell lines (22Rv1 and LNCaP) and hence revealed the anticancer potential of ethanolic root extracts in treating prostate cancer (Kim et al., 2020). Further, the nephron-protective potential of the root extract of *W. somnifera* was also recorded via renal function analysis against the cisplatin-induced nephrotoxicity in the albino Wistar rats (Rasheed et al., 2020).

The recent studies emphasized on the potential role of *W. somnifera* in plant-based drug formulations for the treatment of the pandemic viral disease COVID-19 (caused by the novel coronavirus or SARS-CoV-2). In order to evaluate various bioactive phytochemicals for drug candidacy against SARS-CoV-2 causing the pandemic COVID-19 (coronavirus disease 2019, initiated from Wuhan, China, in 2019), using in silico screening and computational analyses. Out of the total 27 phytochemicals screened against SARS-CoV-2 main protease proteins (MPPs), Nsp9 RNA binding protein, spike receptor binding domain, spike ecto-domain, and HR2 domain using molecular docking, withaferin (a bioactive from *W. somnifera*) along with three other phytochemicals showed maximum binding affinity with all the key proteins in terms of lowest global binding energy (Azim et al., 2020). The binding potential of both withaferin A and withanone (two different bioactives from *W. somnifera*) was investigated by Kumar et al. (2020a) to its highly conserved protein M^{pro}. It was recorded that the withanone can bind to the substrate-binding pocket of SARS-CoV-2 M^{pro} with the efficacy and binding energies equivalent to an already claimed N3 protease inhibitor. Withanone was found to be capable of interacting with the highly conserved residues of the viral proteases. Therefore, it was predicted that withanone may possess the potential to inhibit the functional of SARS-CoV-2 protease and may offer some therapeutic value for the management of COVID-19. When the binding potential of two bioactives from Ashwagandha, i.e., withaferin A and withanone to TPMRSS2 (Transmembrane Protease Serine 2) using molecular docking and molecular dynamics approach, in comparison to its known inhibitor, Camostat mesylate. It was recorded that both withaferin A and withanone may bind and stably interact at the catalytic site of TMPRSS2. However, withanone was found to have a stronger interaction with TMPRSS2 catalytic residues than withaferin A, and it may also be able to induce changes in the allosteric site. Further, a remarkable downregulation of TMPRSS2 mRNA in withanone treated MCF7 cells suggested the dual action of withanone to block SARS-CoV-2 entry to the host cells (Kumar et al., 2020b). Therefore, Ashwagandha might be useful in the management of the pandemic COVID-19. Some of the important therapeutic/medicinal properties of *W. somnifera* are listed in Table 15.1.

15.5 Major challenges in ashwagandha

There are some problems associated with the crop, which limit its cultivation and popularization as herbal medicine. Some of these major challenges have been discussed below and some of the efforts made in overcoming these challenges have also been highlighted.

Table 15.1 List of some important therapeutic properties of Ashwagandha and their effects' studies.

Therapeutic potential	Experimental model	Effects of treatment	References
Tumor-preventing activity	Male albino mice	Oral administration (200 mg/kg daily dose) of Ashwagandha showed tumor-preventing potential by significantly increasing the total leukocyte count and lymphocyte percentage; preventing body weight loss and rate of mortality induced by urethane.	Singh et al. (1986)
Immune-modulatory activity	Babl/c mice	Treatment with root extracts of *Withania somnifera* (20 mg/dose/animal) showed immunomodulatory response through increase of total WBC count, enhancement of the cellularity of bone marrow, increase in positive cell number of α-esterase, increase in enhancement in phagocytic activity (peritoneal macrophages), and inhibition of delayed-type hypersensitivity reaction in mice. Supplementation of antigen along with *W. somnifera* root extract stimulated the circulating antibody titer and increase the count of plaque-forming cells in the mice spleen.	Davis and Kuttan (2000)

Continued

Table 15.1 List of some important therapeutic properties of Ashwagandha and their effects' studies.—*cont'd*

Therapeutic potential	Experimental model	Effects of treatment	References
Hyaluronidase inhibition potential (as antidote to snake bite)	Skin tissues	External application of a glycoprotein purified from *W. somnifera* could inhibit the hyaluronidase activity of *Naja naja* (cobra) and *Daboia russelii* (viper) snake venoms. Hyaluronidase (enzyme) present in snake venoms cause rapid spread of the toxins into the tissues of victim through destruction of extracellular matrix. The glycoprotein present in *W. somnifera* completely inhibited the enzyme activity of snake venoms, thereby demonstrated the use of Ashwagandha extract as an antidote for snake bite victims.	Machiah et al. (2006)
Antitumor effect through immunomodulatory potential	Male Swiss albino mice	Treatment with benzo(a)pyrene induced lung cancer which led to immune dysfunction in affected albino mice. Further, Ashwagandha (root extracts) treatment restored the levels of immunoglobulins, immunocompetent cells, and immune complexes which revealed antitumor potential of Ashwagandha treatment through immunomodulation in albino mice.	Senthilnathan et al. (2006)

Table 15.1 List of some important therapeutic properties of Ashwagandha and their effects' studies.—*cont'd*

Therapeutic potential	Experimental model	Effects of treatment	References
Potential to treat male infertility (improvement of semen quality)	Spermatozoa and seminal plasma from infertile men	*W. somnifera* treatment significantly lowered ROS levels in oligozoospermic (n = 25) and asthenozoospermic (n = 25) men; lowered the apoptosis in normozoospermic (n = 25) and oligozoospermic (n = 25) men. Also, with treatment the concentrations of essential metal ions were increased in seminal plasma of infertile men. Thus, *W. somnifera* treatment improved semen quality and could be used to treat male infertility.	Shukla et al. (2011)
Prophylactic and neuroprotective potential	Male Sprague Dawley rats	Ashwagandha (root extracts) treatments modulated the levels of nitric oxide, acetylcholine, the activity of acetylcholine esterase, and enhanced number of pyknotic cells as well as levels of corticosterone in hippocampal region of rats. Therefore, the root extracts of Ashwagandha ameliorated the memory impairment and neurodegeneration induced by hypobaric hypoxia in male Sprague Dawley rats.	Baitharu et al. (2013)

Continued

Table 15.1 List of some important therapeutic properties of Ashwagandha and their effects' studies.—*cont'd*

Therapeutic potential	Experimental model	Effects of treatment	References
Antioxidant and antiinflammatory agent	Rats	Ashwagandha root powder (600 mg/kg) significantly reduced the severity of arthritis through suppression of its symptoms and effective restoration of motor activity in the arthritic rats.	Gupta and Singh (2014)
Cardiorespiratory potential and improvement of quality of life	Male/female athletic adults	Root extract of Ashwagandha enhanced the levels of maximum oxygen consumption (VO_2 max) resulting in improved cardiovascular dynamics and enhanced cardiorespiratory endurance. Furthermore, the quality of life was improved in healthy male and female adult athletics.	Choudhary et al. (2015)
Neuroprotective and antistress potential; prolonged life expectancy	*Caenorhabditis elegans*	A bioactive component of *W. somnifera*, withanolide A, has antiaging, antistress, neuroprotective potential and enhanced life span, health span, neuron functionality through stimulation of insulin/IGF-1 signaling pathway in *C. elegans*. Furthermore, it also enhanced the transcriptional level of HSF-1 (heat-shock transcription factor) and SKN-1 (antioxidant response transcription factor).	Akhoon et al. (2016)

Table 15.1 List of some important therapeutic properties of Ashwagandha and their effects' studies.—*cont'd*

Therapeutic potential	Experimental model	Effects of treatment	References
Somnogenic (sleep-inducing) potential	Male C57BL/6 mice	Triethylene glycol present in water extracts of *W. somnifera* induced rapid and nonrapid eye movement sleep in mice. The quality of sleep was also enhanced and the sleep-promoting potential of triethylene glycol has been reported in mice.	Kaushik et al. (2017)
Neuroprotective potential	Transgenic mice carrying G93A *sod1* mutant	Neuroprotective properties of *W. somnifera* root extracts have been reported on transgenic mouse model (expressing G93A *sod1* mutant) of amyotrophic lateral sclerosis with TDP-43 proteinopathy. Ashwagandha treatment resulted in enhanced longevity, better motor coordination, regulated expression of cellular chaperons, and improved number of motor neurons situated in the lumbar spinal cord of $SOD1^{G93A}$ mice.	Dutta et al. (2018)
Antistress potential and mood-enhancing properties	Healthy male and female adults aged between 18 and 65 years	Study on efficacy and tolerability of Shoden (standardized Ashwagandha extract) revealed that Shoden treatments relieved the anxiety, hormone production, and stress in healthy male and females under mild stress.	Lopresti et al. (2019)

Continued

Table 15.1 List of some important therapeutic properties of Ashwagandha and their effects' studies.—*cont'd*

Therapeutic potential	Experimental model	Effects of treatment	References
Nephroprotective potential	Albino Wistar rats	Levels of serum urea and creatinine (biomarkers studied to check the renal toxicity) were decreased with the treatment of *W. somnifera* root extracts in nephrotoxic (cisplastin-induced) rats. This study emphasized on the nephroprotective property of Ashwagandha to treat renal dysfunctions.	Rasheed et al. (2020)
Anticancer potential	Human prostate cancer cell lines (22Rv1 and LNCaP)	The ethanolic root extracts of *W. somnifera* inhibited the fatty acid synthesis in human prostate cancer cell lines through downregulation of expression of fatty acid metabolism—related proteins. *Withania* root extracts also suppressed the intracellular amounts of total free fatty acids and decreased the levels of acetyl-CoA in prostate cancer cell lines, revealing the anticancer efficacy of Ashwagandha.	Kim et al. (2020)

15.5.1 Low seed viability

For the commercial crop production under field conditions the high rates of seed germination, seed viability, and seedling survival are required. Although there is an increasing demand of Ashwagandha in the herbal drug industry, the traditional cultivation of Ashwagandha through its seeds has been limited due to low percentage of seed viability, poor germination, and seedling survival; and moreover the dormancy in the seeds is also a limiting factor in its cultivation (Niyaz and Siddiqui, 2014; Shanmugaratnam et al., 2013; Vakeswaran and Krishnasamy, 2003). In addition to these factors, the dependence of germination on both light and temperature (Kambizi et al., 2006; Khanna et al., 2013) further affects its commercial production. To overcome these problems, several researchers tried and optimized various physicochemical treatments, microbial treatments, and the storage conditions to overcome the problem of low seed germination in Ashwagandha. The hydropriming treatment has been found to be effective in enhancing the rate of seed germination (Mehta and Raina, 2016). Siderophore producing *Alcaligenes faecalis* broth promoted the seed germination by 75% (Sayyed et al., 2007). It was also found that inoculation of seed with plant growth–promoting rhizobacteria (PGPR)-selected isolate not only enhanced seed germination under normal circumstances but also under UV-B exposure (Rathaur et al., 2012).

However, the low germination percentage of *W. somnifera* may be attributed to seed dormancy but various treatments have been tried to break the dormancy. Among the different treatments tried to break the seed dormancy including overnight soaking in water and hot water soaking (Shanmugaratnam et al., 2013), a higher germination percentage recorded after hot water soaking (52.75%) than soaking overnight in water (33.75%). Himangini and Thakur (2018) tried various storage conditions and recorded that storing the seeds at 0°C for 4 months is best for germination. In another experiment performed to investigate the effect of different physicochemical treatments, storage, temperature, photoperiod, and growth regulators on the seed germinability of Ashwagandha, it was recorded that the GA_3 treatment (150 μg/mL) was most effective. However, the optimum conditions favoring germination were recorded as 25°C of temperature with continuous light, suggesting a significant role of photoperiod on seed germination (Khanna et al., 2013). Whereas Niyaz and Siddiqui (2014) found that maximum increase in seed germination was at 500 μg/L treatment of GA_3 and the heat treatment at 50°C reduced the germination drastically. The prechilling treatment inhibits the seed germination in Ashwagandha and the content temperature of 25, 35, and 45°C in dark completely inhibits it (Kambizi et al., 2006).

When the effect of various nutritional sources was evaluated on seed germination of *W. somnifera*, the nitrates of sodium, calcium, and potassium were much effective in increasing the rate of seed germination (Ingle and Kareppa, 2012).

15.5.2 Pests and diseases

Despite the immense therapeutic properties of Ashwagandha, it is prone to a number of pests and diseases which not only limit its production but have some implications in the quality of the produce too. The biotic agents pose a big challenge to the quality and quantity of the produce and could be bacterium, fungi, virus, nematode, insect, mite, etc.

One of the most important pathogens is *Alternaria alternata* (Fr.) Keissler, that causes leaf spot disease in *W. somnifera* and one of its related species *W. coagulans* (Pati et al., 2008; Sharma et al., 2013b). This fungal pathogen not only induces some biochemical, physiological, and ultrastructural changes in Ashwagandha but also affects the production of withanolides by lowering the expression of some key genes of withanolide pathway (Sharma et al., 2011, 2014; Singh et al., 2017a,b). Therefore, a sensitive detection system for the presence of *A. alternata* is highly desirable (Sharma, 2013a) for quality control of herbal medicines prepared their off. Since a variability exists in *W. somnifera* germplasm for the resistance toward *A. alternata* (Meena et al., 2019; Mohammad and Shabbir, 2016), the resistant lines can be used as a breeding material to transfer this trait in the elite chemotypes. When different isolates of *A. alternata* were screened for their virulence, one of the isolates, i.e., Chempatti (I6) was found to be most virulent. Further, among the seven fungicides tested, mancozeb (0.2%) was found to be most effective (Kalieswari et al., 2016). Ashwagandha plants treated with the combination of two species of endophytic bacteria viz., *Bacillus amyloliquefaciens* and *Pseudomonas fluorescens*, showed the lowest plant mortality rate in the presence of *A. alternata* stress (Mishra et al., 2018).

Besides *A. alternata*, several diseases of *W. somnifera* have also been reported that are caused by various classes of plant pathogens viz., bacteria, fungi, phytoplasma, viruses, nematodes, etc., that are also listed in Table 15.2.

Table 15.2 List of some common diseases of *Withania somnifera*, caused by various types of plant pathogens.

Nature of the pathogen	Pathogen	Disease	References
Fungi	*Alternaria alternata*	Leaf spot	Pati et al. (2008); Sharma et al. (2011)
	Alternaria dianthicola	Leaf blight disease	Maiti et al. (2007, 2012)
	Alternaria chlamydospora	Leaf blight disease	Vanitha et al. (2006)
	Alternaria solani	–	Alwadi and Baka (2001)
	Fusarium oxysporum	Wilt	Sharma and Trivedi (2010)
	Fusarium solani	Root rot and wilt/leaf spot	Chavan and Korekar (2011); Gupta et al. (2004)
	Cercospora withaniae	Leaf spot	Chavan and Korekar (2011)
	Pseudoercospora withaniae	Leaf spot	Singh (2012)

Table 15.2 List of some common diseases of *Withania somnifera*, caused by various types of plant pathogens.—*cont'd*

Nature of the pathogen	Pathogen	Disease	References
	Pseudocercospora fuligena	Black leaf spot	Saroj et al. (2014)
	Pithomyces chartarum	Leaf spot	Verma et al. (2008)
	Colletotrichum gloeosporioides	Leaf spot	Solanki and Basudeb (2017)
	Puccinia withaniae	Yellow rust	El-Ariqi et al. (2009)
	Choanephora cucurbitarum	Wet rot	Saroj et al. (2012)
	Myrothecium roridum	Leaf spot	Mahrshi (1986)
	Chaetomium globosum	Black leaf lesions	Shah and Daniel (2004)
Phytoplasma	Phytoplasma (16S rVI group)	Witches-broom disease, little leaf symptoms	Khan et al. (2006); Samad et al. (2006); Zaim and Samad (1995)
Virus	Begomovirus (Jatropha mosaic India virus)	Yellow mosaic disease, vein clearing, yellow net or mild leaf curl symptoms	Baghel et al. (2010, 2012)
	Tobacco leaf curl virus	Vain clearing, leaf rolling, and vein banding	Pathak and Raychaudhuri (1967); Singh and Kumar (1998)
	Eggplant mottled dwarf virus	–	Al-Musa and Lockhart (1990)
Nematode	*Meloidogyne incognita*	Root knot disease	Pandey and Kalra (2003); Saikia et al. (2013)
	Meloidogyne javanica	Root knot disease	Bhatti et al. (1974)

In addition to these pathogens, many pest species are also known to affect the health of *W. somnifera* plants. One of the major insect pests is 28-spotted beetle *Henosepilachna vigintioctopunctata*. The larvae of this plant rapidly feed on its leaves, leaving behind a fibrous skeleton, reducing the commercial value of the plants (Sharma and Pati, 2011b). Efforts have been made to identify different parasitoids and fungal biocontrol agents for the eco-friendly management of this pest (Jamwal et al., 2017; Sharma et al., 2012b; Venkatesha, 2006). Cowbug, *Oxyrachis tarandus*, is also an important pest which mainly affects apical portions of the stems, turning them brown, rough, and woody and the apical leaves are shed in severe infestation gradually dried and apical leaves were shed off (Sharma and Pati, 2011a). An invasive mealybug, *Phenacoccus solenopsis*, was also found as one of the important pests of *W. somnifera*, affecting mainly its leaves that can be deformed and shed prematurely (Sharma and Pati, 2013). The carmine red spider mite, *Tetranychus urticae*, is the most important mite species affecting Ashwagandha. It is commonly found on the aerial and more commonly on the apical parts of the infested shoots, commonly feeding on the leaves, turning them shiny white in color, which gradually turn brown, and are shed off at a later stage (Sharma and Pati, 2012).

Besides these important pests mentioned above, several other pests are known to affect the cultivation of Ashwagandha as listed in Table 15.3.

Table 15.3 List of some major pest species infesting *Withania somnifera*.

Pest species	Affected plant part/ symptoms	References
Henosepilachna vigintioctopunctata (Coleoptera: Coccinellidae)	Leaves are fed by both larvae and adults	Sharma and Pati (2011b); Sharma et al. (2012b)
Helicoverpa armigera (Lepidoptera: Noctuidae)	Fruit borer	Rehaman and Pradeep (2016); Rehaman et al. (2018)
Tetranychus urticae (Trombidiformes: Tetranychidae)	Feed on leaves, turning them shiny white in color	Sharma and Pati (2013)
Phalantha phalantha Drary (Nymphalidae: Lepidoptera)	Young leaves are fed by larvae	Gorain et al. (2012)
Phenacoccus solenopsis (Hemiptera: Pseudococcidae)	Leaves are deformed and shed prematurely	Sharma and Pati (2013)
Acherontia styx (Sphingidae: Lepidoptera)	Leaves are affected by caterpillars	Singh and Kumar (1998)
Oxyrachis tarandus (Hemiptera: Membracidae)	Apical portions of the stem affected	Sharma and Pati (2011a)
Acherontia atropos (Lepidoptera: Sphingidae)	Leaves are affected by caterpillars	Akkuzu et al. (2007)
Spilarctia oblique (Lepidoptera: Arctiidae)	–	Chandra (2004); Kumar et al. (2009a)
Tricentrus bicolor (Membracidae: Hemiptera)	–	Kumar et al. (2009c)
Eutetranychus orientalis (Acarina: Tetranychidae)	–	Gupta and Karmakar (2011)
Tricentrus sp. (Membracidae: Hemiptera)	Tender stems are affected, adults and nymphs both suck the plant sap	Singh and Kumar (1998)
Coccidohystrix insolitus (Pseudococcidae: Hemiptera)	Leaves are affected	Ravikumar et al. (2008)

Table 15.3 List of some major pest species infesting *Withania somnifera*.—*cont'd*

Pest species	Affected plant part/ symptoms	References
Drosicha mangiferae (green) (Pseudococcidae: Hemiptera)	–	Bhagat (2004); Kumar et al. (2009b)
Nezara viridula Linn. (Heteroptera: Pentatomidae)	–	Kumar et al. (2009b)
Plautia sp. (Hemiptera: Pentatomidae)	–	Kumar et al. (2009b)
Dysdercus cingulatus (Fabr.) (Hemiptera: Pyrrhocoridae)	–	Kumar et al. (2009b)
Trialeurodes vaporariorum (Westwood) (Homoptera: Aleyrodidae)	–	Kumar et al. (2009b)
Diacrisia oblique (Arctidae: Lepidoptera)	Leaves are affected by caterpillars	Singh and Kumar (1998)
Lygeus equistris (Lygaeidae: Hemiptera)	Leaves are affected, adults and nymphs both suck the plant sap	Singh and Kumar (1998)
Hieroglyphus banian (Acrididae: Orthoptera)	Leaves are eaten by adults and nymphs both	Singh and Kumar (1998)
Trilophida annulata (Acrididae: Orthoptera)	Leaves are eaten by adults and nymphs both	Singh and Kumar (1998)
Aphis craccivora (Aphididae: Hemiptera)	Leaves and tender stems are affected, insects suck the plant sap	Singh and Kumar (1998)
Coccus sp. (Coccidae: Hemiptera)	Leaves are affected, insects suck the plant sap	Singh and Kumar (1998)
Oxycarenus hyalinipennis (Lygaeidae: Hemiptera)	Leaves are affected. Both adults and nymphs suck the plant sap	Singh and Kumar (1998)
Pseudococcus sp. (Pseudococcidae: Hemiptera)	Leaves and other tender shoots are affected. Insects suck the plant sap	Singh and Kumar (1998)
Agromyza sp. (Agromyzidae: Diptera)	Leaves are affected (mined)	Singh and Kumar (1998)
Nisotra striatipennis (Chrysomelidae: Coleoptera)	Leaves are fed by adults	Singh and Kumar (1998)
Polyphagotarsonemuslatus sp. (Tarsonemidae: Acari)	Leaves are affected and defoliation occurs	Singh and Kumar (1998)
Myzus persicae (Hemiptera: Aphididae)	–	Kumar et al. (2009c)
Planococcus citri (Risso) (Hemiptera: Pseudococcidae)	–	Attia and Awadallah (2016)

15.5.3 Quality control in herbal medicine

Because of the use of *W. somnifera* in several herbal drug formulations (as single- and multiherbal products), the stringent measures to ensure the quality of the herbal drugs and the adoption of good manufacturing practices for the preparation of herbal drugs are a challenge. The quality control of medicinal plants has emerged as an important aspect in herbal drug industry and there are strict international guidelines for this too (WHO, 2007).

A study was conducted to analyze the major bioactive constituents like withaferin A, in commercially available herbal formulation derived of Ashwagandha, in the Indian market, a huge variation (more than 70-fold) in the daily intake of withaferin A have been recorded (Sangwan et al., 2004). Age of the plant also significantly affects the metabolic contents of the plant; therefore, the harvest at correct stage is also very important to get a consistent performance of the herbal drugs derived their off and could be responsible for the huge variation in the amount of bioactives. Withaferin A content increases in both the leaves and the roots with the age of plant (up to 12–18 week old), but withaferin A content (along with some other withanolides) is declined on the overmaturation (Dhar et al., 2013; Pal et al., 2011). Further the microbial pathogens like *A. alternata* can reduce the concentration of bioactives present in Ashwagandha too (Pati et al., 2008); therefore, the management of the disease and development of an efficient detection system for the disease becomes even more important.

15.6 Phytochemicals

The phytochemistry of Ashwagandha has been investigated extensively. There are several families of plant secondary metabolite which have been found in it, including the alkaloids, flavonoids, steroidal lactones, tannin, etc. Out of these the steroidal lactones with a basic C-28 framework are the signature bioactive molecules, identified in *W. somnifera*, called as withanolides. The basic structure of withanolides is represented as Fig. 15.1a and structures of three key withanolides (withanolide A, withaferin A, and withanone) are represented as Fig. 15.1b–d. The steroidal lactones present in *W. somnifera* are ashwagandhanolide, withanolide A–Y, withaferin A, withasomidienon, withasomniferin, withanone, withasomniferols, and a novel chlorinated withanolide namely 6a-chloro-5b,17a dihydroxywithaferin A (Dar et al., 2016; Singh et al., 2015; Tripathi et al., 2018). The major phenolics present in Ashwagandha include quercetin, kaempferol, cholorogenic acid, gallic acid, ellagic acid, tannic acid, caffic acid, and rutin. Besides, steroidal lactones and phenolics, *W. somnifera* has glycowithanolides-like sitoindoside (VII–X), sterols, withanol, somnitol, somnisol, cholesterol, diosgenin, stigmastadien, stigmasterol, β-sitosterol, alkaloids, and flavonol glycosides (Bähr and Hänsel, 1982; Bhattacharya et al., 1997; Chatterjee et al., 2010; Chaurasiya et al., 2009, 2012; Ghosal et al., 1988; Mirjalili et al., 2009; Sangwan et al., 2008; Singh et al., 2015; Sivanandhan et al., 2013; Xu et al., 2011).

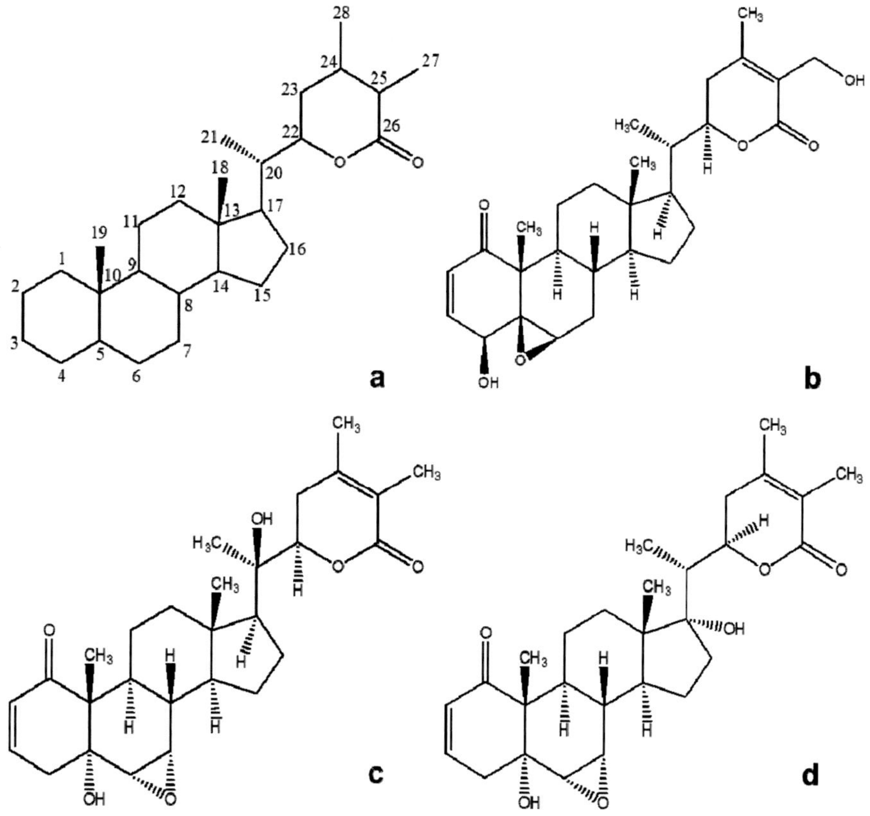

FIGURE 15.1

The chemical structure of (a) withanolide backbone, (b) withaferin A, (c) withanolide A, and (d) withanone.

The 3-α-gloyloxytropane, 3-trigloyloxytropine, withanine, somnine, somniferinine, somniferine, tropine, DL-isopelletierine, anaferine, pseudowithanine, pseudotropine, anhygrine, hygrine, cuscohygrine, choline, withanosomine, and calystegines are the other alkaloids present in Ashwagandha (Dar et al., 2016; Schröter et al., 1966; Schwarting et al., 1963; Singh et al., 2015). However, in *W. somnifera*, a variation of about 0.13%–0.31% has been reported among the total alkaloid content (Tripathi et al., 2018).

Moreover, some other secondary metabolites like anthocyanins, saponins, carotenoids, glycosides, lignins, tannins, flavonoids, phytosterols, and withanamides (A-I), etc., also account for the active biochemical constituents of *W. somnifera* (Anjaneyulu and Rao, 1997; Dhalla et al., 1961; Ghosal et al., 1989; Kirson et al., 1971; Praveen and Murthy, 2010; Pramanick et al., 2008; Takshak and Agrawal, 2014; Xu et al., 2011). The major flavanol glycosides present in *W. somnifera* include 6,8-dihydroxykaempferol 3-rutinoside, quercetin, its 3-O-rutinoside and

3-rutinoside-7-glucoside, etc (Kandil et al., 1994; Singh et al., 2015). Some of the accessory metabolites such as acyl steryl glucosides, starch, dulcitol, hantreacotane, and amino acids (mainly alanine, aspartic acid, glutamic acid, cysteine, cystine, glycine, proline, tryptophan, and tyrosine) are present in *W. somnifera* (Johri et al., 2005; Tripathi et al., 2018).

In the roots of *W. sominfera*, the other medicinally important biochemicals include glycowithanolides, like sitoindosides VII, sitoindosides VIII, sitoindosides IX, and sitoindosides X, etc (Bhattacharya et al., 1997; Ghosal et al., 1989). Whereas in the leaves, withanone, withanolide A, 27-deoxywithaferin A, 7-hydroxywithanolide, trienolide, withanolide Z, withanolide B, and 3α-methoxy-2, 3-dihydro-withaferin are the major withanolides (Anjaneyulu and Rao, 1997; Dhalla et al., 1961; Kirson et al., 1971; Praveen and Murthy, 2010; Pramanick et al., 2008). However, from the seeds of *W. somnifera* two withanolides, i.e., withanolide-WS1 and withanolide-WS2, have also been reported (Xu et al., 2011).

15.6.1 Chemotypic variability

A considerable degree of chemotypic variability exists in the *W. somnifera* germplasm. There is a substantial variability not only in the profile of the different bioactive constituents and other phytochemicals but also in their concentration/accumulation in the plant. The identification of the elite genotypes rich in the phytochemicals having desired bioactive properties may be brought under cultivation and could be used as a breeding material for the crop improvement programs.

Based on the variability in the profile of the different bioactive constituents, Ashwagandha has been classified into various chemotypes. Three distinct chemotypes were recorded from Israel and one each from India and South Africa (Abraham et al., 1968, 1975; Eastwood et al., 1980; Kirson et al., 1971), although the later reports suggest the presence of more than one chemotype in India (Chaurasiya et al., 2009; Kaul et al., 2009; Singh et al., 2017b). The major bioactives of Israel chemotypes I, II, and III were recorded as withaferin A, withanolide D, and withanolide E, respectively (Abraham et al., 1968, 1975) and the major bioactives in Indian chemotype were recorded as withanone and withaferin A (Dhalla et al., 1961; Kirson et al., 1977). Whereas the major bioactives present in the South African chemotype were recorded as withaferin A and withaferin D (Kaul et al., 2009; Kirson et al., 1970). Some new withanolides have also been reported from the hybrids of these chemotypes (discussed in Section 15.8.1). Further, in an attempt to estimate the accumulation of a bioactive, i.e., withanolide A in 25 diverse genotypes of Ashwagandha, grown in India, it was found that withanolide A was maximum in the genotype UWS 59, which was possessing the highest concentration of the alkaloids too (Chauhan et al., 2019).

Besides the chemotypic variability, the variability in the production of bioactives in *W. somnifera*, the leaf ontogenic phase–related variabilities have also been suggested. Using HPLC and radio-labeled precursor feeding experiments, it has been recorded that the de novo biogenesis and accumulation of withanolides is most active in the young leaves. The biogenesis of withanolide may start as early as in primordial stage, reaches maximum in the young leaves, and declines afterward along the maturity (Chaurasiya et al., 2007).

15.6.2 Biosynthetic pathways of major bioactives

Although *W. somnifera* contains a wide range of secondary metabolites distributed differently to different plant parts, but the most important class of bioactives secondary metabolites present in it are withanolides, that have been extensively studied and are credited the most for its widely acclaimed pharmaceutical properties (Dhar et al., 2015). The 24-methylene cholesterol generated via the isoprenoid pathway serves as the precursor for withanolide biosynthesis, that leads to the production of withanolides (Sangwan et al., 2008; Singh et al., 2014). The isoprenoids (structural components of cells or organelle membranes) are naturally synthesized inside the plant cells from the two major precursors viz., dimethylallyl pyrophosphate (DMAPP) and isopentenyl pyrophosphate (IPP) (Hunter, 2007).

The isoprenoid pathway can be biosynthesized through two independent pathways viz., MVA and MEP pathways. The mevalonate (MVA) pathway is localized in the plant cytosol. The alternate pathway is referred to as non-MVA pathway or 2-C-methyl-D-erythritol-4-phosphate (MEP) pathway or 1-deoxy-D-xylulose-5-phosphate (DOXP) pathway or alternative pathway and it occurs in the plastid and has a significant contribution (about 25%) in withanolide production (Chaurasiya et al., 2012; Hunter, 2007). Various plant metabolites like sterols, polyterpenes, triterpenes, and withanolides (C-30 compounds) have been reported to be originated through 24-methylene cholesterol or from 24-methylene lophenol or from some upstream intermediate triterpene pathways (Sangwan et al., 2008). The withanolide biogenesis pathway may be originated through the isoprenogenesis (generation of isoprene units, i.e., IPP and DMAPP) or the isoprenoids biosynthesis pathways (Chaurasiya et al., 2012).

The MVA pathway (cytosol) initiates with the acetyl-CoA C-acetyltransferase (ACT)-catalyzed formation of acetoacetyl-CoA from two units of acetyl-CoA (Hunter, 2007). In the next step, third acetyl-CoA gets condensed with acetoacetyl-CoA by the enzyme hydroxymethylglutaryl-CoA synthase (HMGS), so as to form 3-hydroxy-3-methylglutaryl-CoA (HMG-CoA). In the subsequent step, HMG-CoA is converted (irreversible reaction) to mevalonic acid (MVA) by NADPH-dependent 3-hydroxy-3-methylglutaryl coenzyme A reductase (HMGR). Further, MVA gets phosphorylated to form mevalonate diphosphate sequentially by the enzymes mevalonate kinase (MVAK) and mevalonate phosphate kinase (MVAPK). In the next step, di-phosphomevalonate decarboxylase (MVAPPD) coverts mevalonate diphosphate into IPP and then, some IPP gets converted into DMAPP by IPP isomerase activity.

Besides the classical MVA pathway, alternative MEP pathway can also generate IPP and its isomer DMAPP independently inside the plastids. This pathway starts with the condensation glyceraldehyde 3-phosphate (GA-3-P) and pyruvate to produce 1-deoxy-Dxylulose-5-phosphate (DOXP) and this step is catalyzed in the presence of a cofactor, thiamine pyrophosphate, by the enzyme 1-deoxy-D-xylulose-5-phosphate synthase (DXS). In the next step, DOXP is converted into MEP by the enzyme 1-deoxy-D-xylulose-5-phosphate reductoisomerase (DXR). In subsequent five steps, the MEP is converted into IPP and IPP, the C=C bond gets isomerized to produce DMAPP, and this reaction is catalyzed by the enzyme isopentenyl-diphosphate delta-isomerase (IDI).

In the next step in cytoplasm of plant cell, IPP gets condensed along with its isomer DMAPP by the enzyme geranyl diphosphate Synthase (GPPS) to form geranyl pyrophosphate which is further converted into farnesyl diphosphate by the enzyme farnesyl pyrophosphate synthase (FPSS). Two molecules of farnesyl diphosphate get condensed into squalene (C-30 compound) by the enzyme squalene synthase (SS) and squalene epoxidase (SE) converts squalene into 2,3-epoxysqualene. The withanolide biosynthesis pathway up to SS mediated formation of squalene follows anaerobic reactions (Abe et al., 2007). The first step of oxygenation is catalyzed by SE and after epoxidation of squalene, the pathways are common for biogenesis of other metabolites like sterols, triterpenoids, and brassinosteroids (Sangwan et al., 2008; Sharma et al., 2018).

Further, 2,3-epoxysqualene gets transformed into cycloartenol (plant sterol precursor) by the enzyme cycloartenol synthase (CAS). Cycloartenol is converted into 24-methylene cycloartenol by the enzyme sterol 24-C-methyltransferase (SMT1) and methyl sterol monooxygenase/sterol-4α-methyl oxidase 1 (SMO1) catalyzes the conversion of 24-methylene cycloaretenol into cycloeucalenol (Pal et al., 2019; Thirugnanasambantham and Senthil, 2016). The latter is converted into Obtusifoliol by the enzyme cycloeucalenol cycloisomerase (CEC1) and then, it is converted into Δ 8,14-sterol by the enzyme obtusifoliol 14-demethylase (CPY51G). Δ 8,14-sterol is then converted into 4α-methyl fecosterol by Δ14-sterol reductase (FK) and C-7,8 sterol isomerase (HYD1) catalyzes its conversion into 24-methylene lophenol which is subsequently converted into episterol by methyl sterol monooxygenase/sterol-4α-methyl oxidase 2 (SMO2). Episterol is then converted into dehydroepisterol by the enzyme C-5 sterol desaturase (STE1) and the enzyme 7-dehydro cholesterol reductase (DWF5) catalyzes the conversion of dehydroepisterol into 24-methylene cholesterol. The important intermediate and the branching point of both withanolide biogenesis and sterol biosynthesis pathway is 24-methylene cholesterol which is converted into withanolides by the enzymes sterol glycosyltransferases (SGT) and methyltransferases (MT). However, further detailed pathway where 24-methylene cholesterol gets converted into other withanolides and withaferin A is yet to be elucidated. On the bases of our current understanding of the withanolide pathway, a flow chart has been shown to represent the major steps of the withanolide biosynthesis as Fig. 15.2.

Pathway engineering (genetic engineering of the biosynthetic pathways of secondary metabolites like withanolides in *W. somnifera*) using the recent advancement in in vitro manipulation and cloning and characterization of pathways genes (mainly rate limiting enzymes) can be used in the production of the bioactive metabolites in

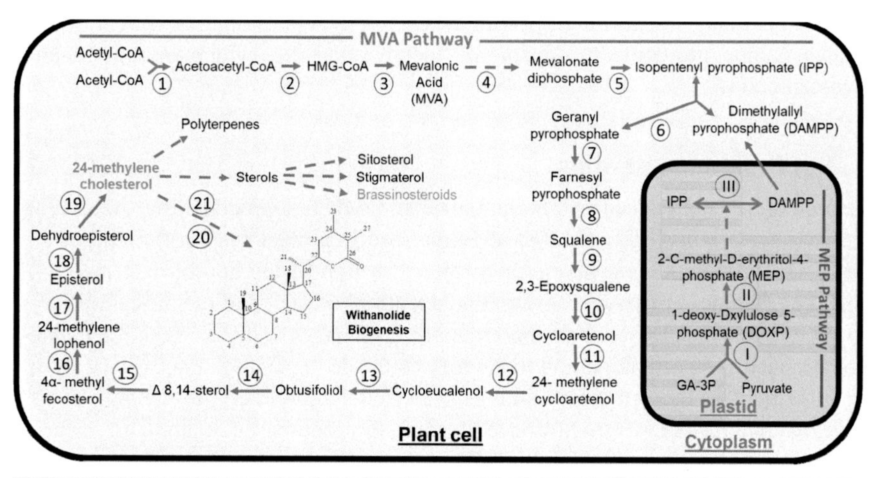

A. Enzymes in the Cytosolic MVA Pathway:

1. Acetyl-CoA C-acetyltransferase (ACT)
2. Hydroxymethylglutaryl-CoA synthase (HMGS)
3. 3-hydroxy-3-methylglutaryl Coenzyme A reductase (HMGR)
4. mevalonate kinase (MVAK)/ mevalonate phosphate kinase (MVAPK)
5. diphosphomevalonate decarboxylase (MVAPPD)
6. Geranyl diphosphate Synthase (GPPS)
7. Farnesyl pyrophosphate Synthase (FPSS)
8. Squalene Synthase (SS)
9. Squalene Epoxidase (SE)
10. Cycloartenol Synthase (CAS)
11. Sterol 24-C-methyltransferase (SMT1)
12. Methyl sterol monooxygenase/Sterol-4α-methyl oxidase 1 (SMO1)
13. Cycloeucalenol cycloisomerase (CEC1)
14. Obtusifoliol 14-demethylase (CPY51G)
15. Δ14-sterol reductase (FK)
16. C-7,8 Sterol isomerase (HYD1)
17. Methyl sterol monooxygenase/Sterol-4α-methyl oxidase 2 (SMO2)
18. C-5 Sterol desaturase (STE1)
19. 7-Dehydro cholesterol reductase (DWF5)
20. Sterol glycosyltransferases (SGT)
21. Methyltransferases (MT)

B. Enzymes in the Plastid-localized MEP Pathway:

I. 1-deoxy-D-xylulose-5- phosphate synthase (DXS)
II. 1-deoxy-D-xylulose-5- phosphate reductase (DXR)
III. Isopentenyl-diphosphate delta-isomerase (IDI)

FIGURE 15.2

The major steps of the withanolide biosynthetic pathway, based on our current understanding. The enzymes of the both MVA and MEP pathway are listed below the flow chart of the biosynthetic pathway.

the medicinal plants. In *W. somnifera* also, the studies have been initiated to genetically engineer the plants to overexpress the key genes of withanolide pathway for the enhanced production of withanolides. Transgenic *W. somnifera* plants overexpressing *WsSQS* (squalene synthase gene from *W. somnifera*) using *Agrobacterium tumefaciens*–mediated genetic transformation showed two- to five-fold increase in *WsSQS* transcripts in the various plant tissues, and two- to three-fold increase in protein expression. In the transformed tissue there was a 1.5- to 2-fold increase in total withanolide content, whereas in leaf alone, the two important withanolides, i.e., withaferin A and withanolide A, by 4- to 4.5-fold (Patel et al., 2015).

15.7 Omics advancements

In the past two decades, a lot of efforts have been in OMICS approaches to understand *W. somnifera* biology. Different OMICS approaches deal with the collective characterization, quantification, and understanding of pools of various biomolecules in relation to the structure, function, and the overall dynamics of an organism, which include genome, proteome, transcriptome, metabolome, *etc*., which are individually discussed below.

15.7.1 Proteomic studies

The proteomics approaches are among the novel OMICS approaches that are used to identify a wide range of genes that are expressed in a plant system (Senthil et al., 2011). The proteomic analysis involves the extraction (using appropriate method), separation (in one- or two-dimensional gel electrophoresis), and identification (prediction of putative role based on the sequence databases), and their functional categorization is an important method to identify their role in plant systems. Every above step is critical in the proteomic studies with better reproducibility. (Singh et al., 2017a) found the TCA (trichloroacetic acid)—acetone-based method of protein extraction is better than both the phenol and the tris buffer—based methods in terms of yield, number of spots (in 2D-electrophoresis), gel resolution, as well as the reproducibility. To study the comparison of the protein profile of *Panax ginseng* (Korean ginseng) and *W. somnifera* (Indian ginseng) root, 2D-electrophoresis was used (Nagappan et al., 2012). Both the plants are comparable to each other due to their health benefits and bioactives (Kulkarni and Dhir, 2008). In order to study the differences in the protein composition of two different tissues, i.e., leaves and seeds, a comparative proteome study was also made using 2D-electrophoresis and mass spectroscopy. In this study, 74 and 70 proteins were identified from the leaves and the seeds tissues, respectively, using protein sequence databases. Besides the common housekeeping genes found in both the tissues, the tissue-specific proteins with specialized metabolic roles were also identified (Dhar et al., 2012). In another experiment the comparative proteomic analysis using 2D-electrophoresis followed by MALDI-TOF-MS (matrix-assisted laser desorption/ionization time of flight—mass spectrometry) was made between in vitro- and in vivo—grown roots and it was found using homology search using MASCOT (http://www.matrixscience.com/) that there is a high similarity in the protein spots of both in vitro and in vivo roots suggesting that a similar developmental process is involved in the development of in vivo and in vitro (independent of exogenous PGRs (plant growth regulators) and independent of in vitro shoots) (Senthil et al., 2011). The proteomics approach was also used for the identification of the differentially expressed proteins in *W. somnifera* in response to the infection by its fungal pathogen *A. alternata* with the aim to identify the proteins involved in the process of pathogenesis. A total of 38 differentially expressed proteins were recorded and identified using MALDI-TOF-MS. The known proteins were categorized into eight different groups based on their function and it was recorded that the majority of them were related to the energy and metabolism, cell structure, stress/defense and RNA/DNA categories; and expression of some key proteins was corresponding to the gene expression at posttranscriptional level also (Singh et al., 2017a). Further, in order to study the structure and properties of the selected enzymes (viz., 3-hydroxy-3-methylglutaryl coenzyme A reductase,

1-deoxy-D-xylulose-5-phosphate synthase, 1-deoxy-dxylulose-5-phosphate reductase, farnesyl pyrophosphate synthase, squalene synthase, squalene epoxidase, and cycloartenol synthase enzymes) involved in withanolide biosynthesis, a bioinformatics analysis was performed and the secondary structure (3d) were predicted and the physicochemical properties (like pI, AI, GRAVY, and instability index) were also studied which may be helpful in predicting their functionality (Sanchita et al., 2014).

15.7.2 Transcriptomics studies: information from ESTs and transcriptome sequencing

The ESTs (Expressed Sequence Tags) are considered as a cost-effective method to identify the genes involved in a particular process/pathway. These ESTs not only provide the gene information but also help in the developments of sequence-based markers like SSRs (simple sequence repeats) from the ESTs of *W. somnifera* or otherwise the cross-species transferability of EST-SSRs has also been exploited (Parmar et al., 2015; Parita et al., 2018).

Due to the availability of NGS (next-generation sequencing) technologies, it is now possible to sequence the transcriptomes and identify the genes being transcribed with the help of the available databases and desired bioinformatic tools. Further, attempts have also been made to identify genes differentially expressed among different conditions (like cell-specific, tissue-specific, organ-specific, development stage–specific, or variety/chemotype-specific, chemical treatment/elicitor-specific, or plant stress–specific manner) in a quick and cost-effective way even in nonmodel plant species. The transcriptome data may also be used for the identification and development of SSRs and SNPs (single-nucleotide polymorphisms) (Jhanwar et al., 2012). Some efforts have also been made in *W. somnifera* with regard to de novo assembly, functional annotation, and comparison in a tissue/organ-specific (Gupta et al., 2013; Senthil et al., 2015; Tripathi et al., 2020), chemotype-specific (Gupta et al., 2015) or chemical treatment–specific (Dasgupta et al., 2014) manner. Identification of the SSRs from the transcriptome data of different chemotypes will be helpful in the marker-assisted breeding in this crop in near future (Gupta et al., 2013). Further, some miRNAs responsible for the regulation of withanolide biosynthesis have also been identified in both root and leaf transcriptomes (Srivastava et al., 2018), which might be helpful in understanding the tight regulatory mechanism of the genes of this pathway.

In order to understand the tissue-specific patterns of withanolide biosynthesis, the transcriptome sequencing was performed in root and leaf tissue synthesizing withanolide A and withaferin A, respectively (Gupta et al., 2015). The sequence data were assembled and, respectively, 89,548 and 1,14,814 unique sequences were identified in leaves and roots of which 47,885 and 54,123 were annotated from leaf and root tissue, respectively, using various databases. Gene Ontology and KEGG analyses were performed to get the detailed information of the enzymes involved in withanolide backbone synthesis. The members of cytochrome P450, glycosyltransferase, and methyltransferase gene families were identified which showed the differential expression in the leaf and root tissue, which might be involved in synthesis of tissue-specific withanolides in Ashwagandha. Another study was made to identify the genes involved in the biosynthesis of specific withanolides in root and leaf tissues of distinct chemotypes of *W. somnifera* and the complete transcriptome analysis of leaf and root tissue was performed. A total of 47,885 and 54,123 unigene sequences were generated from leaf and root tissues, respectively, and have been annotated using the available databases. Based on the sequence homology, these unigenes were categorized into 45 functional groups. On the basis of the annotation, the genes encoding enzymes involved in biosynthesis of triterpenoid backbone (including both MVA and MEP pathways) with their alternatively spliced forms and paralogous were identified. Apart from these, a number of methyltransferases, cytochrome P450s, glycosyltransferase, and transcription factors have also been identified which might be involved in tissue-specific and chemotype-specific diversity in Ashwagandha (Gupta et al., 2015). Recently, the transcriptome sequencing of *W. somnifera* berries was also performed in order to assemble, identify, and annotate the tissue-specific transcripts. Transcripts of nearly all the genes related to the withanolide biosynthetic pathway were obtained. Tissue-wide comparative gene expression analysis suggested almost similar level of transcripts of the genes involved in terpenoid pathway in leaf, root, and berry tissues; however, the transcripts of steroid, phenylpropanoid metabolism, and flavonoid metabolism were relatively more abundant in berries (Tripathi et al., 2020). Attempt was also made to evaluate if the in vitro–raised organ cultures (leaf or adventitious roots) can be a promising source for targeted biosynthesis of bioactive principle, i.e., withanolide using transcriptome sequencing approach. A total of 177,156 transcripts were assembled and it was found that no unique transcripts was there in the in vitro–raised leaves; however, there were about 13% of the transcripts which were unique to in vitro–raised adventitious roots. Furthermore, it was also recorded that the accumulation pattern of withaferin A and withanolide A varied with the tissue type and the duration of culture period (Senthil et al., 2015). In an attempt to analyze the salicylic acid–induced leaf transcriptome of *W. somnifera*, 73,523 unigenes were identified out of which 71,062 unigenes were annotated and 53,424 of them were assigned GO (gene ontology) terms, using available databases. Mapping of the unigenes identified were involved in 182 different biological pathways (Dasgupta et al., 2014).

Although there have been several attempts to study the set of genes involved in the production of bioactives, however, only a meagre information is available on their regulation. Therefore, in order to investigate the miRNA transcriptome responsible for the regulation of withanolide biosynthetic pathway, both in vitro root and the leaf tissue were used. In the root and the leaf tissues, a total of 24 and 39 miRNA families were identified, respectively; out of which 15 and 27 miRNA families were involved in the biological functions of the respective organs. Mainly the endogenous root-miR5140, root-miR159, leaf-miR477, and leaf-miR530 were found to be involved in the regulation of withanolide biosynthetic pathway (Srivastava et al., 2018).

15.7.3 Genomic studies: genetic diversity and beyond

The cytogenetic studies suggest that the chromosome number of *W. somnifera* is variously described as 2n = 24 and 2n = 48 (Ram et al., 2012; Samaddar et al., 2012; Singhal and Kumar, 2008). However, it was suggested that *W. somnifera* might be possessing allopolyploid genome with basic number of x = 12 and there could be more than one cytotypes in natural population (Ram et al., 2012; Sharma, 2013b).

There have been several attempts to study the genetic variability in *W. somnifera* using different kinds of genetic marker in various populations from various geographical regions. The intraspecific variability in the ITS (internal-transcribed spacer region of rDNA) sequences was recorded in Ashwagandha (Mir et al., 2010). This variability in the sequence of ITS regions was also successfully utilized to identify the wild and cultivated genotypes. The amplified PCR product using ITS1 and ITS4 primers, a single 710 bp amplified product was found in the wild genotypes, whereas two separate amplified products of 709 and 552 bp were recorded in the cultivated genotypes. Further a discrete PCR-RFLP fingerprint (using four restriction endonucleases viz., EcoRV, Hinf I, Rfa I, and Hae III to digest amplified PCR products of respective ITS regions) was found in wild versus cultivated genotypes. In another study, 18 RAPD (Random Amplified Polymorphic DNA) primers and 6 AFLP (Amplified Fragment Length Polymorphism) primer combinations were used to study the genetic diversity among 23 accessions which showed 37.82% and 43.94% polymorphism, respectively. AFLP revealed higher levels of polymorphism as compared to the RAPD. The data indicated the existence of two distinct clusters and a clear distinction of cultivated and wild accessions was recorded (Mir et al., 2011). Further, it was recorded that the molecular marker data show a high degree of correlation with the morphological distinctness of various accessions.

The genetic variability was also recorded among the different accessions of Ashwagandha collected from the same geographical region. A very high degree of polymorphism (83.78%) was recorded in the accessions collected from different parts of Tamil Nadu, using RAPD primers (Dharmar and De Britto, 2011). In an attempt to study both the inter- and intra-specific genetic variability in 35 genotypes of *W. somnifera* and five genotypes of a related species *W. coagulans,* using AFLP markers. The similarity matrix value based on Jaccard's coefficient was used to construct a dendrogram showing the genetic relationships, that) distinguished *W. somnifera* from *W. coagulans* in the form of two major clusters. Further, it was found that *W. somnifera* cluster can be subdivided into three subclasses corresponding to Kashmiri and Nagori and an intermediate type (Negi et al., 2000). In a comparative study, SAMPL (Selectively Amplified Microsatellite Polymorphic Loci) assay and AFLP were used to assess the levels of genetic diversity in various Ashwagandha genotypes. SAMPL assay revealed higher degree of polymorphism in comparison to AFLP. The cluster analysis based on SAMPL revealed that there was a higher level of polymorphism among Kashmiri and Nagori genotypes. The divergence of these two genotypes based on their polymorphism clearly indicated a distinct grouping of Kashmiri and Nagori genotypes (Negi et al., 2006). In a study using 12 different genotypes of Ashwagandha using 22 Inter Simple Sequence Repeat (ISSR) markers, it was suggested that there is a wide variation (0.02–0.8333) among the selected genotypes (Bamhania et al., 2013). There are several studies available in literature, which described the utilization of the dominant and random RAPD along with the ISSR marker in different *W. somnifera* to study the genetic variability (Khan and Shah, 2016; Niraj et al., 2011; Tripathi et al., 2012). The analysis of different accessions using dominant AFLP markers with genome-wide coverage and the analysis of the major bioactive constituents suggested a high degree of correlation (Dhar et al., 2006). The genetic diversity in Ashwagandha have also been assessed by using the co-dominant EST-SSR markers (Parmar et al., 2015; Parita et al., 2018).

Recently, 154,386 bp long chloroplast genome of *W. somnifera* was sequenced (Mehmood et al., 2020), that was composed up of one small (18,464 bp long) and one large (85,688 bp long) single copy region separated by a pair of large inverted repeats (25,117 bp long). This chloroplast genome was having 132 genes and was found to be similar in some genomic features (viz., structure, nucleotide content, codon usage, RNA editing sites, SSRs, oligonucleotide repeats, and tandem repeats) with the chloroplast genomes of four other species of the family Solanaceae. In this study 147 and 229 SSRs were identified in the protein-coding regions and non–protein-coding region, respectively.

15.7.4 Metabolomics studies

Metabolomics has become an important tool in the field of pharmacology and phytochemistry of medicinal plants in the past two decades and provided the holistic information about the metabolic profile of several medicinal plants which helped in the quality control of the plant material. The information about the full metabolic profile (not the few predominant bioactives only) and their differential distribution in different plant organs, or during various developmental stages, can also be very helpful in providing a relatively better and clearer picture of the key bioactives as well as their available concentrations in herbal formulations. Further, metabolomic analysis will not only speed up the identification of different chemotypes and the distribution the bioactives (and their variants) in them, but will also help in devicing suitable statergies for enhanced production of bio-actives by modulating the biosynthetic pathways.

Significant qualitative and quantitative differences have been found in the metabolome of both the leaf and root tissue of *W. somnifera*. In an experiment to study metabolic profile of the crude extracts of Ashwagandha leaf and root using NMR and other chromatographic techniques, a total of 62 major and minor primary and secondary metabolites from leaves and 48 from roots have been identified; however, 29 of them were found to be common to both (Chatterjee et al., 2010). In another study when three different organs (leaf, stem, and roots) were analyzed for their metabolic profile using NMR spectroscopy, the leaf was found to exhibit a wide range of metabolites than the other two organs. Further, two major types of withanolides (group I containing 4-OH and 5,6-epoxy groups, withaferin A-like steroids, and group II containing 5-OH and 6,7-epoxy groups (withanolide A-like)), and their ratio was found to be a marker for discriminating leaf samples from the *W. somnifera* plants growing in different areas (Namdeo et al., 2011). In an attempt to study the pattern of the accumulation of three important bioactives viz., withaferin A, 12-deoxywithastramonolide, and withanolide A, using LC-ESI-MS-MS (liquid chromatography—electrospray ionization—tandem mass spectrometry) in four different organs (root, stem, fruits, and leaves) of Ashwagandha, it was found that withaferin A was highest in the leaves, whereas 12-deoxywithastramonolide and withanolide A were highest in the roots (Gajbhiye et al., 2015).

Besides the difference in the metabolic profile of different organs, the difference among the different chemotypes of *W. somnifera* was analyzed using HR-MAS NMR spectroscopy (Bharti et al., 2011). Further, when the withanolide profiles of different *W. somnifera* accessions were examined by scoring as presence and absence of major withanolides and were processed for UPGMA-based dendrogram construction, a high level of phytochemical diversity was recorded in various accessions (as discrete chemotypes). The different accessions were clustered together with respect to their characteristic profile of major withanolides and represented as withaferin A, withanone, withanolide D, or withanolide A rich groups, or the accessions lacking a specific withanolide like withanone minus chemotypes and withaferin A minus chemotypes. Some accessions were found to be rich in 17-hydroxy withaferin A too (Chaurasiya et al., 2009).

Moreover, the differential alterations in profile of primary and secondary metabolites in Ashwagandha fruits at seven different stages of development have also been recorded using both one- and two-dimensional NMR spectroscopy. The fruits from 1 week after fertilization until maturity were classified in seven stages; and a qualitative as well as quantitative analysis of metabolites were made in order to find out the critical stage for harvesting the fruits for obtaining significant amount of the desired bioactive ingredients with the required pharmacological activity. During the early stages of fruit development, a relatively higher concentrations of alanine, aspartate, caffeic acid, choline, phosphocholine, sucrose, and withanolides were observed, whereas during the maturation phase, accumulation of citrate and withanamides was recorded (Sidhu et al., 2011). The study, therefore, suggested a higher metabolic activity during the initial stage of fruit development and a metabolic rerouting in the later stages.

15.8 Crop improvement interventions

Various efforts have been made in this crop to collect, evaluate, and characterize the germplasm and utilize the interspecific variability in view of the crop improvement. The hybridization among different chemotypes has also been tried. Recently, many biotechnological interventions like in vitro manipulations, elicitation (both in vitro and in vivo), and the genetic transformation have also been tried for the enhanced production of bioactives. All these attempts have been discussed below.

15.8.1 Breeding efforts

Furthermore, there have been a few attempts to improve the crop using conventional breeding approaches like development of hybrids, etc. One major step in the classical breeding is characterization/screening of germplasm for the desired characters. In this regard, several attempts have been made to study the chemotypic variability in the germplasm (discussed in Section 15.6.1) from different parts of the world. In India, the systematic germplasm collection efforts for Ashwagandha are in place since inception of the AICRPMAP (All India Coordinated Project on Medicinal and Aromatic Plants) of ICAR (Indian Council of Agricultural Research) in 1969 and a total of 665 accessions of this crop are maintained at different AICRPMAP centers and its headquarters at DMAPR (Directorate of Medicinal and Aromatic Plants Research) (Venugopal et al., 2018). The hybrid between Israel chemotypes and an Indian chemotype of *W. somnifera*, new substituted withanolides, was found in the hybrid plants (Bessalle et al., 1987; Kirson et al., 1977; Nittala and Lavie, 1981). Similarly, three new withanolides were recorded from the hybrids between a South African and an Israeli chemotype (Eastwood et al., 1980). However, Kumar et al. (2011a) found the problems like very low seed set and a poor percentage germination was recorded in the hybrids. Venugopal et al. (2018) have discussed the major breeding objectives of *W. somnifera* viz., development of dwarf, high root yielding plants, uniform crop canopy, nonspreading plant architecture, high withanolide content per unit biomass, resistance toward major pests and pathogens, resistance toward abiotic stresses, increasing starch–fiber ratio, and early maturity.

15.8.2 Biotechnological interventions

One of the major objectives for the biotechnological interventions employed for improvement of *W. somnifera* is the increased production of bioactives (like withanolides, etc.), that can be achieved by a number of in vitro and in vivo manipulations listed in Fig. 15.3 and discussed in detail under the subsequent sections.

15.8.2.1 In vitro propagation of Ashwagandha

Because of the problems associated with the crop (discussed in Section 15.5), like low percentage of seed germination and seed viability, attack by various pests/pathogens, and availability of different chemotypic variants in the crop, the plant tissue culture (PTC) in an important alternative, that can benefit the herbal industry by means of a constant supply of the consistent chemotype, expected to meet with the increasing demand of Ashwagandha. Further, the utilization of in vitro–raised plantlets enhances the productivity per unit area. PTC is, therefore, an important tool for conserving and exploiting the commercially important medicinal plants like Ashwagandha. This plant grows well in Indo–Himalayan region and the conventional methods of harvesting may lead to the habitat destruction; thus, the in vitro methods have been tried extensively for its propagation and conservation. Further, various PTC-based approaches like tissue and organ cultures, cell suspension/callus cultures, somatic embryos, and somaclonal variations/hairy root cultures are considered good alternatives to the conventional methods for getting high value bioactives.

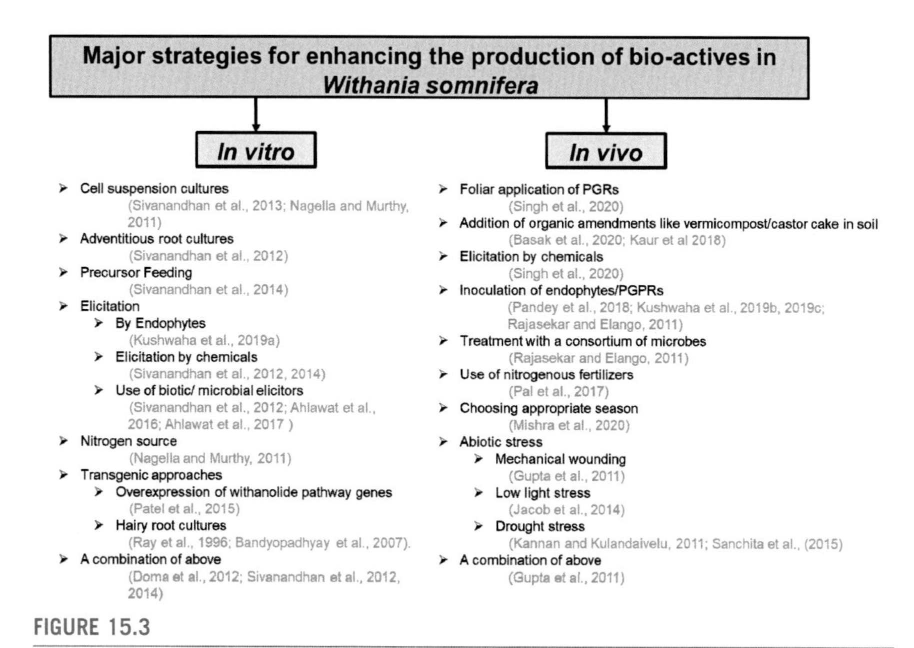

FIGURE 15.3

A flow chart describing the major strategies for enhancing the production of bioactives in *Withania somnifera*, both in vitro and in vivo.

A variety of explants such as seeds, cotyledonary nodes, nodal segments, and axillary buds taken from field-grown mature plants as well as from the in vitro–propagated plantlets have been tested for mass multiplication of *W. somnifera*. Out of tested explants, enhanced caulogenesis have been observed in cultures initiated from nodal segments or shoot tip. Almost all type of basal tissue culture media starting from MS to Gamborg B5, Nitsch medium, WPM (woody plant medium), SH, and their different modifications have been tested by various researchers for the in vitro culture establishment. However, the optimum bud break, shoot proliferation, and maturation have been observed in MS medium (Shasmita and Naik, 2017). In most of the studies conducted in *W. somnifera*, sucrose at a concentration of 3% has given better results for shoot multiplication than the other carbon sources.

The *in vitro* propagation of *W. somnifera* has been reported by many researchers using different explants viz., shoot tips and axillary buds (Autade et al., 2016; Baba et al., 2013; Rani et al., 2014), hypocotyl, cotyledon (Kumar et al., 2013), seed (Supe et al., 2006), cotyledonary leaf segments (Rani et al., 2003a), callus from leaves (Arumugam and Gopinath 2013), and the nodal areas (Kumar et al., 2011b).

Utilization of the axillary shoot buds for multiplication in the micropropagation serves as a safer strategy to reduce the chances of somaclonal variations and confirming the clonal fidelity of regenerated plants. The vital role played by PGRs alone or in combination like cytokinins for breaking the dormancy of axillary buds, and initiating caulogenesis, as well as synergistic combinations of auxins and cytokinins for multiplication, elongation, and proliferation of shoots has been studied and are well established in *W. somnifera*. The high frequency of shoot bud induction was observed using MS medium fortified with BAP (6-benzene aminopurine). Further, the synergistic effect of BAP and NAA (1-nephtalene acetic acid) in the refinement of shoot regeneration efficiency has been obtained in *W. somnifera* (Fatima et al., 2015). The cytokinins viz., BAP, kinetin, 2-iP (iso-pentyl adenine), and TDZ (thidiazuron) alone or in combination with the auxins IAA (indole-3-acetic acid), IBA (indole-3-butyric acid), NAA, and 2, 4-D (2, 4-dichlorophenoxy acetic acid) have been tried for the varied responses in its micropropagation. BAP alone or in combination with other cytokinin or auxins in lower proportions have been found optimum in shoot multiplication from the meristems (Shasmita and Naik, 2017). In a study on the micropropagation of *W. somnifera* involving the use of thidiazuron has resulted in proliferation of shoots but it also established the adverse effect of its longer exposure resulting in distorted shoots, reduced elongation, and even death of primary shoots (Fatima and Anis, 2011). Apart from the use of the above PGRs, which have also been used in various other medicinal plants, the effect of the PGRs that are used relatively lesser frequencies for the micropropagation has been tested as an additive in shoot proliferation and has been resulted in shoot elongation (Sangwan et al., 2008; Sivanesan and Murugesan, 2008). Addition of the other growth factors viz., polyamines, reduced nitrogen sources like amino acids, and coconut milk was reported to have a profound effect on the in vitro caulogenesis (Ray and Jha, 2001; Sivanandhan et al., 2011, 2015). The one-step elongation and rooting has also been established in *W. somnifera* (Kulkarni et al., 2000), by reducing the concentration of sucrose or the PGRs.

Moreover, the effect of type of culture vessel and the sealing material of culture vessel on organogenesis and regeneration potential of different explants has also been evaluated and found to be of utmost importance, for an in vitro regeneration protocol. The glass culture tubes were found to be more efficient in direct organogenesis in comparison to the plastic petri plates, which resulted only in the callus formation. This differential response may be because of the accumulation of ethylene (a PGR) in the glass tubes that could have induced the shoots. Cotton plugs in comparison to parafilm sealing were recorded better sealing agent for the regeneration, as they can allow the free gas exchange (Kulkarni et al., 1996). Indirect organogenesis via callus formation is also a good means of mass multiplication, somatic embryogenesis, and has been well established in *W. somnifera* using the nodal segment, leaf explant, in vitro−grown seedlings, internode, and hypocotyl and epicotyl explants. 2,4-D has been used for the callus initiation either alone or in combination with BAP or kinetin in MS basal medium and after callus maturation shoot induction has been done using BAP alone or in combination with auxins (Manickam et al., 2000; Rani et al., 2003b; Udayakumar et al., 2013b). The suitability of leaf explant for inducing embryogenic callus over internodal segments has been established by Rani et al. (2004) and Sharma et al. (2010). The in vitro rhizogenesis of microshoots has also been a deciding factor in establishing a successful micropropagation protocol and extensively studied in *W. somnifera.* Auxins alone or in combination with cytokinins are having the morphogenic effect in rooting the microshoots, although the cytokinins like BAP have also been found to initiate rooting in *W. somnifera* (Kulkarni et al., 1996, 2000). Various additives like polyamines, adenine sulfate, and ammonium nitrate have the varied effect on in vitro rooting. Polyamines, like putrescine and the additives adenine sulfate and ammonium nitrate, promoted rooting when included in medium. In *W. somnifera* the cost and time of tissue culture protocol have been reduced considerably by introducing ex vitro rooting of microshoots, thereby omitting the in vitro rhizogenesis and acclimatization (Fatima and Anis, 2011).

The tissue culture–raised shoot propagules and the nodal segment of *W. somnifera* have been successfully encapsulated in alginate beads (sodium alginate and calcium chloride based) for the production of synthetic seeds (Fatima et al., 2013; Singh et al., 2006). Encapsulated somatic embryos or vegetative propagules including shoot apices, axillary buds, or nodes, etc., based synthetic seeds may also be used in germplasm conservation of the different chemotypes as well as for exchange of axenic plant material between the laboratories.

Besides the above concise compilation of the efforts made in PTC of Ashwagandha, such as micropropagation, organogenesis, and somatic embryogenesis etc., the detailed reviews on advancements in PTC-based techniques developed in Ashwagandha are also available elsewhere (Singh et al., 2017b; Pandey et al., 2017). However, the attempts made to enhance the productions of bioactives (mainly withanolides) using various PTC-based methods and strategies to enhance their in vitro production are discussed in more details in the subsequent section.

15.8.2.2 In vitro production of bioactives

With the increasing demand for bioactive constituents like withanolides, attempts have been made recently to enhance their in vitro production, which could be a cost-effective and time saving method to produce the adequate amounts. An important approach for the enhancement of bioactives is the in vitro elicitation by some chemical compounds, microbial cultures, and/or microbe free culture filtrates in tissue culture. However, in order to successfully commercialize the in vitro production of plant secondary metabolites are identification of appropriate elicitor, optimization of the concentration of the elicitors, and scaling up of the process for the commercialization of the production of bioactive metabolites at industrial level. Further, some native fungal endophytes (isolated from *W. somnifera* plant, itself) have also been recorded to enhance the *in vitro* production of bioactives. The importance of endophytes is more than the other microbial elicitors since they can be used as effective bioinoculants in the field conditions to increase the production of bioactives, alone or in combination with other biocontrol agents or biofertilizers. In this section, an attempt has been made to discuss the recent advances in this field.

In an experiment to evaluate the elicitation potentiality of a rhizosphere fungi, *Piriformospora indica* with known secondary metabolite–enhancing activities in some other medicinal plants was tried (Ahlawat et al., 2016). In this experiment, when a range of concentrations of the fungal cell homogenate, culture filtrate, and the individual fungal culture discs were added into the cell suspension and callus cultures at different time intervals, it was recorded that the maximum enhancement in withaferin A production was achieved with 3% fungal homogenate, followed by 3% culture filtrate and culture disc, respectively. Further, a concurrent increase in the expression of a number of genes involved in withanolide pathway was recorded, among which the fold increase was highest in *hmgr* gene, considered to be coding for a rate-limiting enzyme HMGR (3-hydroxy-3-methylglutaryl coenzyme A reductase) in the withanolide pathway. Further, when the elicitation potential of the cell homogenates of *P. indica* was compared with that of three other fungi (viz., *A. alternata*, *Fusarium solani*, and *Verticillium dahlia*) in the cell suspension culture of *W. somnifera*, it was found that the 3% *P. indica* cell homogenate was most effective followed by 5% *V. dahliae*, 3% *A. alternata*, and 3% *F. solani* cell homogenate, respectively (Ahlawat et al., 2017). Further, the scaling up of the process was optimized in the study from a shake flask to a lab-scale bioreactor level, thereby leading to the increased production of the three key withanolides (viz., withanolide A, withaferin A, and withanone).

A native endophytic fungus (isolated from *W. somnifera* leaves), *Aspergillus terreus* strain 2aWF, was also found effective in eliciting the production of withanolide A in root cell suspension cultures of *W. somnifera*, when applied as 1% mycelial extract or as 5% culture filtrate with a concurrent upregulation of key genes of the withanolide pathway (Kushwaha et al., 2019a). The culture filtrate was found to be more effective in early and effective elicitation than the mycelial extract. Using the culture filtrate significantly elicited the withanolide A within 6 h, whereas the mycelial extract elicited the maximum withanolide A after 24 h.

When a biotic (chitosan) and an abiotic (aluminum chloride) elicitation was evaluated for their potential to enhance withanolide production in adventitious root cultures, it was found that their responces are variable in magnitude. Under the similar culture conditions, the biotic elicitor (100 mg/L) stimulated higher production of all withanolides in comparison to the abiotic elicitor (Sivanandhan et al., 2012). Further, the nitrogen source plays a critical role in the accumulation of withanolides in vitro. It was found that withanolide A production was maximum when the NH_4^+/NO_3^- ratio was 14.38/37.60 mM (Nagella and Murthy, 2011).

The higher concentrations of total withanolides were obtained in shake-flask culture as well as in the bioreactor using precursor feeding and elicitation, than the control treatments. Maximum withanolides content was recorded in the combined treatment of chitosan (an elicitor) and squalene (withanolide precursor) in the bioreactor (Sivanandhan et al., 2014).

15.8.2.3 Genetic transformation

Genetically transformed plants, plant cells, or the organ cultures are attractive systems to study the production of bioactives and also to get their enhanced production. Several attempts have been made for the genetic transformation for the crop improvement/improved production of bioactives in Ashwagandha. In the recent decades, the hairy root system based on *Agrobacterium rhizogenes* inoculation has been successfully used as an effective method of producing secondary metabolites. The transformed root cultures of Ashwagandha using wild-type *A. rhizogenes* strain LBA 9402 were able to grow axenically in vitro, without exogenous PGRs, which were able to synthesize the bioactives. The withanolide D production in these transformed root cultures was higher than the untransformed cultures (Ray et al., 1996). However, withaferin A was not detected in these transformed cultures, which is usually present in the field-grown plants.

Kumar et al. (2005) also reported that the transformed hairy roots induced by infecting the leaf explants of *W. somnifera* using a wild-type *A. rhizogenes* strain showed the presence of withanolides and the enhancement of the secondary metabolites. This is the first detection of withaferin A in hairy root cultures and was first reported by Bandyopadhyay et al. (2007). The accumulation of withaferin A was recorded in rooty callus line obtained by the transformation using *A. rhizogenes* strains LBA 9402 and A_4 with the maximum accumulation of withaferin A content in the transformed root line WSKHRL-1. Murthy et al. (2008) recorded faster growth, increased biomass, and a 2.7-fold increase of withanolide A content in *A. rhizogenes* strain R1601 transformed roots than the nontransformed cultured roots. Further, the differential performance of different *A. rhizogenes* strains and nature of explant has also been reported for hairy root induction in *W. somnifera* (Saravanakumar et al., 2012). It was also recorded that the addition of acetosyringone is effective in enhancing the transformation frequency. In another experiment, different elicitors were tried to induce the production of bioactives in *W. somnifera*, and it was recorded that both chitosan and nitric oxide increased withaferin A content, acetyl salicylic acid stimulated the accumulation of both withaferin A and withanolide A, and the elicitation by triadimefon highly increased withaferin A content (Doma et al., 2012). The role of heat treatment and sonication in enhancing the rate of *A. rhizogenes*–mediated transformation of leaf segment explant was also reported (Thilip et al., 2015, 2020). In an experiment to try various elicitors for enhancing withaferin A production in hairy root cultures, Thilip et al. (2019) recorded that the elicitation by chitosan (100 mg/L) was most useful and the production was increased by 4.03-folds than the control.

There have been several attempts to optimize the *A. tumefaciens*–mediated methods of genetic transformation, knowledge gain from which can now be utilized in successful metabolic engineering efforts as well as in the functional genomics studies in this medicinal plant. In one of the initial attempts to develop the transformed organ cultures of Ashwagandha using different wild-type strains of *A. tumefaciens*, it was found that some of the galls obtained following infection of *A. tumefaciens* strain N2/73 were able to spontaneously develop shooty teratomas that were grown in unsupplemented basal medium. Further they are able to synthesize more amount of two major native bioactives, i.e., withaferin A and withanolide D, than the untransformed shoot cultures (Ray and Jha, 1999). Transformation using *A. tumefaciens* strain LBA4404 having binary vector pIG121Hm was successfully used by Pandey et al. (2010), the gusA reporter gene with intron under the transcriptional control of the CaMV (Cauliflower Mosaic Virus) 35S promoter. It was recorded that the leaf segments from two-and-a-half-month-old greenhouse–grown seedlings were more efficient in transformation than the in vitro–grown shoots and the second expanded leaf from the shoot tip had the highest efficiency toward transient transformation.

For a reproducible and efficient transformation protocol development for a particular plant, several other factors also require optimization. One of the initial efforts in *W. somnifera* was made by Ishnava et al. (2012), to optimize the infection treatment by *A. tumifaciens*, time duration, PGR treatment, and sucrose concentration with an aim to increase production of bioactives. Further, it was recorded that out of three approaches tried for the development of an efficient transformation system in *W. somnifera*, the microprojectile bombardment–assisted agro-infection was found better than *A. tumefaciens*–mediated transformation and microprojectile bombardment alone (Patel et al., 2014). Transgenic *W. somnifera* plants generated by *A. tumefaciens*–mediated genetic transformation that were overexpressing *WsSQS* (one of the withanolide pathway genes) led to a 1.5- to 2-fold increase in the total withanolide (Patel et al., 2015). Genetic transformation is also an important tool for the function genomics studies. The transgenic lines containing the constructs (in RNAi vector pGSA backbone) which impairs the normal functioning of a *WsCAS* (*W. somnifera* cycloartenol synthase gene) suppressed the withanolide biosynthesis in *W. somnifera* but an enhanced level of withanolide production was recorded in the *CAS* overexpressing transgenic lines. This suggested a key position of CAS in the withanolide biosynthetic pathway (Mishra et al., 2016). Similar gene silencing studies using RNAi were tried to find the possible roles of *SMT1*(sterol-C24-methyltransferase type 1) and *SGTL1* (a sterol glycosyltransferases from class 1) (Pal et al., 2019; Saema et al., 2015).

15.8.2.4 Enhanced production of bioactives in vivo

There are a number of methods reported/suggested for the enhancement of the concentration of bioactives in the Ashwagandha plants in field conditions. They include the foliar application of PGRs, endophytes treatment, elicitor's treatment, imposition of stress conditions, photoperiod, nutritional status, age of the plant at the harvest, season of the harvest, etc.

It has been found that the drought stress induces the concentration of a bioactive, i.e., withaferin A in Ashwagandha (Kannan and Kulandaivelu, 2011). Sanchita et al. (2015) also recorded the increase in the production of different withanolides viz., the withaferin A, 12-withastromonolides, and withanolide A by 42.7%, 78%, and 71%, respectively, under drought stress. However, a decrease in the concentration of these bioactives has been recorded under biotic stress, i.e., leaf spot disease by *A. alternata* (Pati et al., 2008).

Singh et al. (2020) reported the increased production upon the foliar applications of PGRs (jasmonic acid and salicylic acid SA) and chitosan as the elicitors.

Endophytes are also effective in production of secondary metabolites under field conditions. Three native fungal viz., *A. terreus* strain 2aWF (2aWF), *Penicillium oxalicum* strain 5aWF (5aWF), and *Sarocladium kiliense* strain 10aWF (10aWF) were reported to enhance withanolides content in both leaves and roots and a corresponding upregulation of key withanolide biosynthetic pathway genes was also reported (Kushwaha et al., 2019b). When the compatibility of these inherent fungal endophytes was tested with the combined treatment with the biocontrol agent, *Trichoderma viride*, it was recorded that the coinoculation treatments increased the production of secondary metabolite, mainly withanolide A in both leaf and root tissues of *W. somnifera* (Kushwaha et al., 2019c). Treatment with microbial consortium consisting various PGPRs suggested that this treatment increases the withaferin A content of *W. somnifera* (Rajasekar and Elango, 2011).

The production of withanaloids is also influenced by photoperiod. It was recorded that under the long photoperiod-treated condition the concentration of withanolides was higher in Ashwagandha (Sharma and Puri, 2020). However, another study conducted to examine low light stress (25%, 50% and 75% shade along with the control, for 30 days) on withanolide production in *W. somnifera*, the highest withanolide accumulation was recorded under 75% (Jacob et al., 2014). Further, it was found that the expression of withanolide pathway gene *WsFPPS* is enhanced in the response to mechanical injury (Gupta et al., 2011). It has also been recently recorded that the seasonal variations may also influence the withanolide biosynthesis (Mishra et al., 2020); therefore, choosing the appropriate season for the sowing and harvesting of Ashwagandha is useful for the consistent quality of the herbal products thereof.

Moreover, the proper nutrient status of the plant can lead to the enhanced production of the bioactives. It was found that the application of a nitrogenous fertilizer, ammonium sulfate, improved the production of withaferin A (Pal et al., 2017). The use of organic amendments like caster cake and vermicompost along with microbial consortium was able to enhance the total withanolide content in Ashwagandha (Basak et al., 2020). Further, the leaves of Ashwagandha plants cultivated using vermicompost and vermicompost leachate showed the higher concentration of the withanolides viz., withanone, withanolide A, and withaferin A (Kaur et al., 2018). Therefore, the organic farming/adding organic amendments or the integrated nutrient management of this medicinal plant may be good choice to get the enhanced amount of bioactives in field-grown plants.

15.9 Conclusions and future perspectives

Recently, there has been a global shift in consumer behavior to go back to nature in search of the needs of mankind including primary healthcare. Therefore, an increasing number of efforts are being made to look at the potential of the medicinal herbs like Ashwagandha in primary healthcare system in a scientific and systematic way. A number of bioactives with a wide range of pharmaceutical properties have been identified from it. Due to its increasing demand, it has been gradually understood that exploiting the wild-grown plants of Ashwagandha in a way may lead to loss of its genetic variability and even it may lead to habitat destruction. Therefore, efforts were made by different researchers to collect the germplasm, evaluate its chemotypic and genotypic variability in the natural populations, and identify the elite genotopes/chemotypes and promote them for cultivation. Further, researchers have also came across various problems in the cultivation of Ashwagandha and identified some important pests and diseases limiting its cultivation to meet the market demand. However, the harvesting of roots of Ashwagandha from field is a labor-intensive process. Therefore, the in vitro approaches were also tried to meet the demand of bioactives via alternative methods like cell/organ culture (like hairy roots) and other in vitro methods. Further, these methods can be scaled up in bioreactors to meet the requirements of bioactives. Efforts have also been made to increase the production of bioactives through in vitro elicitation and/or precursor feeding. With the increased knowledge of pathway genes accumulate very recently and availability of ESTs and transcriptome data in public domain, more advancement in enhancing the production capacity of bioactives through methods like genetic transformation are also being adopted. Further, the increasing availability of sequence-based markers will help in marker-assisted selection of elite genotypes and will lead to commercialization of improved varieties. In this chapter, we tried to present the efforts made in the improvement of *W. somnifera* using cell culture–based methods with/without the integration of recent knowledge gained through various OMICS approaches (like genomics, transcriptomics, proteomics, and metabolomics) till date and to identify the knowledge gaps hampering the further improvement of the crop. A comprehensive knowledge in a broad holistic view of these diverse strategies will help in paving the way to identify these research gaps in the crop improvement and will motivate the workers to work on them.

Acknowledgments

Authors acknowledge the support from DAV University administration during the preparation of the manuscript. Further, the financial assistance from CSIR (Council for Scientific and Industrial Research), Government of India, is duly acknowledged.

References

Abe, I., Abe, T., Lou, W., Masuoka, T., Noguchi, H., 2007. Site-directed mutagenesis of conserved aromatic residues in rat squalene epoxidase. Biochem. Biophys. Res. Commun. 352 (1), 259–263.

Abraham, A., Kirson, I., Glotter, E., Lavie, D., 1968. A chemotaxonomic study of *Withania somnifera* (L.) dun. Phytochemistry 7 (6), 957–962.

Abraham, A., Kirson, I., Lavie, D., Glotte, E., 1975. The withanolides of *Withania somnifera* chemotypes I and II. Phytochemistry 14 (1), 189–194.

Ahlawat, S., Saxena, P., Ali, A., Abdin, M.Z., 2016. *Piriformospora indica* elicitation of withaferin A biosynthesis and biomass accumulation in cell suspension cultures of *Withania somnifera*. Symbiosis 69 (1), 37–46.

Ahlawat, S., Saxena, P., Ali, A., Khan, S., Abdin, M.Z., 2017. Comparative study of withanolide production and the related transcriptional responses of biosynthetic genes in fungi elicited cell suspension culture of *Withania somnifera* in shake flask and bioreactor. Plant Physiol. Biochem. 114, 19–28.

Akhoon, B.A., Pandey, S., Tiwari, S., Pandey, R., 2016. Withanolide A offers neuroprotection, ameliorates stress resistance and prolongs the life expectancy of *Caenorhabditis elegans*. Exp. Gerontol. 78, 47–56.

Akkuzu, E., Ayberk, H., Inac, S., 2007. Hawk moths (Lepidoptera: Sphingidae) of Turkey and their zoogeographical distribution. J. Environ. Biol. 28, 723–730.

Al-Musa, A.M., Lockhart, B., 1990. Occurrence of Eggplant mottled dwarf virus in Jordan. J. Phytopathol. 128, 283–287.

Alwadi, H.M., Baka, Z.A.M., 2001. Microorganisms associated with *Withania somnifera* leaves. Microbiol. Res. 156, 303–309.

Anjaneyulu, A.S.R., Rao, D.S., 1997. New withanolides from the roots of *Withania somnifera*. Indian J. Chem. Sect. B Org. Chem. Incl. Med. Chem. 36 (5), 424–433.

Arumugam, A., Gopinath, K., 2013. *In vitro* regeneration of an endangered medicinal plant *Withania somnifera* using four different explants. Plant Tissue Cult. Biotechnol. 23 (1), 79–85.

Aslam, S., Raja, N.I., Hussain, M., Iqbal, M., Ejaz, M., Ashfaq, D., Fatima, H., Shah, M.A., Abd-Ur-Rehman, Ehsan, M., 2017. Current status of *Withania somnifera* (L.) Dunal: an endangered medicinal plant from Himalaya. Am. J. Plant Sci. 8 (5), 1159–1169.

Attia, A.R., Awadallah, K.T., 2016. Parasitoids and predators of the *Citrus mealybug*, *Planococcus citri* (Risso)(Hemiptera: Pseudococcidae) infesting the herb, *Withania somnifera*, a new host plant in Egypt. Egypt. J. Biol. Pest Control 26 (2), 245.

Autade, R., Sarika, A.F., Amol, R.S., Sunil, S.G., Choudhary, R.S., Dighe, S.S., 2016. Micropropagation of ashwagandha (*Withania somnifera*). Biosci. Biotechnol. Res. Commun. 9 (1), 88–93.

Azim, K.F., Ahmed, S.R., Banik, A., Khan, M.M.R., Deb, A., Somana, S.R., 2020. Screening and druggability analysis of some plant metabolites against SARS-CoV-2: an integrative computational approach. Inform. Med. Unlocked 20, 100367.

Baba, I.A., Alia, A., Saxena, R.C., Itoo, A., Kumar, S., Ahmed, M., 2013. *In vitro* propagation of *Withania somnifera* (L.) Dunal (Ashwagandha) an endangered medicinal plant. Int. J. Pharm. Sci. Inv. 3 (6), 349–355.

Baghel, G., Jahan, T., Afreen, B., Naqvi, Q.A., Snehi, S.K., Raj, S.K., 2010. Detection of a begomovirus associated with yellow mosaic disease of Ashwagandha (*Withania somnifera* L.) and its impact on biomass yield. Med. Plants - Int. J. Phytomed. Related Ind. 2, 219–223.

Baghel, G., Naqvi, Q.A., Snehi, S.K., Khan, M.S., Raj, S.K., 2012. Molecular identification of three isolates of *Jatropha* mosaic India virus associated with mosaic disease of *Withania somnifera* in India. Arch. Phytopathol. Plant Protect. 45, 2114–2119.

Bähr, V., Hänsel, R., 1982. Immunomodulating properties of 5, 20α (R)-Dihydroxy-6α, 7α-epoxy-1-oxo-(5α)-witha-2, 24-dienolide and solasodine. Planta Med. 44 (01), 32–33.

Baitharu, I., Jain, V., Deep, S.N., Hota, K.B., Hota, S.K., Prasad, D., Ilavazhagan, G., 2013. *Withania somnifera* root extract ameliorates hypobaric hypoxia induced memory impairment in rats. J. Ethnopharmacol. 145 (2), 431–441.

Bamhania, K., Khatakar, S., Punia, A., Yadav, O.P., 2013. Genetic variability analysis using ISSR markers in *Withania somnifera* L. Dunal genotypes from different regions. J. Herbs, Spices, Med. Plants 19, 22–32.

Bandyopadhyay, M., Jha, S., Tepfer, D., 2007. Changes in morphological phenotypes and withanolide composition of Ri-transformed roots of *Withania somnifera*. Plant Cell Rep. 26 (5), 599–609.

Basak, B.B., Saha, A., Gajbhiye, N.A., Manivel, P., 2020. Potential of organic nutrient sources for improving yield and bioactive principle of ashwagandha (*Withania somnifera*) through enhanced soil fertility and biological functions. Commun. Soil Sci. Plant Anal. 51 (6), 779–793.

Bessalle, R., Lavie, D., Frolow, F., 1987. Withanolide Y, a withanolide from a hybrid of *Withania somnifera*. Phytochemistry 26 (6), 1797–1800.

Bhagat, K.C., 2004. Mango mealy bug, *Drosicha mangiferae* (Green) (Margarodidae: Hemiptera) on Ashwagandha - a medicinal plant. Insect Environ. 10, 14.

Bharti, S.K., Bhatia, A., Tewari, S.K., Sidhu, O.P., Roy, R., 2011. Application of HR-MAS NMR spectroscopy for studying chemotype variations of *Withania somnifera* (L.) Dunal. Magn. Reson. Chem. 49 (10), 659–667.

Bharti, V.K., Malik, J.K., Gupta, R.C., 2016. Chapter 52–Ashwagandha: multiple health benefits. In: Nutraceuticals Efficacy, Safety and Toxicity, pp. 717–733.

Bhatt, V.P., Negi, G.C.S., 2006. Ethnomedicinal plant resources of Jaunsari tribe of Garhwal Himalaya, Uttaranchal. Indian J. Tradit. Know. 5 (3), 331–335.

Bhattacharya, S.K., Satyan, K.S., Ghosal, S., 1997. Antioxidant activity of glycowithanolides from *Withania somnifera*. Indian J. Exp. Biol. 35, 236–239.

Bhatti, D.S., Gupta, D.C., Dahiya, R.S., Malham, I., 1974. Additional hosts of the root-knot nematode, *Meloidogyne javanica*. Curr. Sci. 43, 622–623.

Chandra, R., 2004. Status of medicinal plants with respect to infestation of insect pests in and around Chitrakoot, District-Satna (M.P.). Flora Fauna 10, 88–92.

Chatterjee, S., Srivastava, S., Khalid, A., Singh, N., Sangwan, R.S., Sidhu, O.P., Roy, R., Khetrapal, C.L., Tuli, R., 2010. Comprehensive metabolic fingerprinting of *Withania somnifera* leaf and root extracts. Phytochemistry 71 (10), 1085–1094.

Chaudhari, K., Murthy, A.R.V., 2014. Effect of rasayana on mental health-a review study. Int. J. Ayurveda Alternat. Med. 2 (3), 1–7.

Chauhan, S., Joshi, A., Jain, R., Jain, D., 2019. Estimation of withanolide A in diverse genotypes of ashwagandha *Withania somnifera* (L.) dunal. Indian J. Exp. Biol. 57, 212–217.

Chaurasiya, N.D., Gupta, V.K., Sangwan, R.S., 2007. Leaf ontogenic phase-related dynamics of withaferin A and withanone biogenesis in Ashwagandha (*Withania somnifera* Dunal.) - an important medicinal herb. J. Plant Biol. 50, 508–513.

Chaurasiya, N.D., Sangwan, N.S., Sabir, F., Misra, L., Sangwan, R.S., 2012. Withanolide biosynthesis recruits both mevalonate and DOXP pathways of isoprenogenesis in Ashwagandha *Withania somnifera* L. (Dunal). Plant Cell Rep. 31, 1889–1897.

Chaurasiya, N.D., Sangwan, R.S., Misra, L.N., Tuli, R., Sangwan, N.S., 2009. Metabolic clustering of a core collection of Indian ginseng *Withania somnifera* Dunal through DNA, isoenzyme, polypeptide and withanolide profile diversity. Fitoterapia 80, 496–505.

Chavan, S.P., Korekar, S.L., 2011. A survey of some medicinal plants for fungal diseases from Osmanabad district of Maharashtra state. Recent Res. Sci. Technol. 3, 15–16.

Choudhary, B., Shetty, A., Langade, D.G., 2015. Efficacy of Ashwagandha (*Withania somnifera* [L.] Dunal) in improving cardiorespiratory endurance in healthy athletic adults. Ayu 36 (1), 63.

Dar, A.K., ul Hassan, W., Lone, A.H., Haji, A., Manzoor, N., Mir, A.I., 2017. Study to assess high demand and high commercial value medicinal plants of Jammu and Kashmir India-with special focus on routes of procurement and identification. Int. J. Res. Dev. Pharm. Life Sci. 6, 2576–2585.

Dar, P.A., Singh, L.R., Kamal, M.A., Dar, T.A., 2016. Unique medicinal properties of *Withania somnifera*: phytochemical constituents and protein component. Curr. Pharmaceut. Des. 22 (5), 535–540.

Dasgupta, M.G., George, B.S., Bhatia, A., Sidhu, O.P., 2014. Characterization of *Withania somnifera* leaf transcriptome and expression analysis of pathogenesis-related genes during salicylic acid signaling. PLoS One 9 (4) e94803-e94803.

Davis, L., Kuttan, G., 2000. Immunomodulatory activity of *Withania somnifera*. J. Ethnopharmacol. 71 (1–2), 193–200.

Dawa, S., Hussain, A., Tamchos, T., 2013. Medicinal and aromatic plants as a viable livelihood option for the people of Jammu and Kashmir. Indian Hortic. J. 3 (1 and 2), 40–42.

Deb, D.B., 1980. Enumeration, synonymy and distribution of the Solanaceae in India. J. Econ. Taxon. Bot. 1, 33–54.

Dhalla, N.S., Sastry, M.S., Malhotra, C.L., 1961. Chemical studies of the leaves of *Withania somnifera*. J. Pharmaceut. Sci. 50 (10), 876–877.

Dhar, N., Rana, S., Bhat, W.W., Razdan, S., Pandith, S.A., Khan, S., Dutt, P., Dhar, R.S., Vaishnavi, S., Vishwakarma, R., 2013. Dynamics of withanolide biosynthesis in relation to temporal expression pattern of metabolic genes in *Withania somnifera* (L.) Dunal: a comparative study in two morpho-chemovariants. Mol. Biol. Rep. 40, 7007–7016.

Dhar, N., Razdan, S., Rana, S., Bhat, W.W., Vishwakarma, R., Lattoo, S.K., 2015. A decade of molecular understanding of withanolide biosynthesis and in vitro studies in *Withania somnifera* (L.) Dunal: prospects and perspectives for pathway engineering. Front. Plant Sci. 6, 1031.

Dhar, R.S., Gupta, S.B., Singh, P.P., Razdan, S., Bhat, W.W., Rana, S., Latto, S.K., Khan, S., 2012. Identification and characterization of protein composition in *Withania somnifera*-an Indian ginseng. J. Plant Biochem. Biotechnol. 21 (1), 77–87.

Dhar, R.S., Verma, V., Suri, K.A., Sangwan, R.S., Satti, N.K., Kumar, A., Tuli, R., Qazi, G.N., 2006. Phytochemical and genetic analysis in selected chemotypes of *Withania somnifera*. Phytochemistry 67, 2269–2276.

Dharmar, K., De Britto, A.J., 2011. RAPD analysis of genetic variability in wild populations of *Withania somnifera* (L.) Dunal. Int. J. Biosci. Technol. 2, 21–25.

Doma, M., Abhayankar, G., Reddy, V.D., Kishor, P.B., 2012. Carbohydrate and elicitor enhanced withanolide (withaferin A and withanolide A) accumulation in hairy root cultures of *Withania somnifera* (L.). Indian J. Exp. Biol. 50, 484–490.

Dutta, K., Patel, P., Julien, J.P., 2018. Protective effects of *Withania somnifera* extract in $SOD1^{G93A}$ mouse model of amyotrophic lateral sclerosis. Exp. Neurol. 309, 193–204.

Dutta, K., Swarup, V., Julien, J.P., 2017. Potential therapeutic use of *Withania somnifera* for treatment of amyotrophic lateral sclerosis. In: Science of Ashwagandha: Preventive and Therapeutic Potentials. Springer, Cham, pp. 389–415.

Eastwood, F.W., Kirson, I., Lavie, D., Abraham, A., 1980. New withanolides from a cross of a South African chemotype by chemotype II (Israel) in *Withania somnifera*. Phytochemistry 19 (7), 1503–1507.

El-Ariqi, N.S., Al-Sameáe, M.S., El-Moflehi, M.A., 2009. First recording of yellow rust *Puccinia withaniae* Laz. on *Withania somnifera* in Yemen. Arab Gulf J. Sci. Res. 27, 78–82.

Fatima, N., Ahmad, N., Ahmad, I., Anis, M., 2015. Interactive effects of growth regulators, carbon sources, pH on plant regeneration and assessment of genetic fidelity using single primer amplification reaction (SPARS) techniques in *Withania somnifera* L. Appl. Biochem. Biotechnol. 177 (1), 118–136.

Fatima, N., Ahmad, N., Anis, M., Ahmad, I., 2013. An improved in vitro encapsulation protocol, biochemical analysis and genetic integrity using DNA based molecular markers in regenerated plants of *Withania somnifera* L. Ind. Crop. Prod. 50, 468–477.

Fatima, N., Anis, M., 2011. Thidiazuron induced high frequency axillary shoot multiplication in *Withania somnifera* L. Dunal. J. Med. Plants Res. 5 (30), 6681–6687.

Gajbhiye, N.A., Makasana, J., Kumar, S., 2015. Accumulation of three important bioactive compounds in different plant parts of *Withania somnifera* and its determination by the LC-ESI-MS-MS (MRM) method. J. Chromatogr. Sci. 53 (10), 1749–1756.

Ghosal, B., Kaur, R., Shrivastva, R.S., 1988. Sitoindosides IX and X, new glycowithanolides from *Withania somnifera*. Indian J. Nat. Prod. 4, 12–13.

Ghosal, S., Lal, J., Srivastava, R., Bhattacharya, S.K., Upadhyay, S.N., Jaiswal, A.K., Chattopadhyay, U., 1989. Immunomodulatory and CNS effects of sitoindosides IX and X, two new glycowithanolides from *Withania somnifera*. Phytother Res. 3 (5), 201–206.

Gifri, A.N., 1992. Studies in flora of Yemen 2 New records for the flora Aden. Candollea 47, 613–619.

Gorain, M., Ahmed, S.I., Bhandari, R.S., Sharma, N., 2012. Biology of *Phalantha Phalantha* drary (Nymphalidae: Lepidoptera) - an injurious pest of well-known medicinal plant *Withania somnifera* dunal. Biotechnol. Res. 1, 4–6.

Gupta, A., Singh, S., 2014. Evaluation of anti-inflammatory effect of *Withania somnifera* root on collagen-induced arthritis in rats. Pharma. Biol. 52 (3), 308–320.

Gupta, M.L., Misra, H.O., Kalra, A., Khanuja, S.P.S., 2004. Root-rot and wilt: a new disease of ashwagandha (*Withania somnifera*) caused by *Fusarium solani*. J. Med. Aromat. Plant Sci. 26, 285–287.

Gupta, P., Akhtar, N., Tewari, S.K., Sangwan, R.S., Trivedi, P.K., 2011. Differential expression of farnesyl diphosphate synthase gene from *Withania somnifera* in different chemotypes and in response to elicitors. Plant Growth Regul. 65 (1), 93–100.

Gupta, P., Goel, R., Agarwal, A.V., Asif, M.H., Sangwan, N.S., Sangwan, R.S., Trivedi, P.K., 2015. Comparative transcriptome analysis of different chemotypes elucidates withanolide biosynthesis pathway from medicinal plant *Withania somnifera*. Sci. Rep. 5 (1), 1–13.

Gupta, P., Goel, R., Pathak, S., Srivastava, A., Singh, S.P., Sangwan, R.S., Asif, M.H., Trivedi, P.K., 2013. De novo assembly, functional annotation and comparative analysis of *Withania somnifera* leaf and root transcriptomes to identify putative genes involved in the withanolides biosynthesis. PLoS One 8 (5), e62714.

Gupta, R., 1964. Survey record of medicinal and aromatic plants of Chamba forest Division, Himachal Pradeash. Indian For. 90, 454–468.

Gupta, S.K., Karmakar, K., 2011. Diversity of mites (Acari) on medicinal and aromatic plants in India. Zoosymposia 6, 56–61.

Himangini, Thakur, A., 2018. Effect of seed storage conditions on seed germination and vigor of *Withania somnifera*. J. Pharmacogn. Phytochem. 7 (6), 1409–1413.

Hunter, W.N., 2007. The non-mevalonate pathway of isoprenoid precursor biosynthesis. J. Biol. Chem. 282 (30), 21573–21577.

Ingle, S.T., Kareppa, B.M., 2012. Effect of different nutritional sources on seed germination of ashwagandha [(*Withania somnifera* (L.) Dunal.]. BIOINFOLET-A Q. J. Life Sci. 9 (1), 52–53.

Ishnava, K.B., Patel, T., Chauhan, J.B., 2012. Study of genetic transformation of medicinal plants, *Withania somnifera* (L.) Dunal by *Agrobacterium tumefaciens* (MTCC-431). Asian J. Exp. Biol. Sci. 3 (3), 536–542.

Jacob, L., Manju, R.V., Stephen, R., Edison, L.K., 2014. Alterations in withanolide production in *Withania somnifera* (L) Dunal under low light stress. J. Plant Sci. Res. 30, 121–123.

Jamwal, V.V.S., Ahmad, H., Sharma, A., Sharma, D., 2017. Seasonal abundance of *Henosepilachna vigintioctopunctata* (Fab.) on *Solanum melongena* L. and natural occurrence of its two hymenopteran parasitoids. Braz. Arch. Biol. Technol. 60, e160455.

Jhanwar, S., Priya, P., Garg, R., Parida, S.K., Tyagi, A.K., Jain, M., 2012. Transcriptome sequencing of wild chickpea as a rich resource for marker development. Plant Biotechnol. J. 10 (6), 690–702.

Johri, S., Jamwal, U., Rasool, S., Kumar, A., Verma, V., Qazi, G.N., 2005. Purification and characterization of peroxidases from *Withania somnifera* (AGB 002) and their ability to oxidize IAA. Plant Sci. 169 (6), 1014–1021.

Kakaraparthi, P.S., Rajput, D.K., Komaraiah, K., Kumar, N., Kumar, R.R., 2013. Effect of sowing dates on morphological characteristics, root yield and chemical composition of the root of *Withania somnifera* grown in the semi-arid regions of Andhra Pradesh, India. J. Sci. Res. Rep. 2, 121–132.

Kalieswari, N., Raja, I.Y., Devi, M., 2016. Effect of fungicides on the mycelial growth of *Alternaria alternata* causing leaf spot disease in ashwagandha. Int. J. Plant Protect. 9 (1), 153–157.

Kambizi, L., Adebola, P.O., Afolayan, A.J., 2006. Effects of temperature, pre-chilling and light on seed germination of *Withania somnifera*; a high value medicinal plant. South Afr. J. Bot. 72 (1), 11–14.

Kandil, F.E., El Sayed, N.H., Abou-Douh, A.M., Ishak, M.S., Mabry, T.J., 1994. Flavanol glycosides and phenolics from *Withania somnifera*. Phytochemistry 37, 1215–1216.

Kannan, N.D., Kulandaivelu, G., 2011. Drought induced changes in physiological, biochemical and phytochemical properties of *Withania somnifera* Dun. J. Med. Plants Res. 5 (16), 3929–3935.

Kaul, M.K., Kumar, A., Ahuja, A., Mir, B.A., Suri, K.A., Qazi, G.N., 2009. Production dynamics of withaferin A in *Withania somnifera* (L.) Dunal complex. Nat. Prod. Res. 23 (14), 1304–1311.

Kaur, A., Singh, B., Ohri, P., Wang, J., Wadhwa, R., Kaul, S.C., Pati, P.K., Kaur, A., 2018. Organic cultivation of Ashwagandha with improved biomass and high content of active Withanolides: use of Vermicompost. PLoS One 13 (4), e0194314.

Kaushik, M.K., Kaul, S.C., Wadhwa, R., Yanagisawa, M., Urade, Y., 2017. Triethylene glycol, an active component of Ashwagandha (*Withania somnifera*) leaves, is responsible for sleep induction. PLoS One 12 (2).

Khan, J.A., Shrivastva, P., Singh, S.K., 2006. Sensitive detection of a phytoplasma associated with little leaf symptoms in *Withania somnifera*. Eur. J. Plant Pathol. 115, 401–408.

Khan, S., Shah, R.A., 2016. Assessment of genetic diversity among India Ginseng, *Withania somnifera* (L) Dunal using RAPD and ISSR markers. Res. Biotechnol. 7, 1–10.

Khanna, P.K., Chandra, R., Kumar, A., Dogra, N., Gupta, H., Gupta, G., Verma, V., 2014. Correlation between morphological, chemical and RAPD markers for assessing genetic diversity in *Withania somnifera* (L.) Dunal. J. Crop Sci. Biotechnol. 17 (1), 27–34.

Khanna, P.K., Kumar, A., Chandra, R., Verma, V., 2013. Germination behaviour of seeds of *Withania somnifera* (L.) Dunal: a high value medicinal plant. Physiol. Mol. Biol. Plants 19 (3), 449–454.

Kim, S.H., Singh, K.B., Hahm, E.R., Lokeshwar, B.L., Singh, S.V., 2020. *Withania somnifera* root extract inhibits fatty acid synthesis in prostate cancer cells. J. Tradit. Complement. Med. https://doi.org/10.1016/j.jtcme.2020.02.002.

Kirson, I., Abraham, A., Lavie, D., 1977. Chemical analysis of hybrids of *Withania somnifera* L.(Dun.). 1. Chemotypes III (Israel) by Indian I (Delhi). Isr. J. Chem. 16 (1), 20–24.

Kirson, I., Glotter, E., Abraham, A., Lavie, D., 1970. Constituents of *Withania somnifera* dun-XI: the structure of three new withanolides. Tetrahedron 26 (9), 2209–2219.

Kirson, I., Glotter, E., Lavie, D., Abraham, A., 1971. Constitutents of *Withania somnifera* Dun. Part XII. The withanolides of an Indian chemotype. J. Chem. Soc. C Org. 2032–2044.

Kulkarni, A.A., Thengane, S.R., Krishnamurthy, K.V., 1996. Direct *in vitro* regeneration of leaf explants of *Withania somnifera* (L.) Dunal. Plant Sci. 119, 163–168.

Kulkarni, A.A., Thengane, S.R., Krishnamurthy, K.V., 2000. Direct shoot regeneration from node, internode, hypocotyl and embryo explants of *Withania somnifera*. Plant Cell Tissue Organ Culture 62, 203–209.

Kulkarni, S.K., Dhir, A., 2008. *Withania somnifera*: an Indian ginseng. Prog. Neuro-psychopharmacol. Biol. Psychiatry 32 (5), 1093–1105.

Kumar, A., Kaul, M.K., Bhan, M.K., Khanna, P.K., Suri, K.A., 2007. Morphological and chemical variation in 25 collections of the Indian medicinal plant, *Withania somnifera* (L.) Dunal (Solanaceae). Genet. Resour. Crop Evol. 54, 655–660.

Kumar, A., Mir, B.A., Sehgal, D., Dar, T.H., Koul, S., Kaul, M.K., Raina, S.N., Qazi, G.N., 2011a. Utility of a multidisciplinary approach for genome diagnostics of cultivated and wild germplasm resources of medicinal *Withania somnifera*, and the status of new species, *W. ashwagandha*, in the cultivated taxon. Plant Syst. Evol. 291 (3–4), 141–151.

Kumar, A., Sharma, P.C., Mehta, P.K., 2009b. Insect pests on ashwagandha, *Withania somnifera* dunal in Himachal Pradesh. J. Insect Sci. 22, 210–211.

Kumar, A., Singh, C.P., Pandey, R., 2009a. Influence of environmental factors on the population buildup of *Helicoverpa armigera* Hübner and *Epilachna vigintioctopunctata* (Febr.) on Aswagandha (*Withania somnifera* Linn.). J. Entomol. Res. 33, 123–127.

Kumar, A., Singh, C.P., Pandey, R., 2009c. Insect pests of ashwagandha, *Withania somnifera* Linn. In tarai region of Uttarakhand. Entomon 34, 115–118.

Kumar, O.A., Jyothirmayee, G., Tata, S.S., 2011b. *In vitro* plant regeneration from leaf explants of *Withania somnifera* (L) Dunal (Ashwagandha) - an important medicinal plant. Res. Biotechnol. 2 (5), 34–39.

Kumar, O.A., Jyothirmayee, G., Tata, S.S., 2013. *In vitro* conservation of *Withania somnifera* (L) Dunal (Ashwagandha) - a multipurpose medicinal plant. J. Asian Sci. Res. 3 (8), 852–861.

Kumar, V., Dhanjal, J.K., Bhargava, P., Kaul, A., Wang, J., Zhang, H., Kaul, S.C., Wadhwa, R., Sundar, D., 2020b. Withanone and withaferin-A are predicted to interact with transmembrane protease serine 2 (TMPRSS2) and block entry of SARS-CoV-2 into cells. J. Biomol. Struct. Dyn. https://doi.org/10.1080/07391102.2020.1775704 (in press).

Kumar, V., Dhanjal, J.K., Kaul, S.C., Wadhwa, R., Sundar, D., 2020a. Withanone and caffeic acid phenethyl ester are predicted to interact with main protease (Mpro) of SARS-CoV-2 and inhibit its activity. J. Biomol. Struct. Dyn. https://doi.org/10.1080/07391102.2020.1772108 (in press).

Kumar, V., Murthy, K.N.C., Bhamid, S., Sudha, C.G., Ravishankar, G.A., 2005. Genetically modified hairy roots of *Withania somnifera* Dunal: a potent source of rejuvenating principles. Rejuvenation Res. 8 (1), 37–45.

Kushwaha, R.K., Singh, S., Pandey, S.S., Kalra, A., Babu, C.S.V., 2019a. Innate endophytic fungus, *Aspergillus terreus* as biotic elicitor of withanolide A in root cell suspension cultures of *Withania somnifera*. Mol. Biol. Rep. 46 (2), 1895–1908. https://doi.org/10.1007/s1103 3-019-04641-w.

Kushwaha, R.K., Singh, S., Pandey, S.S., Kalra, A., Babu, C.S.V., 2019b. Fungal endophytes attune withanolide biosynthesis in *Withania somnifera*, prime to enhanced withanolide A content in leaves and roots. World J. Microbiol. Biotechnol. 35 (2), 20.

Kushwaha, R.K., Singh, S., Pandey, S.S., Rao, D.V., Nagegowda, D.A., Kalra, A., Babu, C.S.V., 2019c. Compatibility of inherent fungal endophytes of *Withania somnifera* with *Trichoderma viride* and its impact on plant growth and withanolide content. J. Plant Growth Regul. 38 (4), 1228–1242.

Lattoo, S.K., Dhar, R.S., Khan, S., Bamotra, S., Dhar, A.K., 2007. Temporal sexual maturation and incremental staminal movement encourages mixed mating in *Withania somnifera* - an insurance for reproductive success. Curr. Sci. 92, 1390–1399.

Lopresti, A.L., Smith, S.J., Malvi, H., Kodgule, R., 2019. An investigation into the stress-relieving and pharmacological actions of an ashwagandha (*Withania somnifera*) extract: a randomized, double-blind, placebo-controlled study. Medicine 98 (37).

Machiah, D.K., Girish, K.S., Gowda, T.V., 2006. A glycoprotein from a folk medicinal plant, *Withania somnifera*, inhibits hyaluronidase activity of snake venoms. Comp. Biochem. Physiol. C Toxicol. Pharmacol. 143 (2), 158–161.

Mahrshi, R.P., 1986. *Withania somnifera* - a new host for *Myrothecium roridum*. Indian J. Mycol. Plant Pathol. 16, 199.

Maiti, C.K., Sen, S., Paul, A.K., Acharya, K., 2007. First report of *Alternaria dianthicola* causing leaf blight on *Withania somnifera* from India. Plant Dis. 91, 467.

Maiti, C.K., Sen, S., Paul, A.K., Acharya, K., 2012. *Pseudomonas aeruginosa* WS-1 for biological control of leaf blight disease of *Withania somnifera*. Arch. Phytopathol. Plant Protect. 45, 796–805.

Manickam, V.S., Mathavan, E.R., Antonysami, R., 2000. Regeneration of Indian Ginseng plantlets from stem callus. Plant Cell Tissue Organ Culture 62, 181–185.

Meena, R.P., Kalariya, K.A., Saran, P.L., Manivel, P., 2019. Evaluation of ashwagandha (*Withania somnifera* L.) Dunal accessions and breeding lines against leaf spot disease caused by *Alternaria alternata* under subtropical condition of India. J. Appl. Res. Med. Aromat. Plants 14, 100211.

Mehmood, F., Shahzadi, I., Ahmed, I., Waheed, M.T., Mirza, B., 2020. Characterization of *Withania somnifera* chloroplast genome and its comparison with other selected species of Solanaceae. Genomics 112 (2), 1522–1530.

Mehta, A., Raina, R., 2016. Effect of hydropriming on seed germination parameters in different accessions of *Withania somnifera*. Med. Plants-Int. J. Phytomed. Related Ind. 8 (1), 18–23.

Mir, B.A., Khazir, J., Mir, N.A., Hasan, T.U., Koul, S., 2012. Botanical, chemical and pharmacological review of *Withania somnifera* (Indian ginseng): an ayurvedic medicinal plant. Indian J. Drugs Dis. 1, 147–160.

Mir, B.A., Koul, S., Kumar, A., Kaul, M.K., Soodan, A.S., Raina, S.N., 2011. Assessment and characterization of genetic diversity in *Withania somnifera* (L.) Dunal using RAPD and AFLP markers. Afr. J. Biotechnol. 10, 14746–14756.

Mir, B.A., Koul, S., Kumar, A., Kaul, M.K., Soodan, A.S., Raina, S.N., 2010. Intraspecific variation in the internal transcribed spacer (ITS) regions of rDNA in *Withania somnifera* (Linn.) Dunal. Indian J. Biotechnol. 9, 325–328.

Mirjalili, M.H., Moyano, E., Bonfill, M., Cusido, R.M., Palazón, J., 2009. Steroidal lactones from *Withania somnifera*, an ancient plant for novel medicine. Molecules 14, 2373–2393.

Mishra, A., Singh, S.P., Mahfooz, S., Singh, S.P., Bhattacharya, A., Mishra, N., Nautiyal, C.S., 2018. Endophyte-mediated modulation of defense-related genes and systemic resistance in *Withania somnifera* (L.) Dunal under *Alternaria alternata* stress. Appl. Environ. Microbiol. 84 (8) e02845-17.

Mishra, B., Bose, S.K., Sangwan, N.S., 2020. Comparative investigation of therapeutic plant *Withania somnifera* for yield, productivity, withanolide contents, and expression of pathway genes during contrasting seasons. Ind. Crop. Prod. 154, 112508.

Mishra, L.C., Singh, B.B., Dagenais, S., 2000. Scientific basis for the therapeutic use of *Withania somnifera* (ashwagandha): a review. Alternat. Med. Rev. 5, 334–346.

Mishra, S., Bansal, S., Mishra, B., Sangwan, R.S., Jadaun, J.S., Sangwan, N.S., 2016. RNAi and homologous over-expression based functional approaches reveal triterpenoid synthase gene-cycloartenol synthase is involved in downstream withanolide biosynthesis in *Withania somnifera*. PLoS One 11 (2), e0149691.

Mohammad, Z., Shabbir, A., 2016. Screening of *Withania somnifera* L. Germplasm for resistance against leaf spot caused by *Alternaria alternata* (Fr.) Keissler. J. Funct. Environ. Botany 6 (1), 54–57.

Mohanty, I., Arya, D.S., Dinda, A., Talwar, K.K., Joshi, S., Gupta, S.K., 2004. Mechanisms of cardioprotective effect of *Withania somnifera* in experimentally induced myocardial infarction. Basic Clin. Pharmacol. Toxicol. 94, 184–190.

Murthy, H.N., Dijkstra, C., Anthony, P., White, D.A., Davey, M.R., Power, J.B., Hahn, E.J., Paek, K.Y., 2008. Establishment of *Withania somnifera* hairy root cultures for the production of withanolide A. J. Integr. Plant Biol. 50 (8), 975–981.

Nagappan, A., Karunanithi, N., Sentrayaperumal, S., Park, K.I., Park, H.S., Lee do, H., Kang, S.R., Kim, J.A., Senthil, K., Natesan, S., Muthurajan, R., Kim, G.S., 2012. Comparative root protein profiles of Korean ginseng (*Panax ginseng*) and Indian ginseng (*Withania somnifera*). Am. J. Chin. Med. 40 (01), 203–218.

Nagella, P., Murthy, H.N., 2011. Effects of macroelements and nitrogen source on biomass accumulation and withanolide-A production from cell suspension cultures of *Withania somnifera* (L.) Dunal. Plant Cell Tissue Organ Cult. 104 (1), 119–124.

Namdeo, A.G., Sharma, A., Yadav, K.N., Gawande, R., Mahadik, K.R., Lopez-Gresa, M.P., Kim, H.K., Choi, Y.H., Verpoorte, R., 2011. Metabolic characterization of *Withania somnifera* from different regions of India using NMR spectroscopy. Planta Med. 77 (17), 1958–1964.

Nayar, S.L., 1964. Medicinal plants of commercial importance found wild in Uttar Pradesh and their distribution. J. Bombay Nat. Hist. Soc. 61, 651–661.

Negi, M.S., Sabharwal, V., Wilson, N., Lakshmikumaran, M.S., 2006. Comparative analysis of the efficiency of SAMPL and AFLP in assessing genetic relationships among *Withania somnifera* genotypes. Curr. Sci. 91, 464–471.

Negi, M.S., Singh, A., Lakshmikumaran, M., 2000. Genetic variation and relationship among and within *Withania* species as revealed by AFLP markers. Genome 43 (6), 975–980.

Nigam, K.B., 1984. Aswagandha cultivation. Indian Hortic. 28, 39–41.

Niraj, T., Navinder, S., Vandana, M., Patel, P.K., Tiwari, A.B., Sharad, T., 2011. Diversity analysis among *Withania somnifera* genotypes using RAPD and ISSR markers. J. Trop. For. 27 (4), 11–23.

Nittala, S.S., Lavie, D., 1981. Chemistry and genetics of withanolides in *Withania somnifera* hybrids. Phytochemistry 20 (12), 2741–2748.

Niyaz, A., Siddiqui, E.N., 2014. Seed germination of *Withania somnifera* (L.) Dunal. Eur. J. Med. Plants 920–926.

Pal, S., Rastogi, S., Nagegowda, D.A., Gupta, M.M., Shasany, A.K., Chanotiya, C.S., 2019. RNAi of sterol methyl Transferase1 reveals its direct role in diverting intermediates towards withanolide/phytosterol biosynthesis in *Withania somnifera*. Plant Cell Physiol. 60 (3), 672–686.

Pal, S., Singh, S., Ashutosh, K.S., Madan, M.G., Suman, P.K., Ajit, K.S., 2011. Comparative withanolide profiles, gene isolation, and differential gene expression in the leaves and roots of *Withania somnifera*. J. Hortic. Sci. Biotechnol. 86 (4), 391–397.

Pal, S., Yadav, A.K., Singh, A.K., Rastogi, S., Gupta, M.M., Verma, R.K., Nagegowda, D.A., Pal, A., Shasany, A.K., 2017. Nitrogen treatment enhances sterols and withaferin A through transcriptional activation of jasmonate pathway, WRKY transcription factors, and biosynthesis genes in *Withania somnifera* (L.) Dunal. Protoplasma 254 (1), 389–399.

Pandey, R., Kalra, A., 2003. Root knot disease of ashwagandha *Withania somnifera* and its ecofriendly cost-effective management. J. Mycol. Plant Pathol. 33, 240–245.

Pandey, V., Ansari, W.A., Misra, P., Atri, N., 2017. *Withania somnifera*: advances and implementation of molecular and tissue culture techniques to enhance its application. Front. Plant Sci. 8, 1390.

Pandey, V., Misra, P., Chaturvedi, P., Mishra, M.K., Trivedi, P.K., Tuli, R., 2010. *Agrobacterium tumefaciens*-mediated transformation of *Withania somnifera* (L.) Dunal: an important medicinal plant. Plant Cell Rep. 29 (2), 133–141.

Pandey, S.S., Singh, S., Pandey, H., Srivastava, M., Ray, T., Soni, S., Pandey, A., Shanker, K., Babu, C.V., Banerjee, S., Gupta, M.M., 2018. Endophytes of *Withania somnifera* modulate in planta content and the site of withanolide biosynthesis. Sci. Rep. 8 (1), 1–19.

Parita, B., Kumar, S.N., Darshan, D., Karen, P., 2018. Elucidation of genetic diversity among ashwagandha [*Withania somnifera* (L.) Dunal] genotypes using EST-SSR markers. Res. J. Biotechnol. 13 (10), 52–59.

Parmar, E.K., Fougat, R.S., Patel, C.B., Zala, H.N., Patel, M.A., Patel, S.K., Kumar, S., 2015. Validation of dbEST-SSRs and transferability of some other solanaceous species SSR in ashwagandha [*Withania Somnifera* (L.) Dunal]. 3 Biotech 5 (6), 933–938.

Patel, N., Patel, P., Kendurkar, S.V., Thulasiram, H.V., Khan, B.M., 2015. Overexpression of squalene synthase in *Withania somnifera* leads to enhanced withanolide biosynthesis. Plant Cell Tissue Organ Cult. 122 (2), 409–420.

Patel, N., Patel, P., Kumari, U., Kendurkar, S.V., Khan, B.M., 2014. Microprojectile bombardment assisted agroinfection increases transformation efficiency of *Withania somnifera* (L.). Res. Biotechnol. 5 (4), 13–24.

Pathak, H.C., Raychoudhuri, S.P., 1967. *Withania somnifera* - an additional host of tobacco leaf curl virus. Sci. Cult. 33, 234–235.

Pati, P.K., Sharma, M., Salar, R.K., Sharma, A., Gupta, A.P., Singh, B., 2008. Studies on leaf spot disease of *Withania somnifera* and its impact on secondary metabolites. Indian J. Microbiol. 48 (4), 432–437.

Poojari, P., Kiran, K.R., Swathy, P.S., Muthusamy, A., 2019. *Withania somnifera* (L.) dunal: an overview of bioactive molecules, medicinal properties and enhancement of bioactive molecules through breeding strategies. In: In Vitro Plant Breeding towards Novel Agronomic Traits. Springer, Singapore, pp. 1–25.

Pramanick, S., Roy, A., Ghosh, S., Majumder, H.K., Mukhopadhyay, S., 2008. Withanolide Z, a new chlorinated withanolide from *Withania somnifera*. Planta Med. 74 (14), 1745–1748.

Praveen, N., Murthy, H.N., 2010. Production of withanolide-A from adventitious root cultures of *Withania somnifera*. Acta Physiol. Plant. 32 (5), 1017–1022.

Rajasekar, S., Elango, R., 2011. Effect of microbial consortium on plant growth and improvement of alkaloid content in *Withania somnifera* (Ashwagandha). Curr. Bot. 2 (8), 27–30.

Rajeswara Rao, B.R., Rajput, D.K., Nagaraju, G., Adinarayana, G., 2012. Opportunities and challenges in the cultivation of ashwagandha {*Withania somnifera* (L.) dunal}. J. Pharmacogn. 3, 88–91.

Ram, H., Kumar, A., Sharma, S.K., Ojha, A., Rao, S.R., 2012. Meiotic studies in *Withania somnifera* (L.) dunal.: a threatened medicinal herb of Indian thar desert. Am. J. Plant Sci. 3, 185–189.

Rani, G., Arora, S., Nagpal, A., 2003b. Direct rhizogenesis from *in vitro* leaves of *Withania somnifera* (L.) Dunal. J. Herbs, Spices, Med. Plants 10 (3), 47–54.

Rani, G., Virk, G.S., Nagpal, A., 2003a. Plant regeneration in *Withania somnifera* (L.) Dunal. In Vitro Cell. Dev. Biol. Plant 39, 468–474.

Rani, G., Virk, G.S., Nagpal, A., 2004. Somatic embryogenesis in *Withania somnifera* (L.) Dunal. J. Plant Biotechnol. 6 (2), 113–118.

Rani, A., Kumar, M., Kumar, S., 2014. *In vitro* propagation of *Withania somnifera* (L.) Dunal from shoot apex explants. J. Appl. Nat. Sci. 6 (1), 159163.

Rasheed, A., Younus, N., Waseem, N., Badshsah, M., 2020. Protective effect of *Withania somnifera* root extract against cisplatin induced nephrotoxicity through renal function analysis in albino Wistar rats. In Med. Forum 31 (4), 61.

Rathaur, P., Raja, W., Ramteke, P.W., John, S.A., 2012. Effect of UV-B tolerant plant growth promoting rhizobacteria (PGPR) on seed germination and growth of *Withania somnifera*. Adv. Appl. Sci. Res. 3 (3), 1399–1404.

Ravikumar, A., Rajendran, R., Chinniah, C., Irulandi, S., Pandi, R.J., 2008. Evaluation of certain organic nutrient sources against mealy bug, *Coccidohystrix insolitus* (Green) and the spotted leaf beetle, *Epilachna vigintioctopunctata* Fab. on aswagandha, *Withania somnifera* Dunal. J. Biopestic. 1, 28–31.

Ray, S., Ghosh, B., Sen, S., Jha, S., 1996. Withanolide production by root cultures of *Withania somnifera* transformed with *Agrobacterium rhizogenes*. Planta Med. 62 (06), 571–573.

Ray, S., Jha, S., 1999. Withanolide synthesis in cultures of *Withania somnifera* transformed with *Agrobacterium tumefaciens*. Plant Sci. 146 (1), 1–7.

Ray, S., Jha, S., 2001. Production of withaferin A in shoot cultures of *Withania somnifera*. Planta Med. 67, 432–436.

Ray, S., Ray, A., 2015. Medhya Rasayanas in brain function and disease. Med. Chem. 5, 505–511.

Rehaman, S.K., Pradeep, S., 2016. Effect of biopesticides on management of *Helicoverpa armigera* in ashwagandha (*Withania somnifera* Linn.). Pest Manag. Hortic. Ecosyst. 22 (1), 99–102.

Rehaman, S.K., Pradeep, S., Dhanapal, R., 2018. Incidence of fruit Borer *Helicoverpa armigera* (Hubner) on ashwagandha in Shivamogga. Int. J. Curr. Microbiol. App. Sci 7 (3), 1060–1066.

Saema, S., Rahman, L.U., Niranjan, A., Ahmad, I.Z., Misra, P., 2015. RNAi-mediated gene silencing of WsSGTL1 in *W. somnifera* affects growth and glycosylation pattern. Plant Signal. Behav. 10 (12), e1078064.

Saikia, S.K., Tiwari, S., Pandey, R., 2013. Rhizospheric biological weapons for growth enhancement and *Meloidogyne incognita* management in *Withania somnifera* cv. Poshita. Biol. Control 65, 225–234.

Samad, A., Shasany, A.K., Gupta, S., Ajayakumar, P.V., Darokar, M.P., Khanuja, S.P.S., 2006. First report of a 16SrVI group phytoplasma associated with witches'-broom disease on *Withania somnifera*. Plant Dis. 90, 248.

Samaddar, T., Nath, S., Halder, M., Sil, B., Roychowdhury, D., Sen, S., Jha, S., 2012. Karyotype analysis of three important traditional Indian medicinal plants, *Bacopa monnieri, Tylophora indica* and *Withania somnifera*. Nucleus 55 (1), 17–20.

Sanchita, Singh, R., Mishra, A., Dhawan, S.S., Shirke, P.A., Gupta, M.M., Sharma, A., 2015. Physiological performance, secondary metabolite and expression profiling of genes associated with drought tolerance in *Withania somnifera*. Protoplasma 252 (6), 1439–1450.

Sanchita, Singh, S., Sharma, A., 2014. Bioinformatics approaches for structural and functional analysis of proteins in secondary metabolism in *Withania somnifera*. Mol. Biol. Rep. 41 (11), 7323–7330.

Sangwan, R.S., Chaurasiya, N.D., Misra, L.N., Lal, P., Uniyal, G.C., Sharma, R., Sangwan, N.S., Suri, K.A., Qazi, G.N., Tuli, R., 2004. Phytochemical variability in commercial herbal products and preparations of *Withania somnifera* (Ashwagandha). Curr. Sci. 461–465.

Sangwan, R.S., Chaurasiya, N.D., Lal, P., Misra, L., Tuli, R., Sangwan, N.S., 2008. Withanolide A is inherently de novo biosynthesized in roots of medicinal plant Ashwagandha (*Withania somnifera*). Physiol. Plantarum 133 (2), 278–287.

Saravanakumar, A., Aslam, A., Shajahan, A., 2012. Development and optimization of hairy root culture systems in *Withania somnifera* (L.) Dunal for withaferin-A production. Afr. J. Biotechnol. 11 (98), 16412–16420.

Saroj, A., Kumar, A., Qamar, N., Alam, M., Singh, H.N., Khaliq, A., 2012. First report of Wet rot of *Withania somnifera* caused by *Choanephora cucurbitarum* in India. Plant Dis. 96, 293.

Saroj, A., Kumar, A., Srivastava, A.K., Khaliq, A., Absar, N., Alam, M., Samad, A., 2014. New report of black leaf spot mold (*Pseudocercospora fuligena*) on *Withania somnifera* from India. Plant Dis. 98 (9), 1275.

Sayyed, R.Z., Naphade, B.S., Chincholklar, S.B., 2007. Siderophore producing *A. feacalis* promoted the growth of Safed musali and Ashwagandha. J. Med. Aromat. Plant Sci. 29, 1–5.

Schröter, H.B., Neumann, D., Katritzky, A.R., Swinbourne, F.J., 1966. Withasomnine. A pyrazole alkaloid from *Withania somnifera* Dun. Tetrahedron 22 (8), 2895–2897.

Schwarting, A.E., Bobbitt, J.M., Rother, A., Atal, C.K., Khana, K.J., Leary, J.D., Walter, W.G., 1963. The alkaloids from *Withania somnifera*. Lloydia 26, 258–273.

Senthil, K., Jayakodi, M., Thirugnanasambantham, P., Lee, S.C., Duraisamy, P., Purushotham, P.M., Rajasekaran, K., Charles, S.N., Roy, I.M., Nagappan, A.K., Kim, G.S., 2015. Transcriptome analysis reveals in vitro cultured *Withania somnifera* leaf and root tissues as a promising source for targeted withanolide biosynthesis. BMC Genom. 1 (16), 1–16.

Senthil, K., Karunanithi, N., Kim, G.S., Nagappan, A., Sundareswaran, S., Natesan, S., Muthurajan, R., 2011. Proteome analysis of *in vitro* and *in vivo* root tissue of *Withania somnifera*. Afr. J. Biotechnol. 10 (74), 16866–16874.

Senthilnathan, P., Padmavathi, R., Banu, S.M., Sakthisekaran, D., 2006. Enhancement of antitumor effect of paclitaxel in combination with immunomodulatory *Withania somnifera* on benzo (a) pyrene induced experimental lung cancer. Chem. Biol. Interact. 159 (3), 180–185.

Shah, D., Daniel, M., 2004. *Chaetomium globosum* on *Withania somnifera* - a new host record. Indian Phytopathol. 57, 509.

Shanmugaratnam, S., Mikunthan, G., Thurairatnam, S., 2013. Potential of *Withania somnifera* Dunal cultivation as a medicinal crop in jaffna district. Am.-Eurasian J. Agric. Environ. Sci. 13 (3), 357–361.

Sharma, A., 2013a. Study of Leaf Spot Disease in *Withania somnifera* (L.) Dunal Employing Morphological, Biochemical and Molecular Approaches (Ph.D. thesis). Guru Nanak Dev University.

Sharma, A., Pati, P.K., 2011a. First report of *Withania somnifera* (L.) Dunal, as a new host of cowbug (*Oxyrachis tarandus*, Fab.) in plains of Punjab, northern India. World Appl. Sci. J. 14 (9), 1344–1346.

Sharma, A., Pati, P.K., 2011b. First record of 28-spotted ladybird beetle, *Henosepilachna vigintioctopunctata* (F.) infesting *Withania somnifera* (L.) Dunal in Punjab province of Northern India. Pest Technol. 5 (1), 91–92.

Sharma, A., Pati, P.K., 2012. First record of the carmine spider mite, *Tetranychus urticae*, infesting *Withania somnifera* in India. J. Insect Sci. 12 (1).

Sharma, A., Pati, P.K., 2013. First record of Ashwagandha as a new host to the invasive mealybug (*Phenacoccus solenopsis* Tinsley) in India. Entomol. News 123 (1), 59–62.

Sharma, A., Sharma, I., Pati, P.K., 2011. Post-infectional changes associated with the progression of leaf spot disease in *Withania somnifera*. J. Plant Pathol. 93, 397–405.

Sharma, A., Singh, V., Singh, G., Pati, P.K., 2013b. First report of leaf spot disease in *Withania coagulans* caused by *Alternaria alternata* in India. Plant Dis. 97 (3), 420.

Sharma, A., Thakur, A., Kaur, S., Pati, P.K., 2012b. Effect of *Alternaria alternata* on the coccinellid pest *Henosepilachna vigintioctopunctata* and its implications for biological pest management. J. Pest. Sci. 85 (4), 513–518.

Sharma, A., Vats, S.K., Pati, P.K., 2014. Post-infectional dynamics of leaf spot disease in *Withania somnifera*. Ann. Appl. Biol. 165 (3), 429–440.

Sharma, I., Bhardwaj, R., Gautam, V., Kaur, R., Sharma, A., 2018. Brassinosteroids: occurrence, structure and stress protective activities. Front. Nat. Prod. Chem. 4, 204–239.

Sharma, M., Kaur, R., Puri, S., 2013a. Quantification of withanolide A from *Withania somnifera* Dunal. in tropics of himalaya using HPLC with DAD detector. Int. J. Biol. Pharma. Res. 4, 702–705.

Sharma, M., Puri, S., 2020. Physiological performance, secondary metabolite profiling and photo oxidative tolerance in *Withania somnifera*. Med. Plants-Int. J. Phytomed. Related Ind. 12 (1), 41–47.

Sharma, M.M., Ali, D.J., Batra, A., 2010. Plant regeneration through *in vitro* somatic embryogenesis in Ashwagandha (*Withania somnifera* L. Dunal). Researcher 2 (3), 1–6.

Sharma, P., Trivedi, P.C., 2010. Evaluation of different fungal Antagonists against *Fusarium oxysporum* infecting *Withania somnifera* (L.) Dunal. Assam Univ. J. Sci. Technol. 6, 37–41.

Sharma, R.K., Samant, S.S., Sharma, P., Devi, S., 2012a. Evaluation of antioxidant activities of *Withania somnifera* leaves growing in natural habitats of North-west Himalaya, India. J. Med. Plants Res. 6 (5), 657–661.

Sharma, V., 2013b. HPLC-PDA method for quantification of Withaferin-A and Withanolide-A in diploid (n = 12) and tetraploid (n = 24) cytotypes of "Indian Ginseng" *Withania somanifera* (L.) Dunal from North India. Int. J. Indigenous Med. Plants 46 (2), 2051–4263.

Shasmita, M.K.R., Naik, S.K., 2017. Exploring plant tissue culture in *Withania somnifera* (L.) Dunal: *in vitro* propagation and secondary metabolite production. Crit. Rev. Biotechnol. 38 (6), 836–850.

Shukla, K.K., Mahdi, A.A., Mishra, V., Rajender, S., Sankhwar, S.N., Patel, D., Das, M., 2011. *Withania somnifera* improves semen quality by combating oxidative stress and cell death and improving essential metal concentrations. Reprod. Biomed. Online 22 (5), 421–427.

Sidhu, O.P., Annarao, S., Chatterjee, S., Tuli, R., Roy, R., Khetrapal, C.L., 2011. Metabolic alterations of *Withania somnifera* (L.) Dunal fruits at different developmental stages by NMR spectroscopy. Phytochem. Anal. 22 (6), 492–502.

Singh, A., 2012. New report of *Pseudocercospora Speg* on some medicinal plants form Sonebhadra forest U.P. Biol. Forum Int. J. 4, 61–67.

Singh, A.K., Varshney, R., Sharma, M., Agarwal, S.S., Bansal, K.C., 2006. Regeneration of plants from alginate-encapsulated shoot tips of *Withania somnifera* (L.) Dunal, a medicinally important plant species. J. Plant Physiol. 163 (2), 220–223.

Singh, M., Poddar, N.K., Singh, D., Agrawal, S., 2020. Foliar application of elicitors enhanced the yield of withanolide contents in *Withania somnifera* (L.) Dunal (variety, Poshita). 3 Biotech 10 (4), 1–8.

Singh, V., Singh, B., Joshi, R., Jaju, P., Pati, P.K., 2017a. Changes in the leaf proteome profile of *Withania somnifera* (L.) Dunal in response to *Alternaria alternata* infection. PLoS One 12 (6).

Singh, N., Singh, S.P., Nath, R., Singh, D.R., Gupta, M.L., Kohli, R.P., Bhargava, K.P., 1986. Prevention of urethane-induced lung adenomas by *Withania somnifera* (L.) Dunal in albino mice. Int. J. Crude Drug Res. 24 (2), 90–100.

Singh, P., Guleri, R., Angurala, A., Kaur, K., Kaur, K., Kaul, S.C., Wadhwa, R., Pati, P.K., 2017b. Addressing challenges to enhance the bioactives of *Withania somnifera* through organ, tissue, and cell culture based approaches. BioMed Res. Int. 3278494.

Singh, P., Guleri, R., Singh, V., Kaur, G., Kataria, H., Singh, B., Kaur, G., Kaul, S.C., Wadhwa, R., Pati, P.K., 2015. Biotechnological interventions in *Withania somnifera* (L.) dunal. Biotechnol. Genet. Eng. Rev. 31 (1–2), 1–20.

Singh, R.H., 2018. Exploring the medicinal value of green wealth with special reference to neuronutrient Medhya rasayana plant drugs. In: New Age Herbals. Springer, Singapore, pp. 331–346.

Singh, S., Kumar, S., 1998. Withania somnifera: The India Ginseng Ashwagandha. Central Institute of Medicinal and Aromatic Plants (CIMAP), Lucknow, India, pp. 1–293.

Singh, S., Pal, S., Shanker, K., Chanotiya, C.S., Gupta, M.M., Dwivedi, U.N., Shasany, A.K., 2014. Sterol partitioning by HMGR and DXR for routing intermediates toward withanolide biosynthesis. Physiol. Plantarum 152, 617–633.

Singh, V., 2009. Phenology and reproductive biology of *Withania somnifera* (L.) Dunal (Solanaceae). J. Plant Reprod. Biol. 1, 81–86.

Singh, V., Singh, B., Sharma, A., Kaur, K., Gupta, A.P., Salar, R.K., Hallan, V., Pati, P.K., 2017b. Leaf spot disease adversely affects human health-promoting constituents and withanolide biosynthesis in *Withania somnifera* (L.) Dunal. J. Appl. Microbiol. 122 (1), 153–165.

Singhal, V.K., Kumar, P., 2008. Cytomixis during microsporogenesis in the diploid and tetraploid cytotypes of *Withania somnifera* (L.) Dunal, 1852 (Solanaceae). Comp. Cytogenet. 2, 85–92.

Sivanandhan, G., Arun, M., Mayavan, S., Rajesh, M., Mariashibu, T.S., Manickavasagam, M., Selvaraj, N., Ganapathi, A., 2012. Chitosan enhances withanolides production in adventitious root cultures of *Withania somnifera* (L.) Dunal. Ind. Crop. Prod. 37 (1), 124–129.

Sivanandhan, G., Dev, G.K., Jeyaraj, M., Rajesh, M., Muthuselvam, M., Selvaraj, N., Manickavasagam, M., Ganapathi, A., 2013. A promising approach on biomass accumulation and withanolides production in cell suspension culture of *Withania somnifera* (L.) Dunal. Protoplasma 250 (4), 885–898.

Sivanandhan, G., Mariashibu, T.S., Arun, M., Rajesh, M., Kasthurirengan, S., Selvaraj, N., Ganapathi, A., 2011. The effect of polyamines on the efficiency of multiplication and rooting of *Withania somnifera* (L.) Dunal and content of some withanolides in obtained plants. Acta Physiol. Plant. 33 (6), 2279.

Sivanandhan, G., Selvaraj, N., Ganapathi, A., Manickavasagam, M., 2015. Effect of nitrogen and carbon sources on in vitro shoot multiplication, root induction and withanolides content in *Withania somnifera* (L.) Dunal. Acta Physiol. Plant. 37 (2), 12.

Sivanandhan, G., Selvaraj, N., Ganapathi, A., Manickavasagam, M., 2014. Enhanced biosynthesis of withanolides by elicitation and precursor feeding in cell suspension culture of *Withania somnifera* (L.) Dunal in shake-fask culture and bioreactor. PLoS One 9 (8), 1–11.

Sivanesan, I., Murugesan, K., 2008. An efficient regeneration from nodal explants of *Withania somnifera* Dunal. Asian J. Plant Sci. 7, 551–556.

Solanki, S., Basudeb, D., 2017. First report of leaf spot of Aswagandha (*Withania somnifera* Dunal) caused by *Colletotrichum gloeosporioides* from West Bengal, India. J. Mycopathol. Res. 55 (3), 257–259.

Srivastava, S., Singh, R., Srivastava, G., Sharma, A., 2018. Comparative study of Withanolide biosynthesis-related miRNAs in root and leaf tissues of *Withania somnifera*. Appl. Biochem. Biotechnol. 185 (4), 1145–1159.

Supe, U., Dhote, F., Roymon, M.G., 2006. *In vitro* plant regeneration of *Withania somnifera*. Plant Tissue Cult. Biotechnol. 16 (2), 111–115.

Takshak, S., Agrawal, S.B., 2014. Secondary metabolites and phenylpropanoid pathway enzymes as influenced under supplemental ultraviolet-B radiation in *Withania somnifera* Dunal, an indigenous medicinal plant. J. Photochem. Photobiol. B Biol. 140, 332–343.

Thilip, C., Mehaboob, V.M., Varutharaju, K., Faizal, K., Raja, P., Aslam, A., Shajahan, A., 2019. Elicitation of withaferin-A in hairy root culture of *Withania somnifera* (L.) Dunal using natural polysaccharides. Biologia 74 (8), 961–968.

Thilip, C., Raja, P., Rafi, K.M., Faizal, K.P., Thiagu, G., Aslam, A., Shajahan, A., 2020. Genetic transformation using *Agrobacterium rhizogenes* for the production of valuable anticancer compound, withaferin-A from *Withania somnifera* (L.) Dunal. Jamal Acad. Res. J. Interdisc. 1 (1), 1–5.

Thilip, C., Raju, C.S., Varutharaju, K., Aslam, A., Shajahan, A., 2015. Improved *Agrobacterium rhizogenes*-mediated hairy root culture system of *Withania somnifera* (L.) Dunal using sonication and heat treatment. 3 Biotech 5 (6), 949–956.

Thirugnanasambantham, P., Senthil, K., 2016. *In vitro* and omics technologies opens a new avenue for deciphering withanolide metabolism in *Withania somnifera*. Int. J. Pharm. Pharm. Sci. 8 (7), 17–26.

Tripathi, N., Saini, N., Mehto, V., Kumar, S., Tiwari, S., 2012. Assessment of genetic diversity among *Withania somnifera* collected from central India using RAPD and ISSR analysis. Med. Aromat. Plant Sci. Biotechnol. 6 (1), 33–39.

Tripathi, N., Shrivastava, D., Mir, B.A., Kumar, S., Govil, S., Vahedi, M., Bisen, P.S., 2018. Metabolomic and biotechnological approaches to determine therapeutic potential of *Withania somnifera* (L.) Dunal: a review. Phytomedicine 50, 127–136.

Tripathi, S., Sangwan, R.S., Mishra, B., Jadaun, J.S., Sangwan, N.S., 2020. Berry transcriptome: insights into a novel resource to understand development dependent secondary metabolism in *Withania somnifera* (Ashwagandha). Physiol. Plantarum 168 (1), 148–173.

Udayakumar, R., Choi, C.W., Kim, K.T., Kim, S.C., Kasthurirengan, S., Mariashibu, T.S., JJ, S.R., Ganapathi, A., 2013b. *In vitro* plant regeneration from epicotyl explant of *Withania somnifera* (L.) Dunal. J. Med. Plants Res. 7 (1), 43–52.

Udayakumar, R., Kasthurirengan, S., Mariashibu, T.S., Sudhakar, B., Ganapathi, A., Kim, E.J., Jang, K.M., Choi, C.W., Kim, S.C., 2013a. Analysis of genetic variation among populations of *Withania somnifera* (L.) in South India based on RAPD Markers. Eur. J. Med. Plants 266–280.

Vakeswaran, V., Krishnasamy, V., 2003. Improvement in storability of Ashwagandha (*Withania somnifera* Dunal) seeds through pre-storage treatments by triggering their physiological and biochemical properties. Seed Technol. 25, 203.

Vanitha, S., Alice, D., Bharathi, S., 2006. Occurrence of leaf blight disease caused by *Alternaria chlamydospora* in Ashwagandha (*Withania somnifera* Dunal) - a new report. J. Plant Protect. Environ. 3, 115.

Ved, D.K., Goraya, G.S., 2007. Demand and Supply of Medicinal Plants in India. NMPB, New Delhi & FRLHT, Bangalore, India, p. pp18.

Venkatesha, M.G., 2006. Seasonal occurrence of *Henosepilachna vigintioctopunctata* (F.) (Coleoptera: Coccinellidae) and its parasitoid on ashwagandha in India. J. Asia Pac. Entomol. 9 (3), 265–268.

Venugopal, S., Saidaiah, P., Karthik, C.S., 2018. Chapter-6, genetics and breeding of ashwagandha (*Withania somnifera* (L.) dunal.). In: Saidaiah, P. (Ed.), Advances in Genetics and Plant Breeding, vol. 2. AkiNik Publications, New Delhi, pp. 91–106.

Verma, K.S., Thakur, N.S., 2010. Economic analysis of Ashwagandha (*Withania somnifera* L. Dunal) based agroforestry land-use systems in Mid Hills of Western Himalayas. Indian J. Agrofor. 12 (1), 62–70.

Verma, O.P., Gupta, R.B.L., Shivpuri, A., 2008. A new host for *Pithomyces chartarum*, the cause of a leaf spot disease on *Withania somnifera*. Plant Pathol. 57, 385.

Verma, S.K., Shaban, A., Purohit, R., Chimata, M.L., Rai, G., Verma, O.P., 2012. Immunomodulatory activity of *Withania somnifera* (L.). J. Chem. Pharm. Res. 4, 559–561.

Wadhwa, R., Konar, A., Kaul, S.C., 2016. Nootropic potential of Ashwagandha leaves: beyond traditional root extracts. Neurochem. Int. 95, 109–118.

WHO, 2007. WHO Guidelines on Good Manufacturing Practices (GMP) for Herbal Medicines. World Health Organization Press, France, pp. 1–92.

Wood, J.M., 1908. List of flora of natal. Trans. South Afr. Philos. Soc. 18, 197.

Xu, Y.M., Gao, S., Bunting, D.P., Gunatilaka, A.A., 2011. Unusual withanolides from aeroponically grown *Withania somnifera*. Phytochemistry 72, 518–522.

Zaim, M., Samad, A., 1995. Association of phytoplasmas with a witches-broom disease of *Withania somnifera* (L.) Dunal in India. Plant Sci. 109, 225–229.

CHAPTER

Zanthoxylum armatum

16

Mamta Kashyap[1], Varun Garla[2], Aditya Dogra[1]

[1]*Department of Biotechnology, Shoolini Institute of Life Sciences and Business Management, Solan, Himachal Pradesh, India;* [2]*Department of Information Technology, Shoolini Institute of Life Sciences and Business Management, Solan, Himachal Pradesh, India*

16.1 Introduction

The genus *Zanthoxylum* is a traditional medicinal plant belonging to the family Rutaceae, consisting of about 250 species (Arun and Paridhavi, 2012). Eleven species of *Zanthoxylum* have been reported from India: *Zanthoxylum budrunga, Zanthoxylum oxyphyllum, Zanthoxylum ovalifolium, Zanthoxylum acanthopodium, Zanthoxylum planispinum, Zanthoxylum armatum, Zanthoxylum nitidium, Zanthoxylum rhetsa, Zanthoxylum simulans, Zanthoxylum avicennae*, and *Zanthoxylum limonella* (Hooker, 1999; DPR, 2011a, 2016; Rajbhandari et al., 2015). Some of them have been reported in Uttarakhand (*Z. armatum* DC., *Z. acanthopodium* DC., *Zanthoxylum edgew.*, and *Z. budrunga*) (Hajra et al., 1996), and these species (*Zanthoxylum hamiltonianum, Z. rhetsa, Z. oxyphyllum, Zanthoxylum alatum, Z. ovalifolium*, and *Z. acanthopodium*) have been reported in Northeast India (Hajra et al., 1996). Among these species, the most common one is *Z. armatum* DC., which is considered to be one of 30 medicinal plants in the country that is emphasized for cultivation and agrotechnological development (DPR, 2006). It is commonly known as Indian prickly ash, Nepal pepper, and toothache tree. In different regions and languages, it is known by various names. Table 16.1 shows the vernacular names, which are listed subsequently.

16.1.1 Taxonomical arrangement

Domain	:	Eukaryota
Kingdom	:	Plantae
Subkingdom	:	Viridae plantae
Phylum	:	Tracheophyta
Subphylum	:	Euphyllophytina
Infraphylum	:	Radiatopses
Class	:	Magnoliopsida
Subclass	:	Rosidae
Superorder	:	Rutanae

Himalayan Medicinal Plants. https://doi.org/10.1016/B978-0-12-823151-7.00008-8

Table 16.1 Different vernacular names of *Zanthoxylum armatum* in India and other Southeast Asian countries.

Languages	Names
Hindi	*Nepali dhaniya*, *tejphal*, *tumuru*
Sanskrit	*Tejovati*
Nepali	*Timur*, Nepali pepper
English	Bamboo-leaved prickly ash, Nepal pepper, toothache tree, prickly ash
Urdu	*Tamu*
Bengali	*Gaira*
Chinese	*Ci Zhu Ye Hua Jiao*, *Qin Jiao* (Taiwan), *Zhu Ye Jiao*
German	*Nepal pfeffer*
Japanese	*Fuyu Zanshou*, *Fuyu-Sansh*
Manipuri	*Mukthrubi*
Kannada	*Dhiva*, *Tumburudu*, *Jimmi*
Malayalam	*Tumpunal*, *Tumpuni*
Mizoram	*Arhrikreh*
Oriya	*Arhrikreh*, *Ranabelli*
Tamil	*Tumpunalu*

From Manandhar, N.P., 2002. Plants and People of Nepal, Timber Press, Inc., Oregon, USA. Khare, C.P., 2007. Indian Medicinal Plants: An Illustrated Dictionary, first ed. Springer Verlog, New York, p. 730. Bachwani, M., Srivastava, B., Sharma, V., Khandelwal, R., Tomar, L., 2012. An update review on Zanthoxylum armatum *DC. Am. J. Pharm. Technol. Res. 2 (1), 274–285. Bharti, S., Bhushan, B., 2015. Phytochemical and pharmacological activities of* Zanthoxylum armatum *DC: an overview. Res. J. Pharm. Biol. Chem. Sci. 6 (5), 1403–1409.*

Order : Rutales
Suborder : Rutineae
Family : Rutaceae
Genus : *Zanthoxylum*
Species : *Zanthoxylum armatum*

16.2 Distribution

Z. armatum is native to the warm temperate and subtropical region of the world (Pirani, 1993). China, Japan, Korea, Taiwan, Bangladesh, Bhutan, Nepal, Pakistan, Laos, Myanmar, Thailand, Vietnam, and Indonesia can also grow *Z. armatum*. Different states in India such as Andhra Pradesh, Jammu and Kashmir, Manipur, Meghalaya, Nagaland, Orissa, and Uttar Pradesh have different varieties of timur plant (Batool et al., 2010).

16.3 Morphocytological studies

Dhingra et al. (2012) studied different morphological parameters (e.g.,. root or shoot length and diameter, total height of plants; primary, secondary, and tertiary branches;

types of chromosomal anomalies; chromosomal stickiness and clumping; dissolution, condensation, fragmentation, and micronuclei formation; total number of chromosomes in control and treated plants of *Z. armatum*). When mature seeds of *Z. armatum* were treated with different concentrations of colchicine solutions, a 1.0-M concentration resulted in highly retarded growth whereas low concentrations showed accelerated growth of roots. For plant height, a 50% reduction was reported at 1.0 M, and the frequency of retarded height of plants was reported to be higher at 1.0 M and less at 0.50 M concentration of colchicine. Colchiploids showed delayed flower and leaf emergence. The somatic chromosome number in all plants were found to be $2n = 66$ in diploid plants; colchicine-treated plant had a chromosome number of $2n = 132$.

16.3.1 Macroscopic view

Macroscopic studies were carried out by determining the shape, size, color, taste, odor, fracture, and dimensions of the drug. These macroscopic characters are presented in Fig. 16.1 and Table 16.2.

Other macroscopic features of leaf, bark, and fruits were determined. *Z. armatum* is a small aromatic xerophytic tree or large shrub up to 6 m high with three to five pairs of leaflets:

1. Branches are glabrous or rust-colored pubescent, usually armed with straight or slightly compressed reddish brown stipular spines.
2. Petals are absent.
3. Minute and polygamous flowers bear axillary on shoot cymes.
4. Flowers have six to eight sepals. Male flowers have six to eight stamens. Filaments are 2 mm, arranged around a pistillode globose. Female flowers have one to three carpellate ovaries attached to the inner angle of the axis. Fruits are a reddish brown subglobose that split into two parts after ripening. These single rounded, shining black seeds are about 2–3 mm in size (Grierson and Long, 1991; Nair and Nayar, 1997).
5. Spirally arranged leaves originating from the branch were observed. While studying the petiole, narrow extensions of the leaf blade were noticed to continue along each side of petiole from blade base to stem.
6. Incisions of lamina with slight dentate margin in leaves were reported. Venation was unicostate reticulate type.

16.3.2 Microscopic studies

The transverse section of a leaf of *Z. armatum* has a dorsiventral character (Ullah et al., 2014) (Figs. 16.2–16.4). Important microscopic characteristics are shown in Table 16.3.

FIGURE 16.1

(a) Leaves of *Zanthoxylum armatum.* (b) Bark of *Zanthoxylum armatum.* (c) *Seeds of Zanthoxylum armatum.* (d) Fruit of *Zanthoxylum armatum.*

16.4 Biochemical analysis

All of the essential oils have considerable molecular diversity of major and minor constituents in their qualitative and quantitative composition, with specific constituents. Most of the essential oil of leaves of *Z. armatum* was studied by the combination of gas chromatography (GC) and GC−mass spectrometry (MS) . Vast chemical diversity was observed in terms of the qualitative and quantitative makeup of the constituents on different accessions.

Table 16.2 Macroscopic characters of *Zanthoxylum armatum*.

Parameters	Leaf	Bark	Fruit
Color	Dark green in adaxial and light green in abaxial surface	Pale brown	Greenish to brown, round black or brown seeds
Size	20–80 mm long and 12 –25 mm wide	Variable	5 ± 0.2 mm
Shape	Irregular, cylindrical, Lanceolate, acuminate, and imparipinnate	Flat with grayish brown spots owing to removed spines Spines are short with rounded base, irregular, cylindrical curved	Spherical rounded
Taste	Astringent, pungent	Unpleasant	Acrid with pleasant aroma
Odor	Aromatic, pleasant	Characteristics, unpleasant	Pungent, aromatic
Fracture	Short and smooth, brittle when dry	Short bark; when freshly broken, with almost straight edges	Fruit is hard and its envelop breaks into small pieces
Dimensions	Small leaves 2.3–3 cm, large leaves 12–14 cm	2–5 cm long and up to 3 cm wide	6–10 mm in diameter
Powdered drug study			
Color	Greenish brown	Yellow to brown	Dark brown with aroma
Odor	Slight	Slight, characteristic, aromatic	Aromatic
Taste	Oily	Bitter and astringent	Slight bitter but pleasant

From Alam, F., Us Saqib, Q.N., 2015. Pharmacognostic study and development of quality control parameters for fruit, bark and leaves of Zanthoxylum armatum *(Rutaceae). Ancient Sci. Life 34 (3), 147–155. Ullah, B., Ibar, M., Muhammad, N., Khan, A., Khan, S.A., Zafar, S., Jan, S., Riaz, N., Ullah, Z., Farooq, U., Hussain, J., 2017. Pharmacognostic and phytochemical studies of* Zanthoxylum armatum *DC. Pakistan J. Pharm. Sci. 30 (2), 429–438. Jothi, G., Keerthana, K., Sridharan, G., 2019. Pharmacognostic, physicochemical and phytochemical studies on stem, bark of* Zanthoxylum armatum *DC. Asian J. Pharm. Clin. Res. 12 (2), 1–5.*

16.4.1 Qualitative studies

In these studies, the presence of various constituents was reported from different parts of *Z. armatum*. A qualitative analysis of different parts of *Z. armatum* revealed the presence of carbohydrates, alkaloids, phytosterol, terpenoids, phenols, flavonoids, glycosides, sterols, saponins, tannins, fixed oils, fats, and volatile oil in *Z. armatum* leaves, bark, and fruit (Ullah et al., 2013a,b; Upreti et al., 2013; Alam and Saqib 2015; Dabral et al., 2019; Khan et al., 2020). Anthraquinone was found in the leaves and fruits of *Z. armatum* (Alam and Us Saqib, 2015).

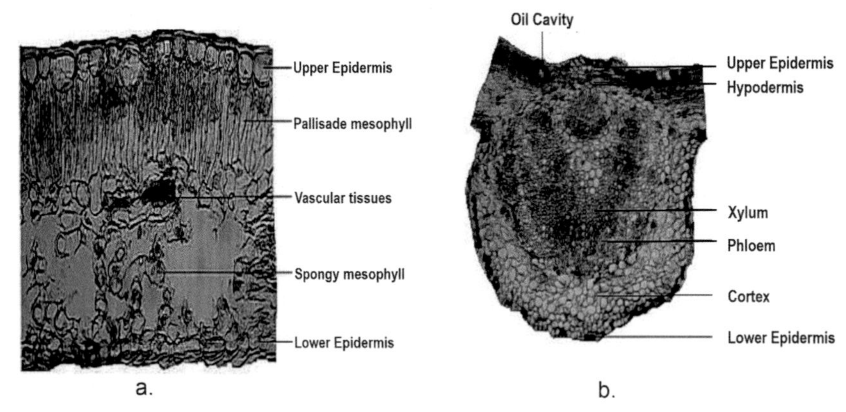

FIGURE 16.2

(a) T. S. leaf of *Zanthoxylum armatum.* (b) T. S. leaf midrib of *Zanthoxylum armatum. T.S.*, transverse section.

From Ullah, B., Ibar, M., Ghulam, J., Ahmad, I., 2014. Leaf, stem bark and fruit anatomy of Zanthoxylum armatum *(Rutaceae). Pakistan J. Bot. 46 (4), 1343–1349.*

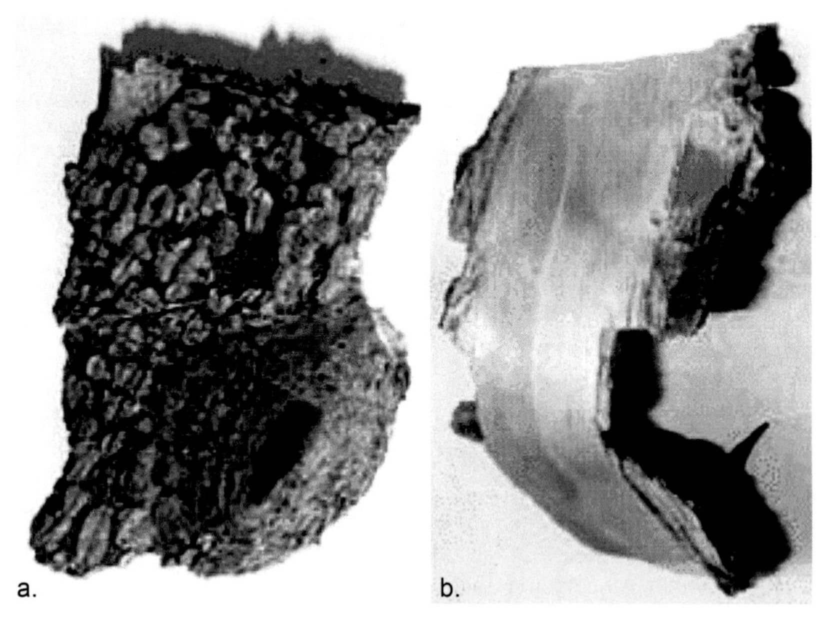

FIGURE 16.3

(a) Adaxial surface of *Zanthoxylum armatum.* (b) Abaxial surface *Zanthoxylum armatum.*

From Ullah, B., Ibar, M., Muhammad, N., Khan, A., Khan, S.A., Zafar, S., Jan, S., Riaz, N., Ullah, Z., Farooq, U., Hussain, J., 2017. Pharmacognostic and phytochemical studies of Zanthoxylum armatum *DC. Pakistan J. Pharm. Sci. 30 (2), 429–438.*

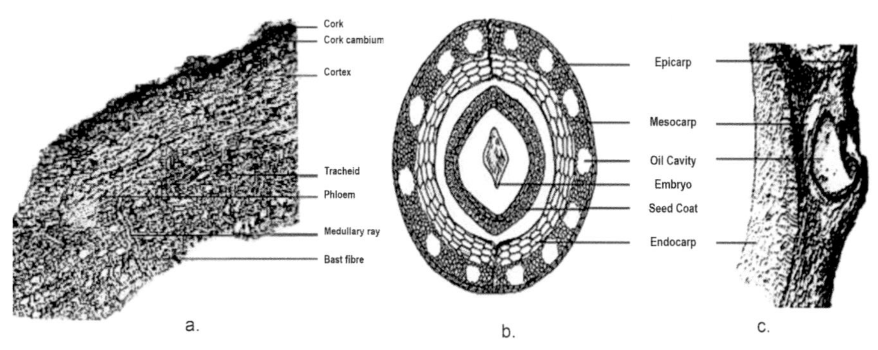

FIGURE 16.4

(a) T. S. of bark of *Zanthoxylum armatum*. (b) T. S. of fruit of *Zanthoxylum armatum*. (c) T. S. of fruit wall of *Zanthoxylum armatum*.

From Ullah, B., Ibar, M., Ghulam, J., Ahmad, I., 2014. Leaf, stem bark and fruit anatomy of Zanthoxylum armatum *(Rutaceae). Pakistan J. Bot. 46 (4), 1343–1349.*

16.4.1.1 Leaf

In *Z. armatum* (leaf) carbohydrates, alkaloids, flavonoids, glycosides, sterols, terpenoids, tannins, and saponins were reported (Upreti et al., 2013; Dabral et al., 2019; Khan et al., 2020). Ranawat et al. (2010) reported that isoquinoline alkaloid, berberine, and flavonoids as well as phenolic compounds were found in the ethanol extract of *Z. armatum*.

16.4.1.2 Fruit and seed

High-performance liquid chromatography (HPLC) was used to assess the identity of phytoconstituents in the aqueous extract of whole fruit (fruit with seeds) of *Z. armatum*. The major constituents were identified as phenolic acids, some flavonoid aglycones, and glycosides in *Z. armatum* fruit extract (Sabir et al., 2017). In an ethanol extract of *Z. armatum* seeds, terpenoids, saponins, steroids, flavonoids, alkaloids, phenols, volatile oils, amino acids, and fatty acids were found (Boggula et al., 2018).

16.4.1.3 Stem and bark

In *Z. armatum* (bark), the existence of compounds such as saponins, tannins, sterols, terpenes, flavonoids, coumarins, lignins, alkaloids, glycosides, carbohydrates, phenols, fixed oils, and fats was reported (Ranawat et al., 2013; Jothi et al., 2019). A new amide was identified as armatamide, along with two lignans, asarinin and fargesin; others were isolated from the bark of *Z. armatum* (e.g., amyrins, lupeol, and sitosterol D-glucoside) (Kalia et al., 1999).

Mukhija et al. (2014) reported three prominent lignans (e.g., sesamin, kobusin, and 4′-O-dimethyl magnolin [novel]) in a petroleum-ether extract of stem-bark of *Z. armatum*. All three lignans were found to have cytotoxic/anticancer activity against A540 and MIA-PaCa cell lines. Among the three lignans, 4′-O-dimethyl

Table 16.3 Microscopic features of *Zanthoxylum armatum* leaf.

Leaf anatomy (Fig. 16.2a)		
Characteristics		**Observation**
Upper epidermis		Nonstomatiferous, thin cuticle, compactly arranged rectangular shaped cells
Palisade mesophyll		Single-layered, cylindrically shaped, compactly arranged cells
Spongy mesophyll		Rounded to somewhat elongated, loosely arranged with large intercellular spaces, schizogenous and lysischizogenous cells
Lower epidermis		Stomatiferous, rectangular, elongated, or irregularly shaped cells
Vascular tissues		Present
Leaf midrib anatomy (Figs. 16.2b, 16.3a and b)		
Adaxial surface		Planoconvex
Abaxial surface		Semicircular
Upper epidermis		Single layered, cuticle, oval to rectangular shaped cells
Hypodermis		Thick-walled collenchymatous cells.
Cortex		Round, thin-walled parenchymatous cells, schizogenous oil cavities
Vascular tissues: Surrounded by iodioblast cells containing Ca-oxalate crystals	xylem	Arc-shaped in which xylem was in adaxial position, appeared in radial rows
	phloem	Arc shaped in which phloem was in abaxial position, rounded cells
Lower epidermis		Rectangular cells
Leaf surface characters		
Palisade		Compact, elongated, and large cells
Vein islets		Prominent distinct, wide, squarish, elongated, polyhedral, thick boundaries, forked and unforked vascular branches and randomly oriented
Vein termination		Arboreous
Types of stomata		Anomocytic, actinocytic, actinostephanocytic, staurocytic, laterocyclocytic, brachyparacytic, brachyparatetracytic, hemiparacytic, and stomatal cluster
Bark anatomy (Fig. 16.4a)		
Cork (outer phellem)		Few layers of brown color, lignified thick-walled, rectangular, or squared cells
Cork cambium (middle phellogen)		Continuous layer of small, elongated, rectangular, thin-walled parenchymatous cells
Cortex (phelloderm)		Compound and simple globular, ovoid starch grains, closely packed large parenchymatous cells, iodioblast cells

Table 16.3 Microscopic features of *Zanthoxylum armatum* leaf.—*cont'd*

Medullary rays			It was passing through the phloem
Phloem			Consist of intact and crushed phloem elements
Bast fibers			Present
Tracheids			Present
Fruit anatomy (Fig. 16.4b and c)			
Fruit wall	Epicarp		Outermost layer, closely arranged thick-walled rectangular cells
	Mesocarp		Middle layer, irregular thin-walled parenchymatous cells, large schizogenous, lysoschizogenous oil cavities
	Endocarp	Outer layer	Small isodiametric or rectangular, thin-walled parenchymatous cells
		Inner layer	Larger, thin-walled rectangular cells
Seed			Oval shaped, outer pigmented layer of testa followed by a layer of thin-walled small cells, nonendospermic and small elongated embryo

From Alam, F., Us Saqib, Q.N., 2015. Pharmacognostic study and development of quality control parameters for fruit, bark and leaves of Zanthoxylum armatum *(Rutaceae). Ancient Sci. Life 34 (3), 147–155. Ullah, B., Ibar, M., Ghulam, J., Ahmad, I., 2014. Leaf, stem bark and fruit anatomy of* Zanthoxylum armatum *(Rutaceae). Pakistan J. Bot. 46 (4), 1343–1349. Ullah, B., Ibar, M., Muhammad, N., Khan, A., Khan, S.A., Zafar, S., Jan, S., Riaz, N., Ullah, Z., Farooq, U., Hussain, J., 2017. Pharmacognostic and phytochemical studies of* Zanthoxylum armatum *DC. Pakistan J. Pharm. Sci. 30 (2), 429–438.*

magnolin was found to have strong cytotoxic activity. An ethyl acetate fraction of *Z. armatum* stem-bark was used for phytochemical and cytotoxic studies. Two pure flavonoids (apigenin and kaempferol-7-O-glucoside) were isolated from column chromatography of an ethyl acetate extract of *Z. armatum.* Apigenin and kaempferol isolated from other plants were shown to have anticancer activity (Muhkija et al.,2015). Flavonoids were reported to inhibit cell growth and proliferation (Adhami et al., 2007) and induce cell toxicity in cancer cells.

HPLC was employed on an ethyl acetate fraction of *Z. armatum* (stem and roots). The researchers reported eight lignans (eudesmin, horsfieldin, fargesin, kobusin, sesamin, asarinin, planispine, and pinoresinol-di-3,3-dimethylallyl) in *Z. armatum.* These lignan constituents showed antiinflammatory properties (Guo et al., 2011). Two new phenolic glycosides were isolated from the stem of *Z. armatum* (e.g., 2-methoxy-4-hydroxylphenyl-1-O-α-L-rhamnopyranosyl-[1″→6″]-β-D-glucopyranoside and threo-3-methoxy-5-hydroxy-phenylpropanetriol-8-O-β-D-glucophyanoside) (Guo et al., 2017a). A new lignan, (7S,8R)-guaiacylglycerol-ferulic acid ether-7-O-β-D-glucopyranoside, along with 12 known compounds (i.e., five known phenylpropanoids (2–6) and seven phenylpropanoid glycosides (7–13), were isolated from the stems of *Z. armatum*, which were analyzed by UV-nuclear magnetic resonance spectrophotometry (Guo et al., 2017b).

16.4.2 Quantitative studies

16.4.2.1 Leaf

Phytochemicals (e.g., sterols, tannins, saponins, alkaloids, flavonoids, and phenols) from Z. *armatum* leaves were investigated quantitatively (Ullah et al., 2013a,b, 2017). Essential oil obtained from *Z. armatum* (syn. *Z. alatum*) leaves was analyzed by GC-MS spectroscopy. The major components identified were bornyl acetate (Negi et al., 2012) and linalool (Ullah et al., 2013b; Guleria et al., 2013). It was revealed that the presence of linalool, 2-undecanone, and β-phellandrene were the major compound not previously found for *Z. armatum* leaves (Bhatt et al., 2017; Singh et al., 2019). The three major components, linalool, limonene, and undecan-2-one, were isolated from the essential oil of *Z. armatum* leaves (Phuyal et al., 2019).

Two major compounds (i.e., 2-undecanone and 2-tridecanone) were identified through GC-MS analysis; an n-hexane extract of *Z. armatum* leaves showed maximum larvicidal activity (Kumar et al., 2015). In phytochemical analysis, fargsin was identified as the major constituent in methanolic and chloroform extracts of *Z. armatum* leaves (Singh et al., 2020).

16.4.2.2 Fruit and seed

The highest percentage of major compounds (e.g., 3-borneol, iso-bornyl acetate and dihydro carveol) was isolated from the essential oil of *Z. armatum* seeds and identified through GC-MS (Waheed et al., 2011). A total of 36 compounds were identified through GC-MS; few of them were reported to be major constituent (i.e., 2-hydroxy cyclopentadecanone followed by palmitic acid and piperitone) (Kayat et al., 2016). Two new compounds were identified by column chromatography and isolated from the methanolic extract of *Z. armatum* fruits and characterized as 2α-methyl-2β-ethylene-3β-isopropyl-cyclohexan-1β, 3α-diol and phenol-O-β-D-arabinopyranosyl-4′-(3″,7″,11″,15″-tetramethyl)-hexadecan-1″-oate (Nooreen et al., 2017). Linalool from Dharchula and Z)-β-ocimene from Pithoragarh were reported to be major compounds isolated from the essential oil of *Z. armatum* seeds. They were identified through the combination of GC and GC-MS (Dhami et al., 2018). *Z. alatum* fruit was reported to have major phenolic acids, flavonoid aglycone, and glycosides, which were analyzed by HPLC; the major compounds were identified as ellagic acid, chlorogenic acid, gallic acid, chrysin, quercetin, and epicatechin. The observed antioxidant activity and protective ability of *Z. alatum* against liver damage results from the phenolic acids (ellagic acid, gallic acid, chlorogenic acid, caffeic acid, and coumaric acid) and flavonoids (chrysin, catechin, epicatechin, rutin, and quercetin) (Sabir et al., 2017). GC-MS analysis of the seed oil of *Z. armatum* (Syn. *alatum*) revealed linalool and limonene to be the main constituents (Jain et al., 2001; Tiwary et al., 2007; Kwon et al., 2011). Singh et al. (2016) performed a proximate quantitative analysis of the total phenolic and flavonoid content of *Z. armatum* seeds.

16.4.2.3 Stem and bark

Two major compounds were characterized as apigenin and kaempferol-7-O-glucoside, which was isolated from the ethyl acetate extract of *Z. alatum* stem bark and identified through column chromatography (Mukhija et al., 2015). Five major compounds were isolated and identified as (E)-anethole, 1.8-cineole, 2-tridecanone, limonene and piperitone from the essential oil of *Z. armatum* twigs through GC and GC-MS analysis (Wang et al., 2015). The two new major compounds (N-[31,41-methylenedioxyphenyl ethyl]-3,4-methylenedioxycinnamoyl amide and N-[31,41-dimethoxyphenylethyl]-3,4- methylenedioxydihydrocinnamoyl amide), were isolated from a methanol extract of *Z. armatum* bark through spectroscopic techniques (Siddhanadham et al., 2017b). α-Pinene and 2-undecanone were isolated from the essential oil of *Z. armatum* bark through GC-MS (Dhami et al.,2019). In a methanol extract of bark of *Z. armatum*, t-butylamine was the dominant compound followed by others such as 1-[(trimethylsilyl) oxy]propan-2-ol, propylene glycol, 2-TMS derivative, glycerol, 1-tert-butyl 3-trimethylsilyl ether, doxepin, acetin, bis-1,3-trimethylsilyl ether, laudanosine, and 2-methyl-1,2-butanediol 2-TMS, respectively. In a chloroform extract of *Z. armatum* bark, benzoxazole, 2-(isobutylamino), was found to be more strongly present than the other constituents (e.g., [Z,Z]-6,9-cis-3,4-epoxy-nonadecadiene, [+]-eudesmin, thujaplicatin, tri-O-methyl, [+]-sesamine, [(2E)-3,7-dimethyl-2,6-octadienyl-2,6-octadienyl] benzene and 1,3,14,16-nonadecatetraene) (Singh et al. 2020).

The secondary metabolites were isolated from *Z. armatum* bark and the total alkaloids were reported (2.73 ± 0.23 mg/g), as were the total flavonoids (0.05 ± 0.02 QE equivalent), terpenoids (153 ± 3.21 mg/g), saponins (0.13 ± 0.01 mg/g), tannins (0.041 ± 0.01 mg/g), and phenols (25.92 ± 1.36 GAE (gallic acid equivalent)) of their respective concentrations (Jothi et al. 2019).

Two different samples (wild and cultivated type) were isolated from seeds of *Z. armatum* for their total phenolic and flavonoid contents. In wild and cultivated seeds of *Z. armatum*, the mean total phenolic content value was reported to be 185.15 and 171.13 mg GAE/g weight of the dried extract, whereas the mean total flavonoid content value was reported to be 91.27 and 111.2 mg QE/g dried weight of the sample (Phuyal et al. 2020a).

16.5 Mineral elemental analysis

Histochemical analysis showed the presence of lignin, calcium, carbonate, mucilage, resins, starch, fats, fatty oils, and volatile oils in *Z. armatum* leaf, bark, and fruit. Cellulose was detected in the bark and fruit of *Z. armatum*, but not in the leaf. Aleurones were detected only in the fruit. The presence of hydroxyanthraquinones was reported only in the leaf and fruit of *Z. armatum*, but not in the bark. HPLC was used for analytical studies; it showed the presence of rutin and gallic acid in a leaf and bark extract of *Z. armatum* (Alam and Saqib, 2015). Crude protein, phosphorus, calcium, zinc, and copper were reported in *Z. armatum* seeds (Singh et al., 2016). The results of a mineral analysis of leaf of *Z. armatum* revealed the presence of calcium, chlorine, potassium, magnesium, silicon, sulfur, phosphorus, iron, sodium, manganese, strontium, zinc, molybdenum, copper, and nickel of their respective concentrations in leaf extract (Khan et al. 2020).

16.6 Genomics

Wang et al. (2019) performed the first study on complete chloroplast sequences of *Z. armatum*. The results showed that *Z. armatum* contains a chloroplast DNA (cpDNA) 158,579 base pairs (bp) long. The cpDNA was found to contain a pair of inverted repeat regions of 27,598 bp between the large single copy region of 85,780 bp and a small copy region of 17,598 bp. Among 133 genes of genome, 88 were protein-coding, eight were ribosomal RNA genes, and 37 were found to be transfer RNA genes. The comprehensive guanine-cytosine content of the whole genome was found 38.5%. Phylogenetic analysis was performed on 18 chloroplast genomes that were native to the family Rutaceae; *Z. armatum* was closely related to *Zanthoxylum schinifolium*.

Fluorescence in situ hybridization was employed to detect (GAA) six loci and ribosomal DNA in *Z. armatum* using oligonucleotide probes. Both (GAA) six loci and 5S ribosomal DNA were detected on pericentromeric regions of chromosomes, but the presence of (GAA) six loci was detected in five chromosome pairs, and ribosomal DNA was detected in another two chromosomal pairs. The variation in densities and locations of (GAA) 6 and 5S rDNA signals was observed by means of individual chromosome. High-intensity (GAA) six signals were detected at the centromeres of two large and two smaller metacentric chromosomes. Strong (GAA) six signals were found at the centromeres of two relatively small metacentric chromosome, whereas weak (GAA) six signals were detected at the centromeres of four large metacentric chromosomes. In the case of 5S rDNA, the strong 5S rDNA signals were detected at the centromere of two other smaller metacentric chromosomes (exclude the small metacentric chromosomes, which were found in strong [GAA] six signals), whereas weak 5S rDNA signals were detected at the centromere of two smaller metacentric chromosomes. No signals were detected in the remaining chromosomes. The approximate 2n number for *Z. armatum* was 128. The range of length of the mitotic metaphase chromosome was recorded as 1.22–2.34 μm (Luo et al., 2018).

Gupta and Kumar (2014) evaluated *Z. armatum* for karyotypic studies. A slight asymmetric 2B type of karyotype in *Z. armatum* was identified. The somatic chromosome number was 66 in the case of *Z. armatum*. The arm's ratio was reported for about 12% of the whole set of chromosomes was 2:1. The total length and volume of whole set of chromosome were 28 and 0.079 μm^3, respectively. The largest to smallest chromosome ratio was 2.8. The range of centromeric index was 26–50 and the variation in total length of chromatin was reported to be 3.43 to 9.86.

This study claim to perform the de novo transcriptome assembly for five major parts of *Z. armatum*, including the roots, stems, leaf buds, mature leaves, and fruit. Comprehensively 111,318 UniGenes were generated from an average length of 1014 bp DNA. GO (gene ontology) and KEGG (Kyoto encyclopedia of genes and genomes) enrichment analysis of *Z. armatum* parts was done by means of organ-specific UniGenes . A total of 53 and 34 UniGenes found from whole-fruit samples were recorded as candidate UniGenes in the case of many biological pathways (i.e., terpenoid biosynthesis or fatty acid biosynthesis) and their elongation and degradation pathways, respectively. The prominent period for accumulating terpenoid compounds was recorded after 40 days of fertilization (Fr4 stage) of the plant, when the development and maturation of fruit of *Z. armatum* were observed. The Fr4 stage was found to be the initial stage that induced the process of fatty acid biosynthesis; catalysis of subsequent reactions was reported up to 62 days after fertilization (Fr6 stage) (Hui et al., 2020).

16.7 Medical significance

Z. armatum demonstrated pharmacological and biological activities that showed highly potent antiinflammatory (Dhami et al., 2019), hepatoprotective (Barua et al., 2018, 2019; Ullah et al., 2011), cytotoxic (Ullah et al., 2013b; Alam et al., 2017), antispasmodic (Ullah et al., 2013a, analgesic (Pathakala et al., 2018), antidiabetic (Khan et al., 2020), and anthelmintic (Singh et al., 2016) (Table 16.4).

16.7.1 Antimicrobial activity

Different explants (fruit, leave, and bark) of *Z. armatum* were employed for the assessment of in vitro antibacterial activity using of agar well diffusion methods against gram-positive bacteria (*Micrococcus luteus*, *Bacillus subtilis*, *Staphylococcus aureus* and *Streptococcus faecalis*) and gram-negative bacteria (*Escherichia coli*, *Proteus vulgaris*, *Klebsiella pneumoniae*, *Streptococcus viridans*, *Pseudomonas multocida*, and *Pseudomonas aeruginosa*). It was found that an acetone extract showed a maximum zone of inhibition against *S. aureus* (42.3 mm), followed by a methanolic extract against *S. aureus* (28.7 mm) as gram-positive bacteria, whereas maximum activity was found in a chloroform extract against *P. vulgaris* (28.3 mm) as gram-negative bacteria (Ullah et al., 2013b; Srivastava et al., 2013). It was concluded that gram-positive bacteria showed more sensitivity to the essential oil than did gram-negative bacteria. For most bacterial species, minimum inhibitory concentration values were 0.65 μg/mL (Ullah et al., 2013b). The ethanol is the most potent extract and had a maximum inhibitory potential at 25% concentration (Dabral et al., 2019).

In another study, among the different extracts used (e.g., petroleum ether, acetone, essential oil, n-hexane, chloroform, ethanol, and methanol), the essential oil of *Z. armatum* was most effective against various tested gram-positive and

Table 16.4 Pharmacological properties of *Zanthoxylum armatum*.

Plant parts	Extracts used	*In vitro/ In vivo*	Standard used	Samples		Value of Detection	Method	References
Antibacterial activity								
Fruits	Crude ethanolic	*In vitro*	Ciprofloxacin	Positive	*Micrococcus luteus*	*21.33 mm	Agar well diffusion	Ullah et al. (2012)
				Negative	*Pasteurella multocida*	*18.33 mm		
Leaves	Essential oil	*In vitro*	Ciprofloxacin	Positive	*Streptococcus faecalis*	*23 mm	Agar well diffusion	Negi et al. (2012)
				Negative	*Klebsiella pneumoniae*	*12 mm		
Leaves	Essential oil	*In vitro*	Ciprofloxacin	Positive	*Micrococcus luteus* *Bacillus subtilis*	*28.45 mm *20.45 mm	Agar well diffusion	Ullah et al. (2013b)
				Negative	*Streptococcus viridians*	*18.54 mm		
Leaves	Essential oil	*In vitro*	Chloramphenicol	Positive	*Bacillus subtilis* *Micrococcus luteus* *Staphylococcus aureus*	*15 mm *21 mm *19 mm	Agar well diffusion	Guleria et al. (2013)
				Negative	*Streptococcus viridians* *Escherichia coli*	*18.54 mm *18 mm		
Leaf	Essential oil	*In vitro*	Chloramphenicol	Positive	*Micrococcus roseus*	*22 mm	Agar well diffusion	Mehmood et al. (2013)
				Negative	*Enterobacter aerogenes*, *Escherichia coli*	*21 mm *22 mm		
Bark	Acetone, methanol, and chloroform	*In vitro*	Tetracycline, streptomycin, and ampicillin	Positive	*Staphylococcus aureus*	*28.7 mm	Agar well diffusion	Srivastava et al. (2013)
				Negative	*Proteus vulgaris*	*28.3 mm		
Leaf	Ethanol	*In vitro*	-	Positive	*Pseudomonas aeruginosa*	*33± 1 mm	Agar well diffusion	Akbar et al. (2014)
				Negative	*Enterococcus faecalis*	*14 ± 1 mm		
Fruits	Crude ethanolic, n-hexane, chloroform, and aqueous methanol	*In vitro*	Imipenem	Positive	*Bacillus subtilis*	**30%	Agar well diffusion	Alam and Saqib (2017)
				Negative	*Escherichia coli*, *Shigella flexnari*, *Pseudomonas aeruginosa*, and *Salmonella Typhi*	***-		

Bark	Methanol	*In vitro*	Benzyl penicillin	Positive	*Staphylococcus aureus*	*13 mm	Agar well dilution method	Siddhanadham et al. (2017a)
				Negative	*Escherichia coli*	*15 mm		
Seeds	Essential oil	*In vitro*	Gentamicin sulfate	Positive	*Staphylococcus aureus*	*15.66±0.58 mm and 6.33±0.58 mm	Agar well diffusion	Dhami et al. (2018)
				Negative	*Escherichia coli*	*14.33±0.58 mm		
Stem bark	Essential oil	*In vitro*	Gentamicin sulfate (20.33±0.58 mm to 40.33±0.58 mm)	Positive	*Staphylococcus aureus*	*14.66±0.58 mm and 15.66±0.58 mm	Agar well diffusion	Dhami et al. (2019)
				Negative	*Escherichia coli*	*14.33±0.58 mm		
Leaves	Petroleum ether and chloroform	*In vitro*		Positive	*Bacillus subtilis* *Staphylococcus aureus*	*18 mm *18 mm	Disk diffusion	Dabral et al. (2019)
				Negative	*Pseudomonas aeruginosa*	*15 mm		
Fruits	Methanolic	*In vitro*	Chloramphenicol	Positive	*Staphylococcus aureus*	*20.72 mm (wild) *18.10 mm (cultivated)	Disk diffusion	Phuyal et al. (2020b)
				Negative	*Proteus vulgaris* *Pseudomonas aeruginosa* *Salmonella typhi* *Shigella dysenteriae*	No Inhibition Zone		
Leaves	Methanol	*In vitro*	Gentamicin sulfate	Positive	*Staphylococcus aureus*	*17.67 mm	Agar well diffusion	Singh et al. (2020)
				Negative	*Escherichia coli*	*17.67 mm		
Antifungal activity								
Bark	Ethanolic	*In vitro*	Miconazole	*Fusarium solani* *Candida albicans* *Microsporum canis*	**66.67% **56.33% **54.33%		Agar dilution method	Ullah et al. (2012)
Leaf	Essential oil	*In vitro*	Miconazole	*Microsporum canis* *Candida albicans* *Candida glabrata*	**84% **83% **79%		Agar dilution method	Ullah et al. (2013b)

Continued

Table 16.4 Pharmacological properties of *Zanthoxylum armatum*.—*cont'd*

Plant parts	Extracts used	*In vitro/ In vivo*	Standard used	Samples	Value of Detection	Method	References
Seeds	Essential oil	*In vitro*		*Aspergillus flavus, Aspergillus terreus, Aspergillus candidus, Aspergillus fumigates, Fusarium nivale, Penicillum italicum,* and *Trichoderma viride*	**100%	Agar dilution method	Prakash et al. (2012)
Leaf	Methanolic	*In vitro*	Miconazole	*Alternaria alternata* *Curvularia lunata*	**47.4% **51.2%	Agar dilution method	Guleria et al. (2013)
Leaf	Essential oil	*In vitro*	Chloramphenicol	*Aspergillus niger*	*28 mm	Agar dilution method	Mehmood et al. (2013)
Dry fruit	Essential oil	*In vitro*	Miconazole	*Cladosporium sp.*	Highest inhibition	Agar dilution method	Prajapati et al. (2015)
Fruits	Crude ethanolic, n-hexane, chloroform, and aqueous methanol	*In vitro*	Miconazole	*Trichophyton longifusus* *Microsporum canis*	**Ethanolic=86% **n-hexnae=90 % **Chloroform=85% **Aqueous-methanol =70 % **Aqueous-methanol =87 %	Agar tube dilution method	Alam and Saqib (2017)
Bark	Methanol	*In vitro*	Fluconazole	*Aspergillus niger*	*12 mm	Agar tube dilution method	Siddhanadham et al. (2017a)
Leaf	Chloroform	*In vitro*	-	*Aspergillus fumigatus*	*12 mm	Disk diffusion method	Dabral et al. (2019)
Antioxidant activity							
DPPH free radical scavenging activity							
Fruits	Ethanolic	*In vitro*	Ascorbic acid	Male Wistar rats	#4.56 ± 1.3 mg/ml	DPPH free radical assay	Batool et al. (2010)
Leaves	Essential oil	*In vitro*	Ascorbic acid	-	#27.0 μg/ml	DPPH free radical assay	Negi et al. (2012)
Seeds	Essential oil	*In vitro*	BHT and BHA	-	#5.6 μl/ml	DPPH free radical assay	Prakash et al. (2012)
Leaves	Methanolic fraction	*In vitro*	BHT and BHA	-	#0.044 mg/ml	DPPH free radical assay	Guleria et al. (2013)
Stem bark	Petroleum ether and ethyl acetate	*In vitro*	Ascorbic acid	-	#85.16 μg/ml #99.25 μg/ml	DPPH free radical assay	Mukhija and Kalia (2014)

Stem bark	Ethanolic	*In vitro*	Ascorbic acid	Male Wistar rats	++50%	DPPH free radical assay	Sati et al. (2011)
Leaf	Methanol	*In vitro*	Ascorbic acid	-	#2.6 μg/ml	DPPH free radical assay	Kanwal et al. (2015)
Leaves	Methanol	*In vitro*	Ascorbic acid	-	#3.63 μg/ml	DPPH free radical assay	Karmakar et al. (2015)
Stem bark	Ethyl acetate	*In vitro*	Ascorbic acid	-	#99.25 ± 2.53 μg/ml	DPPH free radical assay	Mukhija et al. (2015)
Seed coat	Essential oil	*In vitro*	Ascorbic acid	Indian foods	++98.5%	DPPH free radical assay	Joshi et al. (2016)
Seeds	Essential oil	*In vitro*	BHT and Catechin	-	#10.72 μl/ml	DPPH free radical assay	Dhami et al. (2018)
Fruit	Methanol	*In vitro*	Quercetin/Vitamin C	-	#70.4±0.5 μg/ml	DPPH free radical assay	Alam et al. (2019)
Stem bark	Essential oil	*In vitro*	BHT and Catechin	-	#12.58±0.04 μl	DPPH free radical assay	Dhami et al. (2019)
Leaves	Methanol	*In vitro*	Ascorbic acid	-	#57.83 μg/ml	DPPH free radical assay	Khan et al. (2020)
Fruits (cultivated and wild)	Methanol	*In vitro*	Ascorbic acid	-	# 40.62 μg/ml and 45.62 μg/ml	DPPH free radical assay	Phuyal et al. (2020b)
Leaves	Methanol	*In vitro*	BHT and Catechin	-	#50.87 μg/ml	DPPH free radical assay	Singh et al. (2020)
H_2O_2 scavenging activity							
Leaves	Methanol	*In vitro*	Ascorbic acid	-	#79.13 μg/ml	Hydrogen peroxide scavenging assay method	Khan et al. (2020)
Lipoxygenase inhibitory activity							
Fruit	Methanol	*In vitro*	Baicalein	-	# 70.3±2.1 μg/ml	5-LOX assay	Alam et al. (2019)
Metal Chelating activity							
Leaves	Ethyl acetate fraction	*In vitro*	Quercetin	-	++60.1%	Spectrophotometric method	Guleria et al. (2013)
Leaf	Essential oil	*In vitro*	-	-	++70.47%	Spectrophotometric method	Mehmood et al. (2013)
Seeds	Essential oil	*In vitro*	EDTA	-	#18.25 μl/ml	Spectrophotometric method	Dhami et al. (2018)
Stem bark	Essential oil	*In vitro*	EDTA	-	#18.78±0.09 μl and #19.67±0.61 μl	Spectrophotometric method	Dhami et al. (2019)
Leaves	Methanol	*In vitro*	EDTA	-	#19.42 μg/ml	Spectrophotometric method	Singh et al. (2020)

Continued

Table 16.4 Pharmacological properties of *Zanthoxylum armatum*.—*cont'd*

Plant parts	Extracts used	*In vitro/ In vivo*	Standard used	Samples	Value of Detection	Method	References
Nitric oxide scavenging activity							
Leaves	Methanol	*In vitro*	Ascorbic acid	-	@105 ±1.64 μg/ml	Nitric oxide radical scavenging method	Karmakar et al. (2015)
Stem bark	Petroleum ether and ethyl acetate	*In vitro*	Ascorbic acid	-	# 72.39 μg/ml #94.81 μg/ml	Nitric oxide radical scavenging method	Mukhija and Kalia (2014)
Stem bark	Ethyl acetate	*In vitro*	Ascorbic acid	-	#94.81 ± 2.56 μg/ml	Nitric oxide radical scavenging method	Mukhija et al. (2015)
Reducing power assay							
Leaves	Methanolic fraction	*In vitro*	BHT and BHA	-	#0.3 μg/ml	Potassium ferricyanide –ferric chloride method	Guleria et al. (2013)
Leaf	Essential oil	*In vitro*	-	-	@0.094 mM And 0.684 mM	Potassium ferricyanide –ferric chloride method	Mehmood et al. (2013)
Stem bark	Petroleum ether and ethyl acetate	*In vitro*	Ascorbic acid	-	@Petroleum ether=1.537 μg/ml @Ethyl acetate=1.287 μg/ml	Potassium ferricyanide –ferric chloride method	Mukhija and Kalia (2014)
Leaf	Methanol	*In vitro*	Ascorbic acid (2.86nm)	-	@1.411 nm	Potassium ferricyanide –ferric chloride method	Kanwal et al. (2015)
Stem bark	Ethyl acetate	*In vitro*	Ascorbic acid	-	@Ethyl acetate=1.287 ± 0.009 μg/ml	Potassium ferricyanide –ferric chloride method	Mukhija et al. (2015)
Seeds	Essential oil	*In vitro*	Ascorbic acid	-	-RP_{50}=18.45 μl/ml	Potassium ferricyanide –ferric chloride method	Dhami et al. (2018)
Fruit	Methanol	*In vitro*	Quercetin/Vitamin C	-	#14.4±0.6 μg/ml	DPPH free radical assay	Alam et al. (2019)
Stem bark	Essential oil	*In vitro*	Ascorbic acid	-	-RP_{50}= 17.91±0.07 μl) and -RP_{50}= 18.59±0.28 μl	Potassium ferricyanide –ferric chloride method	Dhami et al. (2019)
Leaves	Methanol	*In vitro*	Ascorbic acid	-	-RP_{50} = 28.93 μg/ml	-	Singh et al. (2020)
OH Scavenging Activity							
Fruits	Ethanolic	*In vitro*	-	Male Wistar rats	++47%	Benzoic acid hydroxilation method	Batool et al. (2010)
Leaves	Methanol	*In vitro*	Ascorbic acid	-	@28.10 ± 0.75 μg/ml	Hydroxyl radical	Karmakar et al. (2015)
Leaves	Ethanol	*In vivo*	Ascorbic acid	Albio mice	@8.46 mg/ml	Hydroxyl radical	Rynjah et al. (2017)

Superoxide Radical Scavenging Activity							
Leaves	Methanol	*In vitro*	Ascorbic acid	-	#19.80 ± 0.96 μg/ml	Superoxide	Karmakar et al. (2015)
Phosphomolybdate Assay							
Fruit	Methanol	*In vitro*	Ascorbic acid	-	@0.183 nm	Phosphomolybdenum method	Kanwal et al. (2015)
Anthelmintic activity							
Seeds	Methanol	*In vitro*	Piperazine citrate	*Pheretima posthuma* (Pheritimidae)	-+5.18 ± 0.40 (paralysis) -+15.5 ± 0.40 (deaths)	-	Mehta et al. (2012)
Seeds	Aqueous	*In vitro*	Levamisole	*Haemonchus contortus worms*	-+100 mg/ml (100%)	-	Singh et al. (2016)
Antidiabetic Activity							
Bark	Hydro methanolic	*In vivo*	Glibenclamide	Rats	@264.0 ± 12.88 to 256.0±9.274 gm @282.0± 15.94 to 286.0±13.64 gm	Antidiabetic assay	Karki et al. (2014)
Leaves	Ethanol	*In vivo*	Acarbose	Albio mice	^79.82 %	α-glucosidase inhibition assay	Rynjah et al. (2017)
Leaves and bark	Methanol	*In vitro/ In vivo*	Gibenclamide	Albino mice	^96.61% and 93.58%	Antidiabetic assay	Alam et al. (2018)
Leaves	Methanol	*In vitro*	Acarbose	-	$88.08±0.055%	α-amylase inhibition assay	Khan et al. (2020)
Antiinflammatory activity							
Stem bark	Ethanolic	*In vivo*	Ibropfen (39.68%)	Wistar rats	£19.12%	Direct contact method	Sati et al. (2011)
Stem bark	Aqueous	*In vivo*	Diclofenac Sodium	Wistar rats	£44.43%	Direct contact method	Mukhija et al. (2012)
Seeds	Essential oil	*In vitro*	Diclofenac Sodium (IB_{50} =13.42 μl)	Albumin	©31.96 μl ©38.15±0.01μl	Indirect method	Dhami et al. (2018)
Stem bark	Essential oil	*In vitro*	Diclofenac Sodium(IB_{50} =13.42±0.13)	Albumin	©16.40±0.48 μl	Indirect method	Dhami et al. (2019)
Leaves	Methanol, chloroform	*In vitro*	Diclofenac Sodium (IB_{50} =19.63 μg)	Albumin	©28.53 μg	Indirect method	Singh et al. (2020)
Analgesic activity/Antinociceptive activity							
Leaves	Essential oil	*In vivo*	Ibuprofen	Albio mice	≤58.41% 400 mg/kg ≤54.02%400 mg/kg	Writhing method Formalin-induced flinching	Ibrar et al. (2012)
Bark	Methanolic	*In vivo*	Diclofenac Sodium	Albino mice	≤47.8%	Tail immersion method	Siddhanadham et al. (2017b)
Bark	Methanolic	*In vivo*	Diclofenac sodium	Albino mice	≤39.2%	Acetic acid induced writhing method	

Continued

Table 16.4 Pharmacological properties of *Zanthoxylum armatum*.—*cont'd*

Plant parts	Extracts used	*In vitro/ In vivo*	Standard used	Samples	Value of Detection	Method	References
Bark	Methanolic	*In vivo*	Diclofenac Sodium	Albino mice	≤65.2%	Formalin method	
Seeds	Ethanolic	*In vivo*	Diclofenac Sodium	Wistar rats	μ8.33±0.42 min 200 and 400 mg/kg μ6.3±0.5 200 and 400 mg/kg	Tail clip method Hot plate method	Pathakala et al. (2018)
Fruits	Hydroethanolic	*In vivo*	Diclofenac Sodium	Wistar rats	μ7.01±0.12	Hot plate method	Kour et al. (2019)
Antispasmodic activity							
Stem bark	Ethanolic	*In vivo*	Loperamide (100%)	Mice	≤60%	-	Gilani et al. (2010)
Fruits and leaves	Ethanolic and n-hexnae	*In vitro*	Atropine and without atropine	Rabbit's jejunum	@8.86 mg/ml and 7.49 mg/ml (spontaneous and potassium induced contraction)	-	Ullah et al. (2013a)
Leaves, bark, and fruits	Essential oil	*In vitro*	Ciprofloxacin	Rabbit's jejunum	≤100%	-	Ullah et al. (2013b)
Fruit and leaves	Ethanol	*In vitro*	Verapamil	Rabbit's jejunum	@0.7 mg/ml and 3mg/ml 0.8 mg/ml and 3 mg/ml (spontaneous and potassium induced contraction)	-	Alam and Shah (2019)
Cardiovascular disorders							
Fruit and leaves	Ethanol	*In vitro*	Verapamil	Guinea pig	±2.1 and 3.4 mg/ml	Ca^{++} channel blocking method	Gilani et al. (2010)
Fruit	Hydroethanolic	*In vitro*	Verapamil	Rat	Group-III= TC, FFA, and TG (increased)	Standard methods	Mangalanathan et al. (2017)
Fruit	Ethanol	*In vitro*	Phenylephrine	Rabbit's aortic rings	±1.0 and 8.5 mg/ml	Ca^{++} channel blocking method	Alam and Shah (2019)
Fruit	Hydroethanolic		Verapamil	Rat	Group III=AST, ALT, LDH, CK-MB, and Troponin-T (increased)	Standard methods	Mangalanathan et al. (2019)
Cytotoxic activity							
Fruits	Crude Ethanolic	*In vitro*	Paracetamol	Brine shrimp	®20.00	Brine-Shrimp Microwell Cytotoxicity Assay	Ullah et al. (2011)
Leaves, bark, and fruits	Essential oil	*In vitro*	Paracetamol	Brine shrimp	®15.90	Brine Shrimp Toxicity Assay	Ullah et al. (2013b)
Leaves	Methanol	*In vitro*	-	EAC tumor bearing mouse	#102.3 μg/ml	Trypan blue exclusion method	Karmakar et al. (2015)
Stem bark	Ethyl acetate	*In vitro*	-	Lung cancer and Pancreatic cell lines	#85.33 μg/ml #78.0 μg/ml	Trypan blue exclusion method	Mukhija et al. (2015)
Leaves	Ethanol	*In vitro*	DMSO	Human cervical cell lines	#60 μg/ml	Trypan blue exclusion method	Singh et al. (2015)

Leaves, bark, and fruit	Crude methanol and crude saponin	*In vitro*	Actinomycin-D	Breast cancer cell lines	MTT assay (#200 μg/ml) NRU assay (#100 μg/ml)	Crude methanol (MDA MB 468 cells) Zb=99.76% Crude saponins (MDA MB 468 cells) =ZfSa=95% Crude saponin (MDA MB 468 cells, MCF cells) =ZfSa-100%, ZbSa=100% ZlSa=100% Saponin fraction (MDA MB 468 cells) =ZbSa=95.25%	MTT and NRU method	Alam et al. (2017)
Fruits	Chloroform and Aqueous-methanol	*In vitro*	Etoposide	*Artimia salima*	¥Chloroform and aqueous-methanol=21.4 and 29.6 μg/ml		Brine-Shrimp Microwell Cytotoxicity Assay	Alam and Saqib (2017)
Gastrointestinal disorders								
Aerial parts (stem, leaves, and seeds)	Aqueous-methanol extract	*In vivo*	Verapamil	Rabbit	±2.4 and 0.6 mg/ml		-	Gilani et al. (2010)
Fruit, leaves, and bark	Ethanol	*In vivo*	Verapamil	Rabbit	±**Fruit** = 2.4 and 0.9 mg/ml ±**Leaves** = 1.2 and 3 ±**Bark**= 3.1 and 0.7 mg/ml (carbachol (1μM) and high K + (80 mM) precontractions		-	Alam and Shah (2019)
Hepatoprotective activity								
Leaves	Ethanol	*In vivo*	Silymarin	Wistar albino rats	@500 mg/kg		Acute toxic classic method	Verma and Khosa (2010)
Bark	Ethanol	*In vivo*	Silymarin	Male Wistar rats	@400 mg/kg		Acute toxic classic method	Ranawat et al. (2010)
Leaves and fruits	Crude Ethanolic	*In vivo*	Saline	Albino mice	All survived		Direct contact	Ullah et al. (2011)
Leaves	Essential oil	*In vivo*	Diazepam	Albio mice	≤50%		Direct contact	Ibrar et al. (2012)

Continued

Table 16.4 Pharmacological properties of *Zanthoxylum armatum*.—*cont'd*

Plant parts	Extracts used	*In vitro/ In vivo*	Standard used	Samples	Value of Detection	Method	References
Bark	Ethanol	*In vivo*	Silymarin	Male Wistar rats	@400 mg/kg	Acute toxic classic method	Ranawat and Patel (2013)
Leaves	Ethanol	*In vivo*	Silymarin	Albino rats	@2000 mg/kg (no mortality)	Acute toxic classic method	Oinam et al. (2017)
Leaves	Ethanol	*In vivo*	Metformin	Albio mice	@5000 mg/kg	Karber method	Rynjah et al. (2017)
Fruit	Aqueous	*In vivo*	Paracetamol	Mice	@100 mg/kg	Acute toxic classic method	Sabir et al. (2017)
Leaves	Methanol	*In vivo*	Imipramine	Fasted Mice	@100 and 200 mg/kg	Direct contact	Barua et al. (2018)
Rhizome	Methanol	*In vivo*	Paracetamol	Wistar albino rats	@500 mg/kg	Acute toxic classic method	Talluri et al. (2019)
Fruit, bark, and leaf	Methanol	*In vivo*	Saline	Balb mice	@3g/kg (all survived)	Acute toxic classic method	Alam and Shah (2019)
Seeds	Hydroethanolic	*In vivo*	Imipramine	Fasted Mice	@100 and 200 mg/kg	Direct contact	Barua et al. (2019)
Fruits	Hydroethanolic	*In vivo*	Paracetamol	Female swiss mice and Wistar rats	@2000 mg/kg	Acute toxic classic method	Kour et al. (2019)
Insecticidal activity							
Seeds	Seed oil/vanillin	*In vivo*	DEET	*Aedes aegypti*	-+100%	Direct contact	Kwon et al. (2011)
Leaf	n-hexane	*In vitro*	-	*Plutella xylostella*	¥2988.64ppm	Direct contact	Kumar et al. (2015)
Fruits	Crude Ethanolic, n-hexane, chloroform and Aqueous-methanol	*In vitro*	Coopex	*Rhyzopertha dominica* *Callosobruchus analis*	Crude ethanolic, n-hexane, and chloroform (90% mortality) Crude ethanolic and n-hexane (80 and 90% mortality)	Direct contact	Alam and Saqib (2017)
Pericarp	n-hexane	*In vitro*	-	*Pieris brassiacae*	-+67.92%	Standard soxhlet extraction method	Kaleeswaran et al. (2019)

Phytotoxic activity							
Leaves and fruits	Crude Ethanolic and n- hexane	*In vitro*	Paracetamol	*Lemna minor* L	€ 100%	Lemna assay	Ullah et al. (2011)
Fruits	Crude Ethanolic, n-hexane, chloroform and Aqueous-methanol	*In vitro*	Paraquat	*Lemna minor* L	€90%	Lemna assay	Alam and Saqib (2017)
Sedative-Hypnotic activity							
Leaves	Essential oil	*In vivo*	Bromazepam	Mice	@58 $\pm$ 2.17 300 mg/kg @No. of steps =8.95$\pm$ 2.76 @No. of rearing= 4.76 $\pm$ 2.89 (300 mg/kg) @230 $\pm$ 1.34 400 mg ∞58.06$\pm$ 2.92 (60 min)	Open field method Staircase method Forced swimming method Muscle relaxation method	Muhammad et al. (2013)

= Inhibition zone, += Percent inhibition of bacterial growth, -= not specified, **= percent inhibition of mycelial growth, #= IC_{50}, ++=free radical scavenging activity, @= concentration, -=RP_{50}, ˆ= percent inhibition of α- glucosidase, $ = percent inhibition of α- amylase, * = not specified, -+ = mortality, £ = Carrageenan-induced paw edema, © = IB_{50}, μ = percentage response, ± = EC_{50}, ≤ = percent protection, ¥ = LC_{50}, ® = LD_{50}, € = percent growth inhibition, ∞ = muscle relaxant activity.*

ALT, *alanine aminotransferase;* AST, *aspartate aminotransferase;* BHA, *butylated hydroxyanisole;* BHT, *butylated hydroxytoluene;* CK-MB, *creatinekinase-MB;* DMSO, *dimethyl sulfoxide;* DPPH, *2,2-diphenyl-1-picrylhydrazyl;* EAC, *ehrlich ascites carcinoma;* EDTA, *ethylenediaminetetraacetic acid;* FFA, *free fatty acids;* LDH, *lactate dehydrogenase;* MDA, *MB- 468 and MCF-7- human breast cancer cells;* MTT, *methyl-thiazolyl tetrazolium;* NRU, *neutral red uptake essay;* TC, *total cholesterol;* TG, *Triglycerides;* Zb, *Zanthoxylum bark;* ZFH, *n-hexane extract of fruits;* Zfsa, *saponin extract of fruits.*

gram-negative bacteria (i.e., *Micrococcus luteus*, *S. faecalis*, *B. subtilis*, *E. coli*, *S. aureus*, *K. pneumonia*, *Streptococcus viridans*, *E. coli*, and *Enterobacter aerogenes*. Ciprofloxacin and gentamicin sulfate were used as the standard drug. Ciprofloxacin and gentamicin sulfate were effective against all tested bacterial strains of *S. faecalis*, *M. luteus*, *S. aureus*, *S. viridans*, and *E. coli* (Negi et al., 2012; Ullah et al., 2013b; Dhami et al., 2018, 2019; Singh et al., 2020).

The essential oil exhibited more potent antibacterial activity than methanol, the ethanol fraction, and the ethyl acetate fraction, whereas the hexane fraction was inactive against bacteria (Joshi and Gyawali, 2012). Another study suggested that the ethanol extract of *Z. armatum* leaf has a broad spectrum of antimicrobial activity compared with an aqueous extract. The extracts were tested by the well diffusion method against *E. coli*, *P. aeruginosa*, and *K. pneumonia* (Akbar et al., 2014). Based on the biological activities, it was concluded that the essential oil of *Z. armatum* was the best source for treating various animal, plant and human diseases and could also be greatly significant in the food industry (Alam and Saqib, 2017; Dhami et al., 2018, 2019). *Z. armatum* was used to treat infectious diseases such as dental problems, skin infection, diarrhea, dysentery, and urinary tract infection caused by the pathogens studied (Srivastava et al., 2013; Phuyal et al., 2020b).

16.7.2 Antifungal activity

The essential oil of leaves of *Z. armatum* was evaluated for antimycotic potential against various fungal strains such as *Fusarium solani*, *Microsporum canis*, *Aspergillus flavus*, *Candida glabrata*, *Candida albicans*, and *Trichophyton longifusus* (Ullah et al., 2013b; Prakash et al., 2012; Mehmood et al., 2013; Prajapati et al., 2015). A general pattern of dose dependency was observed (i.e., the effect became more prominent with an increasing concentration of the tested samples). The best antifungal strains were observed against *T. longifusus* (90%) followed by *M. canis* (87%), which was inhibited by fruit n-hexane and aqueous-methanol extract (Alam and Saqib. 2017). Moreover, the leaf essential oil extract inhibited certain fungal strains such as *M. canis* (84%), *C. albicans* (83%), and *C. glabrata* (79%). Most of the results proved that all fungal strains were inhibited by the essential oil obtained from the leaves of *Z. armatum* (Ullah et al., 2013b). Fruit of *Z. armatum* was used as an aromatic tonic to treat fever, dyspepsia, and skin diseases and to expel roundworms (Medhi et al., 2013).

16.7.3 Antioxidant activity

The antioxidant activity of methanol extract of the leaves was evaluated using DPPH (2,2-diphenyl-1-picrylhydrazyl) radical scavenging, reducing power, and phosphomolybdate assay. Ascorbic acid was used as standard. The antioxidant potential evaluated by three assays of different extracts increased in a concentration-dependent manner. The results of the study suggested that *Z. armatum* exhibited remarkable scavenging effects on DPPH and showed maximum reducing ability as an antioxidant capacity and total antioxidant activity (Guleria et al. 2013; Mehmood et al., 2013; Mukhija and Kalia, 2014; Kanwal et al. 2015; Kamakar et al., 2015; Mukhija et al., 2015; Joshi et al., 2016; Dhami et al., 2019). Various authors reported that phenols, carotenoids, flavonoids, and ascorbic acid present in the plants are mostly responsible for strong antioxidant activity. Phytochemical screening of *Z. armatum* indicated the presence of various flavonoids, flavonol glycosides, alkaloids, lignans, phenolics, terpenoids, amino acids, and fatty acids (Batool et al., 2010; Brijwal et al., 2012; Guleria et al. 2013; Mukhija and Kalia, 2014; Mukhija et al., 2015; Akbar et al., 2014).

In this study, petroleum ether, ethyl acetate, methanol and chloroform extracts of the plant were evaluated for their free radical scavenging property using different *in vitro* models. *In vitro* antioxidant potential was evaluated by using DPPH, nitric oxide scavenging assay and ferric reducing power assay. Petroleum ether and ethyl acetate extracts of the plant have shown potent antioxidant potential (Mukhija et al., 2015; Mukhija and Kalia, 2014). Methanolic leaf extract and ascorbic acid exhibited highest superoxide radical scavenging, phosphomolybdate assay, reducing power and antioxidant capacity at the highest concentration (Alam et al., 2019; Guleria et al., 2013; Kanwal et al., 2015; Karmakar et al., 2015; Khan et al., 2020; Phuyal et al., 2020a; Singh et al., 2020). The entire plant was used as a good flavouring agent and good source of phytotherapy as natural herbal antioxidants and highly acceptable in selected Indian food products (Joshi et al., 2016; Phuyal et al., 2020a; Sabir et al., 2017; Singh et al., 2020). It was observed that the methanolic extract of fruit exhibit maximum inhibitory activity and the lipoxygenase enzymes are also associated with the inflammatory conditions like asthma, psoriasis, rheumatoid arthritis, colitis and allergic rhinitis (Alam et al., 2019).

16.7.4 Anthelmintic activity

Aqueous, petroleum ether, and methanol extracts of the seeds of *Z. armatum* were tested against *Pheretima posthuma* (earthworm). Piperazine citrate was taken as the standard. The extracts caused paralysis and death of the worms even at the low concentration of 10 mg/mL. The methanol extract was the most potent (Mehta et al., 2012). The aqueous extract of seeds was tested against *Haemonchus contortus* worms. Levamisole was taken as the standard. The extract showed complete (100%) mortality at a concentration of 100 mg/mL at an 8-h exposure (Singh et al., 2016).

16.7.5 Anticholinergic, antihistaminic, and antiserotonergic activity

The n-hexane extracts of *Z. armatum* seeds induced concentration-dependent inhibition of isolated ileum of rat and guinea pig and fundus of rat. The n-hexane extracts of *Z. armatum* seeds (1000 μg/mL) were significantly higher than the EC_{50} of acetylcholine, 5-hydroxytryptamine, and histamine (Saikia et al., 2017). The essential oil of *Z. armatum* caused bronchorelaxation and demonstrated antiasthmatic properties; this was tested at two dose levels (200 and 400 μL/kg). Dexamethasone is an effective antiasthmatic agent used to treat asthma (Sharma et al., 2018).

16.7.6 Antidiabetic activity

Different extracts (e.g., hydromethanolic, ethanol, and methanol) of the bark, leaf, and fruit of *Z. armatum* were tested for antidiabetic activity in diabetic rats. Among the different extracts used, the methanolic leaf and bark extract were effective in showing maximum α-glucosidase inhibition ($96.61 \pm 2.13\%$ and $93.58 \pm 2.31\%$). Glibenclamide was used as the standard drug (Alam et al., 2018). The methanolic leaf extract of *Z. armatum* showed maximum α-amylase inhibition (89.37 ± 4.68 μg/mL) compared with the standard (acarbose) (Khan et al., 2020). Blood samples were collected from overnight-fasted rats at 7, 14, and 21 days of treatment and analyzed for blood glucose level and lipid profile. The rats were killed on day 21 and the liver and kidney tissues were excised. Oral administration of the extract for 21 days at 200 and 400 mg/kg resulted in a significant reduction in blood glucose, total cholesterol, triglycerides, low-density lipoprotein, and very–low density lipoprotein. There was a significant increase in high-density lipoprotein and body weight of streptozotocin diabetic rats. It was concluded that the plant may have therapeutic value in diabetes and related complications (Karki et al., 2014).

16.7.7 Antiinflammatory activity

The ethanolic and aqueous extract of stem bark of *Z. armatum* and *Z. alatum* was studied for in vivo antiinflammatory activity in Wistar rats using carrageenan-induced paw edema. The ethanolic extract of *Z. armatum* showed inhibition of paw edema by 19.12% at a dose of 250 mg/kg after 4 h of administration; an alkaloid fraction of stem bark of *Z. alatum* at doses of 100 and 150 mg/kg decreased the paw volume, which was dose-dependent. The methanolic extract of fruit showed inhibition of carrageenan that induced paw edema in Wistar rats (Mehta et al., 2012; Sati et al., 2011; Mukhija et al., 2012). Dhami et al. (2019) found maximum (16.40 ± 0.48 μL) antiinflammatory activity in the bark essential oil of Dharchula compared with the seed essential oil of Pithoragarh ($IB_{50} = 31.96$) and leaf methanolic extract ($IB_{50} = 28.53$ μg) of *Z. armatum*. Diclofenac sodium was used as the standard (Dhami et al., 2019; Singh et al., 2020).

16.7.8 Analgesic activity

The leaf essential oil, bark methanolic, and seed ethanolic extract of *Z. armatum* were evaluated for analgesic activities. The leaf essential oil of *Z. armatum* was dose-dependent and showed maximum pain reduction at 400 mg/kg in a acetic acid–induced writhing test compared with formalin induced flinching behavior (Ibrar et al., 2012). Siddhanadham et al. (2017b) performed three different tests (e.g., tail immersion, acetic acid–induced writhing, and formalin test). Among these tests, the formalin test was dose-dependent and maximum inhibition was observed in both phases of the formalin-induced pain response in mice, with a more potent effect in the second phase. It has analgesic activity owing to the presence of lignan components. The seed ethanolic extract of *Z. armatum* showed analgesic activity in the hot plate test (8.33 ± 0.42) compared with the tail clip test (6.3 ± 0.5) (Pathakala et al., 2018). The hydroethanolic extract of fruits of *Z. armatum* was given orally to mice at a dose of 125, 150 and 500 mg/kg and it showed maximum activity at a dose of 500 mg/kg and significantly protected the mice against thermally induced painin mice (Kour et al., 2019).

16.7.9 Antispasmodic activity

The crude extract of *Z. armatum* (100 and 300 mg/kg, orally) showed 20% and 60% protection against castor oil–induced diarrhea in mice and was comparable to the standard drug loperamide (100%) (Gilani et al., 2010). In another experiment, the essential oil of *Z. armatum* leaves was evaluated for possible antidiarrheal activity on spontaneous and potassium chloride induced contracted smooth muscle of isolated rabbit jejunum. The spasmolytic effect of volatile oil started from 0.03 mg/mL and showed 100% effect at a 10-mg/mL dose. The extracts relaxed the contracted muscle, possible owing to calcium channel blocking from the sarcoplasmic reticulum (Ullah et al., 2013b). The samples of ethanolic extract of the fruit and leaves of *Z. armatum* were tested against spontaneous and potassium chloride–induced contracted smooth muscle of the isolated rabbit jejunum. The samples significantly relaxed the contracted smooth muscles and protected against castor oil–induced diarrheal (Ullah et al., 2013a; Alam and Shah, 2019).

16.7.10 Antiallergic activity

Z. armatum extract was evaluated to assess its antiallergic effect against compound β-hexosaminidase enzyme. The effect of the extract on the immediate phase response was credited to the antiphase degranulation from rat basophilic leukemia (RBL)-2H3 cells. The results revealed a significant reduction ($P < .05$) in β-hexosaminidase using *Z. armatum* extract (Shah et al., 2019).

16.7.11 Antileishmanial activity

Crude ethanolic and its n-hexane soluble portion showed good antileishmanial activity against *Leishmania major* (Alam and Saqib, 2017).

16.7.12 Butyrylcholine esterase inhibitory activity

The bark methanolic extract inhibited butyrylcholine esterase with maximum percent inhibition compared with the fruit and leaf methanolic extract. Serine was used as the standard. The fruit, bark, and leaves of *Z. armatum* DC. are considered popular remedies for gastrointestinal, cardiovascular. and respiratory disorders (Alam and Shah, 2019).

16.7.13 Cardiovascular disorders

The crude extract of *Z. armatum* exhibited concentration-dependent relaxation of spontaneous and high K^+ (80 mM)-induced contractions against isolated rabbit jejunum (in vitro). It was more effective against K^+ and suggestive of a Ca^{2+} antagonist effect. The extract also caused the inhibition of both atria force and the rate of spontaneous reactions similar to verapamil in guinea pig atria. It forms the basis of pharmacological activities in the treatment of gastrointestinal, respiratory, and cardiovascular disorders (Gilani et al., 2010).

In isolated rabbit aortic ring, the crude extract of *Z. armatum* has a vasodilator effect in the case of phenylephrine (1 μM) and K^+-induced contractions (Gilani et al., 2010; Alam and Shah, 2019). The hydroethanolic extract of *Z. armatum* (fruit) was reported to treat isoproterenol hydrochloride-induced myocardial infarction in rat. The extract was found to ($P < .05$) normalize the altered level of TC (total cholesterol), TG (triglycerides), FFA (free fatty acids), and PL (phospholipids) significantly in mitochondria and LH (lipid hydroperoxides) and TBARS (thiobarbituric acid reactive substances) in heart tissues (Mangalanathan et al., 2017). The hydroethanolic extract of *Z. armatum* (fruit) was found to revert back to the altered level of serum markers (i.e., AST (aspartate aminotransferase), ALT (alanine aminotransferase), LDH (lactate dehydrogenase), CK-MB (creatine kinase-MB), and troponin-T) in isoproterenol hydrochloride myocardial infarcted rats (Mangalanathan et al., 2019).

16.7.14 Cytotoxic activity

The complete mortality of brine shrimp was recorded at 1000 μg/mL concentration of a leaf extract of *Z. armatum* using a brine shrimp toxicity bioassay (Ullah et al., 2013a,b). The crude extract of *Z. armatum* was significant against brine shrimp larvae (Alam and Saqib 2017). The methanolic extract of *Z. armatum* of leaves, fruit, and bark had better dose-dependent cytotoxic activity (Ullah et al., 2011; Karmakar et al., 2015; Alam et al., 2017). Mukhija et al. (2014) evaluated the petroleum ether fraction of *Z. armatum* and some isolated lignans (sesamin, kobusin, and novel 4′-O-dimethyl magnolin) from it for their cytotoxic activity on A-549 and MIA PaCa cancer cell lines using MTT (methyl tetrazolium assay) assay technique. The petroleum ether fraction was effective in inhibiting A-549 and MIA-PaCa cell lines, whereas lignans showed more significant results against both A-549 and MIA-PaCa cell lines than the petroleum ether fraction of *Z. armatum* (Mukhija et al., 2014). In another study, Mukhija et al. (2015) tested an ethyl acetate fraction of *Z. armatum* bark against cancer cell lines (A-549, MCF- 7, MIA- PaCa and Caco-2) using MTT assay. The ethyl acetate fraction of plant was effective in inhibiting all of these cancer cell lines (Mukhija et al., 2015). The methanol extract of *Z. armatum* and isolated saponins from it were tested on different cancer cell lines: MCF-7, MDA-MB-468 (human breast cell lines), and Caco-2 colorectal cancer cell lines. Significant inhibition was recorded against MCF-7, MDA-MB-468, and Caco-2 cell lines at 200 μg/mL, whereas the saponin fraction of *Z. armatum* fruit has excellent growth inhibition on MCF-7, MDA-MB-468, and Caco-2 cells, respectively, compared with the saponin fraction of the bark and leaf of the plant (Alam et al., 2017). In addition to other biological properties, *Z. armatum* also had excellent anticancer properties and the ability to kill drug-resistant cancer cell lines (Mukhija et al., 2015; Alam et al., 2017; Singh et al., 2015; Alam and Saqib, 2017).

16.7.15 Gastrointestinal disorders

In an isolated rabbit tracheal strip, the crude extract of *Z. armatum* inhibits carbachol (1 μM) and K^+ (80 mM)-induced contractions, which were found to be similar to verapamil (Gilani et al., 2010; Alam and Shah, 2019).

16.7.16 Hepatoprotective activity

An ethanolic extract *Z. armatum* showed hepatoprotective activity against CCl_4-induced liver damage by normalizing the elevated levels of the hepatic enzymes, accelerating the regeneration of parenchyma cells, protecting against membrane fragility, and decreasing the leakage of marker enzymes into the circulation compared with silymarin (Verma and Khosa, 2010; Ranawat et al., 2010; Ranawat and Patel, 2013). An aqueous extract of *Z. alatum* was used to prevent the injurious effect of paracetamol, caused a significant reduction in serum alanine aminotransferase, aspartate amino transferase, and TBARS, and restored the decreased levels of tissue catalase, which supports the hepatoprotective activity of the extract (Sabir et al., 2017; Talluri et al., 2019).

Paracetamol elevated the level of enzymes such as serum glutamate oxalate transaminase, serum glutamate pyruvate transaminase, alkaline phosphatase, and bilirubin in an ethanolic extract of bark of *Z. armatum*, whereas a significant reduction in level of total protein was observed with the administration of paracetamol in rat. The effect of the extract recovered the altered level of serum glutamate oxalate transaminase and serum glutamate pyruvate transaminase, alkaline phosphatase, bilirubin, and total protein. A significant recovery in the altered level of antioxidative class of enzymes (malondialdehyde, glutathione, catalase, and superoxide dismutase) was also recorded with treatment using the ethanolic extract of leaves of *Z. armatum* at an optimum dose in rats (Ranawat and Patel, 2013; Oinam et al., 2017; Kour et al., 2019).

The ethanolic extract of leaves, essential oil of leaves, and methanolic extract of bark, leaves, fruits, and seeds of *Z. armatum* resulted in no mortality or symptoms of toxicity at different dose levels (mg/kg) of body weight in albino mice (Ullah et al., 2011; Ibrar et al., 2012; Alam and Shah, 2019; Oinam et al., 2017; Rynjah et al., 2017; Barua et al., 2018, 2019).

16.7.17 Insecticidal activity

The seed oil, vanillin, crude ethanolic, n-hexane, chloroform, and aqueous methanol of *Z. armatum* demonstrated insecticidal potential against different insects (e.g., *Aedes aegypti*, *Plutella xylostella*, *Rhyzopertha dominica*, *Callosobruchus analis*, and *Pieris brassicae* (Kwon et al., 2011; Kumar et al., 2015; Alam and Saqib, 2017; Kaleeswaran et al., 2019). *Z. armatum* seed oil, vanillin, and the fruit oil of *Z. armatum* in combination were able to enhance repellent activity against female *A. aegypti*. The effect was compared with DEET repellent (Kwon et al., 2011). Alam and Saqib (2017) reported insecticidal potential of *Z. armatum* fruit against different insects. In the case of *R. dominica*, crude, n-hexane and the chloroform extract of fruits of *Z. armatum* showed maximum insecticidal activity, whereas no activity of aqueous methanol fraction was noticed.

16.7.18 Phytotoxic activity

The n-hexane fruit extract of *Z. armatum* was screened for phytotoxic activity using *Lemna minor* L. as a test species; it showed maximum inhibition at 1000 μg/mL. The standard drug used was atrazine, which served as a negative and positive control. Plants were observed on a daily basis; the number of fronds was counted on each seventh day. All parts showed significant dose-dependent photoinhibition. The results of the study suggested that *Z. armatum* has phytotoxic potential and can be a good herbicide or weedicide (Ullah et al., 2011). The hexane extract of fruit of *Z. armatum* was found to have significant phytotoxic activity against *Lemna minor* at a 100-μg/mL concentration of extract (Alam and Saqib, 2017).

16.7.19 Piscicidal activity

An ethanolic extract of *Z. armatum* fruit was evaluated for Mg^{2+}, Na^{+} and K^{+}-ATPase enzyme activity to assess its piscicidal activity against carnivorous air-breathing catfish, *Heteropneustes fossilis*. Maximum inhibition (43%, 40%, and 49%) of Mg^{2+}-ATPase was reported in the brain, muscle, and gill tissues of fish, whereas Na^{+} and K^{+}-ATPase showed 22%, 20%, and 26% of inhibition in the brain, muscle, and gill tissues (Ramanujam and Ratha, 2008).

16.7.20 Sedative-hypnotic activity

The essential oil of *Z. armatum* leaves showed maximum sedative-hypnotic, anxiolytic, and muscle relaxant properties in various animal models. It had a dose-dependent effect (Muhammad et al., 2013).

16.8 In vitro regeneration

Because the seeds of *Z. armatum* have a poor germination rate (Kala, 2010; Purohit et al. 2016) and vegetative propagation through air layering, Purohit et al. (2015) reported a slower speed of seed germination and poor rooting response, owing to which the good quality and quantity of planting material was not obtained. For several years, this important species was harvested continuously from the wild owing to its medicinal properties, which has caused a decline in its natural population and placed *Z. armatum* on the list of endangered plants in the Indian Himalayan region (Samant et al., 2007). These limitations necessitated micropropagation via tissue culture, attracting much attention. It was believed that this method could solve most or at least many problems in propagating *Z. armatum*.

16.8.1 Direct and indirect organogenesis (in vitro)

Purohit et al. (2016) observed the best results for proliferating shoots when nodal explants of *Z. armatum* were inoculated in MS medium with different concentrations of growth regulators (e.g., BAP (6-benzylaminopurine), IAA (indole-3-acetic acid), and GA_3 (gibberellic acxid)). In vitro formed shoots were excised and transfer in half-strength MS containing IBA (50) μM before they were transferred to hormone-free MS medium. When in vitro cultured plants were transferred to pots with soil for acclimatization, the survival rate was 75%. After acclimatization, plantlets were transferred to field conditions, where they started to flower after 1 year. Purohit et al. (2019) reported the in vitro regeneration of shoots by taking leaves as an explant and soaking them in different concentrations of TDZ (thidiazuron) for different times; 15 μM TDZ for 24 h was found to be the best response for pretreatment. Different concentrations of growth regulator (e.g., TDZ either alone or combined with naphthaleneacetic acid (NAA)) were tried; the best result was found with WPM (woody plant medium) supplemented with 6.0 μM TDZ and 0.5 μM NAA after 8 weeks of incubation. Excised small pieces of callus pieces were transferred to different concentrations of BAP, IAA, and GA_3 and the maximum number of shoots was observed in WPM supplemented with BAP (2.0 μM), IAA (1.0 μM), and GA_3 (1.5 μM). Fully developed shoots were transferred to half-strength WPM containing IBA (50.0 μM), and then hormone-free MS medium was used in which 100% root formation was achieved. In vitro formed shoots from callus were transferred to poly bags filled with soil and properly acclimatized. Then, they were placed under sunlight, where an 80% survival rate of plantlets was obtained.

16.8.2 In vivo propagation method

Purohit et al. (2015) developed a propagation method of air layering in *Z. armatum* in which various concentrations of IBA and NAA were applied to intact stem branches during the rainy season. When roots formed, the stems were excised from the mother plant and transferred to soil. When the plant was treated with NAA, the maximum plant survival rate (90%) and number and length of roots was observed at 28 weeks. The Percent rooting decreased at very high concentrations of IBA and NAA; at 1.0 mM, no formation of roots was observed.

Datt et al. (2017) conducted an experiment based on a factorial design with two factors: presowing treatment of seed with six levels (e.g., control, water soaking for

4 days, GA_3, Kin (Kinetin), GA_3 + Kin, and BAP) and seed sources with five levels. The maximum germination (72.5%) of seeds was found with GA_3 + Kin followed by Kin (44.25%) and water soaking (40.00%). The interaction showed a significant difference between variations in seed sources and pretreatments in all measured parameters (e.g., germination starting date, germination closing date, germination period, percent germination, and mean daily germination). It is also useful for regenerating the species to conserve its diversity.

Phuyal et al. (2018) studied the effect of different concentration of IBA and NAA on stem cuttings. The untreated stem cuttings were used as controls. Stem cuttings were planted in three different rooting media (e.g., sand, Neopeat, and mix [containing a mixture of sand, soil, and vermin-compost]). The best response was found with IBA instead of NAA. Among rooting media, Neopeat medium was better than sand and mixed media. The maximum number and length of roots were obtained with IBA (5000 parts per million) in Neopeat medium.

16.9 Conclusion

On the basis of this literature, it can be stated that the *Z. armatum* is a versatile plant that has been used owing to its therapeutic effects in traditional medicine. The study of different morphological parameters such as microscopic and macroscopic characteristics helps identify authenticated parts of *Z. armatum.* Stunted growth and decreased leaflets of plants can be induced by adding colchicine. The plant has been proven to have a number of biological and pharmacological activities. *Z. armatum* consists of linalool, limonene, and 2-undecanone; it is rich in essential oils, few of which are isolated and therapeutically proved by scientific studies. These compounds are mainly marketed and produced by various companies in the form of perfumes, cosmetics, food preservatives, and certain other medicines in curing many diseases. The study showed that extracts of *Z. armatum* have diarrheal protection as well as antispasmodic properties and strong antioxidant and antiinflammatory potential, and can be a source of significant therapeutic agents. *Z. armatum* is a good and safe natural antimicrobial and cytotoxic agent owing to its cytotoxic effect on cancer cell lines isolated from humans by a mechanism involving apoptosis. It could also be an important source in the food industry and for the control of various human, animal, and plant diseases. It also possesses significant antidiabetic activity and appears to be an attractive material for further studies and possible drug development. In vitro regeneration of plants could be a technique for improving the seed germination rate, its threatened status, and agroforestry systems through tissue culture. Through transcriptome data, researchers and breeders can easily detect specific sexual expression patterns in *Z. armatum.* Further studies are required regarding the secondary metabolites of the plant in the search for useful therapeutic agents. The plant can be successfully produced through tissue culture for the commercial production of quality plantlets. It has been found to be one of the best options for therapeutic and agroindustry applications.

References

Adhami, V.M., Malik, A., Zaman, N., Sarfaraz, S., Siddiqui, I.A., Syed, D.N., Afaq, F., Pasha, F.S., Saleem, M., 2007. Combined inhibitory effects of green tea polyphenols and selective cyclooxygenase-2 inhibitors on the growth of human prostate cancer cells both *in vitro* and *in vivo*. Clin. Canc. Res. 13, 1611–1619.

Akbar, S., Majid, A., Hassan, S., Ur Rehman, A., Khan, T., Muhammad, A.J., Ur Rehman, M., 2014. Comparative *in vitro* activity of ethanol and hot water extracts of *Zanthoxylum armatum* to some selective human pathogenic bacterial strains. Int. J. Biosci. 4 (1), 285–291.

Alam, F., Saquib, Q.N., Waheed, A., 2017. Cytotoxic activity of extracts and crude saponins from *Zanthoxylum armatum* DC. against human breast (MCF-7, MDA-MB-468) and colorectal (Caco-2) cancer cell lines. BMC Complement Altern. Med. 17 (368), 1–9.

Alam, F., Shah, A.J., 2019. Butyrlycholine esterase inhibitory activity and effects of extracts (fruit, bark and leaf) from *Zanthoxylum armatum* DC in gut, airways and vascular smooth muscles. BMC Compl. Alternative Med. 19 (180), 1–9.

Alam, F., Us Saqib, Q.N., 2015. Pharmacognostic study and development of quality control parameters for fruit, bark and leaves of *Zanthoxylum armatum* (Rutaceae). Ancient Sci. Life 34 (3), 147–155.

Alam, F., Us Saqib, Q.N., 2017. Evaluation of *Zanthoxylum armatum* Roxb for *in vitro* biological activities. J. Tradit. Complement. Med. 7, 515–518.

Alam, F., Us Saqib, Q.N., Ashraf, M., 2018. *Zanthoxylum armatum* DC extracts from fruit, bark and leaf induce hypolipidemic and hypoglycemic effects in mice *in vivo* and *in vitro* study. BMC Compl. Alternative Med. 18 (68), 1–9.

Alam, F., Us Saqib, Q.N., Ashraf, M., 2019. Phenolic contents, elemental analysis, antioxidant and lipoxygenase inhibitory activities of *Zanthoxylum armatum* DC fruit, leaves and bark extracts. Pak. J. Pharm. Sci. 32 (4), 1703–1708.

Arun, K.K.V., Paridhavi, M., 2012. An ethno botanical phytochemical and pharmacological utilization of widely distributed species Zanthoxylum: a comprehensive overview. IJPI 2 (1).

Bachwani, M., Srivastava, B., Sharma, V., Khandelwal, R., Tomar, L., 2012. An update review on *Zanthoxylum armatum* DC. Am. J. Pharm. Technol. Res. 2 (1), 274–285.

Barua, C.C., Haloi, P., Saikia, B., Sulakhiya, K., Pathak, D.C., Tamuli, S., Rizavi, H., Ren, X., 2018. *Zanthoxylum alatum* abrogates lipopolysaccharide-induced depression-like behaviours in mice by modulating neuroinflammation and monoamine neurotransmitters in the hippocampus. Pharm. Biol. 56 (1), 245–252.

Barua, C.C., Saikia, B., Ren, X., Elancheran, R., Pathak, D.C., Tamuli, S., Barua, A.G., 2019. *Zanthoxylum alatum* attenuates lipopolysaccharide-induced depressive-like behavior in mice Hippocampus. Pharmacogn. Mag. 14 (59), 673–682.

Batool, F., Sabir, S.M., Rocha, J.B.T., Shah, A.H., Saif, Z.S., Ahmed, S.D., 2010. Evaluation of antioxidant and free radical scavenging activities of fruit extract from *Zanthoxylum alatum*: a commonly used spice from Pakistan. Pakistan J. Bot. 42 (6), 4299–4311.

Bharti, S., Bhushan, B., 2015. Phytochemical and pharmacological activities of *Zanthoxylum armatum* DC: an overview. Res. J. Pharm. Biol. Chem. Sci. 6 (5), 1403–1409.

Bhatt, V., Sharma, S., Kumara, N., Sharma, U., Singha, B., 2017. Chemical composition of essential oil among seven populations of *Zanthoxylum armatum* from Himachal Pradesh: chemotypic and seasonal variation. Nat. Prod. Commun. 12 (10), 1643–1646.

Boggula, N., Pathakala, N., Sirigadi, M., Kunduru, V., Bakshi, V., 2018. Assessment of analgesic activity and phytochemical screening of *Zanthoxylum armatum* seeds- an *in vivo* design. J. Global Trends Pharm. Sci. 9 (2), 5378–5389.

Brijwal, L., Pandey, A., Tamta, S., 2012. An overview on phytomedicinal approaches of *Zanthoxylum armatum* DC: an important magical medicinal plant. J. Med. Plants Res. 7 (8), 366–370.

Dabral, H., Singh, G., Singh, R.K., 2019. Phytochemical screening and AntimicrobialActivity of leaf extracts of *Zanthoxylum armatum* against some bacteria and fungi DC: an important medicinal plant. Int. J. Sci. 8 (4), 403–406.

Datt, G., Chauhan, J.S., Ballabha, R., 2017. Influence of pre-sowing treatments on seed germination of various accessions of Timroo (*Zanthoxylum armatum* DC.) in the Garhwal Himalaya. J. Appl. Res. Med. Aromat. Plants 1–6.

Dhami, A., Palariya, D., Singh, A., Kumar, R., Prakash, O., Kumar, R., Pant, A.K., 2018. Chemical composition, antioxidant, *in vitro* anti-inflammatory and antibacterial activity of seeds essential oil of *Zanthoxylum armatum* DC. Collected from two different altitudes of Kumaun region, Uttarakhand. Int. J. Chem. Stud. 6 (6), 363–370.

Dhami, A., Singh, S., Palariya, D., Kumar, R., Prakash, O., Kumar, R., Pant, A.K., 2019. α-Inene rich bark essential oils of *Zanthoxylum armatum* DC. From three different altitudes of Uttarakhand, India and their antioxidant, *in vitro* anti-inflammatory and antibacterial activity. J. Essent. Oil Bear Plant 22 (3).

Dhingra, G.K., Pokhriyal, P., Kumar, M., 2012. A study of foliar epidermal pattern in colchiploid of *Zanthoxylum armatum*. Int. J. Plant Sci. 7 (2), 208–215.

DPR, 2006. Nepal Ko Aarthik Bikaska Lagi Prathamikta Prapta Jadibutiharu. Department of Plant Resources, Ministry of Forest and Soil Conservation, Government of Nepal (in Nepali).

DPR, 2011. Catalogue of Nepalese Flowering Plants-II. Dicotyledons (Ranunculaceae to Dipsaceae). Department of Plant Resources, Ministry of Forest and Soil Conservation. Government of Nepal.

DPR, 2016. Medicinal Plants of Nepal, revised second ed. Department of Plant Resources, Ministry of Forest and Soil Conservation, Government of Nepal.

Gilani, S.N., Khan, A.U., Gilani, A.H., 2010. Pharmacological basis for the medicinal use of *Zanthoxylum armatum* in gut, airways and cardiovascular disorders. Phytother Res. 24, 553–558.

Grierson, A.J.C., Long, D.G., 1991. Flora of Bhutan. 2 Part 1. Royal Botanic Garden, Edinburgh.

Guleria, S., Tiku, A.K., Koul, A., Gupta, S., Singh, G., Razdan, V.K., 2013. Antioxidant and antimicrobial properties of the essential oil and extracts of *Zanthoxylum alatum* grown in north-western Himalaya. Sci. World J. 1–9.

Guo, T., Deng, Y., Xie, H., Yao, C., Cai, C., Pan, S., Wang, Y., 2011. Antinociceptive and anti-inflammatory activities of ethyl acetate fraction from *Zanthoxylum armatum* in mice. Fitoterapia 82, 347–351.

Guo, T., Dai, L., Tang, X., Song, T., Wang, Y., Zhao, A., Cao, Y., Chang, J., 2017a. Two new phenolic glycosides from the stem of *Zanthoxylum armatum* DC. Nat. Prod. Res. 31 (20), 2335–2340.

Guo, T., Tang, X., Chang, J., Wang, Y., 2017b. A new lignan glycoside from the stems of *Zanthoxylum armatum* DC. Nat. Prod. Res. 31 (1), 16–21.

Gupta, S., Kumar, A., 2014. Karyotypic studies in genus *Zanthoxylum armatum* Roxb. – A high value medicinal plant from Uttarakhand. J. Indian Bot. Soc. 93 (3 & 4), 281–283.

Hajra, P.K., Verma, D.M., Giri, G.S., 1996. Materials for the Flora of Arunachal Pradesh, vol. 1. Botanical Survey of India, Kolkata, India, pp. 270–271.

Hooker, J.D., 1999. Flora of British India, vol. 1. Bishen Singh Mahendra Pal Singh, Dehra Dun, India, pp. 492–496.

Hui, W., Zhao, F., Wang, J., Chen, X., Li, J., Zhong, Y., Li, H., Zheng, J., Zhang, L., Que, Q., Wu, A., Gong, W., 2020. *De novo* transcriptome assembly for the five organs of *Zanthoxylum armatum* and the identification of genes involved in terpenoid compound and fatty acid metabolism, 21 (81), 1–15.

Ibrar, M., Muhammad, N., Khan, B.H., Jahan, F., Ashraf, N., 2012. Antinociceptive and anticonvulsant activities of essential oils of *Zanthoxylum armatum*. J. Phytopharmacol. 3 (1), 191–198.

Jain, N., Srivastava, S.K., Aggarwal, K.K., Ramesh, S., Kumar, S., 2001. Essential oil composition of *Zanthoxylum alatum* seeds from northern India. Flavour Fragrance J. 16, 408–410.

Joshi, S., Gyawali, A., 2012. Phytochemical and biological studies on *Zanthoxylum armatum* of Nepal. J. Nepal Chem. Soc. 30, 71–77.

Joshi, P., Raghuvanshi, R.S., Sankhali, A., Sharma, V., 2016. Proximate composition and antioxidant potential of fruits of *Zanthoxylum alatum* roxb. (Tooth ache tree) and its acceptability assessment in food products. Curr. Res. Nutr. Food Sci. 4 (1), 74–79.

Jothi, G., Keerthana, K., Sridharan, G., 2019. Pharmacognostic, physicochemical and phytochemical studies on stem, bark of *Zanthoxylum armatum* DC. Asian J. Pharm. Clin. Res. 12 (2), 1–5.

Kala, C.P., 2010. Assessment of availability and patterns in collection of Timroo (*Zanthoxylum armatum* DC.): a case study of Uttarakhand Himalaya. Med. Plants 2, 91–96.

Kaleeswaran, G., Firake, D.M., Behere, G.T., Challa, G.K., Sanjukta, R.K., Baiswar, P., 2019. Insecticidal potential of traditionally important plant, *Zanthoxylum armatum* DC (Rutaceae) against cabbage butterfly, *Pieris brassicae* (Linnaeus). Indian. J. Tradit. Know. 18 (2), 304–311.

Kalia, N.K., Singh, B., Sood, P., 1999. A new amide from *Zanthoxylum armatum*. J. Nat. Prod. 62 (2), 311–312.

Kanwal, R., Arshad, M., Bibi, Y., Asif, S., Chaudhari, S.K., 2015. Evaluation of ethnopharmacological and antioxidant potential of *Zanthoxylum armatum* DC. J. Chem. 8.

Karki, H., Upadhayay, K., Pal, H., Singh, R., 2014. Antidiabetic potential of *Zanthoxylum armatum* bark extract on streptozotocin-induced diabetic rats. Int. J. Green Pharm. 77–83.

Karmakar, I., Haldar, S., Chakraborty, M., Dewanjee, S., Haldar, P.K., 2015. Antioxidant and cytotoxic activity of different extracts of *Zanthoxylum alatum*. Free. Rad. Antiox. 5 (1), 21–28.

Kayat, H.P., Gautam, S.D., Jha, R.N., 2016. GC-MS analysis of hexane extract of *Zanthoxylum armatum* DC. Fruits. J. Pharmacogn. Phytochem. 5 (2), 58–62.

Khan, K., Richa, Jhamta, R., Kaur, H., 2020. Antidiabetic and antioxidant potential of *Zanthoxylum armatum* DC. leaves (Rutaceae): an endangered medicinal plant. Plant. Sci. Today. 7 (1), 93–100.

Khare, C.P., 2007. Indian Medicinal Plants: An Illustrated Dictionary, first ed. Springer Verlog, New York, p. 730.

Kour, G., Chibber, P., Kaur, S., Kumar, N., Singh, S., Tirpathi, P.K., 2019. Acute toxicity and safety pharmacological study of hydroalcoholic extract of *Zanthoxylum armatum* (ZA-A002). J. Emerg. Technol. Innov. Res. 6 (3), 160–179.

Kumar, V., Reddy, S.G., Chauhan, U., Kumar, N., Singh, B., 2015. Chemical composition and larvicidal activity of *Zanthoxylum armatum* against diamondback moth, Plutella xylostella. Nat. Prod. Res. 1–6.

Kwon, H.W., Kim, S., Chang, K.S., Clark, J.M., Ahn, Y.J., 2011. Enhanced repellency of binary mixtures of *Zanthoxylum armatum* seed oil, vanillin, and their aerosols to mosquitoes under laboratory and field conditions. J. Med. Entomol. 48 (1), 61–66.

Luo, X., Lui, J., Wang, J., Gong, W., Chen, L., Wan, W., 2018. FISH analysis of *Zanthoxylum armatum* based on oligonucleotides for 5S rDNA and $(GAA)_6$. Genome 1–10.

Manandhar, N.P., 2002. Plants and People of Nepal. Timber Press, Inc., Oregon, USA.

Mangalanathan, M., Uthamaramasamy, S., Venkateswaran, R., 2017. Preventive effect of *Zanthoxylum armatum* fruit on mitochondrial lipids alteration in isoproterenol-induced myocardial infracted rats. Int. J. Green Pharm. 11 (4), 721–725.

Mangalanathan, M., Devendhiran, T., Uthamaramasamy, S., Kumarasamy, K., Mohanraj, K., Devendhiran, K., Ilavarasan, L., Lin, M.C., 2019. Efficacy of *Zanthoxylum armatum* fruit on isoproterenol induced myocardial infraction in rats. South. Asian. J. Eng. Tech. 8 (1), 4–11.

Medhi, K., Deka, M., Bhau, B.S., 2013. The genus Zanthoxylum - A stockpile of biological and ethnomedicinal properties. Open Acc. Sci. Rep. 2 (3), 1–8.

Mehmood, F., Muhammad, A., Manzoor, F., Fazal, S., 2013. A comparative study of *in vitro* total antioxidant capacity, *in vivo* antidiabetic and antimicrobial activity of essential oils from leaves and seeds of *Zanthoxylum armatum* DC. Asian J. Chem. 25 (18), 10221–10224.

Mehta, D.K., Das, M.R., Bhandari, A., 2012. *In-vitro* anthelmintic activity of seeds of *Zanthoxylum armatum* DC against *Pheretima Posthuma*. Int. J. Green Pharm. 6, 26–28.

Muhammad, N., Ibrar., B.M., Khan, A.Z., Zia-Lu-Haq, M., Qayum, M., Khan, H., 2013. Behavioural properties of essential oils of *Zanthoxylum armatum* DC leaves: Augmented by chemical profile using GC/GC-MS. J. Chem. Soc. Pak. 35 (6), 1593–1598.

Mukhija, M., Dhar, K.L., Kalia, A.N., 2014. Bioactive lignans from *Zanthoxylum alatum* Roxb. Stem bark with cytotoxic potential. J. Ethnopharmacol. 152, 106–112.

Mukhija, M., Goyal, R., Kalia, A.N., 2012. Alkaloids from *Zanthoxylum alatum* stem bark with anti inflammatory potential in rats against acute and chronic inflammation in rats. Ind. J. Nov. D. Deliv. 4 (2), 139–144.

Mukhija, M., Kalia, A.N., 2014. Antioxidant potential and total phenolic content of *Zanthoxylum alatum* stem bark. J. Appl. Pharmacol. 6 (4), 388–397.

Mukhija, M., Singh, M.P., Dhar, K.L., Kalia, A.N., 2015. Cytotoxic and antioxidant activity of *Zanthoxylum alatum* stem bark and its flavonoid constituents. Int. J. Pharmacogn. Phytochem. 4 (4), 86–92.

Nair, K.N., Nayar, M.P., Rutaceae, 1997. Flora of India (Malpighiceae - Dichapetalaceae), vol. 4. Bot. Surv. Ind., Calcutta, India.

Negi, S., Bisht, V.K., Bhandhari, A.K., Bisht, R., Kandari, N.S., 2012. Major constituents, antioxidant and antibacterial activities of *Zanthoxylum armatum* DC. Essen. Oil Iran. J. Pharm. Ther. 11 (2), 68–72.

Nooreen, Z., Kumarb, A., Bawankuleb, D.U., Tandona, S., Alic, M., Xuand, T.D., Ahmada, A., 2017. New chemical constituents from the fruits of *Zanthoxylum armatum* and its *in vitro* anti-inflammatory profile. Nat. Prod. Res. 1–9.

Oinam, J., Raleng, I., Meitankeishangbam, P., Kumari, B.R., Laishram, S., 2017. A study on hepatoprotective activity of ethanol extract of *Zanthoxylum armatum* Dc (Mukthrubi) leaves in experimental animal. Int. J. Pharm. Sci. Res. 8 (7), 3025–3029.

Pathakala, N., Sirigadi, M., Kunduru, V., Bakshi, V., Boggula, N., 2018. Asessment of analgesic activity and phytochemical screening of *Zanthoxylum armatum* seeds- an *in vivo* design. J. Glob. Trends. Pharm. 9 (2), 5378–5389.

Phuyal, N., Jha, P.K., Raturi, P.P., Gurung, S., Rajbhandary, S., 2018. Effect of growth hormone and growth media on the rooting and shooting of *Zanthoxylum armatum* stem cuttings. Banko Janakari 28 (2), 3–12.

Phuyal, N., Jhaa, P.K., Raturi, P.P., Gurung, S., Rajbhandarya, S., 2019. Essential oil composition of *Zanthoxylum armatum* leaves as a function of growing conditions. Int. J. Food Prop. 22 (1), 1873–1885.

Phuyal, N., Jha, P.K., Raturi, P.P., Rajbhandary, S., 2020a. Total phenolic, Flavonoid contents, and antioxidant activities of fruit, seed, and bark extracts of *Zanthoxylum armatum* DC. Sci. World J. 1–7.

Phuyal, N, Jha, P.K., Raturi, P.P., Rajbhandary, S., 2020b. *In Vitro* antibacterial activities of methanolic extracts of fruits, seeds, and bark of *Zanthoxylum armatum* DC. J. Trop. Med. 7.

Pirani, J.R., 1993. A new species and a new combination in *Zanthoxylum* (Rutaceae) from Brazil. Brittonia 45 (2), 154–158.

Prajapati, N., Ojha, P., Karki, T.B., 2015. Antifungal property of essential oil extracted from *Zanthoxylum armatum* (Timur). J. Nut. Hea. Food. Eng. 3 (1), 1–5.

Prakash, B., Singh, P., Mishra, P.K., Dubey, N.K., 2012. Safety assessment of *Zanthoxylum alatum* Roxb essential oil, its antifungal, antiaflatoxin, antioxidant activity and efficacy as antimicrobial in preservation of *Piper nigrum* L. fruits. Int. J. Food Microbiol. 153, 183–191.

Purohit, S., Bhatt, A., Bhatt, I.D., Nandi, S.K., 2015. Propagation through air layering in *Zanthoxylum armatum* DC: an endangered medicinal plant in the Himalayan region. Proc. Natl. Acad. Sci. U. S.

Purohit, S., Jugran, A.K., Bhatt, I.D., Palni, L.M.S., Bhatt, K., Nandi, S.K., 2016. In vitro approaches for conservation and reducing juvenility of *Zanthoxylum armatum* DC: an endangered medicinal plant of Himalayan region. Trees 31 (3), 1101–1108.

Purohit, S., Joshi, K., Rawat, V., Bhatt, I.D., Nandi, S.K., 2019. Efficient plant regeneration through callus in *Zanthoxylum armatum* DC: an endangered medicinal plant of the Indian Himalayan region. Plant Biosyst. Int. J. Deal. Aspects Plant Biol. 154 (3), 288–294.

Ramanujam, S.N., Ratha, B.K., 2008. Effect of alcohol extract of a natural piscicide — fruits of *Zanthoxylum armatum* DC. on Mg2+- and Na+, K+-ATPase activity in various tissues of a freshwater air-breathing fish, *Heteropneustes fossilis*. Aquaculture 283, 77–82.

Rajbhandari, K.R., Bhatt, G.D., Chhetri, R., Rai, S.K., 2015. Catalogue of Napalese Flowering Plants: Supplement 1. Department of Plant Resources, Ministry of Forest and Soil conservation, Government of Nepal.

Ranawat, L.S., Patel, J., 2013. Antioxidant and hepatoprotective activity of ethanolic extracts of bark of *Zanthoxylum armatum* DC in paracetamol-induced hepatotoxicity. Int. J. Curr. Pharm. Res. 5 (3), 120–124.

Ranawat, L.S., Bhatt, J., Patel, J., 2010. Hepatoprotective activity of ethanolic extracts of bark of *Zanthoxylum armatum* DC in CCl_4 induced hepatic damage in rats. J. Ethnopharmacol. 127 (3), 777–780.

Rynjah, R.V., Devi, N.N., Khongthaw, N., Syiem, D., Majaw, S., 2017. Evaluation of the antidiabetic property of aqueous leaves extract of *Zanthoxylum armatum* DC. using *in vivo* and *in vitro* approaches. J. Tradit. Med. Complement. Ther. 1–7.

Sabir, S.M., Rocha, J.B.T., Boligon, A.A., Athayde, M.L., 2017. Hepatoprotective activity and phenolic profile of *Zanthoxylum alatum* Roxb. fruit extract. Pak. J. Pharm. Sci. 30 (5), 1551–1556.

Saikia, B., Barua, C.C., Haloi, P., Patowary, P., 2017. Anticholinergic, antihistaminic, and antiserotonergic activity of n-hexane extract of *Zanthoxylum alatum* seeds on isolated tissue preparations: an *ex vivo* study. Indian J. Pharmacol. 49 (1), 42–48.

Samant, S.S., Butola, J.S., Sharma, A., 2007. Assessment of conservation diversity: status and preparation distribution of management plan for medicinal plants in the catchment area of Parbati hydroelectric project stage—III in Northwestern Himalaya. J. Mount. Sci. 4, 34–56.

Sati, S.C., Sati, M.D., Raturi, R., Badoni, P., Singh, H., 2011. Anti-inflammatory and antioxidant activities of *Zanthoxylum armatum* Stem Bark. Glob. J. Res. Eng. 11 (5), 19–22.

Shah, M.S., Priya, V.V., Gayathri, R., 2019. Beta-hexosaminidase release-inhibitory activity by *Zanthoxylum armatum* extract. Drug Invent. Today 11 (9), 2318–2321.

Sharma, V., Rasal, V.P., Joshi, R.K., Patil, P.A., 2018. *In vivo* evaluation of antiasthmatic activity of the essential oil of *Zanthoxylum armatum*. Indian J. Pharm. Sci. 80 (2), 383–390.

Siddhanadham, A.S., Prava, R., Alekya, B.B., Pujala, V.K., Tadaka, H., Mantha, S., 2017a. Anti-inflammatory and analgesic activity of methanolic bark extracts of *Zanthoxylum armatum*. World J. Pharm. Res. 5 (10), 29–33.

Siddhanadham, A.S., Yejella, R.P., Prava, R., Sama, J.R., Koduru, A., 2017b. Isolation, characterization and biological evaluation of two new lignans from methanolic extract of bark of *Zanthoxylum armatum*. Int. J. Pharmacogn. Phytochem. Res. 9 (3), 395–399.

Singh, G., Singh, R., Verma, P.K., Singh, R., Anand, A., 2016. Anthelmintic efficacy of aqueous extract of *Zanthoxylum armatum* DC seeds against *Haemonchus contortus* of small ruminants. J. Parasit. Dis. 40 (2), 528–532.

Singh, A., Dhami, A., Palariya, D., Prakash, O., Kumar, R., Kumar, R., Pant, A.K., 2019. Methyl nonyl ketone and linalool rich essential oils from three accessions of *Zanthoxylum armatum* (DC.) and their biological activities. Int. J. Herb. Med. 7 (3), 20–28.

Singh, H.D., Meitei, H.T., SharmaRobinson, A.L.A., Singh, L.S., Singh, T.R., 2015. Anticancer properties and enhancement of therapeutic potential of cisplatin by leaf extract of *Zanthoxylum armatum* DC. Biol. Res. 48, 46.

Singh, A., Palariya, D., Dhami, A., Prakash, O., Kumar, R., Rawat, D.S., Pant, A.K., 2020. Biological activities and phytochemical analysis of *Zanthoxylum armatum* DC leaves and bark extracts collected from Kumaun region, Uttarakhand, India. J. Med. Her. Ethmed. 6, 1–10.

Srivastava, N., Kainthola, A., Bhatt, A.B., 2013. *In vitro* antimicrobial activity of bark extracts of an ethnic plant *Zanthoxylum armatum* DC. Against selected human pathogens in Uttarakhand Himalaya. Int. J. Herb. Med. 1 (3), 21–24.

Talluri, M.R., Gummadi, V.P., Battu, G.R., Killari, K.N., 2019. Evaluation of hepatoprotective activity of *Zanthoxylum armatum* on paracetamol-induced liver toxicity in rats. Indian J. Pharm. Sci. 81 (1), 138–145.

Tiwary, M., Naik, S.N., Tewary, D.K., Mittal, P.K., Yadav, S., 2007. Chemical composition and larvicidal activities of the essential oil of *Zanthoxylum armatum* DC (Rutaceae) against three mosquito vectors. J. Vector Borne Dis. 44, 198–204.

Ullah, B., Ibrar, M., Muhammad, N., 2011. Evaluation of *Zanthoxylum armatum* DC for *in-vitro* and *in-vivo* pharmacological screening. Afr. J. Pharm. Pharmacol. 5 (14), 1718–1723.

Ullah, B., Ibrar, M., Muhammad, N., Tahir, L., 2012. Antimicrobial evaluation, determination of total phenolic and flavonoid contents in *Zanthoxylum armatum* DC. J. Med. Plants Res. 6 (11), 2105–2110.

Ullah, B., Ibrar, M., Ali, N., Muhammad, N., Rehmanullah, 2013a. Antispasmodic potential of leaves, barks and fruits of *Zanthoxylum armatum* DC. Afr. J. Pharm. Pharmacol. 7 (13), 685–693.

Ullah, B., Ibrar, M., Muhammad, N., Ur-Rehman, I., Ur-Rehman, M., Khan, A., 2013b. Chemical composition and biological screening of essential oils of *Zanthoxylum armatum* DC leaves. J. Clin. Toxicol. 3 (5), 1–6.

Ullah, B., Ibar, M., Ghulam, J., Ahmad, I., 2014. Leaf, stem bark and fruit anatomy of *Zanthoxylum armatum* (Rutaceae). Pakistan J. Bot. 46 (4), 1343–1349.

Ullah, B., Ibar, M., Muhammad, N., Khan, A., Khan, S.A., Zafar, S., Jan, S., Riaz, N., Ullah, Z., Farooq, U., Hussain, J., 2017. Pharmacognostic and phytochemical studies of *Zanthoxylum armatum* DC. Pakistan J. Pharm. Sci. 30 (2), 429–438.

Upreti, K., Semwal, A., Upadhyaya, K., Masiwal, M., 2013. Pharmacognostic and phytochemical screening of leaf extract of *Zanthoxylum armatum* DC. Int. J. Herb. Med. 1 (1), 6–11.

Verma, N., Khosa, 2010. Hepatoprotective activity of leaves of *Zanthoxylum armatum* DC in CCl_4 induced hepatotoxicity in rats. Indian J. Biochem. Biophys. 47, 124–127.

Waheed, A., Mahmud, S., Akhtar, M., Nazir, T., 2011. Studies on the components of essential oil of *Zanthoxylum armatum* by GC-MS. Am. J. Chem. 2, 258–261.

Wang, C.F., Zhang, W.J., Y, C.X., Guo, S.S., Geng, Z.F., Fan, L., Du, S.S., Deng, Z.W., Wang, Y.Y., 2015. Insecticidal constituents of esential oil derived from *Zanthoxylum armatum* against two stored-product insects. J. Oleo. Sci. 64 (8), 861–868.

Wang, Y., Hao, J., Yuan, X., Sima, Y., Lu, B., 2019. The complete chloroplast genome sequence of *Zanthoxylum armatum*. Mitochondr. DNA B 4 (2), 2513–2514.

Further reading

Gupta, S., Kumar, A., Paliwal, A.K., 2018. Effect of time and colchicine on the mitotic index of *Zanthoxylum armatum* Roxb. OIIRJ 8, 47–53.

Author Index

'*Note*: Page numbers followed by "f" indicate figures and "t" indicate tables.'

D

H

I

L

M

N

O

P

Q

R

S

T

U

V

W

Z

Subject Index

'*Note*: Page numbers followed by "f" indicate figures and "t" indicate tables.'

Printed in the United States
By Bookmasters